Advances in Intelligent Systems and Computing

Volume 1082

The series "Advances in Intelligent Systems and Computing" contains publications on theory, applications, and design methods of Intelligent Systems and Intelligent Computing. Virtually all disciplines such as engineering, natural sciences, computer and information science, ICT, economics, business, e-commerce, environment, healthcare, life science are covered. The list of topics spans all the areas of modern intelligent systems and computing such as: computational intelligence, soft computing including neural networks, fuzzy systems, evolutionary computing and the fusion of these paradigms, social intelligence, ambient intelligence, computational neuroscience, artificial life, virtual worlds and society, cognitive science and systems, Perception and Vision, DNA and immune based systems, self-organizing and adaptive systems, e-Learning and teaching, human-centered and human-centric computing, recommender systems, intelligent control, robotics and mechatronics including human-machine teaming, knowledge-based paradigms, learning paradigms, machine ethics, intelligent data analysis, knowledge management, intelligent agents, intelligent decision making and support, intelligent network security, trust management, interactive entertainment, Web intelligence and multimedia.

The publications within "Advances in Intelligent Systems and Computing" are primarily proceedings of important conferences, symposia and congresses. They cover significant recent developments in the field, both of a foundational and applicable character. An important characteristic feature of the series is the short publication time and world-wide distribution. This permits a rapid and broad dissemination of research results.

**** Indexing: The books of this series are submitted to ISI Proceedings, EI-Compendex, DBLP, SCOPUS, Google Scholar and Springerlink ****

More information about this series at http://www.springer.com/series/11156

Bibudhendu Pati · Chhabi Rani Panigrahi ·
Rajkumar Buyya · Kuan-Ching Li
Editors

Advanced Computing and Intelligent Engineering

Proceedings of ICACIE 2018, Volume 1

Springer

Editors
Bibudhendu Pati
Department of Computer Science
Rama Devi Women's University
Bhubaneswar, Odisha, India

Chhabi Rani Panigrahi
Department of Computer Science
Rama Devi Women's University
Bhubaneswar, Odisha, India

Rajkumar Buyya
Cloud Computing
The University of Melbourne
Melbourne, VIC, Australia

Kuan-Ching Li
Department of Computer Science and Information Engineering
Providence University
Taichung, Taiwan

ISSN 2194-5357 ISSN 2194-5365 (electronic)
Advances in Intelligent Systems and Computing
ISBN 978-981-15-1080-9 ISBN 978-981-15-1081-6 (eBook)
https://doi.org/10.1007/978-981-15-1081-6

This Springer imprint is published by the registered company Springer Nature Singapore Pte Ltd.
The registered company address is: 152 Beach Road, #21-01/04 Gateway East, Singapore 189721, Singapore

Preface

This volume contains the papers presented at the 3rd International Conference on Advanced Computing and Intelligent Engineering (ICACIE 2018). ICACIE 2018 (www.icacie.com) was held during 22–24 December 2018, at the Institute of Technical Education and Research (ITER), Siksha 'O' Anusandhan Deemed to be University, Bhubaneswar, India, in collaboration with Rama Devi Women's University, Bhubaneswar, India. There were 528 submissions, and each qualified submission was reviewed by a minimum of two Technical Program Committee members using the criteria of relevance, originality, technical quality and presentation. The committee accepted 103 full papers for oral presentation at the conference, and the overall acceptance rate is 20%.

ICACIE 2018 was an initiative taken by the organizers, which focuses on research and applications on topics of advanced computing and intelligent engineering. The focus was also to present state-of-the-art scientific results, to disseminate modern technologies and to promote collaborative research in the field of advanced computing and intelligent engineering.

Researchers presented their work in the conference and had an excellent opportunity to interact with eminent professors, scientists and scholars in their area of research. All participants were benefitted from discussions that facilitated the emergence of innovative ideas and approaches. Many distinguished professors, well-known scholars, industry leaders and young researchers participated in making ICACIE 2018 an immense success.

We also had panel discussion on the emerging topic entitled *Intellectual Property and Standardisation in Smart City* consisting of panellists from software industries like TCS and Infosys, educationalist and entrepreneurs.

We thank all the Technical Program Committee members and all reviewers/sub-reviewers for their timely and thorough participation during the review process.

We express our sincere gratitude to Prof. Manojranjan Nayak, President, S'O'A Deemed to be University, and Prof. Damodar Acharya, Chairman, Advisor Committee, for their endless support towards organizing the conference. We extend our sincere thanks to Honourable Vice-Chancellor, Dr. Amit Banerjee, for allowing us to organize ICACIE 2018 on the campus and also for his timely support. We also express our sincere gratitude to Honourable Patron Vice-Chancellor, Prof. Padmaja Mishra, Rama Devi Women's University, for her moral support and ceaseless encouragement towards successful completion of the conference. We thank Prof. Pradeep Kumar Sahoo, Dean Faculty of Engineering and Technology, for his guidance in local organization of ICACIE 2018. We also thank General Chairs, Prof. P. K. Nanda, Prof. Bibudhendu Pati and Prof. A. K. Nayak, for their valuable guidance during review of papers as well as other aspects of the conference. We appreciate the time and efforts put in by the members of the local organizing team at the Institute of Technical Education and Research (ITER), Siksha 'O' Anusandhan Deemed to be University, Bhubaneswar, India, especially the faculty members of Department of Computer Applications, Computer Science & Information Technology, and Computer Science & Engineering, student volunteers, and administrative and hostel management staff, who dedicated their time and efforts to make ICACIE 2018 successful. We thank Er. Subhashis Das Mohapatra for designing and maintaining ICACIE 2018 Website.

Bhubaneswar, India — Bibudhendu Pati
Bhubaneswar, India — Chhabi Rani Panigrahi
Melbourne, Australia — Rajkumar Buyya
Taichung, Taiwan — Kuan-Ching Li

About This Book

The book focuses on theory, practice and application in the broad areas of advanced computing techniques and intelligent engineering. This two-volume book includes 109 scholarly articles, which have been accepted for presentation from 528 submissions in the 3rd International Conference on Advanced Computing and Intelligent Engineering held at the Institute of Technical Education and Research (ITER), Siksha 'O' Anusandhan Deemed to be University, Bhubaneswar, India, in collaboration with Rama Devi Women's University, Bhubaneswar, India, during 22–24 December 2018. The first volume of this book consists of 54 papers, and the second volume contains 49 papers with a total of 103 papers. This book brings together academic scientists, professors, research scholars and students to share and disseminate their knowledge and scientific research works related to advanced computing and intelligent engineering. It helps to provide a platform for the young researchers to find the practical challenges encountered in these areas of research and the solutions adopted. The book helps to disseminate the knowledge about some innovative and active research directions in the field of advanced computing techniques and intelligent engineering, along with some current issues and applications of related topics.

Contents

About the Editors

Dr. Bibudhendu Pati is an Associate Professor and the Head of the Department of Computer Science at Rama Devi Women's University, Bhubaneswar, India. He received his Ph.D. degree from IIT Kharagpur. Dr. Pati has 21 years of experience in teaching, research, and industry. His areas of research include wireless sensor networks, cloud computing, big data, Internet of Things, and networks virtualization. He is a life member of the Indian Society of Technical Education (ISTE), a senior member of IEEE, member of ACM, and life member of the Computer Society of India (CSI). He has published several papers in leading international journals, conference proceedings and books. He is involved in organizing various international conferences.

Dr. Chhabi Rani Panigrahi is an Assistant Professor at the Department of Computer Science at the Rama Devi Women's University, Bhubaneswar, India. She holds a doctoral degree from the Indian Institute of Technology Kharagpur. Her research interests include software testing and mobile cloud computing. She has more than 18 years of teaching and research experience, and has published numerous articles in leading international journals and conferences. Dr. Panigrahi is the author of a number of books, including a textbook and. She is a life member of the Indian Society of Technical Education (ISTE) and a member of IEEE and the Computer Society of India (CSI).

Dr. Rajkumar Buyya is a Redmond Barry Distinguished Professor of Computer Science and Software Engineering and Director of the Cloud Computing and Distributed Systems (CLOUDS) Laboratory at the University of Melbourne, Australia. He is also serving as the founding CEO of Manjrasoft, a spin-off company of the university, commercializing its innovations in cloud computing. He served as a future fellow of the Australian Research Council during 2012–2016. He has authored over 525 publications and seven textbooks including "Mastering Cloud Computing" published by McGraw Hill, China Machine Press, and Morgan Kaufmann for Indian, Chinese, and international markets, respectively. He has also edited several books including "Cloud Computing: Principles and Paradigms"

(Wiley Press, USA, Feb 2011). He is one of the highly cited authors in computer science and software engineering worldwide (h-index=112, g-index=245, 63,900+ citations). Microsoft Academic Search Index ranked Dr. Buyya as #1 author in the world (2005–2016) for both field rating and citation evaluations in the area of Distributed and Parallel Computing. Recently, Dr. Buyya is recognized as "2016 Web of Science Highly Cited Researcher" by Thomson Reuters.

Prof. Kuan-Ching Li is a Professor in the Department of Computer Science and Information Engineering at the Providence University, Taiwan. He has been the vice-dean for the Office of International and Cross-Strait Affairs (OIA) at the same university since 2014. Prof. Li is the recipient of awards from Nvidia, the Ministry of Education (MOE)/Taiwan and the Ministry of Science and Technology (MOST)/ Taiwan, and also guest professorship from various universities in China. He received his Ph.D. from the University of Sao Paulo, Brazil, in 2001. His areas of research include networked and graphics processing unit (GPU) computing, parallel software design, and performance evaluation and benchmarking. He has edited two books: Cloud Computing and Digital Media and Big Data published by CRC Press. He is a fellow of the IET, senior member of the IEEE, and a member of TACC.

Machine Learning Applications

Effect of Dimensionality Reduction on Classification Accuracy for Protein–Protein Interaction Prediction

Satyajit Mahapatra, Anish Kumar, Animesh Sharma and Sitanshu Sekhar Sahu

Abstract "Large Dimension" of features derived from protein sequences is a major problem in protein–protein interaction (PPI) prediction. Thus, reduction of the feature dimension may increase the classification accuracy. In this paper, particle swarm optimization (PSO) and principal component analysis (PCA) have been used for dimensionality reduction of PPI sequence features. The performance of the algorithm has been assessed using the intraspecies E coli protein–protein interaction database, containing an equal number of positive and negative interacting pairs. Standard sequence-based features such as amino acid composition (AAC), dipeptide composition (Dipep), and conjoint triad composition (CTD) are extracted. From the results, it is seen that the PSO-based dimensionality reduction method provides steady and better performance in terms of accuracy when applied to the features.

Key terms Protein–protein interaction · Feature extraction · Dimensionality reduction · Machine learning

1 Introduction

Interaction between DNA, RNA, and proteins plays significant roles in the execution of cellular processes and cell functions. Among this, protein–protein interaction possesses a vital role in facilitating essential cellular and molecular processes. Proteins

S. Mahapatra (✉) · A. Kumar · A. Sharma · S. S. Sahu
Department of Electronics and Communication Engineering, Birla Institute of Technology Mesra, Ranchi, India
e-mail: satyajit6243@gmail.com

A. Kumar
e-mail: anishkr10052@gmail.com

A. Sharma
e-mail: animeshsharma97@gmail.com

S. S. Sahu
e-mail: sitanshusekhar@gmail.com

B. Pati et al. (eds.), *Advanced Computing and Intelligent Engineering*, Advances in Intelligent Systems and Computing 1082,
https://doi.org/10.1007/978-981-15-1081-6_1

are represented by a polypeptide chain. One protein differs from others due to difference in arrangement of the amino acids in the polypeptide chain. When two or more proteins unite to carry out their respective biological functions, then the process can be termed as protein–protein interaction (PPI). Prior information about the PPIs can be helpful for better understanding of disease mechanism that can become the basis for new methods of remedies for diseases. Although a number of experimental methods have been designed to analyze PPI, they remain expensive and time-consuming. Therefore, computational methods are preferred as an alternative to predict PPI. Results obtained from computational methods can be used as an efficient guide for experimental methods.

Barman et al. have studied intraspecies protein–protein interaction in enteropathogens using features such as domain–domain association (DDA), AAC, Dipep, and degree of amino acid for developing SVM predictor [1]. Zhao et al. have predicted protein–protein interaction using deep neural networks [2]. Cui et al. have studied a human virus interaction model using conjoint triad feature and have used SVM for prediction [3]. Barman et al. have made a comparison study on SVM, random forest, and Naïve Bayes predictor for prediction of host and virus based on the fivefold cross-validation test [4]. The above features are of huge dimensions resulting in the curse of dimensionality. Dimensionality reduction algorithms can be applied on the feature matrix, and can enhance the predictor's accuracy. Some of the dimensionality reduction techniques used for image data analysis and microarray data classification have been discussed below.

Zhuo et al. have made a comparative study of six dimensionality reduction methods. These methods are PCA, LLP, LLE, FLDA, LFDA, and ISOMAP and their study shows that LLP and LLE can effectively reduce the dimension which helps in increasing the accuracy in image retrieval [5]. Sun et al. have proposed an all-dimensional neighborhood (ADN)-based PSO for improving the local search capability around the Gbest in PSO [6]. Sahu and Mishra have proposed a PSO-based filtering technique for high-dimensional microarray data classification [7]. Xue et al. have proposed a multi-objective PSO for feature selection for removing redundant and irrelevant features [8]. Hameed et al. have proposed a GBPSO-SVM algorithm for improving feature selection and classification process for autism spectrum disorder [9]. Han et al. have proposed a feature selection algorithm using binary PSO encoding gene-to-class sensitivity information [10].

This paper is divided into five sections. Motivation and introduction are given in Sect. 1. Section 2 contains information on datasets and algorithm. Section 3 describes support vector machine classifier and performance evaluation parameters. Results and analysis are given in Sect. 4. Finally, conclusion and future scope are reported in Sect. 5.

2 Materials and Methods

To perform protein–protein interaction (PPI) prediction, first, the character sequences are converted into its numeric values, otherwise, called features. These features are then filtered to find out the dominant features. The database is then divided into two parts (80% for training and 20% for independent testing) and is given to the classifier (SVM). Then the accuracy, MCC, sensitivity, and specificity are calculated. Figure 1 shows the implementation of the above PPI prediction method.

2.1 *Dataset*

In this study for the development of a prediction model, intraspecies E coli protein–protein interaction data have been used. The datasets have been collected from EnPPIpred: Enteropathogen Protein–Protein Interactions (PPIs) Prediction (http://bicresources.jcbose.ac.in/ssaha4/EnPPIpred/) [1]. It contains 3886 positive and equal number of negative interacting pairs.

2.2 *Feature Extraction Technique*

2.2.1 Amino Acid Composition (AAC)

This technique provides the normalized frequency of occurrences of amino acids in a polypeptide chain. The numbers of occurrences of amino acids are divided by length of polypeptide chain, for normalization. Combining the features of a pair of protein sequence we get 40-dimensional feature vector. It is given by

$$X(s_i) = \frac{f_1(s_i)}{\sum_{i=1}^{20} f_1(s_i)} \quad i = 1, 2, 3 \ldots\ldots 20 \tag{1}$$

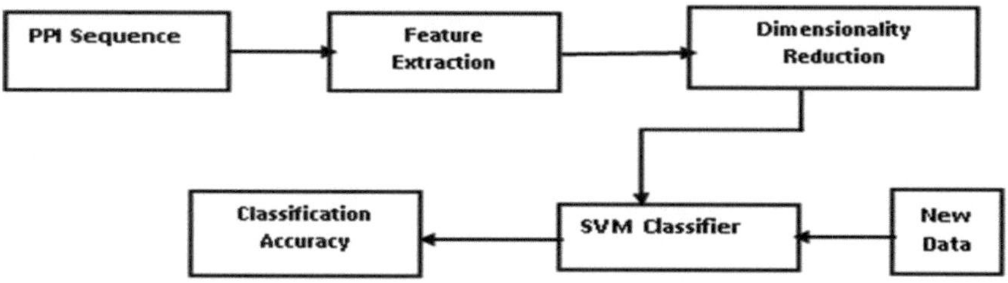

Fig. 1 Implementation of the prediction algorithm

In the above equation, $X(s_i)$ represents sequence features and $f_1(s_i)$ represents the frequency count of ith amino acid [1].

2.2.2 Dipeptide Amino Acid Composition

In this technique, 400 fragments will be generated for each protein sequence. These fragments are the fractions of dipeptides, i.e., AA, AC, AD, ... YV, YW, and YY. Combining the features of a pair of protein sequence we get 800-dimensional feature vector. The dipeptide features are given by

$$X(s_i) = \frac{f_1(s_i)}{\sum_{i=1}^{400} f_1(s_i)} \quad i = 1, 2, 3 \ldots 400 \tag{2}$$

In the above equation, $X(s_i)$ represents the 400 dipeptides and $f_1(s_i)$ is the frequency count of ith dipeptide [1, 11].

2.2.3 Conjoint Triad

In this technique, the 20 amino acids are grouped into seven different classes, i.e., {A,G,V}, {I,L,F,P}, {Y,M,T,S}, {H,N,Q,W}, {R,K}, {D,E}, {C} on the basis of volumes of the side chains and dipoles. Features obtained are normalized frequency of 3-mer in the seven class representation of the protein sequence [13, 14]. Combining the features of a pair of protein sequence we get 686-dimensional feature vector.

2.3 *Dimensionality Reduction*

2.3.1 Principal Component Analysis (PCA)

It is an unsupervised linear dimensionality reduction method, which is used for projection of a data space into smaller dimensional space by using orthogonal transformation. The process used for reduction is given below:

I. At first, find out the mean.
II. Then go for an element by element subtraction of mean from its data, in order to form the covariance matrix.

$$\mathbf{COV} = \sum_{i=1}^{N} (x_i - \mu)(x_i - \mu)^T \tag{3}$$

III. Find the eigenvalues using this covariance matrix.

IV. Now sort the eigenvectors in descending order and extract the required number of vectors according to the required number of dimensions.
V. Principal component scores = input × eigenvectors.

2.3.2 Particle Swarm Optimization (PSO)

I. It is an optimization technique based on population, which evaluates the objective function, here standard deviation, at each particle.
II. The best position is calculated after each iteration and the new velocity of the particles is then evaluated.
Velocity update:

$$\left[V_i(t+1) = W \times V_i(t) + C_1 \times \textbf{rand} \times (\textbf{pbest}(t) - X_i(t) + C_2 \times \textbf{rand} \times (\textbf{gbest}(t) - X_i(t))\right] \quad (4)$$

$$\text{Position update:} X_i(t+1) = X_i(t) + V_i(t+1) \quad (5)$$

III. For dimensionality reduction to a specific number of columns, first, the required numbers of columns are selected randomly. Then the column numbers are updated using PSO to obtain the best cost on the basis of a minimum of standard deviation.
IV. Implementation of PSO-based dimensionality reduction is given below:

```
Step1: Initialization of Constants
Weight
constant1
constant2
Step2: Randomly initialize Position of the particles
Positions = zeros(population,nOfSelection);
fori = 1:population
tempVar = randperm(Boundary(2));
Positions(i,:) = tempVar(1:nOfSelection);
end
Step3: Randomly initialize velocity of the particles
Velocity = ones(population,of selection);
Step4: Initialization of the Global best
Step5: Calculating Fitness Values
Step6: Updating the particle best
Step7: Updating the Global Best
Step8: Velocity Update
Step9: Position Update
Step10: Boundary Checking for Position
Positions(Positions > Boundary(2))
= round(rand(1)*(Boundary(2)-1)) + 1;
```

```
Positions(Positions < Boundary(1))
= round(rand(1) * (Boundary(2)-1)) + 1;
Step11: To avoid repeating same position
if length(unique(Positions(i,:))) ~ = nOfSelection
Step12: New Positions are selected randomly again
End while when the No. of maximum iterations is reached
```

3 Classification and Evaluation

3.1 Support Vector Machines

In this study, support vector machine (SVM) developed by Vapnik has been used for prediction of interacting and noninteracting protein pairs [12]. SVM is supervised learning machines that are associated with learning algorithms for analysis of data. For model development, first, the SVM is trained using a set of training data each tagged with its respective class and then a set of testing data is given whose class is to be predicted by the model.

A hyperplane or margin is necessary that gives the maximum distance of separation between the classes.

The hyperplane is represented by equation below:

$$h(x) = w^T + b \tag{6}$$

where $'w'$ and $'b'$ are the weight and bias vectors, respectively.

Minimizing the cost function, hyperplane is obtained

$$J(\omega) = \frac{1}{2}w^T w = \frac{1}{2}\|w\|^2 \tag{7}$$

$$\text{Subject to the constraints } d_i\left[w^T x_i + b\right] \geq 1 \quad i = 1, 2, \ldots, N \tag{8}$$

In this paper, Scikit-learn package in Python has been used for implementation of SVM.

3.2 Evaluation

Performance is evaluated using fivefold cross-validation tests. Four different parameters, i.e., sensitivity, specificity, accuracy, and MCC are calculated [13, 15].

$$\textbf{Accuracy}(\textbf{Acc}) = \frac{\textbf{Pos}^{+} + \textbf{Neg}^{+}}{\textbf{Pos}^{+} + \textbf{Neg}^{+} + \textbf{Pos}^{-} + \textbf{Neg}^{-}}$$

$$\textbf{Sensitivity}(\textbf{Se}) = \frac{\textbf{Pos}^{+}}{\textbf{Pos}^{+} + \textbf{Pos}^{-}}$$

$$\textbf{Specificity}(\textbf{Sp}) = \frac{\textbf{Neg}^{+}}{\textbf{Neg}^{+} + \textbf{Neg}^{-}}$$

$$\textbf{Mcc} = \frac{(\textbf{Pos}^{+} \times \textbf{Neg}^{+}) - (\textbf{Pos}^{-} \times \textbf{Neg}^{-})}{\sqrt{(\textbf{Pos}^{+} + \textbf{Pos}^{-}) \times (\textbf{Pos}^{+} + \textbf{Neg}^{-}) \times (\textbf{Neg}^{+} + \textbf{Pos}^{-}) \times (\textbf{Neg}^{+} + \textbf{Neg}^{-})}}$$

$(\textbf{Pos}^{+})$ is the count positive samples classified as positive. $(\textbf{Neg}^{-})$ is the count positive samples classified as negative. $(\textbf{Pos}^{-})$ is the count negative samples classified as positive. $(\textbf{Neg}^{+})$ is the count negative samples classified as negative.

4 Result and Discussions

In this paper, for assessment of prediction accuracy, we have used three commonly used PPI sequence information extraction methods such as AAC, Dipep, and CTD. The above feature extraction techniques are applied on a benchmark intraspecies E coli dataset provided by Barman et al. available at EnPPIpred [1]. Then the dimensionality reduction algorithm is applied to reduce the feature dimension keeping the number of samples constant. AAC contains 40 features and upon application of a dimensionality reduction technique, it is reduced to 15. Similarly, Dipep and conjoint triad contain 800 and 686 features which are reduced to 40 after application of dimensionality reduction algorithm.

It is evident from the Andrews plot (Figs. 2 and 3) that after reduction of features, there is a clear difference between the pattern of positive and negative samples.

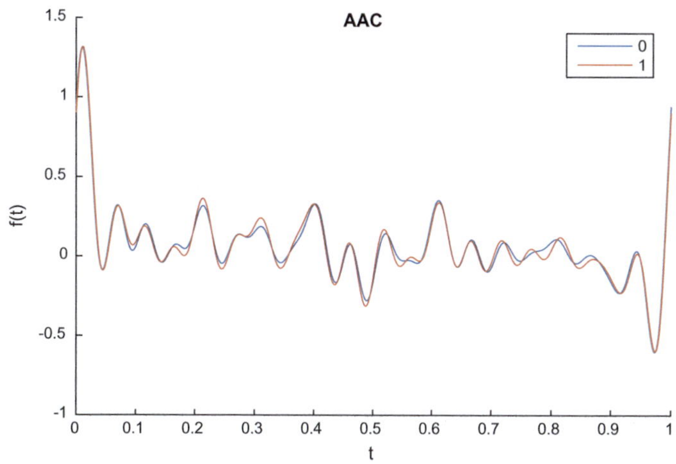

Fig. 2 AAC features without reduction ("0" represents negative samples and "1" represents positive samples)

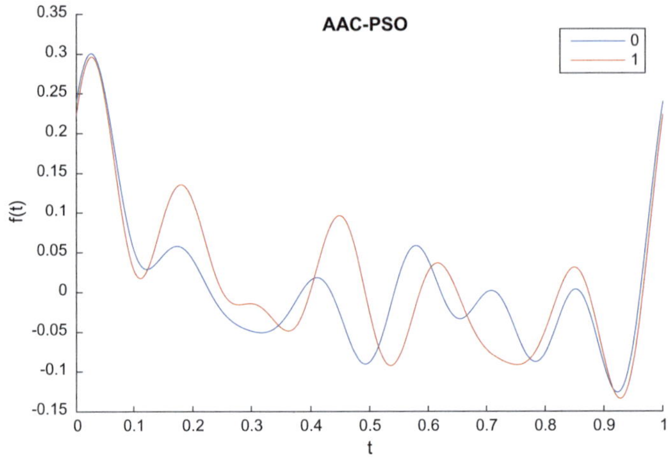

Fig. 3 AAC features reduced using PSO ("0" represents negative samples and "1" represents positive samples)

So, machine learning techniques used on the reduced features will provide better classification accuracy. In the plot shown in Figs. 2 and 3, "0" and "1" represent negative and positive samples, respectively.

4.1 Comparison of Dimensionality Reduction Algorithm on an Intraspecies Dataset

The database has been divided into two parts, i.e., 80% for training by fivefold cross-validation and 20% for independent testing.

The accuracy obtained using AAC features is 79.65, for PCA-AAC is 99.80, for PSO-AAC is 99.42, and for HMS-AAC is 99 as shown in Table 1.

Using dipeptide composition, the accuracy obtained is 83.15, for PCA-DIPEP, the accuracy is 99.80, for PSO-DIPEP, the accuracy is 96.37, and for HMS-DIPEP, the accuracy is 87.43 as shown in Table 2.

Table 1 Performance of SVM using amino acid composition

Features	Accuracy	Sensitivity	Specificity	MCC
AAC (c = 1000, γ = 0.5, rbf)	79.65	76.45	82.85	59.42
PCA-AAC (c = 1000, linear)	99.80	99.74	99.87	99.61
PSO-AAC (c = 1000, γ = 0.5, rbf)	99.42	99.48	99.35	98.83

Table 2 Performance of SVM using dipeptide composition

Features	Accuracy	Sensitivity	Specificity	MCC
DIPEP (c = 100, γ = 0.5, rbf)	83.15	77.74	88.57	66.70
PCA-DIPEP (c = 10, linear)	99.80	99.74	99.87	99.61
PSO-DIPEP (c = 1000, linear)	96.37	96.24	96.49	92.74

Table 3 Performance of SVM using conjoint triad composition

Features	Accuracy	Sensitivity	Specificity	MCC
Conjoint triad (c = 10,000, linear)	80.10	76.32	83.90	60.40
PCA-conjoint triad (c = 10,000, γ = 0.5, rbf)	78.41	72.96	83.96	57.2
PSO-conjoint triad (c = 1000, linear)	98.51	99.60	97.40	97.05

The accuracy obtained using conjoint triad composition is 80.10, for the PCA-conjoint triad, the accuracy is 78.41, for the PSO-conjoint triad, the accuracy is 98.51, for the HMS-conjoint triad, the accuracy is 99.81 as shown in Table 3.

From Tables 1, 2, and 3, it has been observed that PCA provided the best accuracy for AAC and dipeptide features but failed in case of conjoint triad features. Similarly, HMS provided the best accuracy in the case of conjoint triad and comparable in case of AAC features. The accuracy of PSO is comparable with the best accuracy provided by PCA (for AAC and dipeptide) and HMS (for a conjoint triad).

Although PCA gives comparable results in many cases, it fails to provide suitable results in some cases because PCA-based dimensionality reduction provides derived features which sometimes disturb the pattern of the original features. In PSO-based dimensionality reduction, the reduced features are among the original set of features due to which the pattern is conserved.

5 Conclusion

In this paper, dimensionality reduction technique such as PSO and PCA has been used for enhancement of performance of predictor. It has been seen that reduced features provide 15–18% more accuracy. Through exhaustive simulation, it has been found that PSO-based dimensionality reduction provides steady performance in terms of accuracy, sensitivity, and MCC.

In future, this work can be extended for analysis of interspecies protein–protein interaction.

Acknowledgements This work has been partly supported by SERB (ECR/2017/000345) and is carried out at Signal Processing lab of ECE department, BIT Mesra.

References

1. Barman, R.K., Jana, T., Das, S., Saha, S.: Prediction of intra-species protein-protein interactions in enteropathogens facilitating systems biology study. PLoS One **10**, 1–9 (2015). https://doi.org/10.1371/journal.pone.0145648
2. Zhao, Z., Yang, Z., Lin, H., Wang, J.: Aprotein-protein interaction extraction approach based on deep neural network. Int. J. Data Min. Bioinform. **15**, 145–164 (2016)
3. Cui, G., Fang, C., Han, K.: Prediction of protein-protein interactions between viruses and human by an SVM model. BMC Bioinform. **13**, S5 (2012). https://doi.org/10.1186/1471-2105-13-S7-S510
4. Barman, R.K., Saha, S., Das, S.: Prediction of interactions between viral and host proteins using supervised machine learning methods. PLoS One **9** https://doi.org/10.1371/journal.pone.0112034 (2014)
5. Zhuo, L., Cheng, B., Zhang, J.: A comparative study of dimensionality reduction methods for large-scale image retrieval. Neurocomputing **141**, 202–210 (2014). https://doi.org/10.1016/j.neucom.2014.03.014
6. Sun, W., Lin, A., Yu, H., et al.: All-dimension neighborhood-based particle swarm optimization with randomly selected neighbors. Inf. Sci. (NY) **405**, 141–156 (2017). https://doi.org/10.1016/j.ins.2017.04.007
7. Sahu, B., Mishra, D.: A novel feature selection algorithm using particle swarm optimization for cancer microarray data. Procedia Eng **38**, 27–31 (2012). https://doi.org/10.1016/j.proeng.2012.06.005
8. Xue, B., Zhang, M.J., Browne, W.N.: Particle swarm optimization for feature selection in classification: a multi-objective approach. IEEE Trans. Cybern. **43**, 1656–1671 (2013). https://doi.org/10.1109/TSMCB.2012.2227469
9. Hameed, S.S., Hassan, R., Muhammad, F.F.: Selection and classification of gene expression in autism disorder: use of a combination of statistical filters and a GBPSO-SVM algorithm. PLoS One **12**, 1–25 (2017). https://doi.org/10.1371/journal.pone.0187371
10. Han, F., Yang, C., Wu, Y.Q., et al.: A gene selection method for microarray data based on binary PSO encoding gene-to-class sensitivity information. IEEE/ACM Trans. Comput. Biol. Bioinform. **14**, 85–96 (2017). https://doi.org/10.1109/TCBB.2015.2465906
11. Lin, H., Ding, H.: Predicting ion channels and their types by the dipeptide mode of pseudo amino acid composition. J Theor. Biol. **269**, 64–69 (2011). https://doi.org/10.1016/j.jtbi.2010.10.019
12. Vapnik, V. N.: The Nature of Statistical Learning Theory, 2nd edn. Springer, New York (2000)
13. You, Z.-H., Lei, Y.-K., Zhu, L., et al.: Prediction of protein-protein interactions from amino acid sequences with ensemble extreme learning machines and principal component analysis. BMC Bioinform. **14**, S10 (2013). https://doi.org/10.1186/1471-2105-14-S8-S10
14. Shen, J., Zhang, J., Luo, X., et al.: Predicting protein-protein interactions based only on sequences information. Proc. Natl. Acad. Sci. USA **104**, 4337–4341 (2007). https://doi.org/10.1073/pnas.0607879104
15. Pandey, C., Sandeep, R., Priyam, A., Mahapatra, S., SahuS, S.: Predicting protein–RNA interaction using sequence derived features and machine learning approach. Int. J. Data Min Bioinform. **19**(3), 270–282 (2017)

Early Detection of Breast Cancer Using Support Vector Machine With Sequential Minimal Optimization

Kumar Avinash, M. B. Bijoy and P. B. Jayaraj

Abstract Breast cancer is largely occurring cancer disease and one of the leading causes of death among the women in the world. Studies have shown that early detection can bring down the mortality rate significantly. Mammography is the most popular cancer detection technique but it is painful as well as it uses X-ray to capture images which spreads radiation in the body. Radiation is one of the causes of breast cancer. Thermography is a method of detection which is noninvasive, painless, and radiation-free. By seeing these thermal images, doctors can't say whether the patient is having cancer or not that's why they use computer-aided diagnosis (CAD) to see the status by feeding these thermal images to model. The detection of breast cancer from these data is very crucial and needs some sophisticated image processing and machine learning techniques. Modern machine learning technique has become a popular tool for the diagnosis of breast cancer. Among various methods, support vector machine (SVM) classification is one of the most popular supervised learning methods. Performance of SVM classification by using sequential minimal optimization (SMO) algorithm is evaluated and our proposed model is giving better result in terms of accuracy (94.6%), recall (89.5%), and execution time (0.085 s) on Wisconsin data set.

Keywords Breast cancer · Preprocessing · Feature extraction · Classification · Sequential Minimal Optimization

K. Avinash · M. B. Bijoy · P. B. Jayaraj (✉)
Department of Computer Science and Engineering, National Institute of Technology Calicut, Calicut, India
e-mail: jayarajpb@nitc.ac.in

K. Avinash
e-mail: kavinash881@gmail.com

M. B. Bijoy
e-mail: bijoy_p170059cs@nitc.ac.in

B. Pati et al. (eds.), *Advanced Computing and Intelligent Engineering*, Advances in Intelligent Systems and Computing 1082,
https://doi.org/10.1007/978-981-15-1081-6_2

1 Introduction

Breast cancer is a disease which is caused by multiple factors like inheritance, tissue composition, carcinogens, immunity levels, hormones, radiation, etc. When a cell becomes cancerous, it's temperature increases and due to this, the surrounding cells also start becoming cancerous that's why it spreads very fast. Mammography is one of the most popular methods to detect breast cancer. It is the manual method of detecting breast cancer with the help of experts to identify the tumors in breast tissue. This method is painful as well as it involves radiation since the X-ray image is being taken of each breast. There are various methods available like magnetic resonance imaging (MRI) and ultrasonography which assist the CAD system in detecting breast cancer [1]. MRI uses magnetic and electric fields, gradients, and radio waves to generate images of the body parts. In ultrasonography, images of body parts are produced using sound waves. It helps in diagnosis and the causes of pain, swelling, and infection in the body's organs.

Thermography is a noninvasive, painless, noncontact, and non-radiation method to detect breast cancer. Since humans are warm-blooded animal, they emit heat radiation. Thermal image of a human body contains temperature of each cell by using which prediction of the cells as benign or malignant can be done. Different features like contrast, area, radius, perimeter, mean, median, etc., of these thermal images play a crucial role in detecting cancer. Thermography can detect breast cancer 8–10 years earlier than all the other methods mentioned above [2].

Breast cancer detection from the thermographic image data needs some preprocessing like noise removal, image enhancement, etc., before feeding it to the algorithm. This is because of the reason that images may contain some noises due to the local environment and human errors which can cause redundancy later. Then, segmentation will be done for each of these images in order to separate the left breast and right breast so that we can extract the features from each of these images. In the feature extraction process, each feature will be stored as a vector. After getting all the desired features from these images, they will be fed to our proposed algorithm for the classification.

Image segmentation helps in feature extraction. After separating the left breast and right breast from the thermal image redundancy can be easily avoided in the extracted data. With image segmentation, we can also remove the undesired edges and the useless areas of the image which is not going to help in decision making. Figure 1 shows the thermal image of a patient's breast [3]. Figures 2 and 3 show the left and right breast images after noise removal and segmentation have been applied.

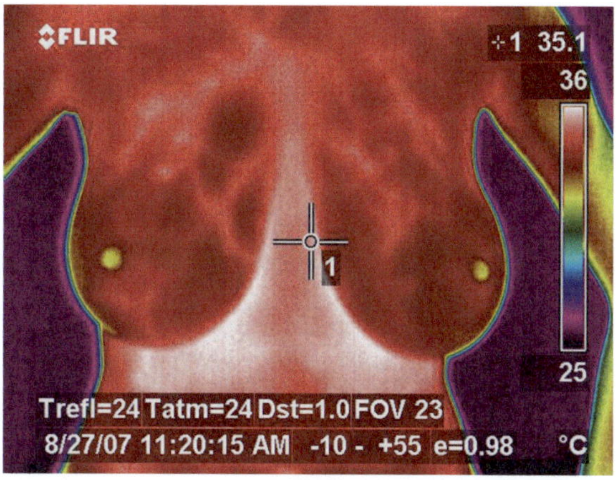

Fig. 1 Thermal image of a patient's breast [3]

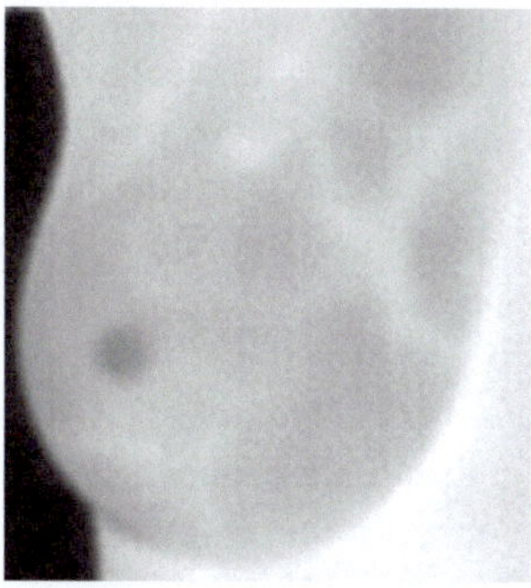

Fig. 2 Left breast image after segmentation

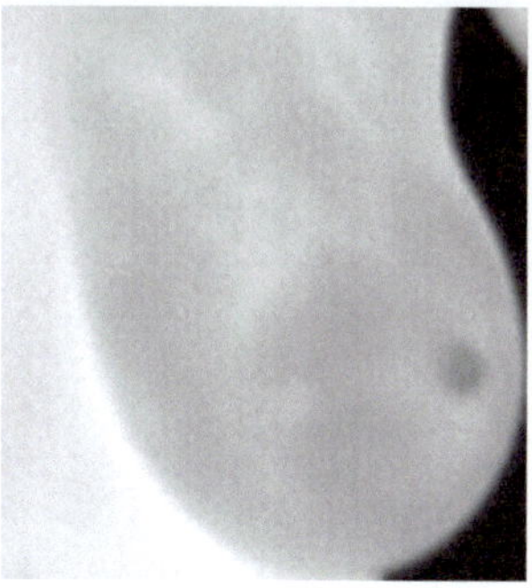

Fig. 3 Right breast image after segmentation

2 Review of Literature

There are many techniques for early detection of breast cancer using various methods. Some of the machine learning related methods are described below.

2.1 Machine Learning Based Related Work for Breast Cancer Detection

In the early detection of breast cancer, support vectors machines play an important role by classifying the data with better precision and accuracy. G. R. Suresh et al., have suggested an approach to detect breast cancer using thermographic color analysis and SVM classifier [1]. The hottest region of breast thermograms was detected using three image segmentation techniques: k-means, fuzzy c-means (FCM), and level set. Experimental results have shown that the level set method has a better performance than other methods as it could highlight the hottest region more precisely.

Ryszard S. Choras suggested an approach in which he has used various image processing techniques like Gabor Wavelets and texture parameter evaluation to detect benign and malignant breast through thermographic grayscale images [4]. Gabor wavelet is a powerful tool to extract texture features and in the spatial domain, is a complex exponential modulated by a Gaussian function.

Rozita Rastghalam et al., have suggested an approach using the spectral probable feature on thermographic images to detect the abnormal pattern in the breast. The gray level co-occurrence matrix is made from image spectrum to obtain spectral co-occurrence feature. This method is not sufficient to extract the spectral probable feature, so they have optimized this matrix and defined it as a feature vector. In this method, normal and abnormal patterns have been separated from each other through the new procedure in which for every image, each of the spectrum features is mentioned [5]. There are many other works which are reported in this area [6–10].

3 Motivation

After the experiment with various classifiers, it is found that SVM is giving better results. The accuracy of SVM classification used in some models is marginal but couldn't exploit the maximization. Our attempt here is to improve the performance of SVM classification by applying SMO for early detection of breast cancer.

4 Problem Definition

To design and develop a machine learning algorithm for early detection of breast cancer using support vector machine with sequential minimal optimization for thermographic images.

5 Proposed Solution

As a first level experiment, implementation of various classification models has been done and accuracy of each model on Wisconsin breast cancer data set is calculated. Figure 4 shows the comparison of accuracy for different models.

It can be observed in the above figure that SVM is giving better results among all other classifiers and that's why SVM is chosen for classification. Since SMO gives the better optimization results for quadratic programming problem, it's being used for optimizing SVM.

5.1 SVM Implementation

Support vectors play an important role in SVM classification. Lagrange duality is used to find the support vectors for the classification purpose. Support vector machine computes a linear classifier of the form

$$f(x) = w^T x + b \tag{1}$$

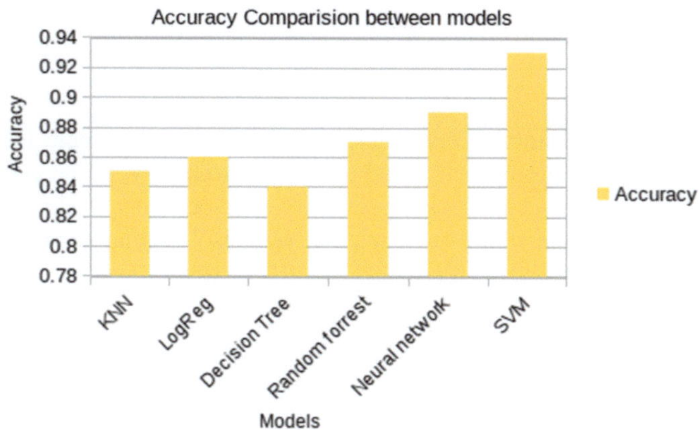

Fig. 4 Accuracy comparison of classification models on WDBC data set

where $f(x)$ is classifier's decision function, w is normal vector to the hyperplane and b is soft margin. In the binary classification of data, SVM will separate the data into either class 1 or class –1 based on the constraints applied to it.

$$y = \begin{cases} 1, & \text{if } f(x) \geq 0 \\ -1, & \text{if} f(x) \leq 0 \end{cases} \tag{2}$$

Considering our problem, the classification with highest margin between data points for different classes is needed. The larger margin will reduce the generalization error. After some mathematical calculations, this optimization problem comes down to $\max_{w,b} \frac{2}{\|\mathbf{w}\|}$ or

$$\min_{w,b} \frac{1}{2} \|\mathbf{w}\|^2 \tag{3}$$

such that

$$y^i(w^T x^i) \geq 1, i = 1, 2, .., n$$

where n is the total number of training examples.

Using Lagrange duality, the above optimization problem can also be written as

$$\max_\alpha W(\alpha) = \sum_{i=1}^{n} \alpha_i - \frac{1}{2} \sum_{i,j=1}^{n} y^i y^j \alpha_i \alpha_j \langle x^i, x^j \rangle \tag{4}$$

such that

$$0 \leq \alpha_i \leq C, i = 1, 2, .., n$$

and

$$\sum_{i=1}^{n} \alpha_i y^i = 0$$

where n is number of training examples, x^i is i-th training example feature vector, $\langle x^i, x^j \rangle$ is the dot product of x^i and x^j feature vectors, y^i is the class for the i-th training example, α_i is the Lagrange multiplier associated with the i-th training example, and C is regularization parameter (C value is $\frac{1}{\propto}$ regularization).

Kernel tricks in SVM When data points are nonlinearly separable, then we can improve the performance of SVM by applying soft margin kernel. For this, we have to map our data points to a different feature space of higher dimension using a mapping function ϕ. So, we can write a kernel function $K(x, z)$ as

$$K(x, z) = \langle \phi(x^i), \phi(x^j) \rangle \tag{5}$$

considering $\phi(x) = x$, then the optimization problem (Eq. 4) can be rewritten as

$$\max_{\alpha} W(\alpha) = \sum_{i=1}^{n} \alpha_i - \frac{1}{2} \sum_{i,j=1}^{n} y^i y^j \alpha_i \alpha_j K(x_i . x_j) \tag{6}$$

Time complexity for computing $\phi(x)$ will be $O(n^2)$ because we have to find the dot product of two feature vectors [11].

Some of the kernel mentioned below are used in our model for comparison purpose.

Linear kernel

$$K(x, y) = \sum_{i=1}^{n} x_i y_i$$

Polynomial kernel

$$K(x, y) = (x^T y + c)^d$$

RBF kernel

$$K(x, y) = e^{-(\frac{\|\mathbf{x}-\mathbf{y}\|^2}{2\sigma^2})}$$

UKF kernel

$$K(x, y) = L(\|\mathbf{x} - \mathbf{y}\|^2 + \sigma^2)^{-\alpha}$$

5.2 *Approach for SMO Implementation*

Previously to optimize SVM's QP problem, most of the algorithms were using numerical optimization functions which need a large number of memories and therefore, they were very slow. In 1998, John Platt developed SMO algorithm to solve SVM optimization problem.

SMO algorithm works by cutting down the SVM optimization problem into many small problems that are easily solvable [11]. In short, the algorithm works as follows:

- Two Lagrange's multipliers α_i and α_j are selected and their values are optimized while keeping all other α values as constant.
- Once those two values are optimized, another two values are chosen again.
- Choosing and optimizing the values will continue until the convergence of the model defined by the constraints.

Heuristics are used to select the two α values to optimize and in turn, it speeds up the convergence rate of the model. The heuristics are based on the error cache that is calculated (b value) and stored during training of the model.

The desired output of the SMO algorithm is a vector of α values which are mostly zeros other than the values which are closest to the decision boundary. These points closest to the decision boundary are nothing but our support vectors (SVs). They should be present near the decision boundary in order to maximize the margin. When algorithm will converge, then it will have very less number of vectors. It means that the resulting decision boundary will only depend on the training examples closest to it. This way adding more examples to the training set which are far from the decision boundary will not change the support vectors [11]. It means that if any unknown data is being added to the feature space, then it is not going to change the support vectors. That's why soft margin SVMs are not sensitive to outliers.

Proposed SVM-SMO algorithms We have implemented the binary class support vector classification algorithm which will be using kernel trick for faster computations. We are using the SMO algorithm to solve the SVM optimization problem.

Algorithm 1 shows the approach of making the SVM classification model using the SMO algorithm. In this algorithm, the feature vectors and the class information of training examples will be fed to SMO.

Algorithm 1: Proposed breast cancer prediction algorithm using SVM with SMO

```
svmSolver(filename)
Read the image data file
for each image img do
    Noise removal    # By applying Mean and Median filter
    Image enhancement    # By applying Fourier transform and IFT
    leftBreastData[ ] = segmentImageLeft(img)
    rightBreastData[ ]=segmentImageRight(img)
    featureVectors[ ]= featureExtraction(leftBreastData[ ], rightBreastData[ ])
end
model ← trainModel(featureVectors)
prediction(model,description)    # Pediction of testing cases
```

Algorithm 2 shows the pseudo-code for trainModel(). The trainModel() function selects the first α to optimize by applying some heuristic and passes this value to examineExample().

Then examineExample() selects the second α to optimize and passes the index of both α values to takeStep().

After getting the support vectors from SMO, these vectors are passed for the prediction on testing instances by calling prediction (model, description). Algorithm 3 shows the pseudo-code for the prediction function.

Algorithm 2: Proposed SMO training algorithm: trainModel()

```
trainModel(featureVectors[ ])
Initialization :
alphaChanged=0
examineAll=1
while (alphaChanged ≥ 0) or (examineAll) do
    if (examineAll) then

        #loop over all training examples
        for i in range(model.alphas) do
            examineRes= examineExample(i)
            alphaChanged = alphaChanged + examineRes
            if (examineRes) then
                supportVector[ ]= alphaChanged
            end
        end
    end
end
return (supportVector)
```

Algorithm 3: Proposed prediction algorithm

```
predictClass(testData[d],description)
Algorithm
lagrangeMultiplier[l]= model.alphas
supVector[v]= model.supportVector
while (testData[d]) ≠ 0 do
    t = 0
    while (supVector[v] ≠ Null) and (lagrangeMultiplier[l] ≠ Null) do
        t = t + lagrangeMultiplier[l]y_i y_j K
    end
    predictedClass[p ] = sgn(b + t)
end
```

6 Results and Discussion

6.1 Dataset Collection and Generation

For detection of breast cancer using thermal images, we require a dataset which contains many images in which both benign as well as malignant tumors are present. We have extracted significant information from these thermal images in order to train our classification model. In Wisconsin data set, each data has a class B or M (benign or malignant) along with 32 feature vectors which have been found by the Wisconsin Hospital. This dataset can be obtained from the UCI machine learning repository. Thermal images which we are using for the training and testing purpose have been taken from DMR Visual laboratory.

6.2 Results

The performance of each classification model is evaluated using four statistical measures: classification accuracy, precision, recall, and F-score [12, 13]. These measures are defined using true positive (TP), true negative (TN), false positive (FP), and false negative (FN). A true positive decision occurs when the positive prediction of the classifier coincided with a positive prediction of the physician. A true negative decision occurs when both the classifier and the physician suggest the absence of a positive prediction. False positive occurs when the system labels the benign case as a malignant one. Finally, false negative occurs when the system labels a malignant case as benign. Accuracy is defined as

$$\text{Accuracy} = (TP + TN)/(TP + TN + FP + FN) \tag{7}$$

precision can be interpreted as the positive predicted value

$$\text{Precision} = TP/(TP + FP) \tag{8}$$

whereas recall can be interpreted as true positive rate.

$$\text{Recall} = TP/(TP + FN) \tag{9}$$

An $F1$ score is a measure of the data that was accurately predicted. If its value is closer to 1, then the prediction is best and if it is closer to 0, it is the worse prediction. It is harmonic mean of precision and recall.

$$F1 - \text{score} = 2 \cdot \frac{\text{Recall} \cdot \text{Precision}}{\text{Recall} + \text{Precision}} \tag{10}$$

After the implementation of the above algorithm on our data set, we have computed all the above parameters and compared the results of SVM classification with and without SMO. Table 1 shows the comparison between the running time of the model using SVM with and without SMO. In the table, for every dataset, increase in speedup when SMO has been applied can be observed.

When our model's accuracy is compared with the existing model of Suresh et al. [1] on the same data set, then it is found that our model is giving better accuracy. Their model has accuracy of 90% whereas ours is having 94% accuracy. Recall in

Table 1 Time comparison of SVM classification with and without SMO

Data set	Time using SVM without SMO (s)	Time using SVM with SMO (s)
Sklearn data	0.43	0.085
WDBC	0.52	0.093
WPBC	0.48	0.091

Table 2 Results comparison with existing model

Results	SVM with SMO	Suresh et al., SVM without SMO
Precision	0.854	0.888
Recall	0.895	0.825
Accuracy	0.946	0.916

Suresh et al. [1] is 83% whereas in our model, it is only 89%. It means our algorithm is returning most of the relevant cases present in the data set. Table 2 shows the comparison between these two model's results.

Increase in accuracy and recall has been achieved because of SMO which reduces the false positive rate significantly. Since the SMO algorithm optimizes the Lagrange's multipliers, the number of resulting support vectors become very less and due to this, the classification time also increases significantly. In SMO, dot products of two feature vector are calculated in order to optimize α values and if there will be less number of vectors, then obviously the time consumption will be less.

7 Conclusion and Future Work

Conclusions

Early detection of breast cancer increases the survival rate of patients. In computer-aided diagnosis (CAD) first, patient's images data is stored in the feature vectors format. For collecting the data, feature extraction method is used to extract the important features from the images. These data have been fed to SVM classifier for the training and testing of the model. After the development of SVM classification model using the SMO algorithm, running time and accuracy is calculated and these results are compared with the existing model. SMO algorithm helps in speeding up the classification process as well as the accuracy of the result. After comparison with the existing model on the same dataset which does not use SMO for optimization of QP problem, our model is giving the better result in terms of recall and accuracy.

Future Work

Following are some of the suggestions for future works

- Parallelization of SVM classification model with SMO can be done using CUDA on GPU [14, 15] for larger datasets.
- Automatic feature selection by principal component analysis (PCA) can be done before feeding the data for training.

Acknowledgements We express our sincere thanks to the Department of Computer Science and Engineering, National Institute of Technology Calicut for their support toward completing the project.

References

1. Wakankar, A.T., Suresh, G.: Automatic diagnosis of breast cancer using thermographic color analysis and SVM classifier. In: The International Symposium on Intelligent Systems Technologies and Applications, pp. 21–32. Springer (2016)
2. Rastghalam, R., Pourghassem, H.: Breast cancer detection using spectral probable feature on thermography images. In: 2013 8th Iranian Conference on Machine Vision and Image Processing (MVIP), pp. 116–120. IEEE (2013)
3. Thermography: http://visual.ic.uff.br/dmi/prontuario/images.php?p=1. Accessed 5 Nov 2017
4. Choras, R.S.: Analysis of breast thermography. In: 2015 2nd World Symposium on Web Applications and Networking (WSWAN), pp. 1–5. IEEE (2015)
5. Venkataramani, K., Mestha, L.K., Ramachandra, L., Prasad, S., Kumar, V., Raja, P.J.: Semi-automated breast cancer tumor detection with thermographic video imaging. In: 2015 37th Annual International Conference of the IEEE Engineering in Medicine and Biology Society (EMBC), pp. 2022–2025. IEEE (2015)
6. Abdelaal, H.M., Elmahdy, A.N., Halawa, A.A., Youness, H.A.: Improve the automatic classification accuracy for arabic tweets using ensemble methods. J. Elect. Syst. Inform. Technol. (2018)
7. Elahi, M., Shahzad, A., Glavin, M., Jones, E., O'Halloran, M.: GPU accelerated confocal microwave imaging algorithms for breast cancer detection. In: 2015 9th European Conference on Antennas and Propagation (EuCAP), pp. 1–2. IEEE (2015)
8. He, H., Jin, H., Chen, J.: Automatic feature selection for classification of health data. In: Australasian Joint Conference on Artificial Intelligence, pp. 910–913. Springer (2005)
9. Ohashi, Y., Uchida, I.: Applying dynamic thermography in the diagnosis of breast cancer. IEEE Eng. Med. Bio. Mag. **19**(3), 42–51 (2000)
10. Sathya, S., Joshi, S., Padmavathi, S.: Classification of breast cancer dataset by different classification algorithms. In: 2017 4th International Conference on Advanced Computing and Communication Systems (ICACCS), pp. 1–4. IEEE (2017)
11. Platt, J.: Sequential minimal optimization: A fast algorithm for training support vector machines (1998)
12. Precision and Recall: https://en.wikipedia.org/wiki/Precision_and_recall. Accessed 5 May 2018
13. Müller, A.C., Guido, S.: Introduction to Machine Learning with Python: A Guide for Data Scientists. O'Reilly Media (2016)
14. Kirk, D.: Programming Massively Parallel Processors: A Hands-on Approach, 1st edn. Morgan Kaufmann Publishers, San Francisco (2010)
15. NVIDIA: Cudazone https://developer.nvidia.com/cuda-zone. Accessed Sept 2017

Clustering Performance Analysis

N. Karthika and B. Janet

Abstract Clustering plays a significant role in identifying the intrinsic structure of data. In this paper, various clustering algorithms are compared on real, numerical, categorical datasets around the cluster size. From the analysis, it is inferred that the repetition of KMeans many times does not bring better significant iterations since it starts randomly. It purely depends on the initial choice of the centroid of clusters. The sum of squared error decreases with increasing cluster size. The Expectation–Maximization (EM) is time-consuming than KMeans.

Keywords Clustering algorithms · KMeans · DBScan · EM

1 Introduction

Clustering is an unsupervised learning method. The goal of clustering is to identify an intrinsic structure/grouping of data in a collection so that the cluster has high intra-cluster similarity as well as low inter-cluster similarity [1]. It plays a foundational role in many areas. It is used to predict the performance of the students in educational research, anomaly detection in social network, and fraudulent transaction detection in crime department. E-retailers and E-commerce gain a lot by applying these clustering approaches in pricing segmentation, customer spending segmentation, branch geographical segmentation, and customer need segmentation. EM plays a vital role in medical image analysis through MRI segmentation, brain tissue segmentation, and also to identify the patterns in images.

Clustering can be broadly segmented into Exclusive clustering, Hierarchical clustering, Overlapping clustering, and Probabilistic clustering [2]. In Exclusive cluster-

N. Karthika (✉) · B. Janet
Department of Computer Applications, National Institute of Technology,
Tiruchirappalli, Tamil Nadu, India
e-mail: bharathikarthika@gmail.com

B. Janet
e-mail: janet@nitt.edu

B. Pati et al. (eds.), *Advanced Computing and Intelligent Engineering*,
Advances in Intelligent Systems and Computing 1082,
https://doi.org/10.1007/978-981-15-1081-6_3

ing, if a particular data item exists in a definite cluster, then it does not accommodate into another cluster, wherein Overlapping cluster approach uses fuzzy sets where a data item may be in one or more clusters with varying degrees of membership. In Hierarchical, the clusters are formed based on combining two closest clusters which start with every data item as an individual cluster and after certain iterations, it reaches the decided clusters. The final one is purely based on probabilistic methods. In this paper, we will examine three popular clustering algorithms namely KMeans, EM, and DBScan with various numerical, real, and categorical datasets.

2 Clustering Algorithms

2.1 Simple KMeans Method

KMeans method [3] is one of the exclusive clustering methods. It is considered as the most powerful, comprehensible, and quick method, but it does not have the ablility to handle noises [4]. The following algorithm 1 shows the functionality of the KMeans method.

Algorithm 1: Simple KMeans Method

Input: X: A database of n data items $\{x_1, x_2, x_3, \ldots, x_n\}$
k: the number of clusters
Output: A set of k clusters

```
  Arbitrarily select k data items as initial cluster centers.
Repeat
  Assign each data item to the cluster center to which the data item is closest.
  Update the cluster center.
Until
  No change in allotment.
```

The time complexity of KMeans is $O(tkN)$ where N is the dataset size, k is the number of clusters, and t is the number of iterations.

2.2 Expectation–Maximization (EM) Clustering

Expectation–Maximization is a two-step iterative method [5]. Though it is highly complex, it has the potential to handle noisy data as well as missing information. The below algorithm 2 shows the step-by-step procedure of the Expectation–Maximization (EM) clustering.

The complexity of Expectation (E) and Maximization (M) step relies both on the number of iterations and time to build up. The computational complexity is $O(dnt)$ where d is the number of input features, n is the number of objects, and t is the number of iterations [6].

Algorithm 2: Expectation–Maximization Method

Input: X: A database of n data items $\{x_1, x_2, x_3, \ldots, x_n\}$
k: the number of clusters
Output: A set of k clusters with maximum log likelihood
Do
Expectation step:
  For each data item $x \in X$, Estimate the membership probability of x in each cluster.
Maximization Step:
  Update the parameter which maximizes the likelihood.
Until
  Converge or Goto step 2

2.3 Density-Based Clustering

Density-based spatial clustering of application with noise (DBScan) is one of the popular density-based clustering algorithms [7]. Density-based algorithms are efficient in handling outliers and are also capable of exploring clusters of random shapes. This algorithm combines the data items according to significant density objective functions. Density is described as the number of data items in a specific neighborhood [8]. The following algorithm 3 gives the procedure of DBScan [7] .

Algorithm 3: DBSCAN Method

Input: X: A database of n data items $\{x_1, x_2, x_3, \ldots, x_n\}$
eps: The radius of neighborhood around a data item
minPts: The minimum number of data items, we wish to have in neighborhood to designate a cluster
Output: A set of k clusters with maximum log likelihood i.e., density
Repeat
  Arbitrarily select a data item which has not been allotted to a cluster or assigned as an outlier.
  Calculate its neighborhood to decide a core data item.
    Then
      Begin a cluster around the data item.
    Else
      Mark as an outlier.
  Extend the cluster by finding the data items which are directly reachable to the cluster.
Until
  All the data items are either assigned to cluster or marked as an outlier.

The complexity of DBScan is $O(N\log N)$, if the spatial index used or else $O(N^2)$ where N is the number of data points.

3 Experimental Results and Discussion

The clustering algorithms examined the following datasets which are downloadable from a UCI machine learning repository [9] using 7th Generation Intel core i7 processor, 16 GB of RAM, and 1 TB HDD for Windows with the help of Weikato Environment for Knowledge Analysis (WEKA) [10]. Table 1 gives the summarization of different datasets analyzed.

Sum of Squared Error:

It is the summation of the squared differences between each data point and its cluster's mean. It can be used to measure the deviation within a cluster. If all the data points within the cluster are identical, then the SSE would be zero. Obviously, the lesser the summation of intracluster distances, the higher the quality of the cluster.

$$\sum_{i=1}^{n}(x_i - \bar{x})^2 \tag{1}$$

All of the above-analyzed datasets show that when the number of clusters increase, then the sum of squared errors will decrease which is depicted in Figs. 1, 2, 3, 4, 5, 6, 7, and 8. The x-axis shows the value of cluster size wherein y-axis is the sum of squared error values. The curve has three different parts: a flat region direct toward the right, a sharp slope region to the left, and a curved transition part in the middle. The elbow bend which is highlighted in all the graphs show the optimal cluster size

Table 1 Summarization of characteristics of various test datasets

Dataset	Count of features	Count of clusters	Count of instances	Attribute description
Iris	4	3	150	Real
Wine recognition	13	3	178	Integer, Real
Wheat seeds	7	3	210	Real
Pima Indians diabetes	8	2	768	Numeric
Contraceptive Method Choice(CMC)	9	3	1473	Integer, Categorical
Ionosphere	34	2	351	Integer, Real
Glass	9	6	214	Real
BreastCancer	9	2	286	Real
Contact lens	4	3	24	Categorical
CPU	9	8	209	Integer
Labor	16	2	57	Integer, Real, Categorical
Soybean	35	19	683	Categorical

Fig. 1 Comparison of sum of squared errors on Wheatseeds data

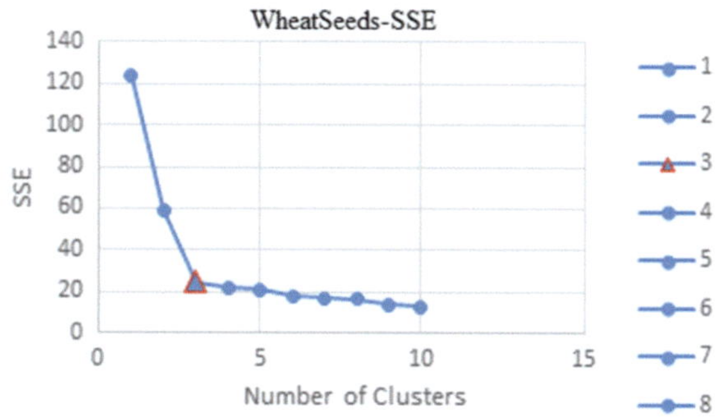

Fig. 2 Comparison of sum of squared errors on Iris data

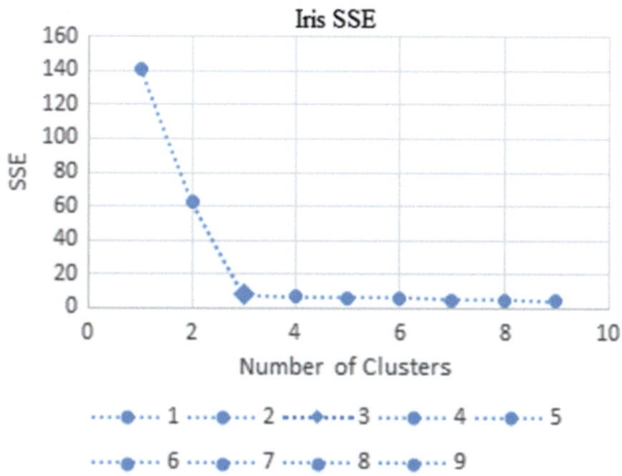

Fig. 3 Comparison of SSE on Pima-Diabetes data

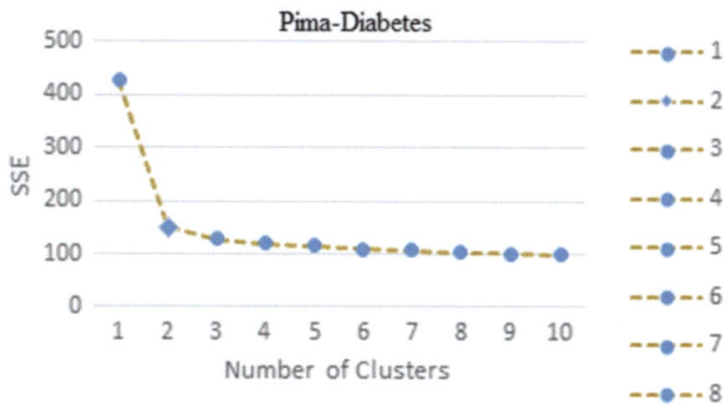

of the dataset, respectively. Wheatseeds, Iris and, Wine Recognition data have three as their optimal cluster size, whereas Pima-diabetes and Ionosphere have two as their significant cluster size, and Lens has four as its optimal size. Table 2 shows the sum of squared errors (SSE) of the KMeans method where the CPU dataset has the least sum of squared errors when compared to other datasets, as it has fewer features with lesser instances, whereas Soybean data has shown the highest sum of squared errors due to the nature of data, i.e., higher instances with higher attributes.

Fig. 4 Comparison of sum of squared errors on Wine Recognition data

Fig. 5 Comparison of SSE on Ionosphere data

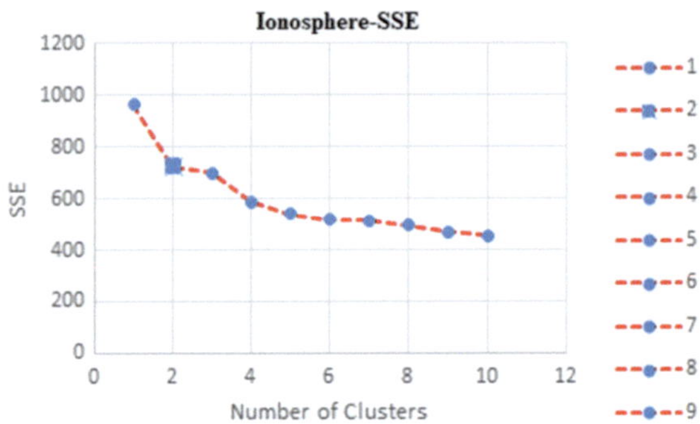

Fig. 6 Comparison of sum of squared errors on Lens data

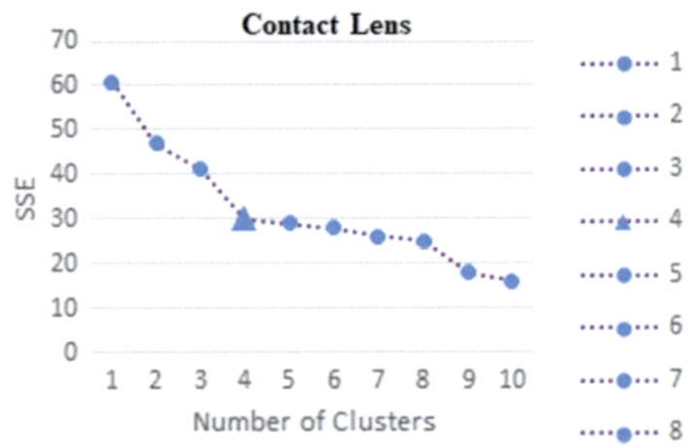

Fig. 7 Comparison of SSE on BreastCancer data

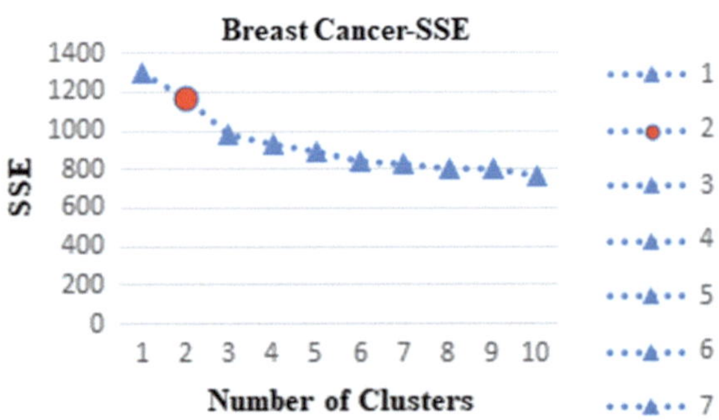

Fig. 8 Comparison of sum of squared errors on CMC data

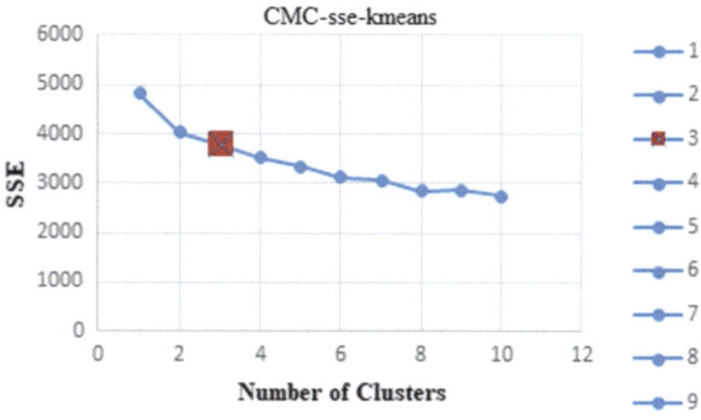

Table 3 depicts the build time utilized by the KMeans model for the examined datasets.

The DBScan method shown in Table 4 shows that the Ionosphere and Soybean data have taken more build time with a higher number of clusters as these data have a higher number of instances with more attributes. Different datasets have shown different execution time over a variety of clusters. CPU data runs almost faster than other datasets against various clusters due to the lesser number of features with small instances. Tables 3 and 4 interpret the time utilized to build the corresponding models. For datasets like Iris, Wheatseeds, Ionosphere, Lens, CPU, Labor consumed equal build time wherein datasets like Wine Recognition, Pima-Diabetes, CMC it is 1% faster and in Glass it is 2% faster than KMeans.

Tables 4 and 5 explore the build time taken by DBScan and EM methods. The values in bold indicate the optimal number of clusters for the corresponding datasets. For Lens data, both methods are equally good. For Wine Recognition data, 1% higher time is taken to yield the optimal clusters in EM than DBScan. In WheatSeeds it is 2% slower, in Pima-Diabetes 4% slower, in CMC 13% slower, in Ionosphere 1% slower, in Glass 2% slower, in BreastCancer 3% slower, in CPU 5% slower, and in Labor 2% slower than DBScan method. It infers that the EM method is time-consuming than DBScan.

Table 6 shows the details of the number of iterations for the varying cluster size. The iterations increase with increase in cluster size for Pima-Diabetes, Lens, Glass, and Wheatseeds datasets whereas there is a fluctuation between iterations and cluster size in the datasets like Iris, Breastcancer, Wine, Ionosphere, and CMC due to the different nature of the data and arbitrary selection of centroid in the data. The Pima-Diabetes and Glass data have shown that the number of iterations increase with the higher cluster size whereas Breast cancer data has taken a less number of iterations with more number of clusters. It explores that the number of iterations does not depend on the number of clusters. Repeating KMeans many times does not bring out better significant iterations since it starts randomly. The results will purely depend on the initial choice of centroid of clusters [11]. This may vary the next time. Thus the build time and iterations change across the different cluster sizes. It does not

Table 2 KMeans—sum of squared error

KMeans—sum of squared errors										
Number of clusters										
Datasets	**1**	**2**	**3**	**4**	**5**	**6**	**7**	**8**	**9**	**10**
BreastCancer	1299	1177	977	932	888	847	826	798	800	766
CMC	4812.479	4025.271	3797.147	3519.121	3328.512	3120.387	3059.08	2836.302	2861.883	2736.5
CPU	33.93	21.179	16.222	12.7338	11.7633	10.4407	8.6837	8.0715	7.7603	7.5988
Glass	195.36	118.203	77.124	75.335	66.136	52.185	49.948	48.473	48.048	38.479
Ionosphere	962.1902	726.103	698.254	585.511	537.736	518.416	512.518	492.882	466.828	452.85
Iris	141.138	62.143	7.817	6.613	6.293	6.131	5.202	4.852	4.672	4.623
Labor	165.483	137.794	119.522	106.3002	99.2303	96.885	96.63	83.534	77.496	70.576
Lens	61	47	41	30	29	28	26	25	18	16
Pima-Diabetes	426.792	149.517	127.72	119.183	115.412	110.235	107.189	103.468	101.103	99.246
Soybean	7161	6209	5765	5693	5262	5061	5051	4941	4923	4353
Wheatseeds	123.676	59.276	24.413	22.154	21.223	17.735	17.025	16.411	13.762	12.834
Wine	121.266	75.811	49.675	45.709	43.425	40.443	38.783	38.116	36.948	35.601

Table 3 Time to build model of various datasets against number of clusters—KMeans

KMeans—time to build model(in Secs)										
Number of clusters										
Datasets	**1**	**2**	**3**	**4**	**5**	**6**	**7**	**8**	**9**	**10**
BreastCancer	0.01	**0.02**	0.02	0.02	0.01	0	0	0	0.02	0.02
Contraceptive method choice	0.01	0.05	**0.05**	0.02	0.02	0.02	0	0.01	0.01	0.02
CPU	0	0.02	0.02	0.02	0	0.02	0.02	**0**	0	0.02
Glass	0.02	0	0	0	0.02	**0.02**	0.01	0	0.02	0.02
Ionosphere	0.01	**0**	0.02	0.02	0.02	0.02	0.03	0.01	0.01	0.01
Iris	0.01	0	**0**	0	0.02	0.02	0	0	0	0.01
Labor	0	**0**	0	0.01	0.02	0	0	0	0	0
Lens	0	0.02	**0**	0	0.01	0	0.01	0	0	0
Pima-Diabetes	0	**0.01**	0.03	0.03	0.03	0.02	0.03	0.02	0.03	0.03
Soybean	0.01	0	0.01	0.02	0.01	0.02	0.01	0.02	0.02	0.03
Wheatseeds	0.01	0	**0**	0	0	0	0	0	0	0.02
Wine	0	0	**0**	0	0	0	0	0.01	0	0

Table 4 Build time of model on tested datasets—DBScan(in Secs)

DBSCAN—build time(in Secs)										
Number of Clusters										
Datasets	**1**	**2**	**3**	**4**	**5**	**6**	**7**	**8**	**9**	**10**
BreastCancer	0.02	**0.02**	0.01	0	0	0	0	0	0	0
Contraceptive method choice	0.02	0.03	**0.02**	0.02	0.01	0.02	0.02	0.02	0	0.02
CPU	0	0	0	0.01	0	0	0	**0**	0	0
Glass	0	0	0	0	0	**0**	0.01	0.01	0.01	0.01
Ionosphere	0.04	**0.07**	0.04	0.03	0.02	0.02	0.01	0.01	0.02	0.02
Iris	0	0.01	**0**	0.01	0.02	0	0.01	0	0	0
Labor	0	**0**	0.02	0	0.02	0	0	0.02	0.01	0
Lens	0	0.01	**0**	0.01	0	0	0.01	0.02	0.01	0
Pima-Diabetes	0	**0.02**	0.01	0.01	0.01	0	0.03	0.02	0.02	0.02
Soybean	0	0.01	0.01	0.01	0.02	0.01	0.01	0.02	0.02	0.02
Wheatseeds	0.01	0	**0**	0.01	0	0	0	0	0.01	0
Wine	0	0	**0.01**	0.01	0.01	0.01	0.02	0.01	0	0.01

Table 5 Time to build model(EM) on various tested datasets

EM build time										
Number of clusters										
Datasets	**1**	**2**	**3**	**4**	**5**	**6**	**7**	**8**	**9**	**10**
BreastCancer	0.02	**0.05**	0.07	0.04	0.08	0.07	0.07	0.08	0.05	0.05
Contraceptive method choice	0.14	0.1	**0.15**	0.23	0.32	0.74	0.69	0.71	0.83	1.24
CPU	0.02	0.15	0.12	0.08	0.05	0.05	0.05	**0.05**	0.05	0.07
Glass	0.05	0.03	0.01	0.03	0.02	**0.02**	0.02	0.02	0.03	0.04
Ionosphere	0.08	0.06	0.04	0.03	0.15	0.09	0.08	**0.07**	0.1	0.1
Iris	0.03	0	**0.02**	0.02	0.02	0.02	0.01	0.05	0.03	0.01
Labor	0	**0.02**	0.01	0	0.02	0	0.01	0	0.02	0.01
Lens	0.01	0.02	**0**	0	0.02	0	0.02	0.02	0	0
Pima-Diabetes	0.04	**0.07**	0.05	0.07	0.1	0.07	0.1	0.14	0.14	0.15
Soybean	0.03	0.22	0.18	0.17	0.2	0.55	0.2	0.35	0.37	0.45
Wheatseeds	0.03	0.02	**0.02**	0.01	0.02	0.02	0.02	0.02	0.02	0.02
Wine	0	0.01	**0.02**	0.02	0.02	0.02	0.02	0.03	0.02	0.01

Table 6 Number of iterations—DBScan

DBScan—number of iterations										
Number of clusters										
Datasets	**1**	**2**	**3**	**4**	**5**	**6**	**7**	**8**	**9**	**10**
BreastCancer	1	**3**	4	4	4	5	7	5	4	4
Contraceptive method choice	1	5	**14**	6	8	5	6	10	6	6
CPU	1	12	10	10	8	8	12	**11**	11	12
Glass	1	9	7	8	7	**7**	11	10	11	16
Ionosphere	1	**8**	9	12	9	15	12	8	8	9
Iris	1	7	**3**	4	4	6	6	6	5	6
Labor	1	**3**	3	5	4	4	3	4	4	4
Lens	1	2	**2**	3	2	2	2	2	5	5
Pima-Diabetes	1	**4**	13	12	16	9	18	19	20	27
Soybean	1	10	10	9	9	7	6	6	6	9
Wheatseeds	1	5	**6**	9	7	7	11	11	13	15
Wine	1	5	**7**	5	6	8	8	13	5	6

Table 7 EM—number of iterations

EM-number of iterations										
Number of clusters										
Datasets	**1**	**2**	**3**	**4**	**5**	**6**	**7**	**8**	**9**	**10**
BreastCancer	1	**29**	39	35	14	36	34	30	19	22
Contraceptive method choice	1	10	**30**	34	37	100	75	66	68	100
CPU	2	42	82	18	24	2	2	**4**	4	1
Glass	2	3	2	2	3	**2**	1	2	1	1
Ionosphere	2	**2**	1	1	22	1	1	1	1	1
Iris	2	2	**1**	16	11	12	21	84	26	0
Labor	2	**7**	1	1	3	3	3	3	2	2
Lens	1	18	**0**	0	0	0	0	0	0	0
Pima-Diabetes	2	**46**	9	9	7	1	1	2	2	2
Soybean	1	6	22	11	14	43	5	13	14	16
Wheatseeds	2	12	**3**	1	2	2	1	1	1	1
Wine	2	11	**1**	1	1	1	1	1	1	1

mean that lower cluster size will lower the number of iterations as the difference in the nature of datasets and also the random choice of centroid.

Tables 6 and 7 depict the number of iterations employed by DBScan and EM to generate the desired number of cluster size. The values are highlighted in boldface. The tables display that Iris has 2% lesser, Wine Recognition has 6% lesser, Wheat-Seeds has 2% lesser, Ionosphere has 4% lesser, Glass has 2.5% lesser, and the CPU data has 1.75% lesser iterations taken by the EM method whereas in Pima-Diabetes it is 91% lesser, in CMC 53% lesser, Breastcancer 89% lesser, Labor data 57% lesser iterations taken by DBScan method.

For all the datasets, the log likelihood is improved with the cluster size. From the DBScan method in Table 8 and the EM approach in Table 9, it is shown that the Ionosphere and Glass data achieved better likelihood than other datasets. In terms of the Sum of Squared Errors, KMeans as well as DBScan methods perform equally for all the datasets. CPU and Lens data attained less sum of squared errors in both the methods due to the lesser number of instances and features.

Tables 8 and 9 display the log likelihood values of DBScan and EM approaches. It has been clear from the data that the Iris is 0.03% better, Wheatseeds is 48.8% better, Wine Recognition is 65% better, Pima-Diabetes is 2% better, CMC is 44.3% better, Ionosphere is 79% better, Breastcancer is 1% better, Lens is 0.4% better, CPU is 25.5% better, and Labor is 3% better than the DBScan method in yielding the desired number of clusters. For all the datasets, EM outperforms DBScan in terms of log likelihood.

For Breast Cancer, CMC, and Iris datasets, the EM method has taken a huge number of iterations than KMeans and DBScan methods. Initially for CPU dataset,

Table 8 DBScan—log likelihood

DBSCAN—log likelihood										
Number of clusters										
Datasets	**1**	**2**	**3**	**4**	**5**	**6**	**7**	**8**	**9**	**10**
BreastCancer	−10.0408	**−9.799**	−9.753	−9.658	−9.634	−9.624	−9.622	−9.627	−9.718	−9.663
CMC	−12.792	−12.4826	**−12.512**	−12.1913	−12.1705	−12.1012	−12.084	−12.099	−12.087	−12.0659
CPU	−47.0724	−42.9172	−41.701	−40.566	−40.145	−39.957	−39.743	**−39.342**	−39.446	−39.141
Glass	−4.577	−1.499	−0.696	0.214	1.662	**1.378**	1.6757	2.118	2.544	2.429
Ionosphere	−13.049	**−5.104**	−4.1612	−1.637	0.706	2.631	3.413	5.082	7.8002	8.168
Iris	−6.0342	−3.063	**−2.21007**	−1.965	−1.948	−1.889	−1.818	−1.725	−1.622	−1.61
Labor	−19.7903	**−18.512**	−17.4007	−17.345	−16.744	−16.648	−16.023	−15.467	−15.596	−15.029
Lens	−4.0995	−3.818	**−3.839**	−3.858	−3.924	−3.9392	−3.957	−3.979	−3.891	−3.907
Pima-Diabetes	−30.5773	**−30.211**	−29.705	−29.459	−29.239	−28.989	−28.919	−28.892	−28.636	−28.639
Soybean	−25.571	−23.238	−21.906	−22.205	−20.231	−19.426	−19.779	−19.257	−19.4206	−17.992
Wheatseeds	−6.644	−3.769	**−1.211**	−0.9404	−0.8501	−1.154	−1.019	−0.676	−0.373	−0.184
Wine	−23.697	−20.925	**−18.238**	−19.014	−18.6148	−18.303	−17.97	−17.899	−17.687	−17.3401

Table 9 Expectation maximization-log likelihood

EM-log likelihood

Number of clusters

Datasets	**1**	**2**	**3**	**4**	**5**	**6**	**7**	**8**	**9**	**10**
BreastCancer	−10.0408	**−9.689**	−9.365	−9.283	−9.417	−9.271	−9.266	−9.265	−9.272	−9.2703
CMC	−12.792	−12.223	**−11.933**	−11.806	−11.736	−11.6628	−11.625	−11.091	−11.549	−11.518
CPU	−47.0724	−42.228	−40.718	−40.014	−39.591	−39.065	−39.227	**−38.872**	−38.761	−38.885
Glass	−4.577	1.153	1.9708	2.362	4.394	**6.6673**	5.305	5.667	5.717	5.252
Ionosphere	−13.049	**1.4901**	−0.327	1.833	3.541	4.4142	4.6609	7.2931	10.845	12.524
Iris	−6.0342	−3.063	**−2.2104**	−2.035	−1.776	−1.623	−1.475	−1.416	−1.353	−1.03
Labor	−19.7903	**−17.896**	−16.706	−16.025	−15.276	−10.78	−9.187	−12.681	−12.015	−11.134
Lens	−4.099	−3.828	**−3.698**	−3.789	−3.747	−3.795	−3.839	−3.829	−3.858	−3.876
Pima-Diabetes	−30.577	**−29.687**	−24.972	−23.145	−27.751	−28.025	−28.054	−25.308	−26.485	−25.493
Soybean	−25.571	−22.985	−21.283	−20.335	−27.751	−28.025	−28.054	−25.308	−26.485	−25.493
Wheatseeds	−6.644	−3.606	**−0.206**	0.216	3.248	3.298	2.411	2.871	2.75	3.46
Wine	−23.697	−19.915	**−11.078**	−16.012	−15.739	−14.836	−14.466	−12.532	−12.161	−12.266

the EM method has taken more number of iterations whereas while increasing the cluster size, the number of iterations are reduced considerably. For Glass dataset, EM takes very fewer iterations than the other two. For Ionosphere, DBScan takes larger iterations than the others. Except for two clusters, the Labor, Lens, Wheat seeds, and Pima-Diabetes take very less iterations than the others. For Soybean, EM takes huge iterations except for two and six clusters.

For Breast cancer data, EM attains the desired output in high build time than DBScan and KMeans and DBScan took less time for more than two clusters wherein CMC, Glass and CPU data, KMeans and DBScan have been achieved in less time for all the variants of cluster size. In Glass data, when the cluster size increases DBScan took less time. In Ionosphere data, KMeans has taken lower build time than the others. Initially, it seems that EM and DBScan have taken more or less equal timings but when the cluster size increases, the build time also increases accordingly, whereas in Iris data, KMeans took less time in most of the cluster's size. In case of Labor, KMeans works faster than the others and in Lens data it is fluctuated between cluster size. In Pima-Diabetes, Wine,Wheatseeds, and Soybean datasets, the EM is slower than DBScan and KMeans. There are a variety of clustering algorithms which are related to various scenarios. Thus, it is difficult to find which method is the best but it is possible to say which algorithm is the most appropriate for a particular scenario.

For Iris data, EM and DBScan performed equally good in small cluster size, but for large cluster size, EM is better than DBScan method. Considering all the datasets, EM outperforms in terms of log likelihood than DBScan.

4 Conclusion

The study covers the performance analysis of different clustering algorithms over various datasets. The analysis is performed in terms of the sum of squared errors (SSE), the number of iterations, and the time taken to produce clusters against a number of clusters. Hence it is inferred from the experimental results that the iterations taken and time utilized will depend on the random initial selection of centroids. In terms of SSE, the performance of both KMeans as well as DBScan are equal. From all the analyzed datasets, it can be concluded that in terms of log likelihood the EM outperforms DBScan but it is time consuming.

References

1. Xu, D., Tian, Y.: A comprehensive survey of clustering algorithms. Ann. Data Sci. **2**(2), 165–193 (2015). 10.1007/s40745-015-0040-1. URL https://doi.org/10.1007/s40745-015-0040-1
2. Jung, Y.G., Kang, M.S., Heo, J.: Clustering performance *k*-means and expectation maximization algorithms. Biotechnol. Biotechnol. Equip. pp. s44–s48. URL http://doi.org/10.1080/13102818.2014.949045

3. Dash, R., Misra, B.B.: Performance analysis of clustering techniques over microarray data: a case study. Phys. A Stat. Mech. Appl. **493**, 162–176 (2018). https://doi.org/10.1016/j.physa.2017.10.032. URL http://www.sciencedirect.com/science/article/pii/S0378437117310427
4. Jain, A.K.: Data clustering: 50 years beyond k-means. In: Joint European Conference on Machine Learning and Knowledge Discovery in Databases, pp. 3–4. Springer (2008)
5. Saxena, A., Prasad, M., Gupta, A., Bharill, N., Patel, O.P., Tiwari, A., Er, M.J., Ding, W., Lin, C.T.: A review of clustering techniques and developments. Neurocomputing **267**, 664–681 (2017). https://doi.org/10.1016/j.neucom.2017.06.053. URL http://www.sciencedirect.com/science/article/pii/S0925231217311815
6. Firdaus, S., Uddin, M.A.: A survey on clustering algorithms and complexity analysis. Int. J. Comput. Sci. Issues (IJCSI) **12**(2), 62 (2015)
7. Boongoen, T., Iam-On, N.: Cluster ensembles: a survey of approaches with recent extensions and applications. Comput. Sci. Rev. **28**, 1–25 (2018). https://doi.org/10.1016/j.cosrev.2018.01.003. URL http://www.sciencedirect.com/science/article/pii/S1574013717300692
8. Rai, P., Singh, S.: A survey of clustering techniques. Int. J. Comput. Appl. **7**(12), 1–5 (2010)
9. https://archive.ics.uci.edu/ml/
10. Witten, I.H., Frank, E., Hall, M.A., Pal C.J. (eds.): Data Mining (Fourth Edition)-Appendix B—The WEKA workbench, 4th edn., pp. 553 – 571. Morgan Kaufmann (2017). https://doi.org/10.1016/B978-0-12-804291-5.00024-6. URL http://www.sciencedirect.com/science/article/pii/B9780128042915000246
11. Shih, M.Y., Jheng, J.W., Lai, L.F.: A two-step method for clustering mixed categroical and numeric data. Tamkang J. Sci. Eng. **13**(1), 11–19 (2010)

Heuristic Algorithm for Resolving Pronominal Anaphora in Hindi Dialects

Seema Mahato, Ani Thomas and Neelam Sahu

Abstract Artificial intelligence is a necessity for today's realistic world knowledge. The facts hidden by the anaphoric expressions can be revealed by anaphora resolution only. The relevance of anaphora resolution could not be avoided as its productivity affects the performance of text summarization, automatic question answering system, information extraction, etc. The paper comes up with implementations of algorithms for first, second, and third-person pronouns for Hindi language, with the ability to tackle intersentential anaphora within the scope of 3–5 sentences and subsequently the algorithms can be tailored-up for more sentences. Approximately, 698 sentences were experimented for feature selection of each type of personal pronoun. The proposed algorithms have been tested on the synthetic datasets of 1059 sentences which contain 712 pronominal anaphors out of 781 anaphoric pronouns. The F-measure evaluation for selected corpora gives promising results, indicating that the algorithms are effective in resolving Hindi pronominal anaphora.

Keywords Pronominal anaphora resolution · Machine learning · Syntactic rules · Morphological knowledge

1 Introduction

Machine learning is a significant phase in artificial intelligence and is related to develop strategies so that a system could expertise itself to learn with data. While coding text manually, our human brain can easily and quickly detect the referring entity but finding equivalence class for anaphora using machine learning in text or discourse needs hard work on perceptive of the necessities for resolution. A machine has

S. Mahato (✉)
Dr. C.V. Raman University, Bilaspur, Chattisgarh, India
e-mail: seema_mahato@yahoo.co.in

A. Thomas
Department of IT, Bhilai Institute of Technology, Durg, Chattisgarh, India

N. Sahu
Department of IT, Dr. C.V. Raman University, Bilaspur, Chattisgarh, India

B. Pati et al. (eds.), *Advanced Computing and Intelligent Engineering*, Advances in Intelligent Systems and Computing 1082,
https://doi.org/10.1007/978-981-15-1081-6_4

to train with massive amount of data to mimic such intellectual as human brain does. Generally, a noun phrase is replaced by a pronoun to avoid its excessive restatement and a pronoun becomes anaphoric when point left to any entity.

For example,

1. सोहन ने गौतम से उसे छाता देने को कहा.

The word उसे is a pronominal anaphor substituted for सोहन and so सोहन is a referent or antecedent. The task of connecting a pronoun with a referring entity occurs in the same or previous clause/sentences deals pronominal anaphora resolution, i.e., to determine the noun phrase (NP) referent for anaphors. Not all pronouns are anaphoric in behavior. The personal pronouns whether intimate, honorary, or distal make a huge list of pronominal anaphors. Time to time different approaches/theory and system have been put forward by researchers in this area but nobody claims to be fully automatic. Anaphora cannot be identified by self as hold none information, so it should be determined to help other NLP applications which have inbuilt AR module such as market intelligence system, machine translation system, reputation monitoring, document summarization, etc. The complexity of process and role of anaphor could be understood by this.

Most of the works on resolving Hindi anaphora are done on written text rather spoken Hindi dialects. In written text, the word order rules are followed but spoken dialogues are unbound with the rules. The presence of personal pronoun or pronominal anaphora in spoken dialogues is found in abundance and their antecedents are mostly NP. Different levels of features such as syntactic, semantic, etc. have been employed by researchers for analyzing Hindi anaphora. The limited literature on Hindi anaphora resolutions show Prasad and Strude [1] as earliest work which was based on salience ranking of candidate antecedent. Sobha et al. [2] developed a rule-based multilingual system "Vasisth" for resolving anaphora in Hindi and Malayalam. Dutta et al. [3] had resolved Hindi anaphora modifying the Hobbs' Naïve Algorithm. Lakhmani et al. [4] implemented Gazetteer method for Hindi pronominal anaphora resolution. Dakwale et al. [5] used dependency structure and relations to employ hybrid approach for identifying referent for Entity pronoun. Mahato and Thomas [6] investigated few pronominal anaphora and achieved remarkable result.

Hindi is a free word ordering language, and therefore pronouns are common in sentences and in which personal pronouns occur very often. As Hindi is an inflectional language, the entity reference by a pronoun also gets shifted depending upon the grammatical case to which a pronoun falls and the attached case marker. Case marker should not be confused with "Vibhakti" in Hindi as former one treated as postposition or suffixes, whereas subsequent one is the tense marker. Paninian grammatical model [7] suggested such six cases and different case markers to identify subject, object, and relation between them which are featured and explored by authors. So, the Hindi parser based on Paninian grammatical model was chosen for initial preprocessing phase for this ongoing research. Shallow parser [8] has been employed to parse sentences and the resultant was captured for further processing. The hybrid approach integrates rule based and machine learning for resolving anaphora analyzing morphological information deeply which is good enough to attain state-of-the-art anaphora resolution results. This pronominal anaphora resolution added 22 new rules for performance improvement. The features used in these rules are based on dependency relation and does not need any domain knowledge. Lexical information such as verb agreement, gender, and number agreement are added too for feature selection. The past imperfection in the algorithms was taken into account and scope for searching an antecedent also extended from intrasentential to intersentential discourse. The hybrid approach presented by the authors is different from other hybrid approaches as named entity recognizer (NER) to categorize the named entity and the tool to check animacy has been not utilized. The trained rules crafted for each category of personal pronoun, however, are not adequate for dealing the occurrences of anaphor in all cases. For example, consider the following sentences in (2):

2. "राबिया खान ने जिया खान की कुछ तस्वीरें बाहर की और बताया कि सूरज पंचोली जिया खान के साथ मारपीट करते थे। जिया उनकी ये बातें सहती थीं क्योंकि वो प्यार में थी। राबिया की मानें तो इस रिश्ते में सूरज से उनकी बेटी को सिवाय गालियां, मारपीट और बेइज़्जती के कुछ नहीं मिला। सूरज उन्हें केवल इस्तेमाल कर रहे थे जबकि जिया प्यार में थी।"

The referential link of these sentences is: (सूरज पंचोली→उनकी), (जिया→वो), (प्यार→इस), (राबिया→उनकी), (जिया→उन्हें). The objective of authors is to identify the equivalence class of each entity in the link disjointedly in the form of mapping table paired in (antecedent→anaphor) format. Few constraints have been enforced to select personal pronouns to be used as pronominal anaphor.

2 Generation of Dataset

The authors have generated corpus by randomly collecting 1757 sentences from different domains which are basically spoken discourse as they reveal unlike temperament of speaker than the texts in print. The texts comprise individual statements, conversations between persons, discussion, from newspaper articles related to about everyday subjects. Mostly, in such discourse, pronouns belong to personal pronouns and their antecedent resides in local domain or bind with the expressions in previous sentences and are NPs. From these sentences, 698 were used for training set and rest for testing purpose.

Test dataset contains all total 781 instances of pronouns and in that 712 are anaphoric pronominal pronoun which contains 233 first-person pronoun (FPP), 101 belongs to second-person pronouns (SPP), 378 are third-person pronouns (TPP), and rest not include as they were other types of anaphors. The first- and third-person pronouns were found to be most frequently occurring in test dataset. A collection of distinctive instances of sub-categories of personal pronouns is shown in Table 1.

Table 1 Statistics of anaphors

Personal pronouns	Count of pronouns sub-category-wise	Total pronouns category-wise	Percentage of sub-category-wise pronouns (%)
First-person singular pronoun	154	233	66
First-person plural pronoun	79		33.9
Second-person intimate pronoun	68	101	67.3
Second-person honorary pronoun	33		32.2
Third-person singular pronoun distal	154	378	40.7
Third-person distal plural pronoun	114		30.1
Third-person singular proximal pronoun	42		11.1
Third-person plural proximal pronoun	68		17.9

3 Architecture of the Proposed System

The proposed system works in two stages: preprocessing phase and anaphora resolution phase. The preprocessing phase deals with the input text and supplies it to the parser to get morphological information. The information so obtained has to undergo feature selection module to filter out the unwanted information and generate the list of anaphors and candidates of antecedents which will then be forwarded to the next phase. The resolution phase activates the module as per the type of anaphor to resolve and determine actual antecedent. Briefings of both the phases are given below.

3.1 Preprocessing Phase

The preprocessing phase includes tasks such as annotation of text, filtration of unwanted NPs, categorization of anaphors, selection of potential candidate, and then advancement of list of potential candidates and anaphors to the next level to take final decision on antecedent. The system is totally free of real-world/domain knowledge but use morphological knowledge unexploited. The proposed system takes the input text which is escalated to Hindi shallow parser [8] for part-of-speech (POS) annotation. The parser parses the text sentence by sentence and annotates them with POS tags and generates parsed text in form of a Treebank in a Shakti standard format (SSF) [9]. The Treebank precisely contains three different levels of information for each word (lexical item): lexical item with boundaries, POS category, and morphological attributes set separated by "," (comma) with allied values in a fixed sequence. All the NP chunks in each sentence preceding the anaphors from the parsed text are collected in an array. Not all NPs are considered as candidates for antecedents rather authors have used a set of term patterns for labeling the NP chunks which may contain a single or group of nouns in it. Term patterns in form of [NP (N, CM)] were defined to identify the NPs that can be the possible candidate where "N" is a noun and "CM" is a case marker which can be "ने","से","को","का", "के", "के द्वारा", or "के लिये" and NPs those following the anaphor are excluded. The rules defined automatically identify these term patterns from the annotated data. Generally, the pronouns are anaphors, but this work has been limited to resolution of pronominal anaphora, so all the pronouns belonging to personal pronouns only were gathered in the arrays category-wise as first, second, or third to maintain their account easier.

The filtration process generates provisional tabular information called filtration table to make an account of all NPs preceding an anaphor in each individual sentence in a corpus as shown in Table 2. Each NP and pronoun (PRP) has been numbered on basis of its occurrence in a sentence and within it, in an intermediate clause. A sentence and clause are identified by sentence id (SID) and clause id (CID). A sentence may have multiple clauses. The decision on selecting intermediate clause has been taken on basis of verb phrase. This information helps selection procedures to focus on preceding NPs while resolution and rejecting demonstrative pronouns which have DEM POS tag. In Table 2, the first column indicates the sentence number and the presence of pronouns in same sentence maintained in second column. These pronouns belong to FPP, SPP, or TPP and occur in particular clause as in column third and fourth. When a sentence has multiple clauses, it may be possible that an

Table 2 Filtration table

SID	PRP in sentence	Category of PRP	CID of PRP	Noun in NPs	CID of NP
S1	nil	nil	nil	लीना, किशोर, शाम	Nil
S2	वह	TPP	C1	लीना,शाम, रात, मिलन, समय	लीना (C2), शाम (C2), रात (C2), मिलन (C2), समय (C3)
S2	मुझे	FPP	C3		
S3	वो	TPP	C1	नगरी, दिल	नगरी(C2), दिल (C3)
S3	ये	DEM	C2		
S3	यहां	TPP	C2		
S4	हम	FPP	C1	शहर	शहर(C1)

antecedent of an anaphor resides in same sentence. In such condition, the clause id helps in selecting the right candidate. The last two columns have the noun in NPs in each sentence and their clause id. Example of a filtration table (Table 2) of a sentence (3) has been given below:

3. लीना के मुताबिक किशोर रोज शाम होते ही कुछ देर के लिए बहुत उदास हो जाया करते थे। वह कहते थे की लीना जब शाम और रात का मिलन होता है, तब मुझे बहुत बेचैनी होती है। कई बार वो कहते कि ये मुंबई मायावी नगरी है और यहां अब दिल नहीं लगता। हम सब छोटे से शहर खंडवा चलते हैं।

After preparing the list of NPs for antecedent candidate and pronominal anaphora to resolve, a selection procedure has been conducted for retrieving a list of potential candidates. Once the list of anaphors to resolve and potential candidate were finalized and prepared, they have been handed over to pronominal anaphora resolver module.

3.1.1 Categorization of Anaphors

Categorization of anaphors has been done through the morphological knowledge available in the parsed text. The "वह", "अब, जब" and "कुछ, किसी, कौन" has been POS tagged as pronouns but sometimes "वह" is a demonstrative pronoun, "अब, जब" are adjective and few others are indefinite pronouns, and so were totally ignored and eliminated from the resolution list. As Hindi language has free word order, some personal pronouns were cataphoric in nature that were realized and rule has been defined to eliminate it. Therefore, if the corpus begins with a pronoun then the respective pronoun is considered as cataphoric.

Hindi pronouns cannot be discerned on basis of gender. For example, वह सितार बजाता है, वह स्कूल मे पढाती है. Here, the verb phrases "बजाता है" indicates about its masculine gender and "पढाती है" indicates feminine gender, so the associated pronoun "वह" has masculine and feminine genders, respectively. So "वह" can be used for any gender. When the sentence "मैंने काम कर दिया" is parsed in Hindi shallow parser, then the Treebank shows "any" gender attribute for the verb which indicates that the associated pronoun can be used for any gender: masculine, feminine, or neuter. Thus, gender of neighboring verb phrase helps in identifying gender of pronoun in Hindi. These features of verb phrase can be use in resolution purpose too.

3.2 *Feature Extraction for Anaphora Resolution*

All the selection rules, constraint, and criteria required for efficient machine learning resolution are represented as set of features applied in the modules. These features

Table 3 Features

Feature name	Description
lineList[]	Array holds unique id for each sentence in a discourse
xq	Count of sentences in a discourse
head[][]	Store head of each sentence
NPwithKA = []	NPs with "ka" case marker
NPwithNE = []	NPs with "ne" case marker
NPwithSE = []	NPs with "se" case marker
NPwithKO = []	NPs with "ko" case marker
NNwithNULL = []	NPs without case marker
AnaphoreLine = []	Total anaphora in a discourse
finalAnaphor = ""	Last anaphor in a discourse
firstAnaphor = ""	First anaphor in a discourse
anaphorInSentence = [][]	Total anaphora in a sentence
lastcatchednoun	Last noun in a sentence
sublist[]	Substitution list
suplist = []	Intermediate substitution sentence
pronoun[][]	Pronouns in each sentence
FPP[]	List of first-person pronoun in a discourse
SPP[]	List of second-person pronoun in a discourse
TPP[]	List of third-person pronoun in a discourse

explain the attributes of pronoun, syntactic property of antecedent candidate, association of case markers with noun or noun phrase, etc. Table 3 listed out some important features.

3.3 Pronominal Anaphora Resolution Phase

The anaphora resolution phase consists of features necessary for extraction and selection of antecedent candidate, algorithms for each category of pronominal pronouns, resolver and substitution module, and finally represents the resultant in form of mapping table. The number and gender agreement have been implemented to check the plurality and gender of each anaphor and compared with candidate antecedent. Different values have assigned if any of the attribute was found matched. The syntactic features of head noun or NPs have also been considered as important features. The resolver module constitutes a set of predefined feature selection and factors and machine learning based algorithms. The resolver module has multiple sub-modules

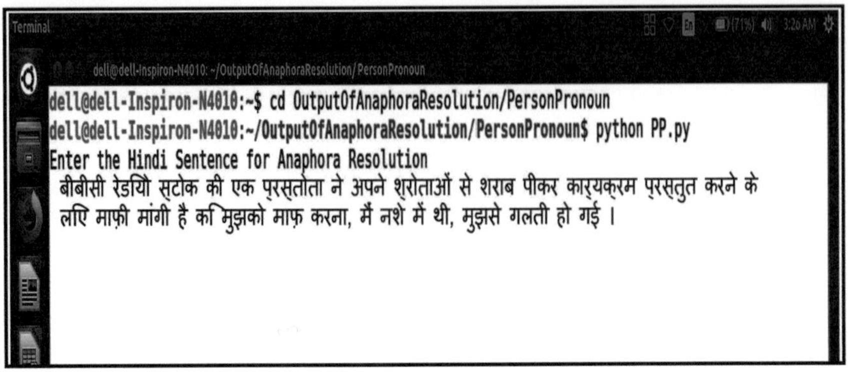

Fig. 1 Input screen

designated for individual pronoun class. The machine learning algorithms are capable of resolving intersentential and intrasentential pronominal anaphora within the scope of three sentences. The substitution module has been linked with all the submodules. The purpose of this module is to make the resolution easier for next iteration and to help the system to construct mapping table automatically.

3.4 Input and Output Screen

Snapshot of input screen of the proposed system for the sentence (4) as input text is given in Fig. 1. On executing the application, the system will prompt user to type or paste the sentence corpus and on pressing enter button the resolution result will get displayed in a browser in form of mapping table. Figure 2 shows the output screen.

4. बीबीसी रेडियो स्टोक की एक प्रस्तोता ने अपने श्रोताओं से शराब पीकर कार्यक्रम प्रस्तुत करने के लिए माफ़ी मांगी है कि मुझको माफ़ करना, मैं नशे में थी, मुझसे गलती हो गई ।

4 Test and Evaluation

Table 4 shows the sample of the information that has been observed and noted from the mapping table generated by the proposed system for overall evaluation.

Performance of anaphora resolution algorithm was examined using precision and recall and the effect of adding more feature selection criteria has been recorded. There are 712 pronominal anaphors captured from the corpus of 1059 sentences. Account of individual category of pronouns is given in Table 5. For these pronominal anaphors, 1780 numbers of antecedent–anaphor pairs have been created initially which were

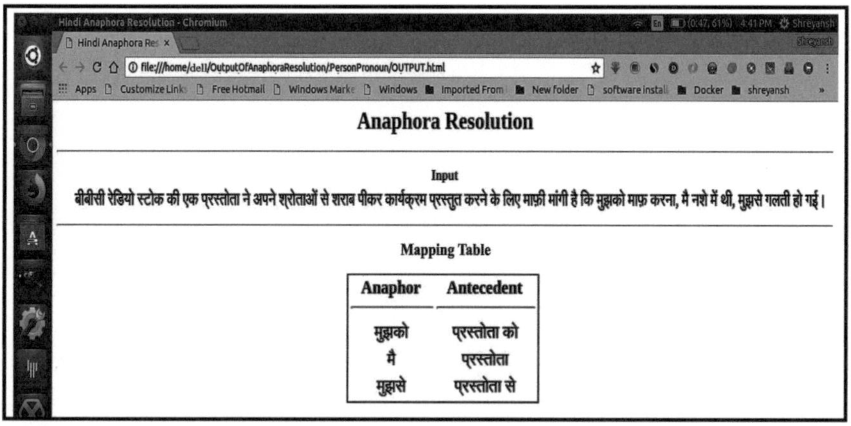

Fig. 2 Output screen

investigated and filtered by feature selection module and thereby reduced to precise identification of 1424 pairs. These pairs were tested and evaluated with respect to the anaphora resolver module based on their morphological categories. Tests were conducted using standard MUC evaluation metrics: precision, recall, and F-score as shown in Table 5.

The efficient performance of these algorithms was due to the integration of advance feature selection set and the modified decision rules in resolver. The result identified 683 pronominal anaphors out of 712 occurrences in which 453 were found resolved correctly. The accuracy rate for first-person pronoun (FPP) is 78%, for second-person pronoun (SPP) is 71%, and for third-person pronoun (TPP) is 52%.

5 Conclusion

The proposed work focuses on the rules imposed on knowledge poor information. The rules crafted and feature selection criteria were prepared by keen observations on training dataset and the effectiveness of each feature added to the accuracy of overall performance. The gap between the achieved accuracy and the utmost was due to the presence of referent in embedded clauses or phrases of the discourse within the context which has been explored and will be included in further research. The authors are also keen to identify the referent for non-pronominal types of anaphora and to compare it with system of other researchers.

Table 4 Information recorded for evaluation purpose

	Sentences	PRP	N	Output from mapping table	R	RC	RNC	UR
[1].	मोदी ने विश्वास जताते हुए कहा कि भाजपा गुजरात में 182 विधानसभा सीट में 151 सीटों पर कब्जा करेगी। उन्होंने कहा, "कितनों को कामराज और देभरभाई यू.एन) देभर) याद हैं जो राष्ट्रीय स्तर पर कांग्रेस के अध्यक्ष थे। जो पार्टी एक परिवार से ज्यादा नहीं सोच सकती, उनसे क्या उम्मीद करेंगे?"	उन्होंने , उनसे	2	उन्होंने → मोदी ने, उनसे → मोदी ने	2	2	0	0
[2].	दिल्ली में रोड शो के दौरान एक व्यक्ति के थप्पड़ जड़ने पर आम आदमी पार्टी के नेता अरविन्द केजरीवाल ने कहा कि हर बार हमला मुझपर ही क्यों?	मुझपर	1	मुझपर → केजरीवाल ने	1	1	0	0
[3].	राजद अध्यक्ष लालू प्रसाद ने कार्यकर्ता व नेताओं को समझाया है कि 'मुझपर संकट आ सकता है, मुझपर कोई भी हमला हो सकता है।	मुझपर, मुझपर	2	मुझपर → प्रसाद ने, मुझपर → प्रसाद ने	2	2	0	0

where N = count of PRP, R = total anaphora resolved, RC = total anaphora resolved correctly, RNC = total anaphora resolved incorrectly, UR = total anaphora unresolved

Table 5 Evaluation results

Algorithms	No. of pronominal anaphora	Total resolved		UR	*P*	*R*	F-score
		Resolved correctly	Resolved incorrectly				
FPP	233	184	37	12	0.83	0.78	0.80
SPP	101	71	13	17	0.84	0.70	0.77
TPP	378	198	138	42	0.58	0.52	0.55

where FPP = first-person singular pronouns, SPP = second-person singular pronouns, TPP = third-person pronoun, UR = total anaphora unresolved, P = precision, R = recall

Acknowledgements The authors gratefully acknowledge the support of Shreyansh Bafna, B.E. (CS), BIT Durg (C.G.) for coding the algorithms in Python and deployment.

References

1. Prasad, R., Strube, M.: Discourse salience and pronoun resolution in Hindi. In: Penn Working Papers in Linguistics, vol. **6**(3), 189–208 (2000)
2. Sobha, L., Patnaik, B.N.: Vasisth: An anaphora resolution system for Malayalam and Hindi. In: Symposium on Translation Support Systems (2002)
3. Dutta, K., Prakash, N., Kaushik, S.: Resolving pronominal anaphora in Hindi using Hobbs algorithm. Web J. Form. Comput. Cognit. Linguist. 10 (2008)
4. Lakhmani, P., Singh, S.: Anaphora resolution in Hindi language. Int. J. Inf. Comput. Technol. **3**(7), 609–616 (2013)
5. Dakwale, P., Mujadia, V., Sharma, D.M.: A hybrid approach for anaphora resolution in Hindi. In: Proceedings of the 6th International Joint Conference on Natural Language Processing, pp. 977–981. Japan (2013)
6. Mahato, S., Thomas, A.: Exploring semantic information from Hindi dependency treebank for resolving pronominal anaphora. Int. J. Comput. Appl. 0975–8887 (2015)
7. Mahato, S., Thomas, A.: Machine learning approach for resolving pronominal anaphora using Hindi dependency treebank. In: Proceedings of BITCON-2015 Innovations for National Development. IJAERS, vol. IV(II), pp. 155–159 (2015)
8. Mannem, P., Bharati, A.: Introduction to the shallow parsing contest for South Asian languages. In: Proceedings of IJCAI Workshop on Shallow Parsing for South Asian Languages (2009)
9. Bharati, A., Sharma, D.M., Husain, S., Bai, L., Begum, R., Sangal, R.: Anncorra: treebanks for Indian languages, guidelines for annotating Hindi treebank (version 2.0) (2009)

End-to-End Reinforcement Learning for Self-driving Car

Rohan Chopra and Sanjiban Sekhar Roy

Abstract Most of the current self-driving cars make use of multiple algorithms to drive. Furthermore, most of the approaches use supervised learning to train a model to drive the car autonomously. This approach leads to human bias being incorporated into the model. We implement the Deep Q-Learning algorithm to control a simulated car, end-to-end, autonomously. The algorithm is based on reinforcement learning which teaches machines what to do through interactions with the environment. The application of reinforcement learning for driving is of high relevance as it is highly dependent on interactions with the environment. Our model incorporates a CNN as the deep Q network. The system was tested on an open-source car-racing simulator called TORCS. The Deep Q-Learning approach allows the system to be more efficient and robust than a system that has been trained solely through supervised training. Our simulation results show that the system is able to drive autonomously and maneuver complex curves.

Keywords Deep learning · Reinforcement learning · TORCS · CNN · Deep Q network

1 Introduction

Road transport is one of the most dangerous means of transport available today, yet everyday millions of people use it to travel. The high rate of accidents is attributed to the slow response time of drivers and fatigue. Jams are caused due to the slow response of human drivers. Autonomous driving can thus reduce the number of accidents worldwide by a significant amount. It will also enable humans to be more efficient and help them spend less time traveling. Currently, autonomous cars employ

R. Chopra (✉) · S. S. Roy
School of Computer Science and Engineering, VIT University, Vellore, India
e-mail: rohan.chopra2014@vit.ac.in

S. S. Roy
e-mail: s.roy@vit.ac.in

B. Pati et al. (eds.), *Advanced Computing and Intelligent Engineering*,
Advances in Intelligent Systems and Computing 1082,
https://doi.org/10.1007/978-981-15-1081-6_5

multiple algorithms to drive the car by decoupling the driving task to multiple small subproblems and then combining all the results to drive the car autonomously. There are two main issues with this approach: first, the subproblems are very complex as depicted in the work of Jo et al. [1, 2]; this approach wastes computation power as well. For example, one of the subproblems is object recognition, the subsystem would recognize and label everything in the frame as opposed to only the objects that are extremely important. Second, combining the results of the subproblems can be tricky and might cause serious performance issues. These issues have pushed researchers to find better alternatives for autonomous driving, which led to the use of Convolutional Neural Networks (CNNs) [3] to build a simplified system to drive cars. Researchers then used expensive hardware such as LIDARs (Light Detection And Ranging), RADARs (RAdio Detection And Ranging), and other sensors like IMU (Inertial Measurement Unit) along with images and trained deep neural networks with it. This approach, called supervised learning, required a large amount of labeled data which requires a lot of human effort. Supervised learning has another major issue where the data that is collected for the use of supervised learning has a human bias. These supervised learning systems had a pretty good performance but, it did not perform better than humans as the data that was used to train it was generated by humans and suffered personal bias. Thus it is hard to create an end-to-end model using supervised learning.

DeepMind's work [4] on deep reinforcement learning opened up new research avenues in problems where the system has to interact with the environment. This deep reinforcement learning proved to give superhuman performance in tasks that did not even perform as good as a human, using previous strategies. Reinforcement learning avoids the problem of collecting a large amount of labeled data and the potential human bias associated with it. The framework of a reinforcement learning system is shown in Fig. 1. The major problem faced by the reinforcement learning algorithm for autonomous cars is the training process which cannot be done on real-life systems as the training would involve many collisions and unpredictable situations. To solve this problem, reinforcement learning models are trained on simulators.

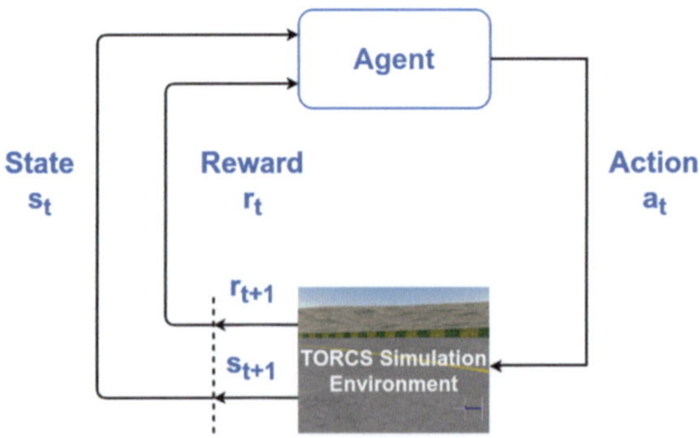

Fig. 1 Framework of reinforcement learning system

Thus, we propose a robust system built using deep reinforcement learning that would drive a car autonomously and give much better efficiency than the previously used methods. To test and train this system, we will use the open-source car-racing simulator called TORCS along with the simulated car-racing add-on which gives us access to different parameters such as velocity and angle of the car with respect to the road.

The remaining part of this paper is as follows: Sect. 2 presents a comparative analysis of currently available systems. Section 3 provides an overview of the proposed system. Experimental results and the discussion of the findings are given in Sect. 4. Section 5 concludes the paper and draws the future lines of research.

2 Related Works

Driving a car autonomously has been approached by several authors, but there has not been much research done in using deep reinforcement learning using raw pixel data to create an end-to-end autonomous car.

Mapping-Based Approach. Jo et al. [1, 2] used multiple different algorithms like perceptron, localization, planning and control to drive the car. This caused their system to be highly complex and difficult to adapt to different scenarios.

CNN-Based Approach. Bojarski et al. [5] used CNNs to drive a car autonomously without the need of a LIDAR; this work was among the first to completely rely on Convolutional Neural Networks to drive a car. This work was different from the traditional methods where highly complex algorithms were used to drive the car. Although their work enabled cars to drive autonomously, it lacked in that the system drove exactly or even worse than the driver who trained it due to the requirement of labeled data for supervised learning which had a human bias. This prevented the system to drive better than humans. Yang et al. [6] used a simulation environment but with a slightly different approach; they trained three CNNs on different sections of the same image from the camera, thus being affected by the issues surrounding supervised learning.

Imitation-Based Approach. Kuefler et al. [7] used a completely different approach where they used Generative Adversarial Networks [8] to mimic human driver behavior. Their system performed quite well over an extended period, but this system would suffer the same issues that the aforementioned pure CNN approaches would.

Deep Reinforcement Based Approach. Tai and Liu [9] used deep reinforcement learning to perform exploration in an unknown environment using a robot simulation. The results showed that the robot was able to successfully avoid obstacles and travel freely in the environment. Sallab et al. [10] incorporated Recurrent Neural Networks with the deep reinforcement learning. Their framework allowed them to create a self-driving car which could successfully drive in a given lane. But, they used parameters provided by the simulation environment TORCS like trackPos as an input to the system instead of the images or video feed of the car. The parameter trackPos is the distance of the car from the center of the track and there is no easy way of calculating this parameter in a real-world scenario. Lange et al. [11] used raw pixel data as an input to the system. They mapped raw input data to a low-dimensional information vector using an autoencoder. The actual control signal is then computed on this low-dimensional information vector, which is also the input to the reinforcement learning framework. Saunders et al. [12] discussed how imitation learning via human intervention, could make training deep reinforcement agents for the self-driving task more data and time efficient, but this approach would impart a human bias to the agent.

The main gap identified in the survey is that there has not been significant research done in using deep reinforcement learning using raw pixel data to create an end-to-end autonomous driving system. Most of the work focuses on using only the sensor readings such as velocity and angle of the car with respect to the direction. A lot of the work that uses deep reinforcement learning uses parameters like trackPos which are hard to recreate in real-world situations.

3 Overview of the Proposed System

Our system consists of a deep Q network which is a Convolutional Neural Network (CNN) with a few fully connected layers at the end. We perform reinforcement learning to make the deep Q network output correct actions for the given input image. We use TORCS, a simulation environment to get the images from a driver's perspective along with other sensor readings. The automation required to drive the car and collect data from the simulation environment is done using a Python script. Moreover, the deep Q network along with the reinforcement training code is written in Python using tensorflow as the deep learning framework.

3.1 Convoluted Neural Networks

Convolutional Neural Networks (CNNs) are neural networks with convolutional filters that assume the input data to be represented as images. CNN's efficiently handle the high dimensionality of raw images due to the use of convolutional filters. Highly complex pattern recognition can be achieved by using a network of neurons. The first few layers are convolution filters where the network applies different filters to the image to find out useful patterns in the image. The activation layer controls which signal flows from one layer to the next, emulating the neurons in our brain. The last few layers in the network are called dense or fully connected, meaning that the neurons of one layer are connected to every neuron in the next layer. The architecture of the CNN that we are using is shown in Fig. 2.

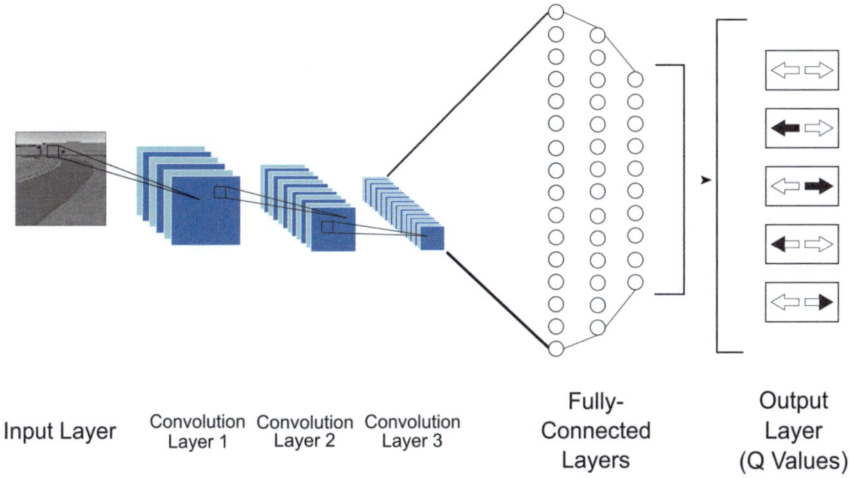

Fig. 2 Architecture of the deep Q network consisting of CNN layers and fully connected or dense layers

3.2 Reinforcement Learning

The Reinforcement Learning framework has been used for a long time for control tasks like manufacturing, inventory management, and delivery management. The Reinforcement Learning framework was formulated by Sutton [13] as a model to provide the best policy, an agent can follow in the given state, such that the total accumulated reward is maximized when the agent follows that policy from the current state to the terminal state. Mnih et al. [4] came up with the idea of deep reinforcement learning which uses a CNN to learn a Q function which played several Atari games with superhuman performance (Fig. 3). The only inputs to the network were images of the game window and the reward values. The basic idea is for the agent to estimate the Q-values whenever an image from the environment is given to it. The Q-values tell the agent which action would lead to the highest reward. Deep reinforcement learning [4] works by maximizing the cumulative future reward. CNNs are used to approximate the optimal action-value function

$$Q_{t+1}(s, a) \leftarrow Q_t(s, a) + \alpha[r + \gamma \arg\max_{a'} Q_t(s', a') - Q_t(s, a)] \quad (1)$$

which is the maximum sum of rewards r, discounted by γ at each timestep t, achievable by a policy π(s) after making an observation s, and taking an action a. The Q-function is formulated as a parameterized function of the states, actions: $Q(s, a, w)$. Deep Q network approximates the Q-function using a Convolutional Neural Network (CNN). The objective of this CNN shall be to minimize the Mean Square Error (MSE) of the Q-values as shown in (3).

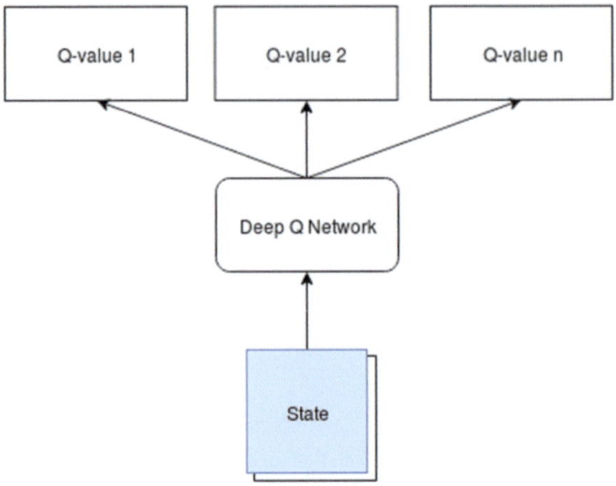

Fig. 3 Optimized architecture of the deep Q network

$$l(w) = E[(r + \gamma \arg\max_{a'} Q_t(s', a', w) - Q_t(s, a, w))^2] \tag{2}$$

$$J(w) = \max_{w} l(w). \tag{3}$$

During training, we used a epsilon-greedy policy which takes a random action with probability ϵ, or else takes the action for the highest Q-value. The epsilon-greedy policy ensures that the agent explores all possible actions and prevents the local maxima problem. We decrease the value of epsilon from 1 to 0.1 over 1 million input images. The action a is given by 4 where a^* is the action according to the highest Q-value.

$$a = \begin{cases} a^* & \text{with probability } 1 - \varepsilon \\ \text{random action} & \text{with probability } \varepsilon \end{cases} \tag{4}$$

We also use replay memory which holds about 600,000 previous states of the environment. Replay memory helps stabilize training of the deep Q network as states are randomly sampled when updating knowledge.

4 Results and Discussion

We use the open-source racing car simulator (TORCS) to simulate and test the proposed deep reinforcement learning framework which was created using tensorflow. To be able to control the car and get sensor values, we use the simulated car racing (SCR) add-on on the Python programming language. This add-on gives access to the car controls, like steering, acceleration, and brakes. The input to the network is the raw pixel data which is the image in the TORCS environment. The reward r is calculated using the speed v of the car and its angle ϕ (Fig. 4) with respect to the road as shown in (5).

$$r = v\cos(\phi) \tag{5}$$

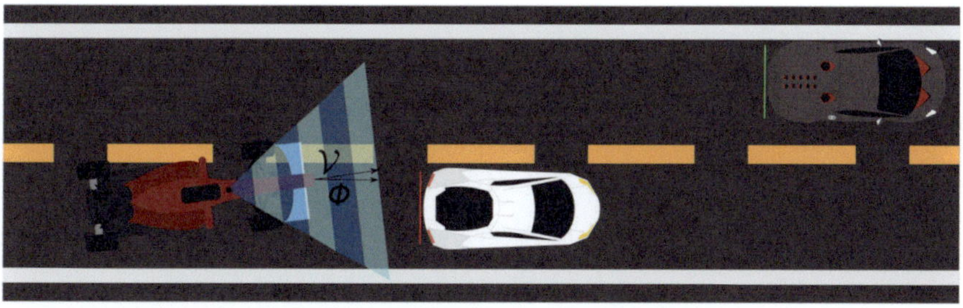

Fig. 4 Simulation environment sensors

Fig. 5 Discretized action space

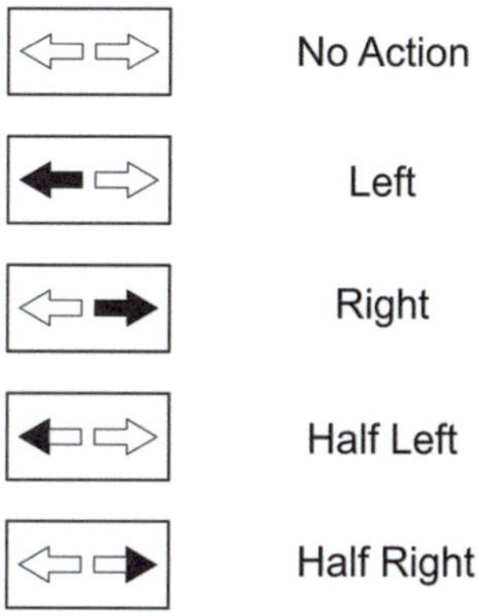

The output is the steering values. The network is trained end-to-end following the same objective of the DQN. In order to make the steering problem as a classification problem, the actions (steer, left, and right) are discretized as shown in Fig. 5. Tile coding of actions makes the steering actions abrupt, but the half left and half right steering actions help make the actions smooth.

The agent learnt to drive through the track successfully after about 1.5 million states. The reward for every state is plotted in Fig. 6. Mean Q value versus state is plotted in Fig. 7.

It took the agent about 90 h to train for 1.5 million states, but in the end it is able to maneuver through the TORCS track successfully. It is important to note that the agent in its current form can maneuver through the track, but not efficiently. The efficiency of the agent, in terms of the time taken to go through the track, can be improved by training the agent for a longer period, as the reward penalizes the agent for having a low speed.

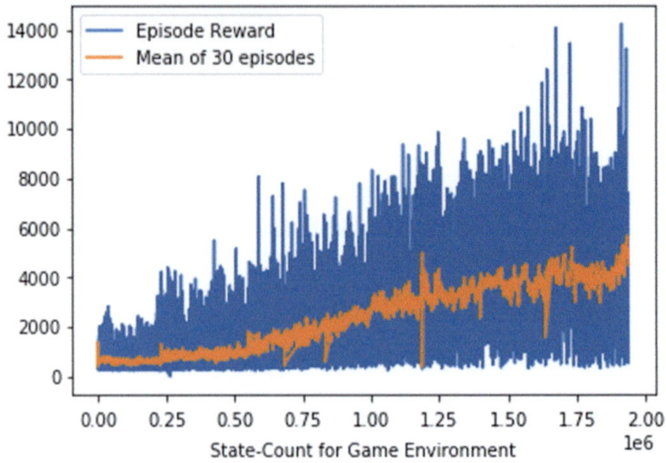

Fig. 6 Rewards versus states

Fig. 7 Mean Q values versus states

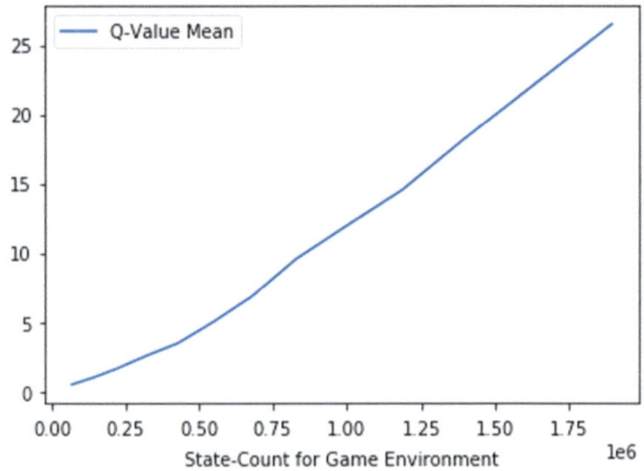

5 Conclusion

In this project, we were able to drive a car inside the simulation environment of TORCS using deep reinforcement learning successfully. We used raw image pixels and rewards calculated using sensor readings to make it form a Q-value function approximator, which we used to steer the car. We made use of certain tricks like epsilon-greedy policy and experience replay to converge the model better. Despite the success of the deep Q network, it takes a lot of time to train. This leads to a high turnaround time. The project has not been tested on real-world systems and might result in unpredictable situations. Our future work includes deploying the proposed framework on a prototype, where the sensors and actuators are part of an actual car. In terms of the machine learning algorithm, we plan on looking into parallel algorithms like A3C [14] which reduce the training time of deep reinforcement learning by a huge amount. We also plan on using imitation learning, to make the algorithm converge faster by training the network beforehand using labeled data and then running reinforcement learning on it. Another area of research would be to use algorithms that output a continuous action space instead of a discrete one.

References

1. Jo, K., Kim, J., Kim, D., Jang, C., Sunwoo, M.: Development of autonomous carpart i: Distributed system architecture and development process. IEEE Trans. Ind. Electron. **61**(12), 7131–7140 (2014)
2. Jo, K., Kim, J., Kim, D., Jang, C., Sunwoo, M.: Development of autonomous carpart ii: a case study on the implementation of an autonomous driving system based on distributed architecture. IEEE Trans. Ind. Electron. **62**(8), 5119–5132 (2015)
3. Krizhevsky, A., Sutskever, I., Hinton, G.E.: Imagenet classification with deep convolutional neural networks. In: Advances in Neural Information Processing Systems, pp. 1097–1105 (2012)
4. Mnih, V., Kavukcuoglu, K., Silver, D., Rusu, A.A., Veness, J., Bellemare, M.G., Graves, A., Riedmiller, M., Fidjeland, A.K., Ostrovski, G., et al.: Human-level control through deep reinforcement learning. Nature **518**(7540), 529 (2015)
5. Bojarski, M., Del Testa, D., Dworakowski, D., Firner, B., Flepp, B., Goyal, P., Jackel, L.D., Monfort, M., Muller, U., Zhang, J., et al.: End to end learning for self-driving cars. arXiv preprint arXiv:1604.07316 (2016)
6. Yang, S., Wang, W., Liu, C., Deng, W., Hedrick, J.K.: Feature analysis and selection for training an end-to-end autonomous vehicle controller using deep learning approach. In: 2017 IEEE Intelligent Vehicles Symposium (IV), pp. 1033–1038. IEEE (2017)
7. Kuefler, A., Morton, J., Wheeler, T., Kochenderfer, M.: Imitating driver behavior with generative adversarial networks. In: 2017 IEEE Intelligent Vehicles Symposium (IV), pp. 204–211. IEEE (2017)
8. Goodfellow, I., Pouget-Abadie, J., Mirza, M., Xu, B., Warde-Farley, D., Ozair, S., Courville, A., Bengio, Y.: Generative adversarial nets. In: Advances in Neural Information Processing Systems, pp. 2672–2680 (2014)
9. Tai, L., Liu, M.: Mobile robots exploration through cnn-based reinforcement learning. Robot. Biomim. **3**(1), 24 (2016)
10. Sallab, A.E., Abdou, M., Perot, E., Yogamani, S.: Deep reinforcement learning framework for autonomous driving. Electron. Imaging **2017**(19), 70–76 (2017)
11. Lange, S., Riedmiller, M., Voigtlander, A.: Autonomous reinforcement learning on raw visual input data in a real world application. In: The 2012 International Joint Conference on Neural Networks (IJCNN), pp. 1–8. IEEE (2012)
12. Saunders, W., Sastry, G., Stuhlmueller, A., Evans, O.: Trial without error: towards safe reinforcement learning via human intervention. In: Proceedings of the 17th International Conference on Autonomous Agents and MultiAgent Systems, pp. 2067–2069. International Foundation for Autonomous Agents and Multiagent Systems (2018)
13. Sutton, R.S.: Learning to predict by the methods of temporal differences. Mach. Learn. **3**(1), 9–44 (1988)
14. Mnih, V., Badia, A.P., Mirza, M., Graves, A., Lillicrap, T., Harley, T., Silver, D., Kavukcuoglu, K.: Asynchronous methods for deep reinforcement learning. In: International Conference on Machine Learning, pp. 1928–1937 (2016)

Understanding Antibiotic Resistance Using Different Machine Learning Approaches

Tanaya Priyadarshini Pradhan, N. K. Debata and Tripti Swarnkar

Abstract Anti-infection resistance is a genuine unrestricted organisms problem. Several microscopic organisms that are fit for causing serious ailments are getting to be impervious to most ordinarily accessible antibiotics. Here we plan a machine learning model which can analyze the clinical information and adequately classify whether a given sample is protected from a particular anti-infection or not. In this paper we have a discussion about various broadly used machine learning strategy including Naive Bayes Classifier, KNN, Multilayer Perceptron, Random Forest, and Decision Tree to characterize the clinical information. We concentrated on the preparation and testing information by 60–40 split and the cross validation (10-fold, 5-fold) for data prepossessing. The model's outcomes are examined considering model predictive accuracy, sensitivity, specificity, MCC, and prevalence for practical outcomes.

Keywords Resistance antibiotic · KNN · Random forest · Multilayer perceptron · Naive bayes classifier · Decision tree

1 Introduction

Antibiotic medications are used to destroy microorganisms which are the cause of illness and diseases. A substance which is delivered by a microorganism that executes or represses the development of another microorganism is known as antibiotics [16].

Organisms have formed a large-scale contribution to human fitness. Microorganisms that are resistant to many anti-infection agents are noted as multiresistant

T. P. Pradhan (✉)
Computer Science and Engineering, S'O'A Deemed to Be University, Bhubaneswar, Odisha, India
e-mail: tanaya.priyadarshini@gmail.com

N. K. Debata
IMS & SUM Hospital, S'O'A Deemed to Be University, Bhubaneswar, Odisha, India
e-mail: nagendebata@soa.ac.in

T. Swarnkar
Computer Application, S'O'A Deemed to Be University, Bhubaneswar, Odisha, India
e-mail: triptiswarnakar@soa.ac.in

B. Pati et al. (eds.), *Advanced Computing and Intelligent Engineering*,
Advances in Intelligent Systems and Computing 1082,
https://doi.org/10.1007/978-981-15-1081-6_6

organisms (MRO) [5]. Antibiotic resistance is a genuine general medical issue, it is the ability of bacteria to resist the effects of an antibiotic. Antimicrobial resistance preserve originator signifies loss of life, costs money, lives and undermines effectiveness of health delivery affairs [3]. Antimicrobial safe microorganisms can stretch out from individual to individual in the general population or from patient to patient in medical center. A careful infection control can diminish spread of these microorganisms in hospitals. Great individual cleanliness can limit spread of these microorganisms in the group. Careful prescribing of anti-infection agents will limit the advancement of more antimicrobial safe strains of microscopic organisms [3].

Machine learning approaches have been used in bioinformatics fields as for different proposition. Here we used the machine learning technique on the clinical data to study if an antibiotic is a resistance or not for a given sample. Machine learning ways are suited to essence information from the clinical dataset and use the acquired facts in the categorization of lately taken sample. We use machine learning (ML) as a form for modeling clinical data by constructing a model suited for research and making predictions.

2 Related Work

Richardson et al. [16] found that resistance to one drug confers the resistance to second drugs thus minimizing the cross-resistance, as a result of one drug sensitizing the bacteria to the second drug. Thus the above processes maximize the collateral sensitivity. The uses of two or more antibiotics synergistically produce more effect and as more point than if each antibiotic is used single.

Blair et al. [4] found that, the extensive widespread usage of antibiotics for the treatment of human beings and other creatures, agriculture, beekeeping as well as other research areas a compulsion inherent therapy. The transformative difficulty for the upgrowth of anti-infection preservation is agreeable. Microscopic organisms have come forward to fight with the activeness of normal antiseptic items for decades.

Berendonk et al. [3] form that the age group of stable identifications forth with an assessment of subsequent flow in antibiotic resistance in the surroundings is shortly incomplete due to the inequality of the research scenario. For excellent management of the display and dispersion of ARBs and ARGs does not recover the need to look for commencing valuable drugs.

Delen et al. [5] learned that machine learning techniques are capable to extract patterns and hidden relation deep into huge clinical datasets, beyond the combination and reaction against the pharmaceutical trainer and owned outcomes. The corresponding research of a distinct model for breast cancer durability utilizes a massive dataset by the side of a 10-fold cross validation provides us the comparative indicator strength of distinct data mining mechanism.

Bellazzi et al. [2] present an entire division on the requirement of the skill of guessing data mining in clinical medication and provide a section of consultation to eliminate data mining manners in the present field . Thus the goal of data division is combined to the embellish intelligence of the instruction that is accommodated in the data as well as focusing sheath that does not approve permanent intelligence or difficulty in the data accumulation behavior.

3 Method and Material

3.1 Dataset Used

Although antibiotic resistance research is establishing currently, still the dispute is to switch against associating data to find whether the drug is resistance or not for a particular sample [9]. The SUM Hospital in Bhubaneswar initiated constructing a database for resistance antibiotics in 2017 gathering pathological information associated with their patients. They have been lifting personal analysis survey with improving individual prescription located on particular consequences.

The clinical dataset is a real-domain dataset which has an explanation of pathological data for patient and collected from different places. Every case in the clinical data is depicted by the binary and the categorical attributes, which simultaneously describes the medical information for an assertive victim. For the reason of this research, 901 samples from the Clinical dataset were worn in their expertise analysis method. The investigation for each case is the cell rank, which determines patient age, gender, organisms, and specimens. Also it contents different bacteria and antibiotics sensitivity patterns for each organism as their features. These sensitivity patterns have been categorized into two parts like sensitive (Infection treatable by normal dosage) and resistant (Not treatable with the agent).

3.2 Data Preparation

The basic phase of data mining is data preparation, the real-world records tend to be unfinished, noisy, and incompatible for a significant analysis. The most widely used data mining algorithms require data to be placed into a tabular form, which contains more significant information for research work. The clinical data being collected for the study is noisy and thus, not in a state to be directly used for the purpose of analysis. We have prepossessed the information to fill in lost values, smooth out commotion and irregularities, so that it can be effectively used for the analysis using data mining techniques.

Prepossessing comprises of two steps, (I) handling missing values, (II) feature representation. For the study purpose we have analyzed the sensitivity patton of the given sample for an AMC (amoxicillin clavulanate) antibiotic. AMC is an antibiotic which is used for the treatment of a number of bacterial infections [5]. The samples are being divided into two classes depending on their sensitivity (resistance or sensitivity) toward the studied antibiotic. The age of every patient has been arranged into a specific range. The patient age is characterized into four types: child (0–15), younger Adult (16–30), middle-age grown-up (31–59), and Senior Adult (60 and above) [9, 13]. Here the child class, younger Adult, middle-age grown-up, and Senior Adult are represented by 1, 2, 3, and 4, respectively. From all the specimens we have taken urine (UR) represented by 1 and blood (BL) represented by 2 for our study. The most utilized organisms in bioinformatics field E. coli (Escherichia coli) and NG (Nasogastric) have been taken here as our organisms feature and most of the sample are form this category. E. coli represent by 1 and NG represent by 2. E. coli has served the E. coli people group, as well as has framed a reason for extrapolation of genes functions. E. coli is a sort of microscopic organisms that regularly lives in your intestines [5]. In that capacity, the accuracy and culmination of the E. coli data is of incredible significance to the group of scientists working in all orders and with all life forms.

3.3 Data Prepossessing

WEKA is open source programming which is uninhibitedly available for mining data and implements an expansive gathering of mining algorithms. It can accept information in various formats and furthermore has converter supported with it. So we have changed over the patient dataset into .csv document. The document was loaded into loaded WEKA explorer. The classify panel is utilized for classification, to evaluate the accuracy of resulting predictive, visualize erroneous predictions, or the model itself.

Initially, we have studied about different classifiers and implemented these on the clinical dataset. WEKA takes the whole dataset as input and divides into training set and testing set (60–40, 70–30). Cross validation is a specific technique for choosing holdout sets, in which a model is created on a preparation set of perceptions and after that precision is evaluated on an autonomous test set. Cross validation error estimation has been very well known for microarray classification [?]. Maybe this is because of the reality that, on average, cross validation error estimates virtually concur with the true errors.

3.4 Classifier Used

Naive Bayes Classifier(NBC)

NBCs is a popular probabilistic method for classification. It prepares the weighted information and helps to prevent over fitting [10]. Experimental results were obtained using standard implementations of binomial Naive Bayes (NB) algorithms [14] provided by the WEKA toolkit.

Multilayer perceptron(MLP):
We used a popular ANN architecture called MLP with back-propagation (a supervised learning algorithm). The MLP is a better function estimated for prediction and classification problems [1]. Its network consists of an arrangement of real components which shapes the input layer, at least one hidden layer of handling components, as well as the output layer is of the processing elements. Here every connection has a related weight, which is calculated adaptively during learning. Here we are using MLP with a single hidden layer.

Instance-based-k-nearest neighbor(KNN):
KNN classifier assigns a class level as the most frequent class level nearest to training samples. On the calculation of the KNN algorithm the Euclidean distance is calculated. Distances from the new vector to all composed vectors are computed as well as the closest k samples are select for classifying the new sample [7].

Decision tree(DT):
DT are the most common used classification techniques which are more popular(ID3) through enhancing of data mining in the field of information systems. Decision tree will be a tree that classifies instances by arranging them in light of feature values.

Random forest(RF):
The view of RF is like a bootstrapping algorithm among Decision tree (CART) appear. RF is a collection of methods that use bagging to get better classification performance via combination of the yield of a number of classifiers [14]. The key scheme for ensemble methods is that a huge number of weak learners can be used to produce a strong learner. The present forecast can fundamentally be the mean of all possibilities.

3.5 Performance Measure

The performance of the different classifiers is compared based on numerical performance indices such as accuracy, sensitivity, specificity, prevalence, and Matthews' connection coefficient (MCC) as our accuracy measure.

MCC is a relationship coefficient among actual and predicted classifications. It takes values between -1 and 1. MCC is utilized as a part of machine learning as a measure of the nature of paired (two class) groupings [6]. It considers true and false positives and negatives and is for the most part viewed as a balanced measure which can be utilized regardless of whether the classes are of altogether different sizes [11].

$$\mathrm{MCC} = \frac{((TP \times TN) - (FP \times FN))}{\sqrt{((TP + FP)(TP + FN)(TN + FP)(TN + FN))}}$$

Sensitivity and specificity are factual measures of the performance of a binary classification test, additionally referred to in insights as classification function: [17] Sensitivity measures the extent of positives that are accurately recognized all things considered. Specificity measures the extent of negatives that are effectively distinguished.

$$\text{Sensitivity} = TP/TP + FP \tag{1}$$

$$\text{Specificity} = TN/TN + FN \tag{2}$$

Prevalence, here and there referred to as prevalence rate, is the extent of people in a populace who have a specific sickness or characteristic at a predetermined point in time or over a predefined time frame [18].

$$\text{Prevalence} = TP + FP/(N) = TP + FP/(TP + FP + FN + TN) \tag{3}$$

where *TP* is the number of true positive samples, *TN* is the number of true negative samples, *FP* is the number of false positive samples, *FN* is the number of false negative samples. Accuracy is a depiction of methodical errors, a measure of factual inclination; as these reasons a distinction between an outcome and a "true" esteem. Accuracy is the weighted normal of a test's sensitivity and specificity, where sensitivity is weighted by prevalence and specificity is weighted by the supplement of prevalence [11].

$$\text{Accuracy} = (\text{Prevalence} \times \text{Sensitivity}) + (1 - \text{Prevalence} \times \text{Specificity}) \tag{4}$$

3.6 Proposed Model

Figure 1 represents the schematic block diagram of the proposed model. In this section, we specify the trial procedure and evaluate the different parameters of the tests. The total model is partitioned into three primary sections, i.e., initialization step, classification, and evolution. Under the initialization step we work for data collection, data preparation, and prepossessing. The data is initially collected from the hospital. The collected dataset isn't set up for execution so first we have prepared the data in a specific execution format by handling missing data and feature representation technique. Now the prepared data is processed using WEKA for further analysis. The data is being divided into training, as well as test dataset using two different validation techniques, i.e., cross validation (10-fold and 5-fold) and data split (60–40). All the prepossessed data is classified using different benchmark classifiers. After

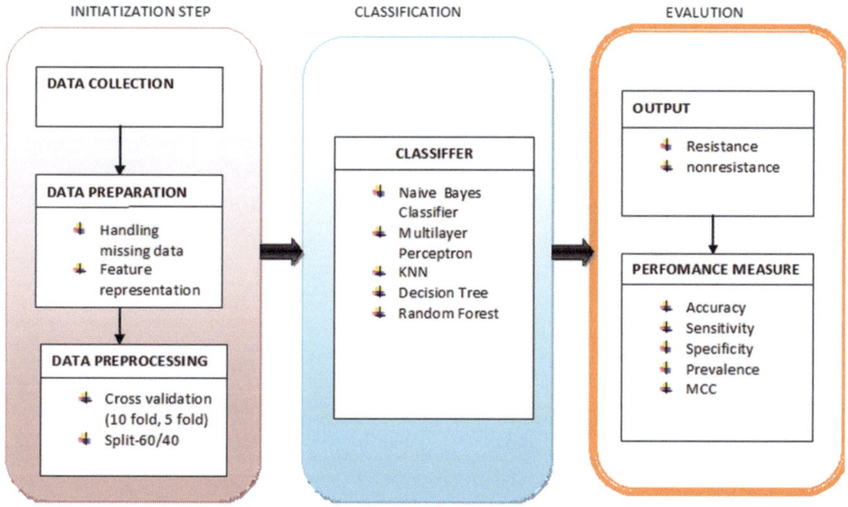

Fig. 1 Proposed model

classification we have discovered the sample which is either resistance or sensitive for a specific antibiotic and measured the performance using different model accuracy measures.

4 Result Discussion

On the above study, we evaluate the proposed model based on the accuracy measures (accuracy, sensitivity, specificity, MCC, and prevalence). The achieved result was found from analysis of validation technique (10-fold, 5-fold) and data split (60–40) for every classifier.

Tables 1, 2, and 3 summarize the qualities for every execution measure acquired from 5 benchmark classification methods for all data prepossessing approaches connected to the clinical information. The result is being discussed in light of standard performance measures; overall accuracy, specificity, sensitivity, prevalence, and

Table 1 Performance measure for E. coli in 5-fold cross validation

Classifiers	MCC	Sens	Spec	Prev	Acc
NBC	0.482	0.889	0.625	0.836	0.845
MLP	0.453	0.875	0.658	0.87	0.846
KNN	0.422	0.863	0.650	0.882	0.834
DT	0.482	0.889	0.625	0.836	0.845
RF	0.421	0.871	0.614	0.864	0.836

Table 2 Performance measure for E. coli in 10-fold cross validation

Classifiers	MCC	Sens	Spec	Prev	Acc
NBC	0.482	0.889	0.625	0.836	0.845
MLP	0.402	0.861	0.646	0.89	0.837
KNN	0.416	0.865	0.647	0.883	0.839
DT	0.482	0.889	0.625	0.836	0.845
RF	0.416	0.87	0.611	0.865	0.835

Table 3 Performance measure for E. coli in 60–40 split

Classifiers	MCC	Sens	Spec	Prev	Acc
NBC	0.532	0.875	0.729	0.843	0.852
MLP	0.302	0.81	0.75	0.947	0.806
KNN	0.449	0.841	0.79	0.905	0.836
DT	0.535	0.878	0.72	0.836	0.852
RF	0.437	0.84	0.76	0.901	0.832

MCC. Here it gives information about E. coli organism on basis of two specimens, i.e., UR and BL which are collected from Wards. Calculation of specificity and prevalence is more for KNN on basis of E. coli organism.

Our proposed comparative learning model discusses the different performance measures with an objective of classifying the given sample with specific organisms found in it is sensitive or resistance to a specific antibiotic listed in the database. From the Tables 1, 2, and 3 DT give the best outcome on the execution of MCC and sensitivity.

Calculation of specificity and prevalence is more for KNN on basis of E. coli organism. In terms of accuracy DT also gives best results for every validation technique for E. coli organisms.

Similarly, Tables 4, 5, and 6 summarize the qualities for every execution measure acquired from 5 benchmark classification methods for 5-fold cross validation, 10-fold cross validation, and 60–40 split connected to the clinical information. Here it provides data about NG organisms on premise of two same specimens, i.e., UR and

Table 4 Performance measure for NG in 5-fold cross validation

Classifiers	MCC	Sens	Spec	Prev	Acc
NBC	0.482	0.889	0.625	0.836	0.845
MLP	0.47	0.877	0.675	0.87	0.85
KNN	0.463	0.875	0.675	0.873	0.849
DT	0.482	0.889	0.625	0.836	0.845
RF	0.478	0.878	0.683	0.87	0.852

Table 5 Performance measure for NG in 10-fold cross validation

Classifiers	MCC	Sens	Spec	Prev	Acc
NBC	0.482	0.889	0.625	0.836	0.845
MLP	0.421	0.865	0.663	0.887	0.843
KNN	0.458	0.874	0.672	0.874	0.848
DT	0.482	0.889	0.625	0.836	0.845
RF	0.461	0.876	0.666	0.87	0.848

Table 6 Performance measure for NG in 60–40 split

Classifiers	MCC	Sens	Spec	Prev	Acc
NBC	0.531	0.89	0.689	0.838	0.857
MLP	0.545	0.884	0.755	0.863	0.866
KNN	0.438	0.853	0.757	0.908	0.844
DT	0.531	0.89	0.689	0.838	0.857
RF	0.517	0.873	0.767	0.88	0.86

BL. From the Tables 4, 5, and 6 DT, NBC give the best outcomes on the execution of MCC and sensitivity. Calculation of specificity and prevalence is more for KNN on basis of E. coli organism. Accuracy of RF also gives best results for every validation technique for NG organisms. MLP also have a maximum accuracy value, so that it correctly classifies (86.66%) our dataset and provides the most excellent end result. From the above tables, in most extreme cases NBC and DT provide the best result lying on the implementation of MCC and sensitivity for 5-fold and 10-fold cross validation. Overall specificity and prevalence measure KNN algorithm and MLP give greater value, respectively.

Figures 2 and 3 show the accuracy measure of different organisms by using stripping techniques. From the graphical view we found that the split method gives better results as compare to validation techniques for all the studied benchmark classifiers. Although in case of E. coli the MLP 10-fold shows better accuracy in compression

Fig. 2 Accuracy measure of E. Coli organisms using stripping techniques

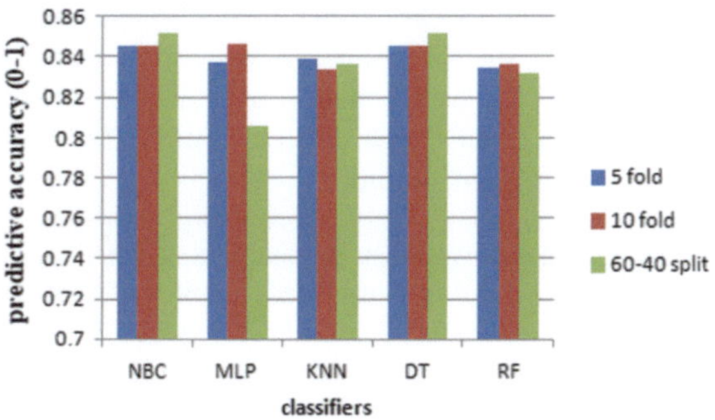

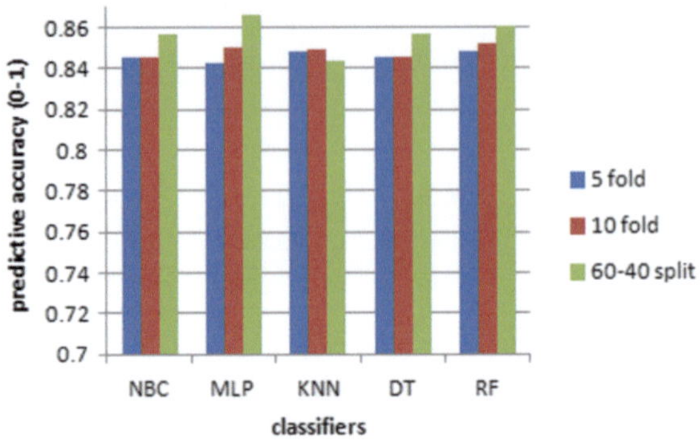

Fig. 3 Accuracy measure of NG organisms using stripping techniques

to other two methods whereas RF result is comparable for all the three methods. Similarly in case of NG we have seen from Fig. 3 in KNN model 10-fold gives better results.

5 Conclusion

In this work we examine about a machine learning model which can dissect the clinical information and viably order whether a given example is either resistance or not to a particular anti-infection. Also, moreover discover the distinctive execution measure for each machine learning model. Each model gives precise values after execution of each learning model. On the off chance that we discuss the MCC esteem NBC and DT have same outcome for 5-fold and 10-fold cross validation in both specimens. KNN algorithm provides better execution for specificity measures. For prevalence measure MLP performed better from different algorithms. MLP likewise have the greatest precision esteem, with the goal that it effectively characterizes (86.66%) of our information collection and gives the most phenomenal final product.

NBC and DT provide best results for overall calculation, as the number of correctly classify samples and incorrectly classify samples are equal in both the cases. From the above work we found that the child and the young adults have more Resistance body than the middle age grown-up people and senior adult, after the age of fifty the people are lost their Resistance power and they have a sensitive body.

The work can be expanded in analyzing the effect of the presence of the specific organism in a given sample collected from different places in regard to its sensitivity toward the different antibiotics.

Acknowledgements We thank to Professor S. K. Kar (skkar@soa.ac.in) (Director Research IMS SUM Hospital), ethical committee head of Institute of medical science (IMS). It has been approved in his presence and thanks for giving the approval from the ethical committee.

References

1. An, A., Cercone, N.: Discretization of continuous attributes for learning classification rules. In: Pacific-Asia Conference on Knowledge Discovery and Data Mining, pp. 509–514. Springer (1999)
2. Bellazzi, R., Zupan, B.: Predictive data mining in clinical medicine: current issues and guidelines. Int. J. Med. Inform. **77**(2), 81–97 (2008)
3. Berendonk, T.U., Manaia, C.M., Merlin, C., Fatta-Kassinos, D., Cytryn, E., Walsh, F., Bürgmann, H., Sørum, H., Norstörm, M., Pons, M.-P., et al.: Tackling antibiotic resistance: the environmental framework. Nat. Rev. Microbiol. **13**(5), 310 (2015)
4. Blair, J.M.A., Bavro, V.N., Ricci, V., Modi, N., Cacciotto, P., Kleinekathfer, U., Ruggerone, P., Vargiu, A.V., Baylay, A.J., Smith, H.E., et al.: Acrb drug-binding pocket substitution confers clinically relevant resistance and altered substrate specificity. In: Proceedings of the National Academy of Sciences **112**(11), 3511–3516 (2015)
5. Delen, D., Walker, G., Kadam, A.: Predicting breast cancer survivability: a comparison of three data mining methods. Artif. Intell. Med. **34**(2), 113–127 (2005)
6. Guilfoile, G., Alcamo, I.E.: Antibiotic-resistant Bacteria. Infobase Publishing (2007)
7. Gupta, V., Mittal, M.: KNN and PCA classifier with autoregressive modelling during different ECG signal interpretation. Procedia Comput. Sci. **125**, 18–24 (2018)
8. Guzella, T.S., Caminhas, W.M.: A review of machine learning approaches to spam filtering. Expert Syst. Appl. 36(7), 10206–10222 (2009)
9. Hall, M.A.: Correlation-based feature selection for machine learning (1999)
10. Hasan, M., Kotov, A., Carcone, A.I., Dong, M., Naar, S., Brogan Hartlieb, K.: A study of the effectiveness of machine learning methods for classification of clinical interview fragments into a large number of categories. J. Biomed. Inform. **62**:21–31 (2016)
11. Kotsiantis, S.B., Zaharakis, I., Pintelas, P.: Supervised machine learning: a review of classification techniques. Emerg. Artif. Intell. Appl. Comput. Eng. **160**, 3–24 (2007)
12. Mehdiyev, N., Enke, D., F, P., Loos, Peter: Evaluating forecasting methods by considering different accuracy measures. Procedia Comput. Sci. **95**, 264–271 (2016)
13. Pham, B.T., Bui, D.T., Pourghasemi, H.R., Indra, P., Dholakia, M.B.: Landslide susceptibility assesssment in the Uttarakhand area (India) using GIS: a comparison study of prediction capability of naïve bayes, multilayer perceptron neural networks, and functional trees methods. Theor. Appl. Climatol. **128**(1–2), 55–273 (2017)
14. Quinlan, J.R.: Induction of decision trees. Mach. Learn. **1**(1), 81–106 (1986)
15. Quinlan, J.R.: C4. 5: Programs for Machine Learning. Elsevier (2014)
16. Richardson, L.A.: Understanding and overcoming antibiotic resistance. PLoS Biol. **15**(8), e2003775 (2017)
17. Riley, M., Abe, T., Arnaud, M.B., Berlyn M.K.B., Blattner F.R., Chaudhuri R.R., Glasner, J.D., Horiuchi, T., Keseler, I.M., Kosuge, T., et al.: Escherichia coli k-12: a cooperatively developed annotation snapshot 2005. Nucl. Acids Res. **34**(1), 1–9 (2006)
18. Sokolova, M., Lapalme, G.: A systematic analysis of performance measures for classification tasks. Inf. Process. Manag. **45**(4), 427–437 (2009)
19. Swarnkar, T., Mitra, P.: Graph-based unsupervised feature selection and multiview clustering for microarray data. J. Biosci. **40**(4), 755–767 (2015)
20. Wang, J., Zucker, J.-D.: Solving multiple-instance problem: a lazy learning approach (2000)

Computer Vision Aided Study for Melanoma Detection: A Deep Learning Versus Conventional Supervised Learning Approach

S. S. Poorna, M. Ravi Kiran Reddy, Nukala Akhil, Suraj Kamath, Lekshmi Mohan, K. Anuraj and Haripriya S. Pradeep

Abstract Melanoma is one of the most fatal type of skin cancer. Among the 2–3 million skin cancer diagnosed around the world each year, around 5% is affected with melanoma. Early detection of melanoma can save a life. A computer vision aided system with reasonable accuracy was developed for the early diagnosis of melanoma. The analysis was done using dermoscopic images downloaded from publically available database. After preprocessing, the features capable of melanoma identification, viz., ABCD parameters: Asymmetry, Border, Color, and Diameter are extracted. The analysis includes a comparative study between conventional machine learning techniques and deep learning. The learning techniques: Total Dermoscopic Score, K-Nearest Neighbor, Support Vector Machine and Convolutional Neural Networks were used for classification. The results of the study showed that deep learning-based method gives more accurate and precise detection of melanoma compared to conventional supervised learning techniques.

Keywords Melanoma · ABCD parameters · KNN · SVM · CNN · Accuracy · Precision · Recall

1 Introduction

In the fast pacing world of modern technology, along with the most modern techniques and equipment's for medical diagnosis, we can see an increase in the number of patients also. One of the most widespread malady as per the reports is cancer. Melanoma type skin cancer is one of the most common types. Statistical surveys predict that in the U.S, 178,560 cases of melanoma will be diagnosed in 2018 [1]. Cancer becomes life threatening or causes mortality when it is diagnosed at a later stage or when the lesion prediction accuracy is less. Skin Cancer is due to uncontrollable growth of cells in the skin. The skin has three layers, the first layer squamous

S. S. Poorna (✉) · M. R. K. Reddy · N. Akhil · S. Kamath · L. Mohan · K. Anuraj · H. S. Pradeep
Department of Electronics and Communication Engineering, Amrita Vishwa Vidyapeetham, Amritapuri, India
e-mail: poorna.surendran@gmail.com

B. Pati et al. (eds.), *Advanced Computing and Intelligent Engineering*,
Advances in Intelligent Systems and Computing 1082,
https://doi.org/10.1007/978-981-15-1081-6_7

cells, second layer basal cells, and third layer melanocyte cells. UV rays and tanning beds are the main reason for skin cancer. According to American Society of Dermatology, skin cancer is diagnosed under three cases viz. Basal Cell Carcinoma (BCC), Squamous Cell Carcinoma (SCC), and Melanoma [2]. Among the three, melanoma which occurs in melanocytes, is the most dangerous and fast growing. It is highly curable and can reduce the risk of spreading into other parts of body, if detected and treated in early stages. Normally biopsy is used for the diagnosis of melanoma. It involves removing samples of tissue from patient's suspected part of the skin and analyzing it in the laboratory. As the sample has to undergo histopathological analysis, this diagnostic method is time consuming. On the other hand, by exploiting computer assisted methods using image processing followed by supervised learning, helps in fast and accurate diagnosis.

The paper aims at using various machine learning and deep learning techniques, for diagnosis and an earlier detection of the lesion, which in turn leads to the better lives of human beings. This work comprises of preprocessing, segmentation, feature extraction, and classification of Melanoma using a threshold-based technique-TDS calculation, using supervised conventional machine learning methods and by deep learning using CNN.

2 Related Works

A lot of literatures are available on computer-aided detection on melanoma from the past few decades. The survey paper by Sathiya et al. [3] gives a brief review on preprocessing and segmenting techniques applied to raw images, feature extraction, feature selection, and classification methods for computer assisted diagnosis of melanoma. Dubai et al. [4] developed a system using ABCD rule (Asymmetry, Border, Color, Diameter) for feature extraction. They segmented images using Otsu's thresholding algorithm. In their work researchers classified a set of 463 digital images as benign or malignant, using feed forward Artificial Neural Network (ANN). Three variants of the back-proportion algorithms were used to train the model. Among these, Bayesian Regularization algorithm achieved the best accuracy of 76.9%.

Hiam et al. [5] adopted a noninvasive method using image processing for the detection, extraction, and classification of the skin lesions. The segmentation of the lesion region was done using Otsu's thresholding method. For the extraction of statistical features, Gray Level Co-Occurrence Matrix (GLCM) algorithm was used along with ABCD rule for the extraction of dermoscopic features in determining skin lesion types. The extracted features were then subjected to feature selection using PCA and classified with SVM. The system achieved an accuracy of 92.1%. Sabouri et al. [3] proposed a cascade classification method for Melanoma diagnosis. In this paper parallel image operations (image denoising and segmentation) are performed simultaneously with a highly efficient GPU. Color and texture features are extracted from the segmented images for determining the class of skin lesion. The obtained features are grouped into five different categories, viz. rgb + texture, hsv + texture, rgb, hsv, and texture. Their performance was evaluated with five different classifiers consisting of KNN, Naïve Bayes, multilayer perceptron, random forest, and SVM. Under the proposed cascade classifier scheme, a sensitivity of 83.06% and specificity of 90.05% was obtained.

In recent years, deep learning has proven extremely powerful in solving high computational problems. The use of deep learning approaches in detecting tumors has become lifesaving tool for many. Nasr-Esfahani et al. [6] employed deep learning approach using Convolutional neural network (CNN) to classify the image dataset into malignant or benign. Using the k-Means clustering, segmentation was done to identify the lesion. For smoothening purpose Gaussian filtering was used. The main advantage of using deep learning techniques lies in the fact that automatic feature extraction will be employed for better efficiency. The database had 6120 images, collected from UMCG, including both normal as well as synthesized. The performance analysis using CNN exhibited an accuracy of 81%.

3 Methodology

The methodology followed in the proposed work included data collection, preprocessing, segmentation, feature extraction, and classification. The performance of these systems is compared using confusion matrix, accuracy, precision, and recall. The ensuing sections give explanations on different processing techniques involved.

3.1 DataBase

Dermoscopic images from the publicly available database, The International Skin Imaging Collaboration (ISIC) [7] was used. The database with 1600 images which comprised 800 malignant and 800 benign images were considered for analysis. The sample subset of malignant and benign images in the database is shown in Fig. 1.

3.2 Preprocessing and Feature Extraction

After Otsu segmentation [8], the raw images were resized to a dimension 180 $\times$ 180, in order to obtain the region of interest. The images are then enhanced and also subjected to noise removal using median filtering. Feature extraction is the most important part used for determination of skin cancer. Here ABCD [9] parameters are employed for distinguishing the benign from malignant lesion. ABCD represents Asymmetry (A), Border (B), Color (C), and Diameter (D). The features are extracted as follows:

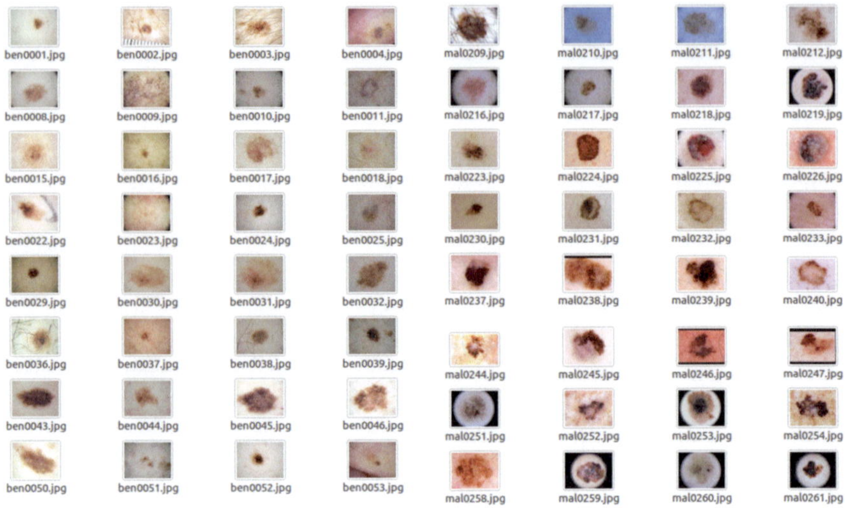

Fig. 1 Sample images in ISIC database

As the malignant lesions are generally asymmetrical in nature, asymmetry can serve as an essential feature for distinguishing benign from malignant. For the calculation of asymmetry index, the image is first divided over its centroid and then each nonoverlapping area is obtained by calculating the difference between the sum of the pixels of image area and lesion. The ratio is given by dividing this difference (ΔA) with the total area of the lesion (A) as in Eq. 1.

$$\text{Asymmetry Index(AI)} = \frac{\Delta A}{A} \tag{1}$$

Malignant lesions are defined by much contrasted borders whereas that of benign are generally defined by clear boundaries. Border irregularity is quantified using two features: compactness index and fractal dimension. Compactness index indicates how disperse or compact the shape is. It is calculated for a lesion with perimeter P and area A as given in Eq. 2.

$$C = \frac{P^2}{4\pi A} \tag{2}$$

Fractal dimension is a reliable tool for the detection of the lesion structure. It exhibits self-similarity and each section has a fractal geometry which has a different scale and same nature with that of the whole fractal. Integer valued dimension size is another characteristic. The growth pattern of the skin cancer cells is measured by using Box-counting method. The Hausdorff dimension calculation method is employed for obtaining the fractal dimension of the image.

Different colors in the pigmented area represent the cancerous cells. Often, malignant cells have black or brown color. In order to get the color variance of an image, HSV (Hue Saturation Value) color map is obtained and the ratio of number of pixels for each color (based on HSV) to the actual number of pixels is taken for each color. For finding the diameter, segmented image is complemented so that the dark areas become lighter and light areas become darker. Then the length of the labeled regions (major axis and minor axis) in pixel is computed and minimum length gives the diameter. It is observed that the moles that are greater than 6 mm in diameter would be malignant in nature.

3.3 Classification

After the feature extraction process, classification using the extracted features is performed. The proposed method compares the three different techniques to classify dermoscopic images. First one is the threshold-based method where Total Dermoscopic Score (TDS)-based classification. The second is based on classification using conventional machine learning techniques viz. Support Vector Machine (SVM) and K-Nearest Neighbor (KNN). These methods are compared against classification using a deep learning-based technique, i.e., Conventional Neural Network.

3.3.1 Threshold-Based Classification: Total Dermoscopic Score (TDS)

This method of classification involves weighting the ABCD parameters and fixing a threshold as the TDS score [8], as given in Eq. 3. The images with TDS less than 4.75 are classified under benign and those with a score more than 5.25, are included under malignant. Suspicious lesions are identified by a TDS in between these two values.

$$\mathrm{TDS} = 1.3A + 0.1B + 0.5C + 0.5D \tag{3}$$

3.3.2 Machine Learning-Based Classification: Using KNN and SVM

The supervised learning techniques, viz., K-Nearest Neighbor (KNN) and Support Vector Machine (SVM) are used for classifying the data as malignant and benign. KNN is a clustering algorithm, which uses Euclidean distance measure to find the nearest neighbors. The testing entity is labeled based on largest number of nearest neighbors [10]. SVM uses the concept of hyperplane for classification. Margin between the classes is maximized with the help of support vectors [11]. The hyperplanes can be either linear or nonlinear. Kernel functions are employed for fitting nonlinear hyperplanes. A sample plot of SVM classification, with the help of Radial

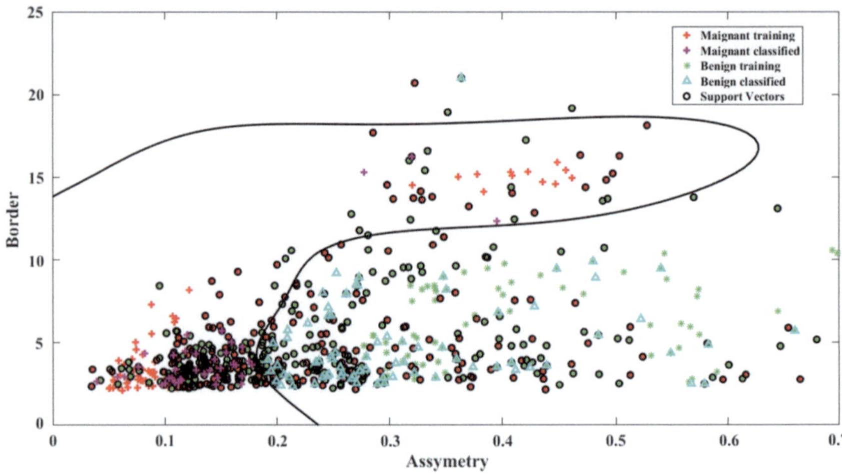

Fig. 2 A plot of SVM classification, with RBF kernel function

Basis Function (RBF) kernel, using the attributes asymmetry versus border, is as shown in Fig. 2.

3.3.3 Deep Learning-Based Classification: Using CNN

Convolutional neural network is a category of neural network approach under deep learning methodologies for classification. CNN consists of different convolutional layers which are capable of adjusting the weights and biases, from appropriate learning. The neurons at the input side receive several inputs, finds their dot product and feeds it to an activation function in order to reply with an output. It starts with the extraction of input image pixels on the front end of the input layer for the calculation of class membership score at the end layer.

AlexNet [12, 13] is the architecture adopted in this work. This architecture used five convolution, three pooling layers, three normalization layers, and three fully connected layers. The training images with dimension $56 \times 56 \times 3$ are fed into the input layer. The convolution layer moves over the input image to extract the features from it. The first convolution layer has 64 filters of dimension $5 \times 5 \times 3$, with a mask of dimension 2 and $28 \times 28 \times 64$ dimension output feature map layer (since stride/mask is 2). ReLU is used as the activation function here. Figure 3 shows a sample malignant input and feature map of 64 filter outputs corresponding to the image. Similarly second convolutional layer has input dimension $28 \times 28 \times 64$, pooling layer dimension $14 \times 14 \times 64$ with 128 filters and output dimension $14 \times 14 \times 128$ and so on. Following this, the architecture has a fully connected layer with 4096 neurons. The last layer does the classification part and it has only two classes at the end. Since outputs are in terms of probability distributions, the probabilities of

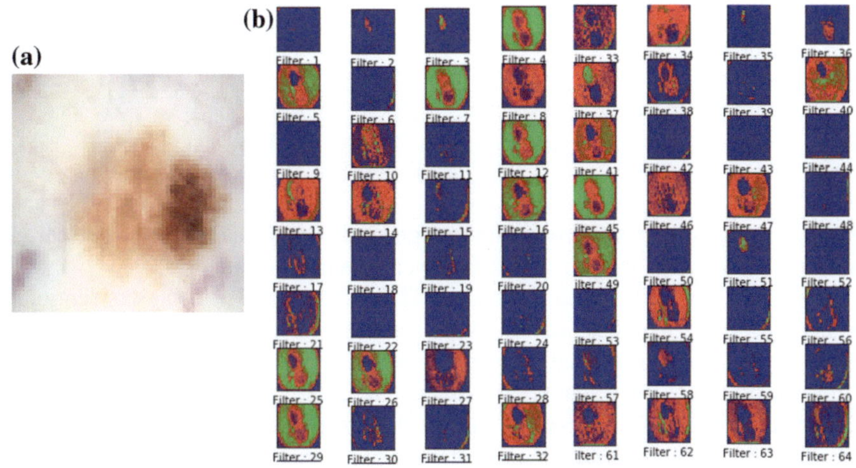

Fig. 3 **a** Malignant input **b** feature map of 64 filter outputs corresponding to the image

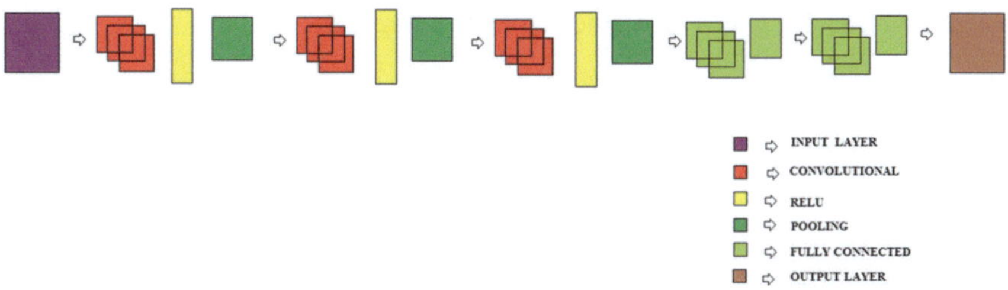

Fig. 4 Blocks in CNN architecture

last layer aggregates to unity and a soft max is used at the output. The blocks used in the architecture of CNN in this work is as shown in Fig. 4.

4 Results

The melanoma skin cancer detection exploiting computer vision-based techniques are implemented using three types of classification methods, viz., a threshold-based technique using TDS, using conventional machine learning techniques: KNN and SVM, and by employing a deep learning technique: CNN. Out of 1600 images taken for analysis from the database, 1120 images are used for training and the rest for testing and validation. Training data is used to train the model whereas the testing data is used to evaluate the performance of the model. The performance measures used for comparison are accuracy, precision, and recall values. Table 1 gives the mathematical definitions for the performance measures. In table 1, MD, MND, BD,

Table 1 Mathematical definitions for the performance measures

Performance measure	Definition
Accuracy	$\frac{MD+BND}{MD+MND+BD+BND}$
Precision	$\frac{MD}{MD+BD}$
Recall	$\frac{MD}{MD+MND}$

and BND represent, respectively, the malignant diseased, malignant not diseased, benign diseased, and benign not diseased elements in classification.

Our results of analysis: case (i) using TDS score, case (ii) using conventional machine learning methods: KNN, SVM, and case (iii) using CNN are shown in Table 2. The table shows that among all the techniques used, CNN proved to be highly accurate, viz., 90% compared to others. The accuracies for TDS-based method, KNN and SVM are 57.3%, 66.9%, and 69.35%, respectively. Further the precision (83.78%) and recall (96.8%) using deep learning techniques are high when compared to conventional machine learning techniques and the threshold-based technique. The confusion matrix for melanoma detection using CNN is given in Fig. 5. According to the color map obtained, deep learning applied to the image dataset gives high true positive rate, giving yellow and green along the diagonal.

Table 2 Performance analysis using three types of classification techniques

Measures in (%)	Case (ii)			
	TDS	KNN	SVM	CNN
Accuracy	57.3	66.9	69.35	90
Precision	58.57	58.16	60.45	83.78
Recall	58.7	61.3	64.36	96.8

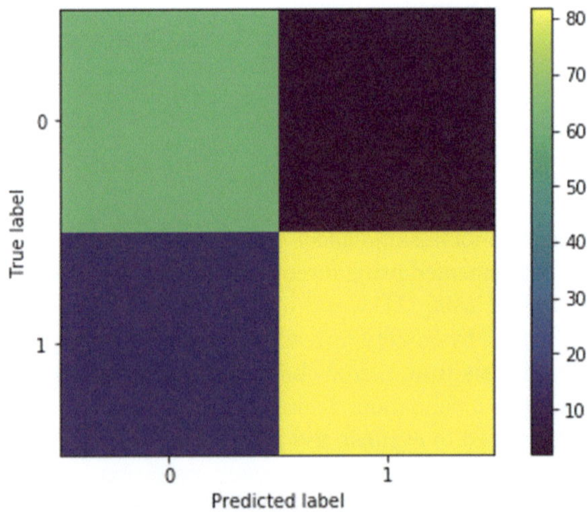

Fig. 5 Confusion matrix color-map for Melanoma detection using CNN

5 Conclusion

A comparative analysis of computer vision aided methods for melanoma skin cancer detection is done in this paper. As per our observation shows that the deep learning-based system using CNN is highly accurate, with 90% accuracy, in detecting melanoma. Apart from features used in ABCD algorithm, analysis using 7 point checklist and GLCM features can be done as an extension of this work. A high-end GPU-based analysis may further increase accuracy rate.

References

1. https://www.cancer.org/cancer/melanoma-skin-cancer/about/what-is-melanoma.html
2. https://www.aad.org/public/spot-skin-cancer/learn-about-skin-cancer/types-of-skin-cancer
3. Sathiya, S., Binu, S., Kumar, S., Prabin, A.: A survey on recent computer-aided diagnosis of melanoma. In: 2014 International Conference on ICCICCT, pp. 1387–1392. IEEE (2014)
4. Dubai, P., Bhatt, S., Joglekar, C., Patii, S.: Skin cancer detection and classification. In: ICEEI, pp. 1–6. IEEE (2017)
5. Alquran, H., Qasmieh, I.A., Alqudah, A.M., Alhammouri, S., Alawneh, E., Abughazaleh, A., Hasayen, F.: The melanoma skin cancer detection and classification using support vector machine. In: 2017 IEEE Jordan Conference on AEECT, pp. 1–5. IEEE (2017)
6. Nasr-Esfahani, E., Samavi, S., Karimi, N., Soroushmehr, S.M.R., Jafari, M.H., Ward, K., Najarian, K.: Melanoma detection by analysis of clinical images using convolutional neural network. In: EMBC, pp. 1373–1376. IEEE (2016)
7. https://isic-archive.com
8. Murumkar, O.S., Gumaste, P.P.: Feature extraction for skin cancer lesion detection. Int. J. Sci., Eng. Technol. Res. **4**(5) (2015)
9. Monisha, M., Suresh, A., Bapu, B.R.T., Rashmi, M.R.: Classification of Malignant Melanoma and Benign Skin Lesion by Using Back Propagation Neural Network and ABCD Rule in Cluster Computing, pp. 1–11. Springer, New York LLC (2018)
10. Poorna, S.S., Anuraj, K., Nair, G.J.: A weight based approach for emotion recognition from speech: an analysis using South Indian Languages. In: International Conference on Soft Computing Systems, Springer CCIS, Kollam on the 19th & 20th of April 2018
11. Franc, V., Hlavac, V.: Multi class support vector machine. In: Proceedings of IEEE International Conference on Pattern Recognition (2002)
12. Krizhevsky, A., Sutskever, I., Hinton, G.E.: Image net classification with deep convolutional neural networks. In: NIPS (2012)
13. Sze, V., Chen, Y.-H., Yang, T.-J., Emer, J.: Efficient Processing of Deep Neural Networks: A Tutorial and Survey (2017) arXiv: 1703.09039v2 [cs.CV]

Clustering of Association Rules on Microarray Gene Expression Data

S. Alagukumar, C. Devi Arockia Vanitha and R. Lawrance

Abstract Association rule mining is one of the most important procedures in data mining. In microarray gene expression analysis, often, large number rules are discovered. The rules have to be reduced significantly by techniques such as interest measures or clustering to knowledge mining. In this paper, a novel method for clustering association rules derived from microarray gene expression data is proposed. The gene expression data are converted into gene expression intervals using discretization technique. The association rules are extracted from the maximal frequent itemsets of the gene expression based on the support and confidence. The Euclidean distance measure is used to calculate similarity matrix for the derived association rules. Finally, the single linkage agglomerative clustering algorithm is employed to cluster the association rules based on the similarity matrix. The results obtained using the proposed method outperforms other methods like complete linkage and average linkage.

Keywords Agglomerative clustering, association rules · Clustering · Discretization · Gene expression

1 Introduction

Microarray technology is a powerful tool by which the expression patterns of thousands of genes can be monitored concurrently. The application of gene expression analysis ranges from cancer diagnosis to drug response. Microarray experiments

S. Alagukumar
Assistant Professor, Department of Computer Applications, Ayya Nadar Janaki Ammal College, Sivakasi 626124, Tamil Nadu, India

C. D. A. Vanitha (✉)
Assistant Professor, Department of Computer Science, The Standard Fireworks Rajaratnam College for Women, Sivakasi 626123, Tamil Nadu, India

R. Lawrance
Director, Department of Computer Applications, Ayya Nadar Janaki Ammal College, Sivakasi 626124, Tamil Nadu, India

B. Pati et al. (eds.), *Advanced Computing and Intelligent Engineering*, Advances in Intelligent Systems and Computing 1082,
https://doi.org/10.1007/978-981-15-1081-6_8

produce huge volume of data, which is main challenge. The clustering of association rules is helpful for discovering the knowledge from the large amount or volume of gene expression data.

Gupta et al. [1] presented a normalized distance metric to group association rules and clustered the association rules. The comparative statement of results obtained using different distance measures for clustering proved that the normalized distance measure outperformed the conventional methods. Giugno et al. [2] reviewed various discretization methods and classified the gene expression data using association rules mining and developed a classification model using gene expression intervals.

Alagukumar and Lawrance [3] reviewed the selective analysis of microarray gene expression data using association rule mining and discussed about the large amount of rule generated.

Alagukumar and Lawrance [4], in this paper, discretized gene expressions into intervals and analyzed microarray gene expression data using apriori association rule mining. Alagukumar and Lawrance [5] analyzed microarray gene expression data using various data discretization methods in their article AprioriClose algorithm that has been applied to generate the class association.

Usman et al. [6] developed the multilevel mining of association rules from warehouse schema compared to the rules discovered from the original data without schema imposed on it. Akben [7] introduced a novel clustering method, selfish data clustering (SDC), for clustering of biological signal datasets containing batched outliers.

Plasse et al. [8] stated a new methodology to an industrial application from the automotive industry with large datasets. Tan et al. [9] analyzed various interest measures that are used to find the significant rules for small set of patterns. Lent et al. [10] developed an approach for clustering and association rules to identify generalized segments in large datasets. Kosters et al. [11] proposed an algorithm for clustering in large database to extract information about the products purchased by the customers.

CDA Vanitha et al. [12] proposed a novel method for identifying clusters of different shapes in a dataset and have shown that the clustered genes have biological significance. The structure of the paper is as follows. Section 2 explained the datasets used in this work. Section 3 proposed the methodology of clustering of association rules on microarray gene expression data. In Sect. 4, the simulation results have been discussed. Finally, in Sect. 5, the approach has been concluded.

2 Materials

The microarray gene expression dataset is formulated based on the $X \times G$ matrix of expression values, where the row represents samples $X = \{x1, x2, x3, \ldots, xn\}$ and column represents genes $G = \{g1, g2, g3, \ldots, gn\}$. The sample microarray data format is shown in Table 1. The gene expression data is highly voluminous and the data mining techniques are used to extract useful knowledge and hidden pattern from the data [13–15].

Table 1 Microarray dataset format

Samples	Gene1	Gene2	…	Gene m
*X*1	*G*[13]	*G*[7, 13]	…	*G*[1, *m*]
…	…	…	…	…
Xn	*G*[*n*, 1]	*G*[*n*, 2]	…	*G*[*n*, *m*]

3 Methodology

The proposed methodology has four phases, such as discretization, mining maximal frequent itemsets, association rule generation, and clustering of rules. The schematic diagram of the proposed approach is shown in Fig. 1.

Fig. 1 Schematic diagram clustering of association rules

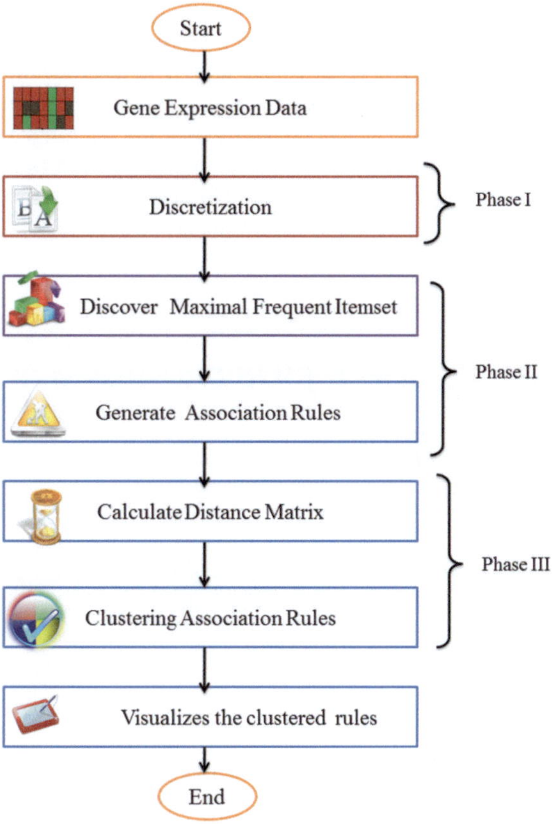

3.1 Discretization

Discretization is the process of transferring categorical or continuous values into discrete values. Discretization techniques can be classified into either supervised methods or unsupervised methods depending on the datasets used. The supervised discretization methods used class label when converting continuous attributes into discrete attributes. But, unsupervised discretization techniques discretize continuous attributes without the class label. The supervised techniques can be further characterized as entropy-based and statistics-based method. The equal-width binning method and equal frequency binning method are categorized into unsupervised discretization methods. Liu et al. [14] have defined a discretization process, which follows four steps, sorting the continuous values, calculating cut points for splitting intervals or merging intervals, based on some condition or criterion, and finally stopping at some point based on the splitting or merging intervals [14].

The equal-width interval bin (EWIB) discretization is one of the easiest methods that divide the observed values for genes expression into m equal-sized bins, where m is a parameter, given by the user. The discretization process involves sorting the continuous values of gene expression and finding the minimum and maximum values. The interval can be calculated by dividing the range into m equally sized bins using the given formula (1). The gene expression values with their specific intervals are separated by the cutting points. For example, $g1$ [$-$inf, cut-point] to $g2$ [cut-point, $+$ inf]. In this paper, equal-width interval bin (EWIB) discretization method has been used to discretize the gene expression data.

3.2 Algorithm 1: EWIB Discretization

Input: Gene expression data.
Output: Discretized gene expression.

Begin
Step 1: Read the genes expression data.
Step 2: Sort the gene expression values.
Step 3: Pass the m value to partitioning the data.
Step 4: Finding the minimum and maximum values.
Step 5: Spilt the range of continuous gene expression values according to cut point.

$$\text{Interval} = \frac{\text{V}_{\text{max}} - \text{V}_{\text{min}}}{m} \tag{1}$$

Step 6: Repeat step 5 until the stopping criteria is satisfied and discretize all the continuous values.
End.

3.3 Association Rule Mining

The fundamental process is of discovering an association rule and extracting interesting relations among the set of items [13, 16]. For example, an association rule between genes in the form of gene1 [−inf:4.03] → gene2 [0.3, 0.73], gene3 [0.14, 0.69] infers that if gene1 is expressed, then both gene2 and gene3 will also be expressed. Let $I = \{i1, i2, i3, \ldots, im\}$ be a set of m elements called items [14, 17]. Support of the rule $A \rightarrow B$ contains in the transaction set $|T|$ with support value s, where s is the number of transactions in $|T|$ that contains $A \cup B$.

$$S = \frac{\text{support } (A \cup B)}{|T|} \tag{2}$$

A rule is defined in the form $A \rightarrow B$, where $A, B \subseteq \text{I}$, and $A \cap B = \emptyset$. The left-hand side of the rule is called as antecedent itemsets and right-hand side of the rule is called as consequent itemsets.

3.4 Maximal Frequent Itemsets

Burdick et al. [18] proved that the set maximal frequent itemset is smaller than the frequent closed itemset, and also smaller than the set frequent itemset. Let D is a set of items, then $A \subseteq D$ be an itemset, and S be a collection of itemsets. In this work, we denote by support (A) the percentage of itemsets $B \in S$ such that $A \subseteq B$. If support$(A) \geq$ minimum support, then A is a frequent itemset. If A is frequent and there is no frequent super set of A, then A is a maximal frequent itemset.

3.5 Algorithm 2: Extraction of Association Rules

Input: Gene expression with samples and gene values with an interval.
Output: Association rules.

Begin
Step 1: Discover candidate itemsets.
Step 2: Find the frequent itemsets using support count.
Step 3: Remove the infrequent itemsets from the frequent itemsets.
Step 4: Discover of all maximal frequent itemset.
Step 5: Generate set of rules that have confidence above the minimum confidence threshold from maximal frequent items $R_k = \{r_1, \ldots, r_n\}$.
End.

3.6 Extract Association Rule Mining

If support $(A \rightarrow B) \geq$ minimum support value and confident $(A \rightarrow B) \geq$ minimum confident, the rule $A \rightarrow B$ is said to be strong or confident association rules, where minimum confident threshold is defined by user. The minimum support confidence value is set as 100% by the user to generate significant rules.

3.7 Clustering Association Rules

Lent et al. [10] introduced the idea of clustering association rules. Rokach et al. [17] analyzed various clustering methods and hierarchical clustering is creating clusters that have a predetermined ordering from top to bottom-up design. The hierarchical clustering is classified into two types known as divisive and agglomerative approaches [17]. In this paper, an agglomerative clustering algorithm is used to group the association rules derived from the gene expression data. Agglomerative hierarchical clustering algorithm works by clustering the data based on the nearest distance measure between the data objects until there is only a single cluster. In this paper, the single linkage hierarchical clustering method has been used to identify the distance between two clusters. The distance measure is used to define the cluster's distance of two points in each cluster. Given formulae ai and bi are represented in the single linkage method as follows to identify the distance of two cluster objects.

$$L = \text{minimum}(D(x_{ai}, x_{bi})) \tag{3}$$

3.8 Algorithm 3: Agglomerative Clustering Algorithm

Input: A set of rules $R = \{r_1, \ldots, r_n\}$ (n is a number of rules).
Output: Number of clustered association rules.

Begin
Step 1: Let $R = \{r_1, \ldots, r_n\}$ be the set of association rules.
Step 2: Start with disjoint cluster level 0.
Step 3: Find the least distance pair of clusters using Euclidean distance formula.

$$d\{c_i, c_j\} == \sqrt{|x_{i1} - x_{j2}|^2 + |x_{i2} - x_{j1}|^2 + \ldots + |x_{in} - x_{jn}|^2} \tag{4}$$

Step 4: Merge clusters into single cluster.
Step 5: Repeat steps 3 and 4 and update the distance matrix D. Finally, form a single cluster using single linkage algorithm $C = c_1, \ldots, c_n$.
End.

4 Result and Discussion

In this research work, the microarray gene expressions are used to evaluate the performance of the clustering of rule mining. The microarray gene expression data are taken from the research paper [2]. The dataset [19] is available at the National Centre for Bioinformatics (NCBI) website, such as http://www.ncbi.nlm.nih.gov, and the dataset has been downloaded and used in this work. The simulations are performed in a computer with a configuration of processor clock rate of 2.8 GHz and 4 GB of main memory. The algorithm is written using R statistical programming language version 3.3.1. R language is used for implementation, because it is one of the effective open-source software to work with the statistical data. R is data analysis tool used by data scientists, statisticians, and big data analysis. Breast cancer dataset is downloaded from National Centre for Bioinformatics (NCBI). The dataset contains 60 numbers of samples and 16 numbers of gene expression conditions. The following Table 2 shows the sample gene expression for primary breast cancer gene expression data. Where row represents the samples and column represents the genes. The numerical values represent the gene expression level.

Table 3 shows that the discrete transaction dataset and the gene expressions are discretized into four intervals using EWIB discretization algorithm, where the binning value is passed by the user.

Table 4 represents transaction itemsets, where first column of transaction represents the transaction id and second column of transaction represents the itemsets. The frequent itemsets are generated using the maximal frequent itemset.

Table 2 Microarray gene expression dataset

	ABCC11	HOXB13	CHDH	EST_3	IL17BR
S1	8.09	−2.78	0.45	0.54	−1.52
S2	4.63	−2.34	−0.28	0.39	−2.02
S3	5.86	−0.57	1.59	0.81	0.84
S4	4.91	1.22	−0.63	−0.1	−1.82

Table 3 Data discretization (bin value is 4)

	ABCC11	HOXB13	CHDH	EST_3	IL17BR
*S*1	[7.22, 8.09]	[−2.78, −1.78]	[−0.075, 0.480]	[0.355, 0.583]	[−2.020, −1.305]
*S*2	[4.63, 5.50]	[−2.78, −1.78]	[−0.630, −0.075]	[0.355, 0.583]	[−2.020, −1.305]
*S*3	[5.50, 6.36]	[−0.78, 0.22]	[1.035, 1.590]	[0.583, 0.810]	[0.125, 0.840]
*S*4	[4.63, 5.50]	[0.22, 1.22]	[−0.630, −0.075]	[−0.100, 0.128]	[−2.020, −1.305]

Table 4 Transaction dataset

tID	Itemsets
*S*1	ABCC11 [7.22, 8.09], HOXB13 [−2.78, −1.78], CHDH[−0.075, 0.480], EST_3 [0.355, 0.583], IL17BR [−2.020, −1.305]}
*S*2	{ABCC11 [4.63, 5.50], HOXB13 [−2.78, −1.78], CHDH [−0.630, −0.075], EST_3 [0.355, 0.583], IL17BR [−2.020, −1.305]}
*S*3	{ABCC11 [5.50, 6.36], HOXB13 [−0.78, 0.22], CHDH [1.035, 1.590], EST_3 [0.583, 0.810], IL17BR [0.125, 0.840]}
*S*4	{ABCC11 [4.63, 5.50], HOXB13 [0.22, 1.22], CHDH[−0.630, −0.075], EST_3 [−0.100, 0.128], IL17BR[−2.020, −1.305]}

Table 5 represents the generated maximal frequent itemset. From the maximal frequent itemset, the association rules are extracted. Table 6 shows the association rules, where the association rules are extracted with the user threshold value minimum support which is 50% and confidence is 100%. The similarity matrixes are calculated for the clustering of association rules using Euclidean distance methods. Table 7 shows the similarity matrix for the clustering of association rules.

The hierarchical clustering method is used to cluster the rules. Figure 2 represented the dendrogram for the clustered association rules. Figure 2 depicts the agglomerative clustering tree on unique splitting points in a different cluster set. For example, splitting at height of 1.0 would give the following cluster sets: (Rule-4), (Rule-3), (Rule-1, Rule-2), (Rule-9, Rule-10, Rule-11), (Rule-6), (Rule-7), (Rule-5, Rule-6), and (Rule-12, Rule-13, Rule-14). Rule-12, Rule-13, and Rule-14 are in the same

Table 5 Generated maximal frequent itemset

#	Frequent items	Support (%)
1	{HOXB13 = [−2.78, −1.78]}	50
2	{EST_3 = [0.355, 0.583]}	50
3	{ABCC11 = [4.63, 5.50]}	50
4	{CHDH = [−0.630, −0.075]}	50
5	{IL17BR = [−2.020, −1.305]}	75
6	{HOXB13 = [−2.78, −1.78], EST_3 = [0.355, 0.583]}	50
7	{HOXB13 = [−2.78, −1.78], IL17BR = [−2.020, −1.305]}	50
8	{EST_3 = [0.355, 0.583], IL17BR = [−2.020, −1.305]}	50
9	{ABCC11 = [4.63, 5.50], CHDH = [−0.630, −0.075]}	50
10	{ABCC11 = [4.63, 5.50], IL17BR = [−2.020, −1.305]}	50
11	{CHDH = [−0.630, −0.075], IL17BR = [−2.020, −1.305]}	50
12	{HOXB13 = [−2.78, −1.78], EST_3 = [0.355, 0.583], IL17BR = [−2.020, 1.305]}	50
13	{ABCC11 = [4.63, 5.50], CHDH = [−0.630, −0.075], IL17BR = [−2.020, −1.305]}	50

Table 6 Association rules

Rules	LHS	RHS	Support (%)	Confidence (%)
Rule1	{EST_3 = [0.355, 0.583]}	{HOXB13 = [−2.78, −1.78]}	50	100
Rule2	{HOXB13 = [−2.78, −1.78]}	{EST_3 = [0.355, 0.583]}	50	100
Rule3	{HOXB13 = [−2.78, −1.78]}	{IL17BR = [−2.020, −1.305]}	50	100
Rule4	{EST_3 = [0.355, 0.583]}	{IL17BR = [−2.020, −1.305]}	50	100
Rule5	{CHDH = [−0.630, −0.075]}	{ABCC11 = [4.63, 5.50]}	50	100
Rule6	{ABCC11 = [4.63, 5.50]}	{CHDH = [−0.630, −0.075]}	50	100
Rule7	{ABCC11 = [4.63, 5.50]}	{IL17BR = [−2.020, −1.305]}	50	100
Rule8	{CHDH = [−0.630, −0.075]}	{IL17BR = [−2.020, −1.305]}	50	100
Rule9	{EST_3 [0.355, 0.583], IL17BR[−2.020, −1.305]}	{HOXB13 = [−2.78, −1.78]}	50	100
Rule10	{HOXB13 [−2.78, −1.78], IL17BR[−2.020, −1.305]}	{EST_3 = [0.355, 0.583]}	50	100
Rule11	{HOXB13 [−2.78, −1.78], EST_3 [0.355, 0.583]}	{IL17BR = [−2.020, −1.305]}	50	100
Rule12	{CHDH[−0.630, −0.075], IL17BR[−2.020, −1.305]}	{ABCC11 = [4.63, 5.50]}	50	100
Rule13	{ABCC11 [4.63, 5.50], IL17BR[−2.020, −1.305]}	{CHDH = [−0.630, −0.075]}	50	100
Rule14	{ABCC11 [4.63, 5.50], CHDH[−0.630, −0.075]}	{IL17BR = [−2.020, −1.305]}	50	100

clusters, ABCC1, IL17BR, and CHDH are underexpressed genes and these genes are related to breast cancer.

The Leukemia gene expression dataset shown in Table 8 is used for demonstrating the performance of the proposed approach. The dataset is [19] available at http://www.broadinstitute.org/cgi-bin/cancer/datasets.cgi. The dataset contains the 72 samples and 7129 genes with a class attribute of acute lymphoblastic leukemia (ALL) and acute myeloid leukemia (AML). In this paper, the class label removed data has been

Table 7 Similarity distance matrix

	Rule-1	Rule 2	Rule-3	Rule-4	Rule-5	Rule-6	Rule-7	Rule 8	Rule 9	Rule-10	Rule-11	Rule-12	Rule-13	Rule-14
Rule-1	0.00	0.00	1.41	1.41	2.00	2.00	2.00	2.00	1.00	1.00	1.00	2.24	2.24	2.24
Rule-2	0.00	0.00	1.41	1.41	2.00	2.00	2.00	2.00	1.00	1.00	1.00	2.24	2.24	2.24
Rule-3	1.41	1.41	0.00	1.41	2.00	2.00	1.41	1.41	1.00	1.00	1.00	1.73	1.73	1.73
Rule-4	1.41	1.41	1.41	0.00	2.00	2.00	1.41	1.41	1.00	1.00	1.00	1.73	1.73	1.73
Rule-5	2.00	2.00	2.00	2.00	0.00	0.00	1.41	1.41	2.24	2.24	2.24	1.00	1.00	1.00
Rule-6	2.00	2.00	2.00	2.00	0.00	0.00	1.41	1.41	2.24	2.24	2.24	1.00	1.00	1.00
Rule-7	2.00	2.00	1.41	1.41	1.41	1.41	0.00	1.41	1.73	1.73	1.73	1.00	1.00	1.00
Rule-8	2.00	2.00	1.41	1.41	1.41	1.41	1.41	0.00	1.73	1.73	1.73	1.00	1.00	1.00
Rule-9	1.00	1.00	1.00	1.00	2.24	2.24	1.73	1.73	0.00	0.00	0.00	2.00	2.00	2.00
Rule-10	1.00	1.00	1.00	1.00	2.24	2.24	1.73	1.73	0.00	0.00	0.00	2.00	2.00	2.00
Rule-11	1.00	1.00	1.00	1.00	2.24	2.24	1.73	1.73	0.00	0.00	0.00	2.00	2.00	2.00
Rule-12	2.24	2.24	1.73	1.73	1.00	1.00	1.00	1.00	2.00	2.00	2.00	0.00	0.00	0.00
Rule-13	2.24	2.24	1.73	1.73	1.00	1.00	1.00	1.00	2.00	2.00	2.00	0.00	0.00	0.00
Rule-14	2.24	2.24	1.73	1.73	1.00	1.00	1.00	1.00	2.00	2.00	2.00	0.00	0.00	0.00

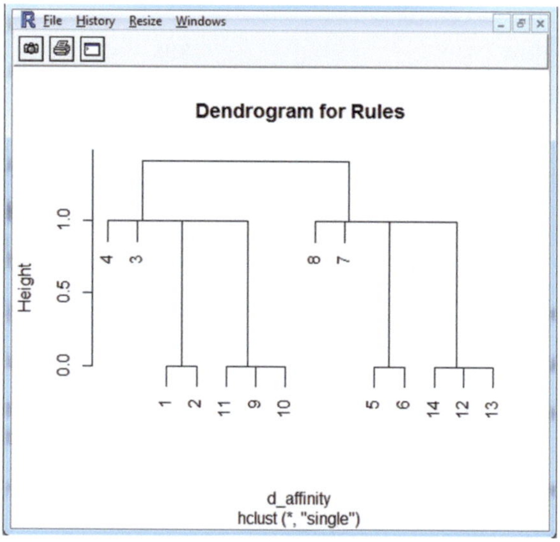

Fig. 2 Single linkage hierarchical clustering

Table 8 Sample leukemia dataset

Sample	AFFX-ioB-5	AFFX-BioB-M	AFFX-BioB-3	AFFX-BioC-5	AFFX-BioC-3
*S*1	−214	−153	−58	88	−295
*S*2	−139	−73	−1	283	−264
*S*3	−76	−49	−307	309	−376
*S*4	−135	−114	265	12	−419
*S*5	−106	−125	−76	168	−230
*S*6	−138	−85	215	71	−272
*S*7	−72	−144	238	55	−399
*S*8	−413	−260	7	−2	−541
*S*9	5	−127	106	268	−210
*S*10	−88	−105	42	219	−178

used for the clustering analysis and also for comparative analysis. The following Table 8 shows the sample gene expression for Leukemia cancer gene expression data. The Leukemia gene expressions are discretized into intervals using EWIB discretization algorithm. Table 9 shows the association rules which are extracted using maximal frequent itemsets.

The hierarchical clustering method is used to cluster the rules and the clustered sets are (Rule-1), (Rule-2), and (Rule-3, Rule-4 and Rule-5).

Table 9 Association rules

Rules	LHS	RHS	Support (%)	Confidence (%)
R1	AFFX-BioB-3 [−116.3, −21.0]	AFFX-BioB-M [−154.5, −119.3]	20	100
R2	AFFX-BioB-M [−84.2, −49.0]	AFFX-BioC-5 [257.2, 309.0]	20	100
R3	AFFX-BioB-M [−119.3, −84.2], AFFX-BioB-3 [169.7, 265.0]	AFFX-BioB-5 [−204.0, −134.3]	20	100
R4	AFFX-BioB-5 [−204.0, −134.3], AFFX-BioB-3 [169.7, 265.0]	AFFX-BioB-M [−119.3, −84.2]	20	100
R5	AFFX-BioB-5 [−204.0, −134.3], AFFX-BioB-M [−119.3, −84.2]	AFFX-BioB-3 [169.7, 265.0]	20	100

5 Conclusion

The proposed methodology of Clustering of Association Rule Mining is used to extract interesting patterns within the small group of rules. Habel et al. [21] studies have shown that a two gene ratio HOXB13 and IL17BR genes are related to the breast cancer. The clustered gene expression shows the most significant genes that cause breast cancer. The proposed novel methodology is compared with leukemia cancer dataset. The results of this proposed method can be used by the drug designers to provide targeting treatments.

References

1. Gupta, G.K., Strehl, A., Ghosh, J.: Distance based clustering of association rules. In: Proceedings of ANNIE Intelligent Engineering Systems Through Artificial Neural Networks, vol. 9, pp. 759–764 (1999)
2. Giugno, R., Pulvirenti, A., Cascione, L., Pigola, G., Ferro, A.: MIDClass: microarray data classification by association rules and gene expression intervals. PloS one **8** (2013)
3. Alagukumar, S., Lawrance, R.: A selective analysis of microarray data using association rule mining. Procedia Comput. Sci **47**, 3–12 (2015)
4. Alagukumar, S., Lawrance, R.: Algorithm for microarray cancer data analysis using frequent pattern mining and gene intervals. Int. J. Comput. Appl. **1**, 9–14 (2015)
5. Alagukumar, S., Lawrance, R.: Classification of microarray gene expression data using associative classification. In: IEEE International Conference on Computing Technologies and Intelligent Data Engineering (ICCTIDE), pp. 1–8 (2016)
6. Usman, M.: Multi-level mining of association rules from warehouse schema. Kuwait J. Sci. **44** (2017)
7. Akben, S.B.: A novel clustering method suitable for clustering of biological signal datasets containing batched outliers. Kuwait J. Sci. **44** (2017)
8. Plasse, M., Niang, N., Saporta, G., Villeminot, A., Leblond, L.: Combined use of association rules mining and clustering methods to find relevant links between binary rare attributes in a large data set. Comput. Stat. Data Anal. **52**, 596–613 (2007)

9. Tan, P.N., Kumar, V., Srivastava, J.: Selecting the right objective measure for association analysis. Inf. Syst. **29**, 293–313 (2004)
10. Lent, B., Swami, A., Widom, J.: Clustering association rules. In: 13-th IEEE International Conference on Data Engineering, pp. 220–23 (1997)
11. Kosters, W., Marchiori, E., Oerlemans, A.: Mining clusters with association rules. In: Advances in Intelligent Data Analysis, pp. 39–50 (1999)
12. Devi ArockiaVanitha C., Devaraj, D., Venkatesulu, M.: Real coded genetic algorithm for development of optimal GK clustering algorithm. In: International Conference on Swarm, Evolutionary, and Memetic Computing, Springer, Cham, pp. 264–274 (2014)
13. Agarwal, R., Srikant, R.: Fast algorithm for mining association rules in large data bases. In: Proceedings of the 20th International Conference on Very Large Data Base (VLDB'94), Santiago, Chile, pp. 487–499 (1994)
14. Liu, H., Hussain, F., Tan, C.L., Dash, M.: Discretization: an enabling technique. Data Min. Knowl. Disc. **6**(4), 393–423 (2002)
15. Tuimala, J., Laine, M.M.: DNA Microarray Data Analysis, 2nd edn. PicasetOy, Helsinki (2005)
16. Han, J., Kamber, M.: Data Mining: Concepts and Techniques. Morgan Kaufmann Publishers, Elsevier (2002)
17. Rokach, L., Maimon, O.: Clustering methods. In: Data Mining and Knowledge Discovery Handbook, pp. 321–352. Springer US (2005)
18. Burdick, D., Calimlim, M., Flannick, J., Gehrke, J., Yiu, T.: MAFIA: a maximal frequent itemset algorithm. IEEE Trans. Knowl. Data Eng. **17**, 1490–1504 (2005)
19. http://www.broadinstitute.org/cgi-bin/cancer/datasets.cgi
20. Golub, T.R., Slonim, D.K., Tamayo, P., Huard, C., Gaasenbeek, M., Mesirov, J.P., Bloomfield, C.D.: Molecular classification of cancer: class discovery and class prediction by gene expression monitoring. Science **286**(5439), 531–537 (1999)
21. Habel, L.A., Sakoda, L.C., Achacoso, N., Ma, X.J., Erlander, M.G., Sgroi, D.C., Quesenberry, C.P.: HOXB13: IL17BR and molecular grade index and risk of breast cancer death among patients with lymph node-negative invasive disease. Breast Cancer Res. **15**, R24 (2013)
22. http://www.ncbi.nlm.nih.gov

Boolean Association Rule Mining on Microarray Gene Expression Data

R. Vengateshkumar, S. Alagukumar and R. Lawrance

Abstract Microarray gene expression contains a dense amount of data from which analyzing and extracting interesting knowledge is a tedious task, hence the data mining techniques are used. In this research, a novel gene association rule algorithm called Boolean association rule (BAR) Mining has been proposed, where t-test has been used to filter the non-informative genes, *k*-means clustering is used to discretize the gene expressions, and the Boolean Association Technique has been implemented to generate frequent gene expressions. It also exposes the relation between normal and abnormal gene expression data and provides to take decisions for gene targeting treatment.

Keywords Gene expression · Gene selection · Discretization · Boolean association rule mining

1 Introduction

Microarray technology provides the way to calculate the expression level of genes in cells. Gene expression level is the procedure of transcribing Deoxyribonucleic Acid sequence to messenger Ribonucleic Acid sequence which known as the protein sequence [11]. A number of created versions from Ribonucleic Acid are described as gene expressions level [5–7]. Microarray technology helps to discover the new genes, to identify gene functions and expression levels. Microarray technology assists

R. Vengateshkumar (✉)
Research and Development Centre, Bharathiar University, Coimbatore, Tamil Nadu, India
e-mail: vengatesh.kumar@gmail.com

S. Alagukumar · R. Lawrance
Department of Computer Applications, Ayya Nadar Janaki Ammal College, Sivakasi, Tamil Nadu, India
e-mail: alagukumarmca@gmail.com

R. Lawrance
e-mail: lawranceraj@gmail.com

B. Pati et al. (eds.), *Advanced Computing and Intelligent Engineering*, Advances in Intelligent Systems and Computing 1082,
https://doi.org/10.1007/978-981-15-1081-6_9

researchers to study cancer, dengue diseases and to develop effective drug. Microarray gene association analysis is describing biological knowledge from the relationship between the genes. The researchers use this information to make drugs [11].

In this research paper, the BAR algorithm focuses on gene association rule mining to discover the interesting relationship between gene expression data by using class label data. The remaining part of this paper is structured as given below. The relevant works are evaluated in Chap. 2. The proposed work of microarray gene association analysis is described in Chap. 3. The experiment results are revealed in Chap. 4 and the conclusion of research is described in Chap. 5.

2 Reviewed Papers

The various techniques have been reviewed and studied to extract the biological knowledge of microarray data. Discovering the rules for normal and abnormal microarray gene expression, the researcher has to study the gene selection, data transformation, and association rule mining.

Jeanmougin et al. [7] have stated that statistical methods are very important in biological and biomedical research. They conducted a survey on gene expression data using various statistical approaches, such as the Welch's *t*-test, Analysis of variance, and Significance analysis of microarray. These approaches have been used to identify the differentially expressed genes. Cui and Churchill [2] have stated that discovering biological information from microarray data needs statistical methods. They have used *t*-test method for detecting differential expression of genes.

Garcia et al. [3] have made a survey of data transformation techniques which is used in preprocessing steps to convert continuous values into categorical values.

Liu et al. [8] have discussed and analyzed the role of discrete values in continuous data. The discrete values are easy to use and extract knowledge better than continuous values.

McIntosh et al. [10] have proposed the MAXCONF algorithm. The experiment shows that MAXCONF outperforms support-based rule mining and rule extraction. Wur et al. [12] have presented Boolean association rule mining in market basket data analysis. This approach reduces the number of scans to generate frequent item sets over the Apriori algorithm.

Martinez et al. [9] have developed a novel association rule for the analysis of genomic data whic is called GenMiner. GenMiner implements normalization and discretization algorithm to discover efficient nonredundant association rules. The experiment shows that GenMiner is significantly smaller than the Apriori-based algorithm in execution time and memory usage.

Giugno et al. [4] have presented a MIDClass classification method for microarray data, which is based on association rules. Finally, they have made an experiment with various microarray gene expression data and compared them with various classification methods.

Alagukumar et al. [1] have made a survey of mining association rules and reviewed microarray gene association analysis. Apriori and FP-growth association rule mining algorithms are applied to microarray data and they compared Apriori and FP-growth algorithms. The results showed that FP-growth algorithm took less memory usage and time complexity.

From these research works, it has been finalized that microarray gene expression data contains a dense amount of data. The statistical method has to be used to select significant gene expression data. The gene expression data having continuous values, hence the gene expression values need to be converted into discrete values. The microarray association mining is used to discover the relationship between microarray gene data. The existing microarray gene association mining algorithm consumes more time for generating frequent gene patterns for each class and extracting the rules for relationships between frequent gene patterns.

The BAR algorithm is implemented to select the important genes, discretize the gene expression, and generate frequent gene expressions without candidate generation and discovered association rules, which takes less memory and low computational time.

3 Methodology

Microarray gene association mining finds normal and abnormal frequent gene sets whose occurrences go above the user-defined value. It discovers the relationship between normal and abnormal frequent gene sets with minimum support and minimum confidence values. The BAR algorithm finds frequent normal and abnormal gene patterns and discovers the rules for normal and abnormal genes. These rules are helpful for making decisions for treatment. The BAR algorithm consists of a microarray preprocessing phase and microarray gene association phase. The procedure of the BAR algorithm is shown in Fig. 1.

3.1 Preprocessing

The preprocessing technique is used for data cleaning, data selection, and data transformation process. Data cleaning is a process to handle the missing data. Data selection is a process to select informative data. Data transformation is a process to transform continuous data into discrete data.

Begin

(1) Preprocessing
    (a) Call Gene Selection (Data D)
    (b) Call Discretization (Filtered Data D)
(2) Converts the Discretized data into the transaction format
(3) Generation of frequent k-itemsets
    (a) Call frequent(support S)
(4) Discover the Boolean association rules
    (c) Call Association(support S and confidence C)

End

Fig. 1 BAR algorithm

3.1.1 Gene Selection

In this paper, *t*-test is used to remove the non-informative genes and to select the informative genes expression data. The gene selection algorithm uses *t*-test to provide estimates of differentially expressed genes [7].

Input: Gene Expression Data, *D*
Output: Selection of Significant Genes

Procedure **Gene Selection (**Data *D***)**
Begin
for each normal and abnormal gene group
Calculate mean value for normal (N) and abnormal (A) genes
Calculate standard deviation and t-distribution with a degree of freedom

$$T = \frac{\text{mean(N)} - \text{mean(A)}}{\sqrt{\frac{\text{sd(N)}^2}{\text{m}} + \frac{\text{sd(A)}^2}{\text{n}}}} \ldots\ldots\ldots\ldots\ldots (1)$$

Calculate p value from t-distribution
if p value < 0.05
Set the gene as significant
else
Set the gene as in-significant
end

where N and A represent mean values of normal and abnormal genes, $\text{sd}(N)^2$ and $\text{sd}(A)^2$ represent standard deviation for normal and abnormal genes, and T represents *t*-distribution value. The *t*-distribution value is calculated using the mean value of normal and abnormal genes and standard deviations of normal and abnormal genes.

The t-distribution value is used to identify the p-value. The p-value which is less than 0.05 is used to select genes that are significantly different from normal gene expression data. Finally, the p-value which is greater than 0.05 is commonly assigned as insignificant genes and is filtered out from the gene set.

3.1.2 Discretization

Data discretization is the preprocessing technique. It transforms the continuous data into distinct discrete intervals [3]. Later, the data are analyzed in the knowledge representation and discrete values are used to simplify the data representation process. In this paper, the k-means discretization has been used to discretize continuous gene expression data. The k-means discretization technique is an unsupervised technique. The k-means discretization process is based on the following steps.

Input: Gene Expression with continuous data, D
Output: Gene Expression with discretized data

Procedure **Discretization** (Filtered Data D)
Begin

(a) Read the significant Genes
(b) Clusters the data D into n groups where n is user defined
(c) Choose n points randomly as cluster centers
(d) Allocate objects to nearest clusters center based on Euclidean distance method
(e) Calculate the centroid for all the objects in each cluster
(f) Repeat steps (c), (d) and (e) upto the same center points are allocated to each cluster in next iterations
(g) Sort center points for each cluster
(h) Each center position is computed as the $\text{Ti} = \frac{n_i + n_{i+1}}{2}$ of its associated points
(i) Discretized value is set by minimum value, Ti, maximum value
(j) Replace the continuous values into discrete values

End

3.2 *Gene Association Analysis*

Gene association rule mining finds frequent gene sets from the transaction gene set based on the user-defined threshold value and Boolean gene association rules are discovered based on minimum support and confidence.

Support: A rule presents in gene sets G with support S_p, where S_p is the number of genes represented in gene sets G [9].

Confidence: A rule has confidence C_p in gene set G, where C_p is the number of genes represented in gene sets G [9].

Gene Association Rules

A gene association rule is designed as $G \rightarrow G1$, where $G, G1 \subseteq D$ and $G \cap G1 = \emptyset$ [9]. In a rule "G" represents the antecedent and "$G1$" represents the consequent. The gene association analysis obtains frequent genes and rules from gene transaction data by using the BAR method. The BAR method generates frequent gene sets without candidate sets and extracts Boolean gene rules in the following phases.

Phase1: using logical OR and logical AND, the frequent gene sets are generated.

Phase2: using logical AND and logical XOR, the Boolean gene association rules are discovered.

Input: Gene Expression Data *D*, minimum support *S* and confidence *C*

Output: Boolean Gene Association Rules

Procedure **frequent** (support *S*)

Begin

(a) Initiating the gene table I_k and sample table T_k from the transaction database
(b) Concatenate the gene table $\mathrm{I_k}$, sample table $\mathrm{T_k}$and column vector table $\mathrm{V_k}$ as the $\mathrm{ITV_k}$and generate up to k-item set
(c) Obtain a pair of various row vectors in item table $\mathrm{I_{k-1}}$
(d) Perform logical-OR on those two row vectors to obtain a new k-item set
(e) Perform logical-AND on those two row vectors to get $\mathrm{T_{k-1}}$ upto final frequent itemset
(f) The result of k-item set, logical-AND result, vector count are stored in $\mathrm{I_k}$, $\mathrm{T_k}$and $\mathrm{V_k}$ table.
(g) Repeat until no new pairs of rows in $\mathrm{I_{k-1}}$

End

Procedure **Association** (support *S* and confidence*C*)

Begin

(a) Read table calledITV_k
(b) Remove the row vectors of 1-item sets from the last iteration which have no antecedent
(c) Perform logical-AND on two row vectors called x and z of $\mathrm{I_k}$in ITV
(d) Perform logical-XOR on two row vectors are selected in step (c) to get results x $\longrightarrow$ y
(e) if$\frac{\mathrm{Support(Z)}}{\mathrm{Support(x)}} \geq$ minimum confidence then

Boolean gene association rule x $\longrightarrow$ y is discovered

(f) Repeat the step (c) and (e) for all grouping of x and z until no new rules are created

End

The gene association to be analyzed is illustrated using the following tables. Table 1 represents the sample gene database. Table 2 shows the initialization of the gene and transaction data. Table 3 represents the gene association rules.

Table 1 Sample database

Tid	Gene set
TT1	Ge1, Ge2
TT2	Ge2, Ge3, Ge5, Ge6
TT3	Ge1, Ge2, Ge3, Ge5, Ge6
TT4	Ge3, Ge5
TT5	Ge1, Ge4

Table 2 Initialization of transaction table

	Transaction table T_k					Vector table V_k
	TT1	TT2	TT3	TT4	TT5	Count
Ge1	1	0	1	0	1	3
Ge2	1	1	1	0	0	3
Ge3	0	1	1	1	0	3
Ge4	0	0	0	0	1	1
Ge5	0	1	1	1	0	3
Ge6	0	1	1	0	0	2

Table 3 Boolean association rules

Rules	LHS	RHS	Support (%)	Confidence (%)
Rule1	Ge1	Ge3	50	100
Rule2	Ge2	Ge5	75	100
Rule3	Ge5	Ge2	75	100
Rule4	Ge2, Ge3	Ge5	50	100
Rule5	Ge3, Ge5	Ge2	50	100

4 Experimental Results

The experiment was carried out by the system with the hardware specification of the Intel Core i3 processor and 4 GB RAM. The BAR algorithm was executed in Java.

The microarray gene expression breast cancer 2 data are collected from National Centre for Biotechnology Information [13]. Table 4 shows sample breast cancer 2 type gene expression data. The t-test method was applied to select significant gene expression.

Table 5 represents the filtered microarray gene expression data. The filtered continuous microarray gene expressions are converted into distinct discrete gene expression using the *k*-means discretization algorithm. The discretized gene expression data are used to mine the frequent gene sets.

Table 4 Sample gene expression data

Sample	GE1	GE2	GE3	GE4	GE5	GE6	GE7	GE8	GE9	GE10
Class label	Abnorm	Abnorm	Abnorm	Abnorm	Abnorm	Norm	Norm	Norm	Norm	Norm
LYPD6	0.50	−1.50	−0.50	0.70	0.10	1.35	1.73	0.49	3.32	2.43
PTGER3	1.70	2.30	1.20	4.00	3.60	2.68	5.27	3.87	2.56	3.06
EST2	−0.50	−0.40	−0.60	−0.50	−0.40	0.04	−0.26	−0.07	−0.24	0.32
CHDH	0.90	1.20	0.30	1.00	−1.80	0.55	2.51	2.04	0.90	3.38
EST3	0.60	0.50	−0.30	−0.03	−1.20	1.12	1.89	2.11	0.95	2.34
…	…	…	…	…	…	…	…	…	…	…

Table 5 Filtered gene expression

Sample	GE1	GE2	GE3	GE4	GE5	GE6	GE7	GE8	GE9	GE10
Class label	Abnorm	Abnorm	Abnorm	Abnorm	Abnorm	Norm	Norm	Norm	Norm	Norm
EST2	−0.50	−0.40	−0.60	−0.50	−0.40	0.04	−0.26	−0.07	−0.24	0.32
LYPD6	0.50	−1.50	−0.50	0.70	0.10	1.35	1.73	0.49	3.32	2.43
ABCC11	6.00	6.60	4.40	6.80	6.00	3.67	0.36	0.28	1.19	1.24
EST3	0.60	0.50	−0.30	−0.03	−1.20	1.12	1.89	2.11	0.95	2.34
IL1R2	1.40	1.70	0.50	1.40	1.10	0.53	0.34	0.45	0.02	−0.14
IL17BR	−0.70	−1.10	−2.20	−0.60	−3.60	0.76	1.47	1.93	−1.21	1.12

Table 6 Transaction data set for an abnormal gene set

Transaction ID	Items
GE1	LYPD6 [−0.30, 0.70], EST2 [−0.60, −0.50], EST3 [−0.50, 0.70] IL17BR [−1.90, −0.60], IL1R2 [1.00, 1.70], ABCC11 [5.30, 6.80]
GE2	LYPD6 [−1.50, −0.30], EST2 [−0.50, −0.40], EST3 [−0.50, 0.70] IL17BR [−1.90, −0.60], IL1R2 [1.00, 1.70], ABCC11 [5.30, 6.80]
GE3	LYPD6 [−1.50, −0.30], EST2 [−0.60, −0.50], EST3 [−0.50, 0.70] IL17BR [−3.60, −1.80], IL1R2 [0.60, 1.00], ABCC11 [4.4, 5.30]
GE4	LYPD6 [−0.30, 0.70], EST2 [−0.60, −0.50], EST3 [−0.50, 0.7] IL17BR [−1.80, −0.60], IL1R2 [1.00, 1.70], ABCC11 [5.30, 6.80]
GE5	LYPD6 [−0.30, 0.70], EST2 [−0.50, −0.40], EST3 [−1.20, −0.50] IL17BR [−3.60, −1.80], IL1R2 [1.00, 1.70], ABCC11 [5.30, 6.80]

Table 7 Transaction data set for a normal gene set

Transaction ID	Items
GE6	LYPD6 [0.49, 2.03], EST2 [0.00, 0.32], EST3 [0.95, 1.57] IL17BR [0.05, 1.93], IL1R2 [0.19, 0.53], ABCC11 [2.22, 3.67]
GE7	LYPD6 [0.49, 2.03], EST2 [−0.26, 0.00], EST3 [1.57, 2.34] IL17BR [0.05, 1.93], IL1R2 [0.19, 0.53], ABCC11 [0.28, 2.22]
GE8	LYPD6 [0.49, 2.03], EST2 [−0.26, 0.00], EST3 [1.57, 2.34] IL17BR [0.05, 1.93], IL1R2 [0.19, 0.53], ABCC11 [0.28, 2.22]
GE9	LYPD6 [2.03, 3.32], EST2 [−0.26, 0.00], EST3 [0.95, 1.57] IL17BR [−1.21, 0.05], IL1R2 [−0.14, 0.19], ABCC11 [0.28, 2.22]
GE10	LYPD6 [2.03, 3.32], EST2 [0.00, 0.32], EST3 [1.57, 2.34] IL17BR [0.05, 1.93], IL1R2 [−0.14, 0.19], ABCC11 [0.28, 2.22]

Table 6 represents the transaction data set for abnormal gene set. Table 7 represents the transaction data set for normal gene set. Table 8 represents a normal frequent gene set. Table 9 represents an abnormal frequent gene set.

The gene expression knowledge is discovered from the Boolean gene association rules. Table 10 represents the Boolean association rule mining for normal genes. Table 11 represents the Boolean association rule mining for abnormal genes.

For example, the rule for normal gene LYPD6 [0.49: 2.03] → IL1R2 [0.19: 0.53] shows that the gene expressions expressed positively, the rule for an abnormal gene LYPD6 [–0.30: 0.70] → IL1R2 [1.00: 1.70] shows that the gene expression is expressed negatively. It provides the decisions for gene-targeting cancer treatments.

Table 8 Normal frequent gene set

Frequent Item Set Class for Normal
LYPD6 [0.49: 2.03]
IL17BR [0.055: 1.93]
EST2 [−0.260: −0.005]
ABCC11 [0.28: 2.22]
IL1R2 [0.19: 0.53]
LYPD6 [0.49: 2.03] IL17BR [0.055: 1.930]
LYPD6 [0.49: 2.03] IL1R2 [0.19: 0.53]
IL1R2 [0.19: 0.53] IL17BR [0.055: 1.930]
EST2 [−0.260: −0.005] ABCC11 [0.28: 2.22]
EST3 [1.57: 2.34] IL17BR [0.055: 1.930]
EST3 [1.57: 2.34] ABCC11 [0.28: 2.22]
LYPD6 [0.49: 2.03] IL17BR [0.055: 1.930] IL1R2 [0.19: 0.53]
EST3 [1.57: 2.34] IL17BR [0.055: 1.930] ABCC11 [0.28: 2.22]

Table 9 Abnormal frequent gene set

Frequent Item Set for Abnormal
LYPD6 [−0.30: 0.70]
IL1R2 [1.00: 1.70]
ABCC11 [5.30: 6.80]
EST2 [−0.60: −0.50]
EST3 [−0.60: 0.70]
IL17BR [−1.90: −0.60]
LYPD6 [−0.30: 0.70] IL1R2 [1.00: 1.70]
LYPD6 [−0.30: 0.70] ABCC11 [5.30: 6.80]
IL1R2 [1.00: 1.70] ABCC11 [5.30: 6.80]
EST2 [−0.60: −0.50] EST3 [−0.60: 0.70]
IL17BR [−1.80: −0.60] EST3 [−0.50: 0.60]
IL17BR [−1.80: −0.60] ABCC11 [5.30: 6.80]
LYPD6 [−0.30: 0.70] IL1R2 [1.00: 1.70] ABCC11 [5.30: 6.80]
IL17BR [−1.80: −0.60] EST3 [−0.60: 0.70] IL1R2 [1.00: 1.70]
IL17BR [−1.80: −0.60] IL1R2 [1.00: 1.70] ABCC11 [5.30: 6.80]
EST3 [−0.50: 0.60] IL1R2 [1.00: 1.70] ABCC11 [5.30: 6.80]
LYPD6 [−0.3: 0.70] IL1R2 [1.00: 1.70] ABCC11 [5.30: 6.80]
EST3 [−0.50: 0.60] IL17BR [−1.80: −0.60] IL1R2 [1.00: 1.70]
EST3 [−0.50: 0.60] IL17BR [−1.80: −0.60] ABCC11 [5.30: 6.80]
EST3 [−0.50: 0.60] IL17BR [−1.80: −0.60] IL1R2 [1.0: 1.7] ABCC11 [5.30: 6.80]

Table 10 Boolean association rule mining—normal genes

No.	Antecedent	Consequent	Support (%)	Confidence (%)
1	LYPD6 [0.49: 2.03]	IL1R2 [0.19: 0.53]	60	100
2	EST3 [1.57: 2.34]	IL17BR [0.055: 1.930]	60	100
3	EST3 [1.57: 2.34]	IL17BR [0.055: 1.930], ABCC11 [0.28: 2.22]	60	100

Table 11 Boolean association rule mining—abnormal genes

No.	Antecedent	Consequent	Support (%)	Confidence (%)
1	LYPD6 [−0.30: 0.70]	IL1R2 [1.00: 1.70]	60	100
2	IL17BR [−1.80: −0.60]	EST3 [−0.50: 0.60]	60	100
3	EST3 [−0.50: 0.70], IL17BR [−1.80: −0.60]	ABCC11 [5.30: 6.80]	60	100

5 Conclusion

The experiment was made by using breast cancer 2 microarray gene expression data set. The BAR algorithm obtains frequent normal and abnormal gene sets without candidate gene set generation. It scans the gene expression data set only one time. It reduces the time complexity to expose the relationship between normal and abnormal gene set. The BAR algorithm generates frequent gene expression intervals for normal and cancer occurrence of gene expression data. Association rules discover significant relations among normal and cancer occurrence genes. The result of the BAR algorithm finds the crucial diseased gene expression and provides decisions for diseased gene treatments.

References

1. Alagukumar, S., Lawrance, R.: A selective analysis of microarray data using association rule mining. Procedia Comput. Sci. **47**, 3–12 (2015)
2. Cui, X., Churchill, G.A.: Statistical tests for differential expression in cDNA microarray experiments. Genome Biol. **4**(4), 210 (2003)
3. Garcia, S., Luengo, J., Saez, J.A., Herrera, F.: A survey of discretization techniques: taxonomy and empirical analysis in supervised learning. Knowl. Data Eng. IEEE Trans. **25**(4), 734–750 (2013)
4. Giugno, R., Pulvirenti, A., Cascione, L., Pigola, G., Ferro, A.: MIDClass: microarray data classification by association rules and gene expression intervals. PLoS ONE **8**(8), 216–231 (2013)
5. Han, J., Kamber, M.: Data Mining: Concepts and Techniques, Morgan Kaufmann Publishers, Elsevier (2002)
6. Hu, Y., Aram C., Norbert E., Stephanie, M.: The Drosophila gene expression tool (DGET) for expression analyses. BMC Bioinform. (1) (2017)
7. Jeanmougin, M.: Should we abandon the *t*-test in the analysis of gene expression microarray data: a comparison of variance modeling strategies. PloS one **5**(9) (2010)
8. Liu, H., Hussain, F., Tan, C.L., Dash, M.: Discretization: an enabling technique. Data Min. Knowl. Disc. **6**(4), 393–423 (2002)
9. Martinez, R., Nicolas, P., Claude, P.: GenMiner: mining non-redundant association rules from integrated gene expression data and annotations. Bioinformatics **24**(22), 2643–2644 (2008)
10. McIntosh, T., Sanjay, C.: High confidence rule mining for microarray analysis. IEEE/ACM Trans. Comput. Biol. Bioinf. **4**(4), 611–623 (2007)

11. Vengateshkumar, R., Alagukumar, S., Lawrance, R.: Analysis of microarray gene expression data using boolean association rule mining. Int. J. Innov. Technol. Creat. Eng. **7**(5), 412–416 (2017)
12. Wur, S.Y., Leu, Y.: An effective boolean algorithm for mining association rules in large databases. In: Proceedings on Database Systems for Advanced Applications, IEEE Transactions, pp. 179–186 (1999)
13. www.ncbi.nlm.nih.gov/geo/query/acc.cgi?acc=GSE1379

Early Detection of Alzheimer's Disease Using Multi-feature Fusion and an Ensemble of Classifiers

G. Janakasudha and P. Jayashree

Abstract Computerized detection of Alzheimer's disease (AD) can assist medical practitioners to a greater extent in the early diagnosis of the onset of Dementia in elderly people. Structural Magnetic Resonance Imaging (MRI) of the human brain is a key modality that aids in the early diagnosis of the disease. We have come up with an automatic identification of the AD examining the structural MRI using machine learning. The proposed approach aids to identify persons with Alzheimer's disease using a multi-feature fusion approach. Multi-feature fusion is performed by Support Vector Machine using Feature elimination in a recursive manner where an optimal subset of features is obtained. An ensemble of classifiers with Multi-Layer Perceptron (MLP), Support Vector Machine (SVM) and J48 are used. The performance of the individual classifiers and the ensemble of classifiers with combined features are measured. A better performance is achieved through the multi-feature approach with ensemble of classifiers. The CAD system provided a maximum accuracy of 93.8%.

Keywords Alzheimer's disease · Ensemble learning · Multiple feature fusion

1 Introduction

Alzheimer's disease is a chronic disease that causes degeneration of neurons in the brain and leads to dementia in elderly persons. It is considered as one of the death causing diseases next to heart diseases and cancer. Early detection of the disease aids in preventing the progression and in treating the disease. As no cure for Alzheimer's disease exists, the early diagnosis of the disease is required to delay the onset of the disease. It also reduces the psychological impact on the patients. Alzheimer's

G. Janakasudha (✉)
Sri Venkateswara College of Engineering, Sriperumbudur, India
e-mail: jsudha@svce.ac.in

P. Jayashree
Madras Institute of Technology Campus, Anna University, Chennai, India
e-mail: pjshree12@gmail.com

B. Pati et al. (eds.), *Advanced Computing and Intelligent Engineering*, Advances in Intelligent Systems and Computing 1082,
https://doi.org/10.1007/978-981-15-1081-6_10

disease causes a slow decline in cognitive, behavioral, psychological, reasoning and thinking skills.

More than 47 million people are affected by Alzheimer's disease throughout the globe. The progression of the disease differs from one patient to the other. The definite diagnosis of the presence of AD can be done only with the detection of plaques and tangles in the human brain. It is a most challenging task to perform an early detection of the presence of AD which can lead to effective treatment to slow down the clinical progression of the AD. For early examination of AD, it is important to study the symptoms of the onset of the disease. This state of decrease in cognitive functions is called as Mild Cognitive Impairment (MCI). MCI is a state where the patient is able to notice the changes in their thinking abilities. Patients with MCI have a greater chance of converting to AD than persons without MCI. The MCI is a state where the impairment could be due to the normal aging of the person or it could lead to AD eventually causing Dementia. The prognosis of the changeover from MCI to Alzheimer's disease is an open issue. Therefore an accurate diagnosis of AD or MCI at an early stage is very significant.

There are a varied number of neuroimaging techniques like Magnetic Resonance Imaging (MRI) which includes Structural, functional and diffusion tensor imaging and Positron Emission Tomography (PET) are used for the examination of Alzheimer's disease. Out of these the Structural MRI is more standardized imaging modality in detecting the clinical progress of AD compared to other modalities. The diagnosis of AD involves the neurogenetic testing measurements, amount of amyloid and tau, neuronal injuries, amount of cerebrospinal fluid (CSF) and cerebral atrophy.

Machine learning and Pattern classification are used widely nowadays to develop a computer-aided diagnosis of AD with different modalities. Our work considers the Structural MRI as it is more promising than other modalities. Several features can be extracted from a Structural MRI image of the whole brain. This includes intensity, gray-matter intensity, white-matter intensity, size of the hippocampus, cerebral atrophy, cortical thickness, morphometry, texture-based features. A combination or fusion of features is used to improved accuracy of the diagnosis of AD and MCI rather using a single feature for diagnosis.

2 Background

Imaging has played important role in the studies on Alzheimer disease over the past 40 years. Computed tomography (CT) was used at the early stages and later Magnetic Resonance Imaging (MRI) is used to investigate the sources of neurodegenerative diseases [1]. Out of the available imaging modalities, each one has its own advantages and disadvantages and no single modality can be used to serve all purposes. The efficient combination of imaging biomarkers will aid in diagnosis, disease phasing and can be used to track the progress of the disease using which the therapies could be fine-tuned specific to the patients.

There are various works contributing to the literature of computerized diagnosis of Alzheimer's disease. In [2] a Voxel-based method is used and a Welch's t-test is done where the mean and standard deviation are determined for the voxels chosen. SVM and classification trees are used to classify AD-affected brain from normal brain. A feature extraction method using landmarks is proposed in [3, 4]. In [3] a longitudinal structural MR image is presented. This approach avoids time-consuming steps during application stage registration and tissue segmentation and provides efficient diagnosis. The consolidated feature representations for patients are extracted from different longitudinal scans. In [4] a landmark detection scheme based on regression forests constrained on shape is proposed. Experiments show that landmark based feature extraction outperformed ROI-based and Voxel-based approaches and has better accuracy of diagnosis and it is approximately 50 times faster than the region-based methods.

A fast filter approach is used in [5] to select voxels commonly triggered for the AD affected and normal brain. The method was evaluated using statistical tests. It is considered a powerful method to find out the dissimilarities in several brain regions.

Selection of desirable features plays a major role in improving the computer-aided investigation of AD. In [6], a significance map or p-map is constructed to quantify the importance of individual features. Two feature selection methods from gray-matter morphometry based on significance maps using a filter (direct method) and a wrapper (indirect method) are proposed. The methods are evaluated and compared based on t-statistics, weight vectors of SVM and domain expert knowledge. In [7] the classifiers support vector machine (SVM), a regularized extreme learning machine (RELM) and an import vector machine (IVM), were used. Feature selection technique using a greedy approach is used to choose decisive feature vectors. A discriminative approach based on kernels is chosen to deal with composite data distributions. RELM is found to give a better accuracy compared to the other two methods.

A lot of works have been carried out on the importance of the feature selection methods and using them to classify AD patients from normal subjects using Support Vector Machine with different kernels and using other classifiers like Multi-Layer Perceptron (MLP), Decision Trees, Naive Bayes and Logistic Regression. The size of brain structures like whole brain, hippocampus, cortical gray matter, white matter, medial temporal lobe are considered as the major biomarkers for different classifications such as AD, MCI, and healthy in various literatures like [8].

In [9] various feature reduction methods are combined on both brain MRI and blood plasma proteins. A model is built using Support Vector Machine (SVM) using the selected set of features and texture descriptors. A set of SVM is further integrated by a weighted sum rule.

Automatic segmentation of various tissues in the brain is performed for every image to obtain the dissemination of gray matter (GM) and white matter (WM) tissues in [10]. SVM with different kernels like linear or radial is used to categorize persons as normal or affected by AD. The efficiency of the classifier is evaluated. The system based on Partial Least Squares (PLS) for extracting features with a linear SVM classifier extracted notable information and performed better than system based on the Principal Component Analysis (PCA) method for extracting features.

An ensemble of classifiers (SVM, MLP, and J48), is employed [11] to overcome the deficiencies of an individual classifier. The decision of the ensemble class is made based on the largest total score or vote from individual classifiers. However in [11] the features are considered individually and then considered together as combined features.

There are also some literatures available on combining multiple features to get a better performance of the classifiers. The volume of the gray-matter, gray-level co-occurrence matrix, and Gabor feature obtained from the brain image adds up to the features available for study [12]. Combining multiple features, results in a better performance of the classifier. SVM-RFE algorithm with co-variance method is used for selecting the multiple features and it is found that multi-feature method outperforms the single-feature method.

In [13] the cortical and sub cortical thickness and volumes and the sulcal measures are found. Suitable combinations of the individual measures are used with different classifiers like Naive Bayes, Logistic Regression and Support Vector Machine and their performances are compared. For classification, the sulcal measures are superior or equal to the other measures. The thickness of the sylvian fissure is one of the important discerning measures and is one of the significant biomarker for detecting early stage of AD.

Apart from the biomarkers considered earlier, a method proposed in [14] combines the Scale Invariant Feature Transform (SIFT) and Histogram of Oriented Gradients (HOG) descriptor in anatomical structure. The methods have a good representational ability to distinguish patients with AD from healthy people.

In [15] hippocampus segmentation using algorithms such as hierarchical AdaBoost, SVM and hierarchical SVM with automated feature selection (Ada-SVM) are studied. In [16] a two-step feature selection procedure is used. This includes statistical and anatomical features. These features along with knowledge based region of interests for Alzheimer's disease in the human brain were used to classify AD and MCI using Support Vector Machines.

In [17] the structural differences at key regions in the brain of different groups of people using region-based method are discussed. The machine learning approach proposed in [17] is used to distinguish patients with Alzheimer's disease or mild impairment from the healthy persons and a prediction of transformation of mild impairment to AD is also made. A smooth and symmetric registration is used to identify the shape deformations between different groups of people. The ability to discriminate people into various groups has been increased using the deformation-based methods.

Therefore, in the proposed method, both Morphometric and Texture-based features are extracted and using the SVM-RFE algorithm an optimal set of features are obtained and those features are used for ensemble classification to improve accuracy compared to the existing methods.

3 Methods

In this section, the proposed work and the methods used are described in detail. The proposed classification method is shown in Fig. 1.

3.1 Image Acquisition

We have obtained the MRI scan images from Alzheimer's Disease Neuroimaging Initiative (ADNI). We used the images in the ADNI1 Annual2Yr_3T dataset. It consists of 200 Normal controls, 400 MCI and 200 AD. The objective of ADNI is to test if MRI, PET, neuropsychological assessments, clinical tests and other biomarkers can be unified to analyze the progression of MCI and early AD. Useful results out of these markers can be used by researchers and clinicians to diagnose early AD. It also aids in developing new treatments, observe the efficiency, reduces the time duration and expenditures of the clinical examinations. The dataset can be downloaded from http://adni.loni.usc.edu/.

3.2 Preprocessing

Image processing algorithms rely heavily on the quality of the images. The process of MRI acquisition involves certain artifacts. Image preprocessing is a significant process which can improve the reliability of the image to a greater extent and enhances certain details in the image for efficient analysis. As MRI acquisition is a time-consuming process and it demands patients to be in complete still state. But due to

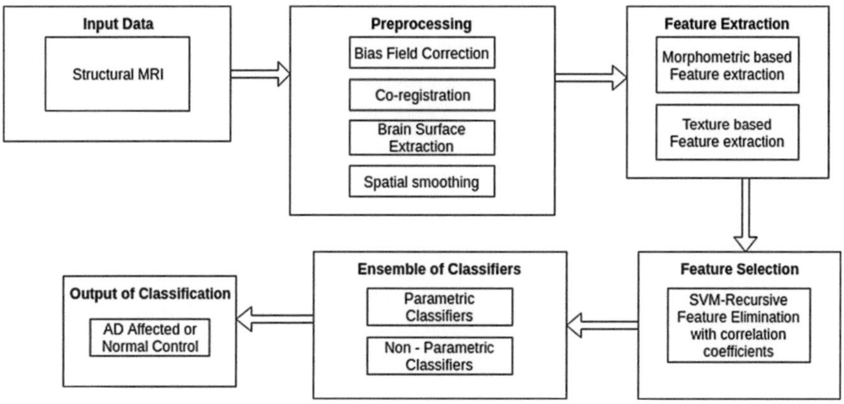

Fig. 1 Workflow of proposed classification system

voluntary and involuntary motion caused by the patients the MRI acquired will have distortions. To remove this artifact, we use motion correction. There is also a possibility of inhomogeneity due to magnetic field or variation in magnetic susceptibility. This could lead to poor feature extraction and segmentation. A bias-field correction is done to remove this artifact. Brain scans differ in their size and shape for individual subjects. Therefore, a co-registration of the acquired image with the standard brain template aids in the better identification and comparison of anatomical structures across subjects. Brain tissues may overlap with other neighboring tissues of neck, bone and skin. As they are not required and also might reduce the reliability of the process, they are removed using brain surface extraction where the non-brain tissues are removed from the MRI image. Finally, spatial smoothing is applied to remove any minute matching error and to enhance the signal-to-noise ratio.

3.3 *Segmentation and Feature Extraction*

Neuroimaging helps in distinguishing efficiently the Healthy persons from Alzheimer's disease affected patients. The major concern is the huge size of the MRI images and processing them takes a lot of computational time. Moreover not all the information available in the images is relevant for the classification of AD. Feature extraction is a useful method using which more relevant and discriminative features can be obtained from the input image and it helps in obtaining better classification accuracy. There are a variety of features that can be obtained from an MRI image. They could be texture-based or morphological-based.

3.3.1 Morphometry Based Features

The morphometry based features like amount of Gray Matter (GM), White Matter (WM), CerebroSpinal Fluid (CSF) and size of the hippocampus are obtained. Many research studies have presented a reduction in the amount of Gray matter in the presence of AD. The tissue volumes are obtained through FSL software package. A threshold value is used to classify the GM, WM and CSF. Intensity values above the threshold are WM, intensity values below the threshold are CSF and intensity value equal to threshold is GM. Figures 2 and 3 shows the segmentation of tissues for Normal Control and Alzheimer's disease.

Hippocampus is present in the limbic system of the human brain. It is important for the consolidation of memory information. With the progression of AD, the hippocampus shrinks and gets reduces in size which is usually termed as atrophy. The atrophy measurement of the hippocampus can be used as a significant biomarker for identifying as well as finding the progression of AD. Measuring the size of the hippocampus requires a segmentation of the region from the other regions of the brain.

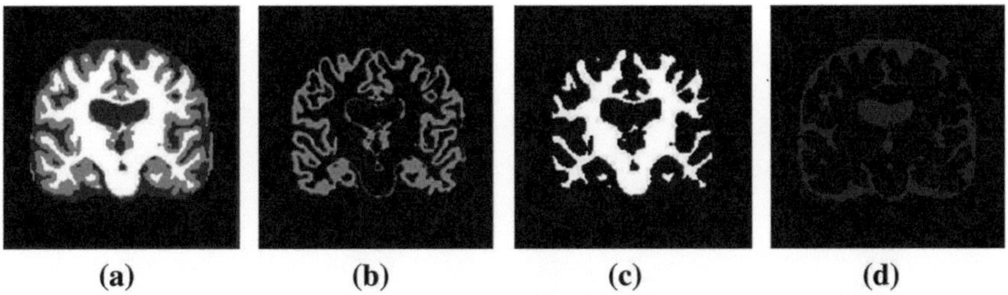

Fig. 2 **a** Normal control—original image, **b** Segmented Gray matter, **c** White matter and **d** CSF

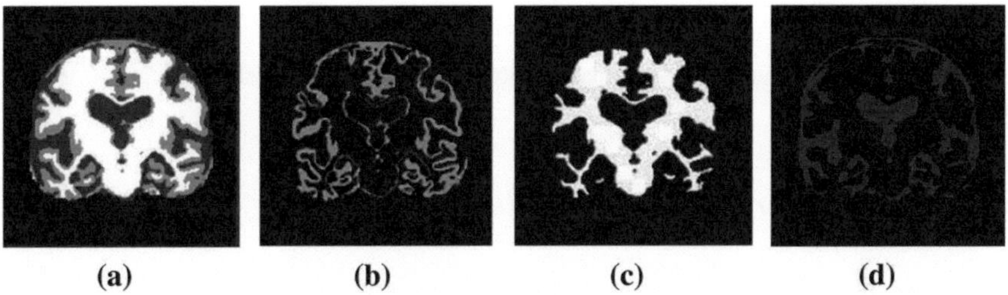

Fig. 3 **a** AD—original image, **b** Segmented Gray matter, **c** White Matter and **d** CSF

As human brain is a complex structure, a low intensity difference between different regions and the absence of well defined lineament, makes segmentation of the hippocampus a challenging task. We propose to make use of ROI mask mapping, ROI extraction, denoising, region trimming, hippocampus region extraction and hippocampus size measurement.

After this, the hippocampus is obtained with the high intensity values as compared to low intensity values of CSF in surrounding regions. The size values of the left and right hippocampi are lesser for the AD patients compared to healthy people. Figure 4 shows the actual and estimated location of the hippocampus.

3.3.2 Texture-based Features

The texture features are low level features that can be used for revealing different characteristics in the image. Gray-level co-occurrence matrix (GLCM) and the Gabor features are the effective features representing the textures are used. From GLCM's 28 feature descriptors can be obtained. Out of them correlation, contrast, sum of squares, sum of averages, sum of variances are more significant. Gabor filters perform suitably in both frequency and spatial domains. A 2D Gabor filter in different direction and frequency is used to get the texture features.

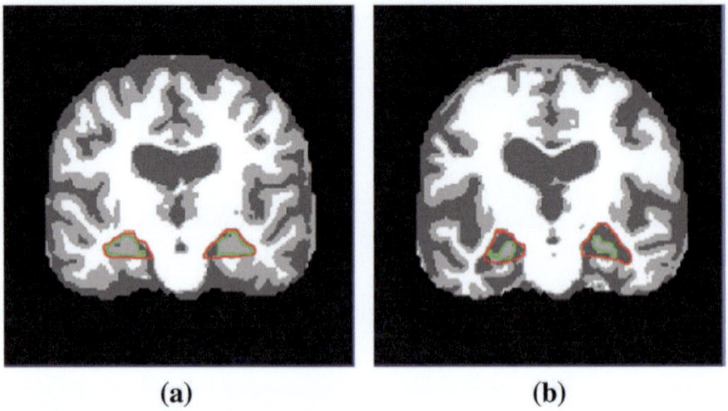

Fig. 4 Brain of **a** NC and **b** AD. Estimated (shown in red) and actual (shown in green) left and right hippocampus

3.4 Multi-feature Fusion and Classification

An optimal feature subset is obtained through the SVM-RFE and co-variance method [12]. This feature subset is used for better classification accuracy than using individual features. SVM-RFE is applied on parameters selected from SVM using the training data. Co-variance matrix is calculated for all the features and then combined with the results of SVM-RFE. The optimal subset is found using sequential forward selection where a feature is added to an empty set one by one based on their rank. The feature with highest rank is selected from unselected feature list initially. More features are added to the optimal subset either based on their rank or its close relationship with the highest ranked features.

There are a variety of classifiers available for classification. Among them three different classifiers: SVM, MLP, J48 are chosen to evaluate the accuracy of the classification on the optimal feature subset obtained. An ensemble of the above mentioned classifiers is used to improve the classification further. The accuracy of the ensemble of classifiers is found to be higher than that of the accuracy of individual base classifiers. The ensemble classification technique used here is the majority voting technique. Each individual classifier's result is considered as a vote and the ensemble classification is based on the classifier that has majority votes.

$$\text{Ensemble\_Class} = \begin{cases} 1, & \text{if} \sum_{i=1}^{n} \text{Base\_Classifier\_Class} > \frac{n}{2}, \\ -1 & \text{Otherwise}, \end{cases} \quad (1)$$

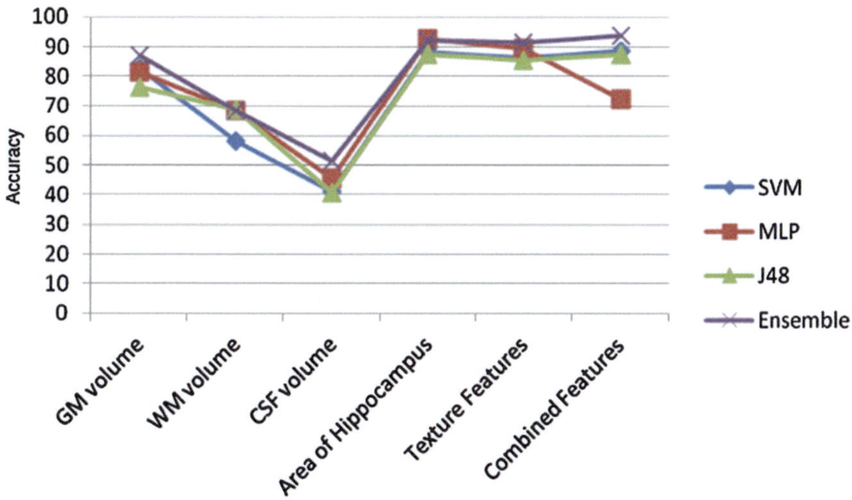

Fig. 5 Comparison of accuracies (%) of classifiers based on different features

4 Results

The implementation of the proposed classification scheme on ADNI Annual 2yr 3T containing 306 subjects is given along with the results obtained. Figure 5 shows the accuracy obtained with the individual base classifiers and ensemble of classifiers with multi-feature fusion.

The features considered are Volume of Gray Matter, White Matter, CerebroSpinal Fluid, Area of Hippocampus and Texture-based features. The combined features are selected from the set of individual features using SVM-RFE algorithm. The ensemble of classifiers is found to outperform the individual classifiers for different features.

5 Conclusion

In this paper, we have analyzed computerized classification of Alzheimer's disease from the MRI scan images. This will aid medical practitioners and clinicians in identifying AD and also in monitoring the progress of AD. The proposed method is based on two types of feature extraction based on Voxel-based morphometric features and Texture-based features. An optimal subset of these features is obtained using SVM-RFE. These features are then used to train an Ensemble of classifiers comprising of SVM, J48 and MLP. It is found that the ensemble of classifiers outperform individual base classifiers. It is also found that SVM and J48 perform better than MLP. In future, this work can be extended to diagnose the conversion of MCI to AD with the help of the data obtained from the same subjects over two to three years of time span.

References

1. Johnson, K.A., Fox, N.C., Sperling, R.A., Klunk, W.E.: Brain imaging in Alzheimer disease. Cold Spring Harb. Perspect. Med. **2**, a006213 (2012)
2. Salas-Gonzalez, D., G´orriz, J.M., Ram´ırez, J., L´opez, M., Alvarez, I., Segovia, F., Chaves, R., Puntonet, C.G.: Computer-aided diagnosis of Alzheimer's disease using support vector machines and classification trees. Phys. Med. Biol. **55**, 2807–2817 (2010)
3. Zhang, J., Liu, M., An, L., Gao, Y., Shen, D.: Alzheimer's disease diagnosis using landmark-based features from longitudinal structural MR images. IEEE J. Biomed. Health Inform. **21**(6), 1607–1616 (2017)
4. Zhang, J., Gao, Y., Gao, Y., Munsell, B.C., Shen, D.: Detecting anatomical landmarks for fast Alzheimer's disease diagnosis. IEEE Trans. Med. Imaging **35**(12), 2524–2533 (2016)
5. Armañanzas, R., Iglesias, M., Morales, D.A., Alonso-Nanclares, L.: Voxel-based diagnosis of Alzheimer's disease using classifier ensembles. IEEE J. Biomed. Health Inform. **21**(3), 778–784 (2017)
6. Bron, E.E., Smits, M., Niessen, W.J., Klein, S.: For the Alzheimer's disease neuroimaging initiative: feature selection based on the SVM weight vector for classification of dementia. IEEE J. Biomed. Health Inform. **19**(5), 1617–1626 (2015)
7. Lama, R.K., Gwak, J., Park, J.-S., Lee, S.-W.: Diagnosis of Alzheimer's disease based on structural MRI images using a regularized extreme learning machine and PCA features. J. Healthc. Eng. (Article ID 5485080) (2017)
8. Kruthika, K.R., Rajeswari, A.P., Maheshappa, H.D.: Alzheimer's disease neuroimaging initiative: classification of Alzheimer and MCI phenotypes on MRI data using SVM. In: Advances in Signal Processing and Intelligent Recognition Systems, Advances in Intelligent Systems and Computing, 678, pp. 263–275. Springer (2018)
9. Nanni, L., Salvatore, C., Cerasa, A., Castiglioni, I.: The Alzheimer's disease neuroimaging initiative: combining multiple approaches for the early diagnosis of Alzheimer's disease. Pattern Recognit. Lett. 84 (2016) 259–266
10. Khedher, L., Ramírez, J., Górriz, J.M., Brahim, A., Segovia, F.: The Alzheimer's disease neuroimaging initiative: early diagnosis of Alzheimer's disease based on partial least squares, principal component analysis and support vector machine using segmented MRI images. Neurocomputing **151**, 139–150 (2015)
11. Farhan, S., Fahiem, M.A., Tauseef, H.: An ensemble-of-classifiers based approach for early diagnosis of Alzheimer's disease: classification using structural features of brain images. Comput. Math. Method. Med. (Article ID 862307) (2014)
12. Xiao, Z., Ding, Y., Lan, T., Zhang, C., Luo, C., Qin, Z.: Brain MR image classification for Alzheimer's disease diagnosis based on multifeature fusion. Comput. Math. Method. Med. (Article ID 1952373) (2017)
13. Cai, K., Xu, H., Guan, H., Zhu, W., Jiang, J., Cui, Y., Zhang, J., Liu, T., Wen, W.: Identification of early-stage Alzheimer's disease using sulcal morphology and other common neuroimaging indices. PLoS One **12**(1), e0170875 (2017). https://doi.org/10.1371/journal.pone.0170875
14. Li, T., Li, W., Yang, Y., Zhang, W.: Classification of brain disease in magnetic resonance images using two-stage local feature fusion. PLoS One **12**(2), e0171749 (2017). https://doi.org/10.1371/journal.pone.0171749
15. Morra, J.H., Tu, Z., Apostolova, L.G., Green, A.E., Toga, A.W., Thompson, P.M.: Comparison of AdaBoost and support vector machines for detecting Alzheimer's disease through automated hippocampal segmentation. IEEE Trans. Med. Imaging **29**(1), 30–43 (2010)

16. Li, M., Oishi, K., He, X., Qin, Y., Gao, F., Mori, S.: For the Alzheimer's disease neuroimaging initiative: an efficient approach for differentiating Alzheimer's disease from normal elderly based on multicenter MRI using gray-level invariant features. PLoS One **9**(8), e105563 (2014). https://doi.org/10.1371/journal.pone.0105563
17. Long, X., Chen, L., Jiang, C., Zhang, L.: Alzheimer's disease neuroimaging initiative: prediction and classification of Alzheimer disease based on quantification of MRI deformation. PLoS One **12**(3), e0173372 (2017). https://doi.org/10.1371/journal.pone.0173372

Data Mining Applications

An SNN-DBSCAN Based Clustering Algorithm for Big Data

Sriniwas Pandey, Mamata Samal and Sraban Kumar Mohanty

Abstract Clustering is a technique to partition data into different groups in such a way that data items in a group are more similar to each other than the data points in any other group. The assumption of infinite main memory is very usual while designing most of the clustering algorithms but this assumption fails when the size of data set starts increasing. In this scenario, data needs to be stored in the secondary memory and time spent in the input/outputs (I/O) dominates the actual computational time. Therefore by reducing the I/O, the efficiency of the clustering techniques can be improved. In this paper, one shared near neighbor based algorithm is devised by minimizing its I/O complexity to make it suitable for the Big Data in external memory model proposed by Aggarwal and Vitter. There is no change in the computational steps, hence cluster quality remains the same. We implement the algorithm in the STXXL library to show its efficacy for Big Data sets.

Keywords Clustering · SNN clustering · External memory algorithms · Big data clustering

1 Introduction

Clustering is an approach of partitioning data into groups according to some similarity criteria. A standard for clustering is the difference of inter-cluster distance and intra-cluster difference. In today's scenario when each and every application is

S. Pandey · S. K. Mohanty (✉)
Computer Science and Engineering,
PDPM Indian Institute of Information Technology, Design and Manufacturing, Jabalpur,
Jabalpur 482005, Madhya Pradesh, India
e-mail: sraban@iiitdmj.ac.in

S. Pandey
e-mail: snp.kecian@gmail.com

M. Samal
Jabalpur, India

B. Pati et al. (eds.), *Advanced Computing and Intelligent Engineering*,
Advances in Intelligent Systems and Computing 1082,
https://doi.org/10.1007/978-981-15-1081-6_11

generating large data, it is a challenging task to understand and analyze that data. Hence, clustering is emerging as the most essential part of data mining, machine learning, and many other applications like document mining, web analysis, medical science, and information retrieval.

1.1 Big Data Clustering

Any algorithm should be scalable to the data set size, but most of the algorithms could not satisfy this requirement and their performance starts degrading as the data set grows. Though the term "Large" or "Big" is relative, it is a kind of agreement among researchers that the rate of increase of data size is remarkably higher than the increase in hardware and technologies required to maintain and process it. Most of the clustering algorithms are not developed considering such heavy load of data. Decomposition [21], incremental [5], parallel implementation [20], summarization [3], approximation [15], and distribution [23] are some techniques in the literature to process big data. Besides several benefits of these approaches, the challenges and limitations attached to these approaches bar them from being universally accepted. To further illustrate our point, we can take a few examples like the incremental approach which is affected by the order of input but the benefit is that the complete data set is not required beforehand [17, 21]. Similarly, approximation techniques are complex in terms of actual practice and data structure usage [22]. If we talk about the parallelization technique, it is not always possible and feasible to parallel the modules of an algorithm [16]. Another approach used for large data is distributed clustering which uses the concept of local and global clustering. This approach fails when there is data dependency between different sites [13].

Balanced Iterative Reducing and Clustering Using Hierarchies (BIRCH) [25], Fuzzy C-Means (FCM) [4], CLARANS [19], DENCLUE [11], OptiGrid [12], Clustering Using Representatives (CURE) [10], Scalable K-Means++ [2], and EM [7] are a few methods developed for big data. Inclination to fixed-size and spherical-shape clusters, reliance on initial parameters, parameter tweaking, no guaranteed convergence, and dependence on the size of main memory are some problems associated with these algorithms. In addition to these issues, nonuniversal applicability makes these algorithms less interesting.

In-core algorithms are designed on the RAM model, which assumes infinite main or internal memory and uniform access to all its addresses, but the reality is different. There are several levels of memory having different cost and performance characteristics. When the size of a data set increases, the main memory fails to accommodate complete data in itself and data needs to be stored in the secondary memory. As the disk access is too slower (in the ratio of millions) as compared to main memory access, rather than measuring the performance of an algorithm in terms of time complexity it will be a good idea to calculate the I/O complexity or in other words number of inputs and outputs from the secondary memory. In 1988, Aggarwal and Vitter proposed a new model called the external memory model. This model makes use of the concept of locality of reference and hierarchical memory structure directly in algorithm design [1]. It is also called I/O model and it considers a computer as a combination of a processor, main memory (M), and secondary memory (disk). The external memory (EM) is assumed to be unlimited in size and data transfer takes place in a bunch of data items (B) called blocks. Transfer of a block of B data items between EM and RAM is counted as one I/O.

1.2 Contribution of the Paper

Shared Near Neighbor (*SNN*) is a similarity metric usually used in the clustering methods that measure the likeness of two points based on the number of common neighbors. But one major problem in the shared near neighbor technique proposed by Jarvis and Patrick [14] is setting the threshold; improper threshold value causes under or over clustering. For addressing all these issues, one new clustering algorithm was proposed by Ertoz et al. [9]. This algorithm uses the shared nearest neighbor count as similarity measure and DBSCAN's core point concept for clustering. The algorithm works well even if clusters are of varying shapes, sizes, and densities and data are noisy and high dimensional. This algorithm is used in various applications like word clustering, time-series data clustering, and document clustering [8].

The algorithm proposed in [9] is suitable for average-size data sets but when it is applied on very large data sets that cannot reside in the main memory, I/O time dominates the time consumed in the computation. Its time and I/O complexities are the same, hence it is not I/O efficient. This paper makes some contribution toward external memory algorithms by designing the algorithm in the external memory model to work on large data. Since the computational steps remain the same, the proposed algorithm generates the same set of clusters. The claim is proven by showing that the I/O complexity of our algorithm is less than the original algorithm by BM factor, where B is the block size and M is the main memory size. STXXL [6] (a software library for external memory) is used to simulate the experiment and theoretical claims are substantiated by the experimental results.

1.3 Organization of the Paper

The structure of the paper is as follows: Sect. 1 contains the proposed scalable SNN-DBSCAN based clustering algorithm with the complexity analysis. The experimental results and observations are described in Sect. 3. The future scope and conclusion of the paper are given in Sect. 4.

2 Proposed SNN-DBSCAN Based Clustering Algorithm for Large Data

Shared near neighbor (SNN) is a similarity measure which calculates closeness of two objects based on the number of common neighbors shared between the two entities [9]. This is a better similarity measure in comparison to other distance measures as it considers both size and density of data points.

The algorithm has two parameters: k (number of neighbors we are considering) and θ (similarity threshold). According to the value of these parameters, the performance of the algorithm varies. An analytic process exists to tune these parameters [18].

2.1 Existing SNN Algorithm

The algorithm proposed in [9] can be divided into three modules. In the first step, calculate the k-nearest neighbor (kNN) of all points. Any distance measure can be used to calculate similarity/dissimilarity between two points. Only k nearest neighbors of a point in non-descending order are considered. First entry of each neighborhood row is the label of that point, i.e., initially same as the point and later it will indicate the label of the cluster or core point to which it is assigned.

In the next step, the common neighbors of each pair of data items are counted and SNN density of each point is calculated. According to this SNN density, core points are recognized. Using the generated kNN matrix, the number of shared near neighbors between each pair of points is counted. Based on this count, SNN density of each point is calculated. SNN density of a point is the number of points with whom the point shares more than Eps $(similarity\ threshold)$ neighbors. If the SNN density of a point is greater than $MinPts$ $(density\ threshold)$, the point is considered as "core point". If a core point shares more than Eps neighbors with any existing core point, its label is replaced with that core point label.

In the last step non-core points are assigned to the nearest core point. Once core points are recognized, next step is to assign all other points (non-core points) to the nearest core point. Here nearest core point means the core point with which a point shares the maximum neighbors and both points include each other in their neighborhood. If a point does not share more than Eps neighbors with any core point, the point is assumed as noise.

The computational complexity of the algorithm is $O(N^2k^2)$. For almost each computational step, an I/O is required in general, hence the I/O complexity of the algorithm $O(N^2k^2)$ which is equal to time complexity.

2.2 Proposed Algorithm

The algorithm is not efficient due to its high I/Os, hence it is not feasible to use these algorithms for large data. In this section, we have made this traditional algorithm [9] scalable by redesigning it on the external memory model . Computational steps are kept unchanged hence clustering quality remains unaffected.

Calculation of KNN Matrix In the first step, the kNN matrix is generated with I/O efficiently. The external memory version of this step is provided in [24]. We are going to use the same procedure. The $N \times D$ data set can be assumed as a collection of $t \times D$ size N/t blocks, where t varies according to the available main memory size. Two data blocks S_i and S_j are brought into memory. For each point pair in these two blocks, distance is measured using some distance metric like Euclidean. Based on the distance, appropriate labels are stored in knn_i matrix block of size $t \times k + 1$. The knn matrix is responsible for holding the k-neighbors and the label of the point, that is why the width of knn matrix is $k + 1$. After computation, knn_i matrix block is written to disk. The same procedure is repeated for all N/t blocks.

It can be observed that only two blocks of size $t \times D$ and two blocks of size $t \times k + 1$ are required simultaneously in the main memory. Hence t can take value up to $\Theta(M/(D + k))$. Hence I/O complexity of generating the knn matrix is $O((ND/B)N/t) = O((N^2D/B)((D + k)/M)) = O((N^2Dk + N^2D^2)/BM)$.

Core Point Recognition The knn matrix, generated in the first step, is the input for the second step, i.e., core point recognition. This matrix can be seen as blocks of size $t \times k + 1$ each. A block knn_i is read and all other blocks knn_j such that $i \neq j$ are read one by one, then all the point pairs are compared. In this way count of shared neighbors of all points of block knn_i with all other points is known. According to this count core points can be decided by comparing with the density threshold $MinPts$. The above procedure is repeated N/t times. This step generates a vector containing all core points and updates the label as well if two core points share more than Eps neighbors.

At a time two blocks of size $t \times k + 1$ reside in the main memory. Hence $M = 2t(k + 1)$, i.e., $t = \Theta(M/k)$. I/O requirement of this step is $O(((Nk + N)/B)N/t) = O((N^2k + N^2)/tB) = O(N^2k^2/BM)$. The method is described in Algorithm 1.

Clustering Step The output of the second step, vector containing core points, is input for this step. Again knn matrix can be assumed as partitioned into N/t blocks of $t \times k + 1$ size each. A block knn_i is read into the main memory and all blocks containing core points are read one by one to the main memory. For each point p in the block knn_i count of shared neighbors with each core point is calculated and point $p's$ label is changed with the nearest core point. The N/t iterations are required.

In this step only two matrix blocks of size $t \times k + 1$ are required at a time in the main memory. Hence $M = 2t * (k + 1)$, i.e., $t = \Theta(M/k)$. I/O requirement of this step is $O(((t \times (k + 1)/B) \times C) \times N/t) = O((C \times k \times N)/B)$, where C represents core points count. The method is described in Algorithm 2.

Algorithm 1 Core Point Recognition

```
Input: Eps // threshold.
  knn[N][k + 1] //k neighborhood matrix
   Output: CorePoints //A vector containing all core points.
t = M/k
for i = 0 to N/t - 1 do
   Read matrix block knn_i
   for j = 0 to N/t - 1 do
      Read matrix block knn_j
      Do the following computations in main memory
      for r = (i)t to (i + 1)t - 1 do
         for l = (i)t to (i + 1)t - 1 do
            for m = 1 to k + 1 do
               if knn[r][1] == knn[l][m] then
                  flag = 1
            for m = 1 to k + 1 do
               if knn[l][1] == knn[r][m] then
                  flagt + +
            if flag == 2 then
               for p = 2 to k + 1 do
                  for q = 2 to k + 1 do
                     if knn[r][p] == knn[l][q] then
                        n + +
            if n > Eps then
               SNN[r/t]++
               if l is a core point then
                  if knn[r][0] == r then
                     knn[r][0] = l
                     changed[r/t] = 1
            flag=0,n=0
      for p = 0 to t - 1 do
         if SNN[p] > MinPts then
            Add point i * t + p to vector CorePoints
         else if changed[p]==1 then
            knn[i * t + p][0] = i * t + p
   Write matrix block knn_i to disk
```

2.3 I/O Analysis

We have calculated the I/Os required by all the three modules of the algorithm. Combining all the three steps, the total number of I/Os required by all the phases of the algorithm is $O((N^2Dk + N^2D^2 + N^2k^2)/BM + CkN/B)$. The constants like D can be ignored. CkN depends on the thresholds (Eps and $MinPts$). If the number of core points C is less as compared to N, this term CkN/B can be ignored, hence the number of I/Os needed for the algorithm is $O(N^2k^2/BM)$, i.e., there is BM factor improvement over the in-core algorithm.

Algorithm 2 Clustering Step

Input: $CorePoints$ //A vector containing all core points.
$knn[N][k+1]$ //k neighborhood matrix.
Output: knn[N][k+1] //kNN matrix with updated labels.

```
t = M/k
for i = 0 to N/t − 1 do
    Read matrix block knn_i.
    for c = 0 to length(CorePoints) do
        j = CorePoint[c]
        Read matrix block knn_c
        //knn_c is the matrix block containing core point j
        Do following in main memory
        for r = (i)t to (i + 1)t − 1 do
            for m = 1 to k + 1 do
                if knn[r][1] == knn[j][m] then
                    flag = 1
            for m = 1 to k + 1 do
                if knn[j][1] == knn[r][m] then
                    flagt ++
            if flag == 2 then
                for p = 2 to k + 1 do
                    for q = 2 to k + 1 do
                        if knn[r][p] == knn[j][q] then
                            n ++
            if n > Eps then
                if n ≥ SNN[r/t] then
                    knn[r][0] = knn[j][0]
                    SNN[r/t] = n
    Write matrix block knn_i to disk
```

3 Experimental Results

Standard Template Library for Extra Large Data Sets (STXXL) [6] is used to implement both the proposed and the in-core algorithms. Data set is generated randomly. For study, data set size is varied from 5000 to 320000. Dimension of data is set 80 and the neighborhood size(k) is set 51. Main memory size is varied from 200 KB to 1 GB. The experiment is run on a system having Ubuntu 12.04 with Intel core 2 Duo 2 GhZ processor with 180 GB disk size.

The first experiment is run on a fixed-size main memory (200 KB), and the traditional in core methods fails to give results for 160000 data items in 5 days. Increasing main memory size does not make some remarkable improvement in in-core algorithm as there is no relation between its I/O complexity and the main memory size.

In the second experiment, the main memory size is varied up to 1 GB. We are running our experiment in a limited environment to exhibit the improvement in I/O complexity. Despite the fact that 8 GB/16 GB memory is common in the vast majority of the frameworks nowadays, however, in the event that we utilize the entire memory or even 50% of the primary memory to run our program, it will be out of CPU before getting out of memory. As an example, suppose 16 GB memory is available and we are utilizing half (8 GB) of it to compute kNN. The number of data point, having 64 features and each dimension takes 4 bytes, that can reside in this memory is ≈ 31 million $(8 \times 10^9 \div (64 \times 4))$. The pair-wise distance calculation takes $\approx 10^{17}$, i.e., 100 $PetaFlop$. This number is not in the capacity of typical personal computers. Henceforth, we cannot run our analysis utilizing the entire memory accessible. On the off chance that the CPUs with such capacities are present, at that point, the analysis can be kept running utilizing the aggregate accessible fundamental memory.

In this section, we discuss the effectiveness of the proposed algorithm. Let us assume that a system has unlimited processing capability but amount of data too large as compared to the internal memory. Assume that unit of B and M is in KBs and GBs respectively, we will get an improvement of order $2^{40}(BM = 2^{30} \times 2^{10})$ in the number of I/Os for proposed algorithm. For the same amount of data (defined in the previous paragraph), there will be a reduction of I/Os from Pis to Kis (Peta vs. Kilo). The viability of the proposed calculation can be comprehended in an accompanying way. Assume a system with no confinement on processing ability is accessible and data is large to the point that it cannot be fit in the main memory and we need to go for out of core usage. As the standard block size(B) in ordinary PCs is in KBs and RAM(M) is in GBs, we will get an improvement of order $2^{40}(BM = 2^{30} \times 2^{10})$ in the count of I/Os for proposed calculation. For the similar size data set as the last experiment, the quantity of I/Os will be diminished from Pis to Kis (Peta vs. Kilo).[1]

The experimental study is divided into two categories:

1. Comparison of the proposed out of core with the in-core algorithm.
2. Impact of internal memory size on proposed algorithm.

3.1 Comparison of the Proposed Out-of-Core and In-Core Algorithms

The first experiment was performed to compare the proposed and the in-core algorithm in terms of number of I/Os, number of reads, and total data read/written. The results are shown in Fig. 1a–d. We have run our experiment for different data sets having size 5000 to 320000, but the algorithm implemented in the traditional way fails to give result after 5 days for 160000 data items. We can see a very clear improvement in the number of I/Os and other characteristics in proposed SNN clustering algorithm. It substantiates our theoretical claim that our proposed algorithm outperforms the traditional SNN algorithm in terms of I/O complexity.

If we analyze the Fig. 1a, we can see that although the number of I/Os is increasing in quadratic order in both in-core and proposed out of core algorithm, the number of I/Os is reduced remarkably. For example, when the data set size is 80,000, the number of I/Os is reduced from 200 million to 20 million approximately. A similar

[1]https://en.wikipedia.org/?title=Binary_prefix.

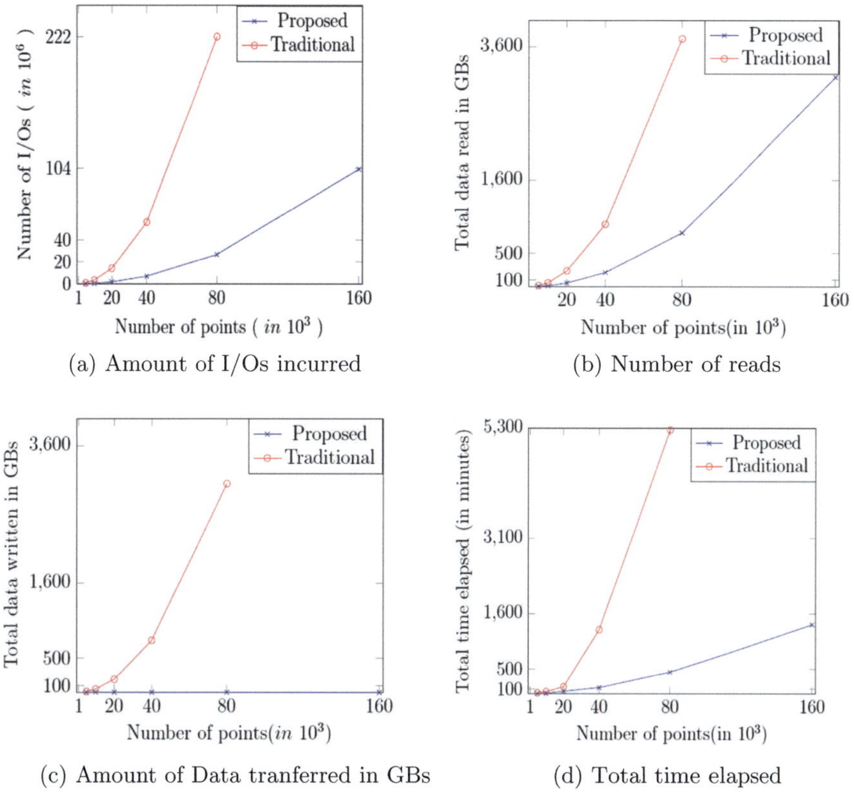

Fig. 1 Comparison of the proposed and in-core techniques with the requirement of data and I/Os

pattern is observed in the amount of data read. Figure 1c shows an interesting pattern in the amount of total written data. Though it depends on the particular data set that how many blocks of data are required to be written, in proposed algorithm it is almost linear. But in traditional algorithm almost all the data read is written back to disk. Figure 1c shows the difference in both algorithms in terms of data written.

3.2 *Effect of Size of the Internal Memory on the Proposed Algorithm*

In this experiment, we study the effect of the main memory size on the proposed algorithm. We vary the main memory size from 200 KB to 1 GB, and we observe that when the main memory size and size required to store the kNN matrix are approximately equal, number of I/Os is almost negligible. This supports our claim that the proposed algorithm shows a BM factor improvement over the conventional algorithm. The results are shown in Fig. 2a–c.

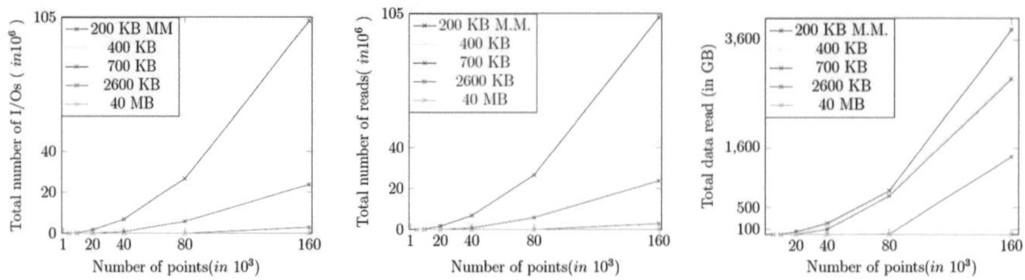

(a) Number of data reads (b) Number of data writes (c) Total number of I/Os

Fig. 2 Algorithm performance versus main memory size

In Fig. 2a, we can see that for 40 MB main memory, the number of I/Os is almost zero for 160000 points, because in this case the total main memory 40 MB is almost equal to size required for kNN matrix ($\approx$ 50 MB). Similar effects are visible for the number of reads/writes and total data read/written. We run our experiment for memory size up to 1 GB, but for data sets having size up to 160000 the number of I/Os is zero, because the main memory is large enough to hold so many data points. Hence we are displaying results only for main memory size less than 40 MB.

4 Conclusion

This paper demonstrated that by redesigning the in-core algorithms in external memory model, large data sets can be handled. The algorithm for *Finding clusters of varying shapes, sizes, and densities in noisy, high-dimensional data* [9] is designed in the external memory model. It is shown that by modifying the data access pattern, the algorithm can be made suitable for large data sets without any change in the algorithm's output. If some I/O efficient data structures can be designed to efficiently compute the k-neighborhood computation, then many existing in-core algorithms can be made scalable.

References

1. Aggarwal, A., Vitter, J.: The input/output complexity of sorting and related problems. Commun. ACM **31**(9), 1116–1127 (1988)
2. Bahmani, B., Moseley, B., Vattani, A., Kumar, R., Vassilvitskii, S.: Scalable k-means++. Proc. VLDB Endow. **5**(7), 622–633 (2012)
3. Ball, G.H., Hall, D.J.: A clustering technique for summarizing multivariate data. Behav. Sci. **12**(2), 153–155 (1967)
4. Bezdek, J.C., Ehrlich, R., Full, W.: Fcm: the fuzzy c-means clustering algorithm. Comput. Geosci. **10**(2), 191–203 (1984)

5. Charikar, M., Chekuri, C., Feder, T., Motwani, R.: Incremental clustering and dynamic information retrieval. In: Proceedings of the Twenty-ninth Annual ACM Symposium on Theory of Computing, pp. 626–635. ACM (1997)
6. Dementiev, R., Kettner, L., Sanders, P.: STXXL: standard template library for xxl data sets. Softw. Pract. Exp. **38**(6), 589–637 (2008)
7. Dempster, A.P., Laird, N.M., Rubin, D.B.: Maximum likelihood from incomplete data via the em algorithm. J. R. Stat. Soc. Ser B (Methodol), 1–38 (1977)
8. Ertoz, L., Steinbach, M., Kumar, V.: A new shared nearest neighbor clustering algorithm and its applications. In: 2nd International Conference on Data Mining, Clustering High Dimensional Data and its Applications, pp. 105–115. SIAM (2002)
9. Ertöz, L., Steinbach, M., Kumar, V.: Finding clusters of different sizes, shapes, and densities in noisy, high dimensional data. In: SDM, pp. 47–58. SIAM (2003)
10. Guha, S., Rastogi, R., Shim, K.: CURE: an efficient clustering algorithm for large databases. In: ACM SIGMOD Record, vol. 27, pp. 73–84. ACM (1998)
11. Hinneburg, A., Keim, D.A.: An efficient approach to clustering in large multimedia databases with noise. In: Proceedings of KDD'98, pp. 58–65 (1998)
12. Hinneburg, A., Keim, D.A.: Optimal grid-clustering: towards breaking the curse of dimensionality in high-dimensional clustering. In: Proceedings of the 25th International Conference on Very Large Data Bases VLDB'99, pp. 506–517 (1999)
13. Januzaj, E., Kriegel, H.P., Pfeifle, M.: Dbdc: density based distributed clustering. In: Advances in Database Technology—EDBT 2004, Lecture Notes in Computer Science, vol. 2992, pp. 88–105 (2004)
14. Jarvis, R.A., Patrick, E.A.: Clustering using a similarity measure based on shared near neighbors. IEEE Trans. Comput. C **22**(11), 1025–1034 (1973)
15. Kanungo, T., Mount, D.M., Netanyahu, N.S., Piatko, C.D., Silverman, R., Wu, A.Y.: A local search approximation algorithm for k-means clustering. In: Proceedings of the Eighteenth Annual Symposium on Computational Geometry, pp. 10–18. ACM (2002)
16. Kim, W.: Parallel clustering algorithms: survey (2009). http://www.cs.gsu.edu/~wkim/indexfiles/SurveyParallelClustering.pdf
17. Liu, Y., Guo, Q., Yang, L., Li, Y.: Research on incremental clustering. In: 2nd International Conference on Consumer Electronics, Communications and Networks (CECNet), 2012, pp. 2803–2806 (April 2012)
18. Moreira, G., Santos, M.Y., Moura-Pires, J.: SNN input parameters: how are they related? In: International Conference on Parallel and Distributed Systems (ICPADS), pp. 492–497. IEEE (2013)
19. Ng, R.T., Jiawei, H.: CLARANS: a method for clustering objects for spatial data mining. IEEE Trans. Knowl. Data Eng. **14**(5), 1003–1016 (2002)
20. Olson, C.F.: Parallel algorithms for hierarchical clustering. Parallel Comput. **21**(8), 1313–1325 (1995)
21. Rokach, L., Maimon, O.: Clustering methods. In: Data Mining and Knowledge Discovery Handbook, pp. 321–352. Springer (2005)
22. Wikipedia.: Approximation algorithm, online (2015). Accessed June 2015
23. Xu, X., Ester, M., Kriegel, H.P., Sander, J.: A distribution-based clustering algorithm for mining in large spatial databases. In: Proceedings of 14th International Conference on Data Engineering, 1998, pp. 324–331. IEEE (1998)
24. Yadav, P.K., Pandey, S., Samal, M., Mohanty, S.K.: Nearest neighbor-based clustering algorithm for large data sets. In: Bhatia, S., Tiwari, S., Mishra, K., Trivedi, M. (eds.) Advances in Computer Communication and Computational Sciences. Advances in Intelligent Systems and Computing, vol. 760. Springer, Singapore (2018)
25. Zhang, T., Ramakrishnan, R., Livny, M.: BIRCH: an efficient data clustering method for very large databases. In: ACM SIGMOD Record, vol. 25, pp. 103–114. ACM (1996)

Wavelet Transform Domain Methods for Resolution Enhancement of Satellite Images

Mansing Rathod, Jayashree Khanapuri and Dilendra Hiran

Abstract In today's world, many applications require satellite images, but resolution is the major weakness of these images. Many researchers have initiated to determine the resolution problem by using transform domain methods like Stationary Wavelet Transform (SWT), Discrete Wavelet Transform (DWT), SWT&DWT and proposed method which integrates DWT, SWT & Integer Wavelet Transform (IWT) that is implemented to overcome the resolution problem of mentioned images. In all these methods, the low resolution (LR) image is broken down into four sub-band images namely Low-Low (LL), Low-High (LH), High-Low (HL) and High-High (HH). The Bi-Cubic Interpolation factor 2 is used on high-frequency sub-band images for resizing the sub-band images. Resolutions of images are enhanced by applying Inverse Transform. For comparison of transform domain methods, the satellite images- LANDSAT8, LANDSAT7 and LANDSAT5 are used in this paper. Objective of the proposed method is to integrate the features of IWT, DWT and SWT in order to improve the resolution. The results demonstrate the supremacy of the suggested method. The Peak Signal to Noise Ratio (PSNR), Root Mean Square Error (RMSE), Mean Absolute Error (MAE) and Time in seconds (s) provide improved result by using the proposed technique.

Keywords Resolution enhancement · Integer wavelet transform · Stationary wavelet transform · Discrete wavelet transform

M. Rathod (✉)
Information Technology Department, K.J.S.I.E/IT, Mumbai 400022, India
e-mail: rathodm@somaiya.edu

M. Rathod · D. Hiran
Pacific University, Udaipur 313003, India
e-mail: sigmapawan72@gamil.com

J. Khanapuri
Electronics and Telecommunication Department, K.J.S.I.E/IT, Mumbai 400022, India
e-mail: jayashreek@somaiya.edu

B. Pati et al. (eds.), *Advanced Computing and Intelligent Engineering*,
Advances in Intelligent Systems and Computing 1082,
https://doi.org/10.1007/978-981-15-1081-6_12

1 Introduction

Resolution of satellite images is very important in many research activities. Some of the research areas use satellite images. For example, geographical information, geosciences, astronomy, etc. Resolution enhancement is an important area of image processing that produces better resolution of mentioned images [11–14]. Various spatial domain methods used for the enhancement of resolution are nearest neighbor, bilinear, and bi-cubic interpolation, but the disadvantage with them is computational problem, due to increased interpolation factor as the also new pixel values are calculated through surrounding pixels. Bi-cubic interpolation is more efficient method as compared to other two interpolation methods, because Bi-cubic interpolation produces more detailed image than other two methods. The transform domain methods produce excellent sharp resolution of satellite images because they directly work on the coefficients of image. Wavelet is one of the techniques that produce good resolution of image. This is one of the best signal processing tools that has been successfully implemented in various fields. The wavelet is a tiny wave with finite energy. Fourier transform loses the time information so wavelet removes this drawback, because with this time and frequency information will simultaneously analyze the amount of frequency with time precision. Numerous researchers have applied transform domain techniques to get better quality resolution of images. This is made by determining the coefficient of wavelets. DWT [1–3], DWT&SWT [4–12] and the proposed method based on DWT&SWT&IWT [15, 16] are compared. These methods have been tested with different sets of satellite images. The evaluation parameters are determined to establish the superiority of methods.

2 Literature Survey

A lot of literature surveys have been studied about the transform domain methods. Various researchers have suggested different methods for improving the resolution of satellite images. Brightness of the image is improved by combining Singular Value Decomposition with DWT features [1]. Enhancement and brightness both the things are carried out in this paper. Paper [2] has presented the concept of DWT and it has divided the input image into down sampled images. The drawback of DWT is that it requires interpolation factor 2 in order to modify the down sampled images but also provides sharpness to the image. Paper [3] has compared transform and spatial domain. Authors have concluded that the transform domain methods are better than spatial domain methods because images are smoothened but it does not sharpen it. Paper [4] has discussed SWT method and it overcomes the drawbacks of DWT. Down sampled image which is a feature of DWT is not present in SWT. Detailed edge information in high frequency components is provided by SWT. Paper [5] has presented wavelet domain with interpolation and suggested that combination of both i.e. interpolation and wavelet gives excellent resolution of satellite image. Paper [6] has implemented the integration of SWT and DWT wavelet transforms. This resolve the disadvantage of DWT, because high frequency components are generated due to adding the features of both transforms. Noise and blur has been removed for satellite image with DT-CWT transform, because good directional and shift invariance is the property of this method [7]. Few wavelet transform domain techniques are introduced & compared [8]. The output of DT-CWT method is superior to DWT, SWT, SWT, and DWT methods. Paper [9] has combined the characteristics of Singular Value Decomposition (SVD) and DWT. It enhances the resolution and contrast of color and gray-level images. Paper [10] has compared local and general histogram equalization by using DWT and SVD method. It improved the contrast and resolution of images. Paper [11] has determined the silent object of images by using model. This model clears the noise and blurring of the image. Paper [12] has integrated the properties of DWT&SWT and is compared the result with SWT transform. It concluded that integrated method gives good result as compared to SWT transform. Paper [13] has discussed about DWT & Discrete Cosine Transform (DCT) and compared both the methods. It is suggested that DWT gives very good PSNR compared to DCT. Paper [14] has surveyed the DWT, SWT and DT-CWT transform techniques and enlisted their drawbacks as well as the advantages. Paper [15, 16] has merged the features of IWT and DWT transforms. It has improved the resolution of image compared to DWT. The performances of implemented methods are evaluated by using PSNR, RMSE, MAE, and TIME in second.

3 Satellite Image Resolution Enhancement Methods

1. **DWT**: This method is used to increase the power of image's resolution enhancement. After the decomposition, it uses the sub band images and LR image which do not have equal size. DWT provides half of the LR image. The parameter of interpolation is used for modifying the input LR images. The LR image is divided into 4 different sub band images. Bi-Cubic interpolation parameter 2 is used on output of LR image. Difference image is determined with LL sub band image and LR image. The difference image is merged with three sub band images of high frequency to reconstruct the estimated images. α/2 as an Interpolation factor is used on estimated images and LR image. Inverse DWT is applied to reconstruct or enlarge the selected LR image (Fig. 1).

   In DWT, the shifting and scaling of the mother wavelet is done by powers of two.

$$\psi_{j,k} = \frac{1}{\sqrt{2^j}}\psi\left(\frac{t - k2^j}{2^j}\right) \tag{1}$$

   Shift parameter is k, scale parameter is j and mother wavelet is $\psi(t)$.

2. **SWT&DWT**: The stationary wavelet and discrete wavelet transform methods are combined to reduce the loss in frequency components. The LR input image is decomposed by mentioned transform. The technique produces four frequency sub band images. Interpolation parameter 2 is used in sub band images of DWT because it produces half of the LR image. There is no need for interpolation in SWT sub band images, because it produces same size of LR image. The high frequency components are projected by merging the high frequency components of mentioned technique. The parameter of Bi-cubic interpolation $\alpha/2$ is utilized on estimated and LR images. Inverse DWT is applied for merging LR input image and estimated images to reconstruct the enhanced image. The PSNR improves in this method as compared to DWT method (Fig. 2).
3. **Proposed method SWT&DWT&IWT**: The proposed method has integrated the characteristics of three wavelet frequency domain methods, i.e., SWT, DWT, and IWT. This method focuses on generating sublime high frequency components compared to the above mentioned methods. The input is divided into four sub-band images, with the help of DWT, SWT, and IWT. Since they use down sampled image, Interpolation factor 2 is applied on DWT and IWT sub band images. The above mentioned method is not applicable for the output of SWT, as same size sub band images are generated as a result after decomposition of input image. Desired images are evaluated by collating the high frequency components of mentioned transforms. By applying Interpolation factor $\alpha/2$ on both images i.e. the input (LR image) and desired sub-band images. Expected high resolution image is produced by merging sub-band and input images using Inverse Lifting Wavelet Transform (ILWT) (Fig. 3).

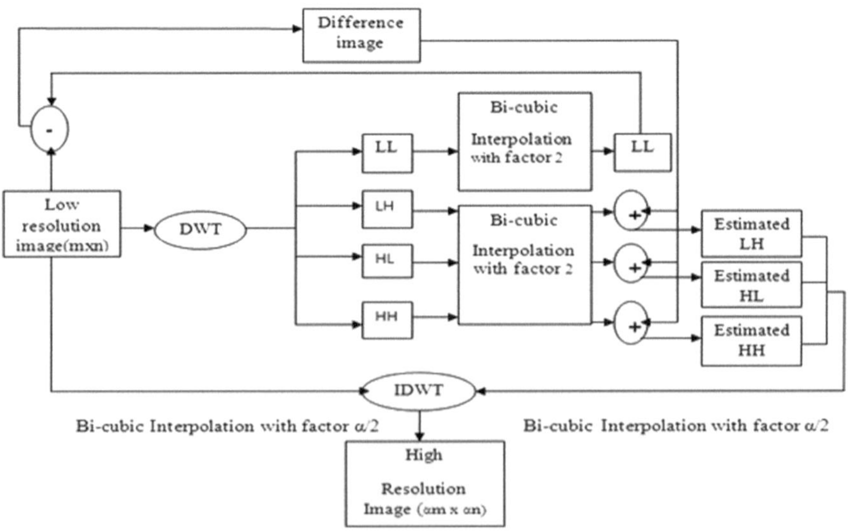

Fig. 1 Block diagram for resolution enhancement with DWT method

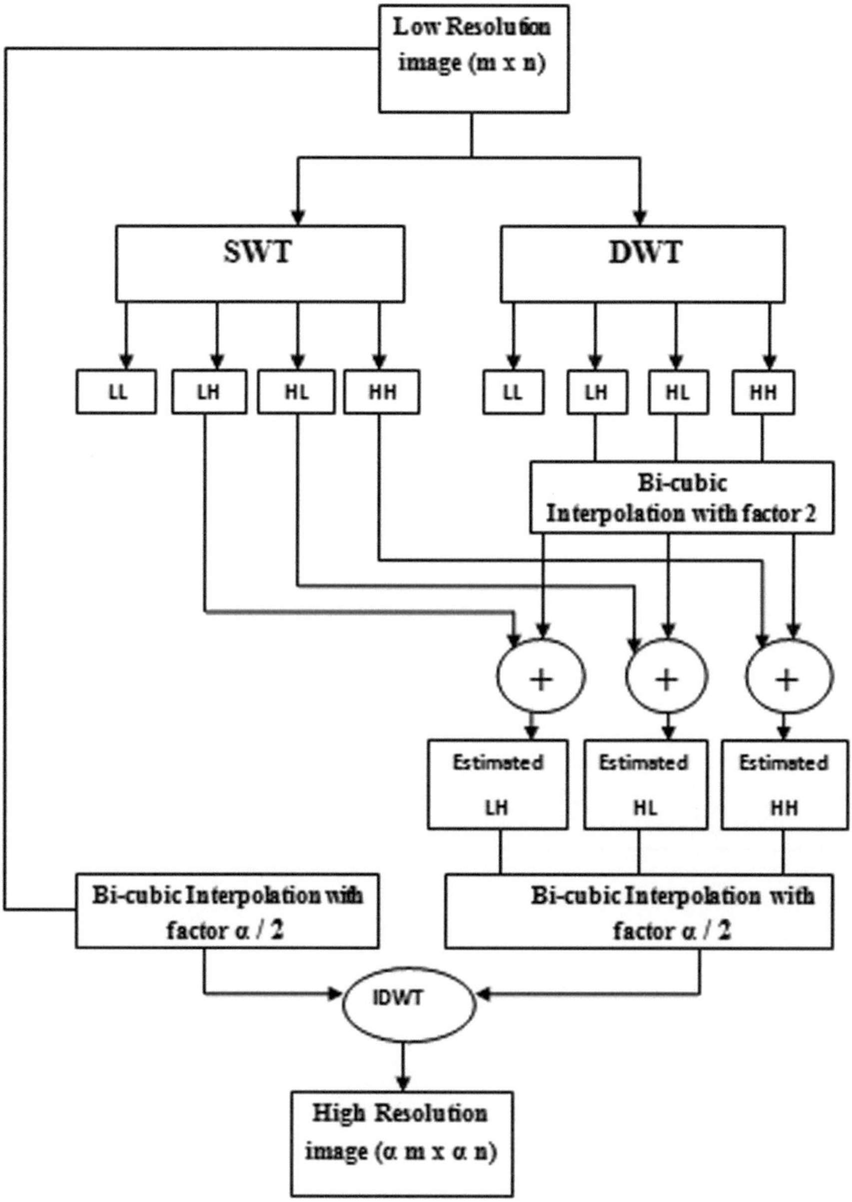

Fig. 2 Block diagram for SWT and DWT method

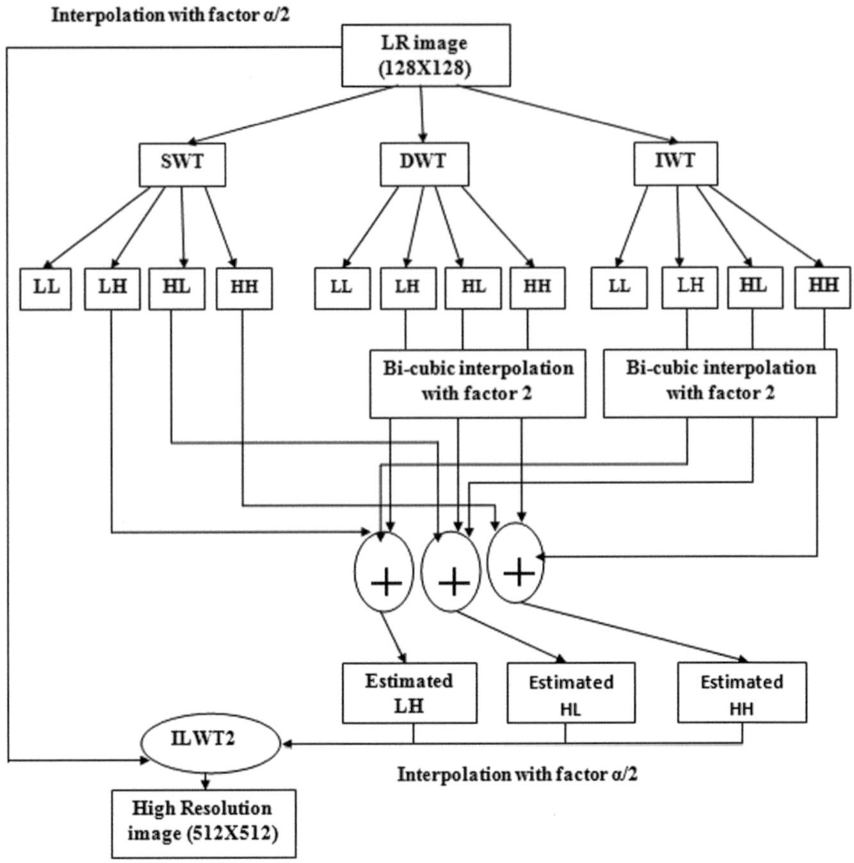

Fig. 3 Proposed method block diagram

The advantage of ILWT is that it is not dependent on Fourier transform and calculation is not as complex as in traditional methods. Integer coefficient from IWT and floating point coefficient from DWT and SWT produce efficient reconstruction. The result obtained by proposed method is excellent in-terms of PSNR, RMSE and MAE and compared to other two methods. Various sets of satellite images namely LANDSAT8, LANDSAT7 and LANDSAT5 are taken into consideration to gauge the result of proposed method and other techniques.

4 Evaluation Parameters

The performances of implemented methods are evaluated by using following parameters that are PSNR, RMSE, MAE, and TIME(s).

1. **MSE**: It is one of the parameter for evaluating the performance of implemented methods. By using the LR image (I_{in}) and the original image (I_{org}) we are calculating the MSE between them. The size of the images is MXN.

$$MSE = \frac{\sum_{i,j} \left(I_{in}(i, j) - I_{org}(i, j)\right)^2}{MXN} \tag{2}$$

2. **RMSE**: This is another parameter to test quality of the mentioned methods. RMSE measures the reference image's pixels to produce enhanced high resolution image (H_r).

$$RMSE = \sqrt{\frac{1}{MN}\sum_{i=1}^{M}\sum_{j=1}^{N}(H_r(i, j) - H(i, j))^2} \tag{3}$$

3. **MAE**: Difference between reference image and enhanced high resolution image (H_r) is determined by this parameter.

$$MAE = \frac{1}{MN}\sum_{i=1}^{M}\sum_{j=1}^{N}|H_r(i, j) - H(i, j)| \tag{4}$$

4. **PSNR**: The ratio of original image to reconstructed image is determined by PSNR. R is the maximum fluctuation of input image.

$$PSNR = 10\log_{10}\left(\frac{R^2}{MSE}\right) \tag{5}$$

5. **TIME (second)**: It determines the computational time for implemented methods.

5 The Result and Discussion

The existing wavelet transform methods and proposed method are implemented, i.e., combination of DWT&SWT&IWT, to overcome the resolution drawback of mentioned images. The method that is proposed produces more efficient result than others. The performance is evaluated by evaluation parameters. Those parameters prove the supremacy of the proposed method. We have considered different sets of satellite images such as LANDSAT8, LANDSAT7 and LANDSAT5 to determine the quality of mentioned methods. The input image of size 128 × 128 is low resolution which is decomposed using mentioned transforms. Inverse transform is applied to reconstruct the image with size 512 × 512 (Figs. 4, 5, 6, 7, 8, 9, 10, 11 and 12 and Tables 1 and 2).

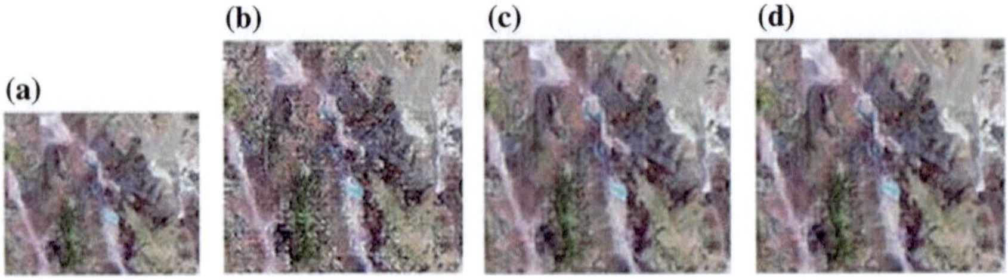

Fig. 4 **a** Landsat8 LR image, **b** DWT resolution enhanced image, **c** DWT&SWT resolution enhanced image, **d** Proposed method (DWT&SWT&IWT) resolution enhanced image

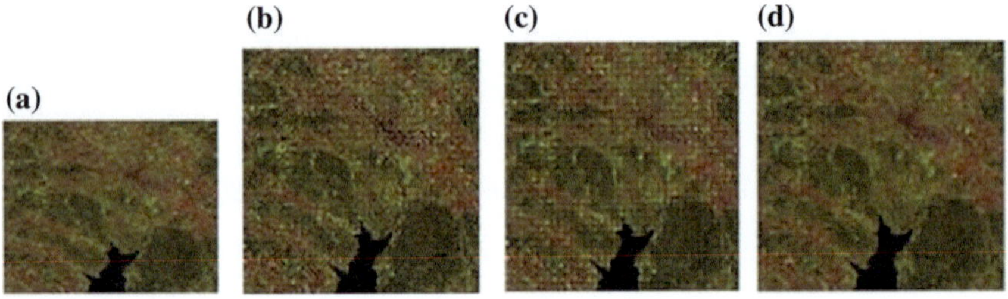

Fig. 5 **a** Landsat7 LR image, **b** DWT resolution enhanced image, **c** DWT&SWT resolution enhanced image, **d** Proposed method (DWT&SWT&IWT) resolution enhanced image

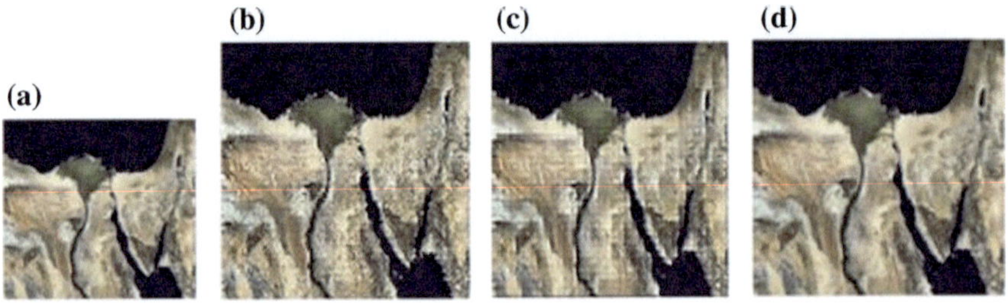

Fig. 6 **a** Landsat5 LR image, **b** DWT resolution enhanced image, **c** DWT&SWT resolution enhanced image, **d** Proposed method (DWT&SWT&IWT) resolution enhanced image

6 Conclusion

Transform domain methods DWT, DWT&SWT and the proposed methods are implemented to remove the resolution problems for satellite images. The proposed and other methods are tested with well-known estimation parameters that are PSNR, RMSE, MAE and TIME in seconds. Various categories of satellite images are carried out for testing the above methods. The proposed method is excellent in-terms of PSNR, RMSE. The graphs and tables show the quality of proposed method and other implemented methods. Thus, the method that is proposed has a more computational time and PSNR.

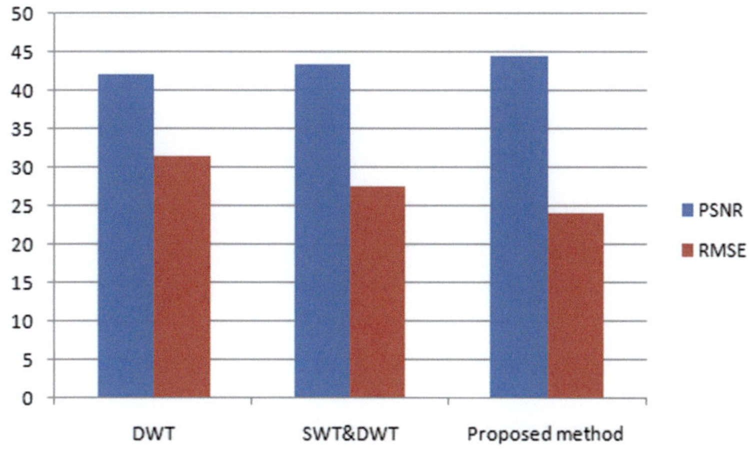

Fig. 7 Graph of PSNR and RMSE for Landsat8 satellite image

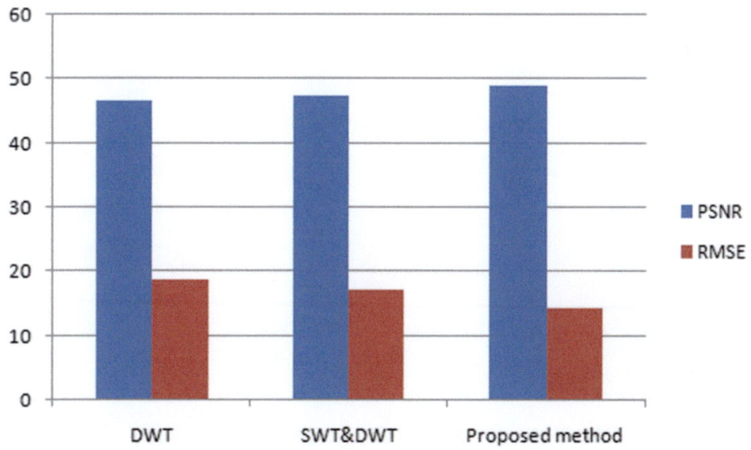

Fig. 8 Graph of PSNR and RMSE for Landsat7 satellite image

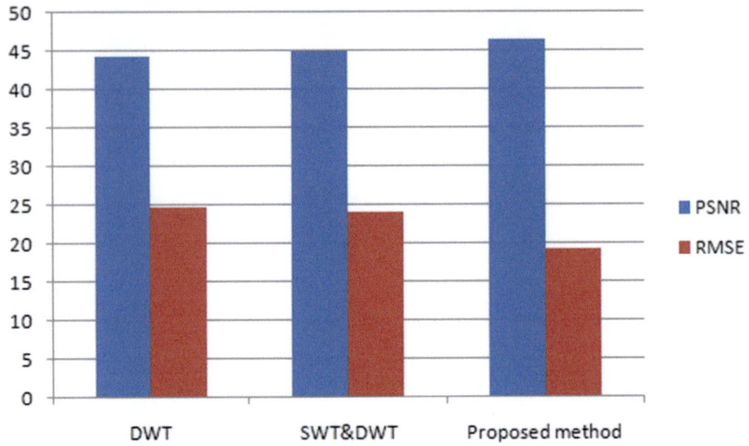

Fig. 9 Graph of PSNR and RMSE for Landsat5 satellite image

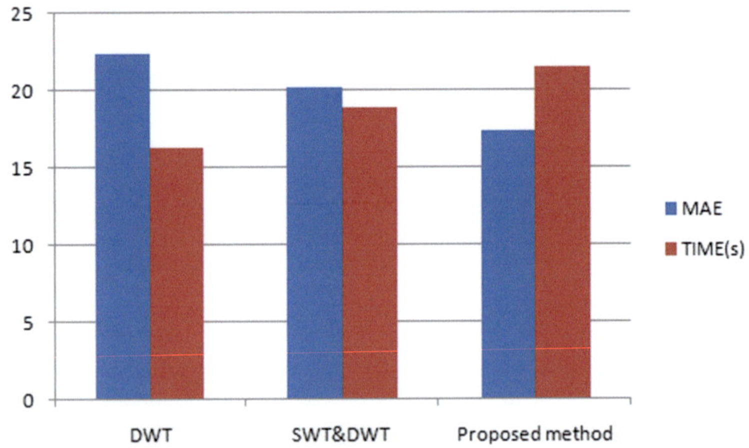

Fig. 10 Graph of MAE and TIME for Landsat8 satellite Image

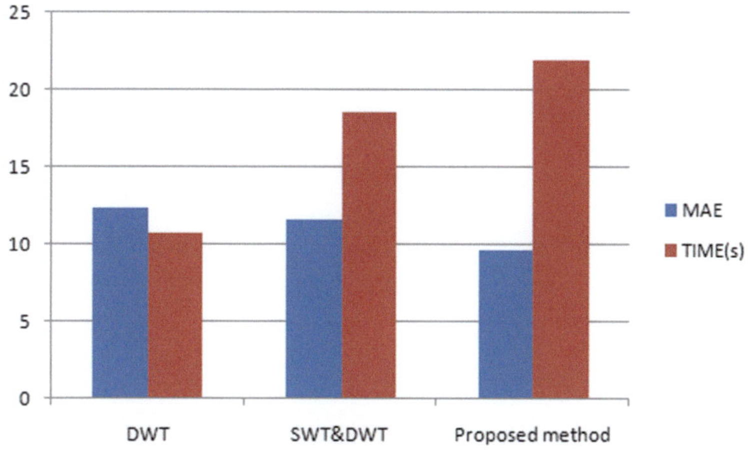

Fig. 11 Graph of MAE and TIME for Landsat7 satellite image

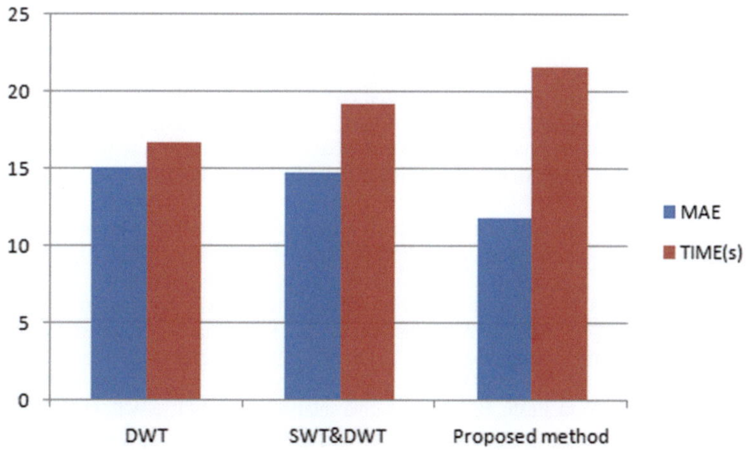

Fig. 12 Graph of MAE and TIME for Landsat5 satellite image

Table 1 Shows the result obtained by DWT, SWT&DWT and Proposed method i.e. DWT&SWT&IWT in-term of PSNR and RMSE. The resolution is enlarged from 128 × 128 to 512 × 512 with ($\alpha = 4$)

Methods/evaluation parameters	Landsat8 satellite image		Landsat7 satellite image		Landsat5 satellite image	
	PSNR	RMSE	PSNR	RMSE	PSNR	RMSE
DWT	42.28	31.49	46.74	18.84	44.36	24.78
SWT&DWT	43.40	27.67	47.53	17.21	45.00	24.07
Proposed method (DWT, SWT&IWT)	44.61	24.07	49.03	14.47	46.54	19.27

Table 2 Shows the result obtained by DWT, SWT&DWT and Proposed method i.e. DWT&SWT&IWT in-term of MAE and TIME(s). The resolution is enlarged from 128 × 128 to 512 × 512 with (α = 4)

Methods/evaluation parameters	Landsat8 satellite image		Landsat7 satellite image		Landast5 satellite image	
	MAE	TIME(s)	MAE	TIME(s)	MAE	TIME(s)
DWT	22.42	16.33	12.42	10.72	15.07	16.69
SWT&DWT	20.25	18.88	11.58	18.58	14.82	19.24
Proposed method (DWT, SWT&IWT)	17.34	21.49	9.66	21.94	11.81	21.67

References

1. Mathew, A.A., Kamatchi, S.: Brightness and resolution enhancement of satellite images using SVD and DWT. Int. J. Eng. Trends Technol. (IJETT) **4**(4), 712–718 (2013)
2. Demirel, H., Anbarjafari, G.: Discrete wavelet transform-based satellite image resolution enhancement. IEEE Geosci. Remote Sens. **49**, 1997–2004 (2011)
3. Shekokar, R.U., Pawar, Y.S.: Resolution enhancement of image captured by satellite using DWT. Int. J. Multidiscip. Res. Dev. **2**(7), 238–234 (2015)
4. Hasan, D., Gholamreza, F.: Image resolution enhancement by using discrete and stationary wavelet decomposition. IEEE Trans. Image Proc. **20**(5), 1458–1460 (2011)
5. Gupta, A.: Miss Sonika: image resolution enhancement technique by interpolation in wavelet domain. Int. J. Prog. Eng. Manag. Sci. Humanit. (IJPE) **1**(3), 2395–7794 (2015). ISSN: 2395-7786
6. Karadge Supriya, S.: An efficient technique for image resolution enhancement using discrete and stationary wavelet transform. Int. Res. J. Eng. Technol. **6**(1), 738–740 (2016). ISSN: 2395-0072
7. Jayanthi, P., Jagadeesh, P.: Image resolution enhancement based on edge directed interpolation using dual tree-complex wavelet transform. In: IEEE-International Conference on Recent Trends in Information Technology (ICRTIT), pp. 759–763, 978-1-4577-0590-8/11/$26.00 ©2011. IEEE (2011)
8. Bala Srinivas, P., Venkatesh, B.: Comparative analysis of DWT, SWT, DWT&SWT and DT-CWT-based satellite image resolution enhancement. Int. J. Electron. Commun. Technol. (IJECT) **5**(4), 137–141 (2014)
9. Bidwai, P., Tuptewar, D.J.: Resolution and contrast enhancement techniques for grey level, color image and satellite image. In: International Conference on Information Processing (ICIP), pp. 511–515. IEEE 978-1-4673-7758-4 (2015)
10. Sharma, A., Khunteta, A.: Satellite image contrast and resolution enhancement using discrete wavelet transform and singular value decomposition. In: International Conference on Emerging Trends in Electrical, Electronics and Sustainable Energy System (ICETEESES), pp. 01–05. IEEE 978-1-5090-2118-5 (2016)
11. Mazhar, A., Hassan, F., Anjum, M.R, Maria, S., Muhammad, A.S.: High resolution image processing for remote sensing application. In: International Conference on Computing Technology (INTECH), pp. 302–305. IEEE 978-1-5090-2000-3 (2016)
12. Rathod, M., Khanapuri, J.: Satellite image resolution enhancement using stationary wavelet transform (SWT) and discrete wavelet transform (DWT). In: International Conference on Nascent Technologies in the Engineering Field (ICNTE), pp. 01–05. IEEE, ISBN-978-1-5090-2794-8 (2017)
13. Rathod, M., Khanapuri, J.: Resolution enhancement of satellite image using discrete cosign transform (DCT) and discrete wavelet transform (DWT). Asian J. Converg. Technol. (AJCT) **3**(3), 67–71 (2017). ISSN: 2350-1146

14. Rathod, M., Khanapuri, J.: A comparative study of transform domain methods for resolution enhancement of satellite image. In: International Conference on Intelligent System Control (ISCO), pp. 287–291. IEEE, ISBN: 978-1-5090-2718-7 (2017)
15. Sagarumar, Ramakrishnaaiah, T.: Satellite image resolution enhancement technique using DWT and IWT. Int. J. Comput. Appl. Technol. Res. **4**(1), 70–76 (2015). ISSN: 2319-8656
16. Moses, Ch., Prasad, P.M.K.: Image enhancement using stationary wavelet transform. Int. J. Comput. Math. Sci. **6**(9), 84–88. ISSN: 2347-8527 (2017)

Improving Query Results in Ontology-Based Case-Based Reasoning by Dynamic Assignment of Feature Weights

J. Navin Chandar and G. Kavitha

Abstract Ontology-Based Case-Based Reasoning combines the effectiveness of memory-based problem-solving technique and semantic reasoning capabilities by integrating with domain ontologies. The knowledge base consists of cases where we search based on the given user query for case retrieval. The accuracy of the ranked results depends on the global similarity of the cases which in turn depends on the local similarity of case features. We transform user input into a query and conduct k-nearest neighbor search. However, the challenge is the default search mechanism relies on the static weights assigned to the various features of the case. This paper will focus on building a model to dynamically determine features weights and similarity functions based on the query provided by the user. With a camera features data set used for illustration, the results showed significant improvement in the ranking of results obtained by using dynamic weights as against the static ones. Finally, we discuss prospective techniques to further improve the ranking of the results.

Keywords Case-Based reasoning · Ontology · Web ontology language · Protégé · jCOLIBRI

1 Introduction

Case-Based Reasoning (CBR) [1] is a memory-based reasoning technique where we capture past experiences of users as cases in the knowledge base. The knowledge captured in the system becomes more structured when the semantics aspect is also considered. This has created a wide interest in Ontology-Based Case-Based reasoning where case representations are instances of a defined ontology. Ontology is a formal specification of shared conceptualization in the domain of interest [2]. Ontologies have applications in various domains like law [3] for legal reasoning. In the medical domain, we use a CBR type called conversational CBR [4] for interactive mode of

J. Navin Chandar (✉) · G. Kavitha
Department of Information Technology, B. S. Abdur Rahman
Crescent Institute of Science & Technology, Chennai 600048, India
e-mail: ncjnavin@gmail.com

B. Pati et al. (eds.), *Advanced Computing and Intelligent Engineering*,
Advances in Intelligent Systems and Computing 1082,
https://doi.org/10.1007/978-981-15-1081-6_13

information retrieval. In this paper, the focus is on applying Ontology-Based Case-Based Reasoning more specifically structural CBR on a sample data set from Kaggle [5]. The data set has camera features with over 1000 data instances.

A CBR system maintains the knowledge base in the form of cases against which we use similarity techniques to compute ranked relevant cases based on user query. The challenge is the retrieval efficiency is highly dependent on the query formalization which in turn characterizes the case. Typically, we search the feature described in the user query against the case base and ranking the search results by utilizing similarity functions. The feature set might have a default weight as part of the defined domain ontology. In the current work, we discuss how dynamically determining the weight and local similarity function of case attributes based on user input can improve the ranking of the search results.

We use camera attributes data set to evaluate the proposed methodology. By comparing the results before and after the application of the proposed technique we showcase how the dynamic assignment of feature weight and local similarity function of case features alter the global similarity of the retrieved cases leading to an improved ranking of search results.

The rest of the paper will be organized as follows: Sect. 2 provides an overview of related works on Ontology-Based Case-Based reasoning with the focus on retrieval challenges and proposed methodology. Section 3 presents the developed case study with the sample data set and the interpretation of results. Finally in Sect. 4 we summarize our conclusion and prospects of future work.

2 Literature Review

In this section, the fundamental concepts of Case-Based reasoning, Ontology, Knowledge Management will be briefly discussed. We will conclude by discussing how the proposed methodology can improve the relevance of the case search results.

2.1 Case-Based Reasoning

A Case Base consists of a set of cases. Each case consists of Case Description and Case Solution. Given a problem which is typically described by a set of features of the case, the first step is to search in the case base. The system returns the case solution when a perfect match is found else we retrieve all the nearest matched cases based on similarity values. These cases would be revised based on the similarity value which is typically called Case Adaptation. We retain the adapted case in the case base as a new learned case based on feedback from the end-user. Figure 1 show the case retrieval cycle.

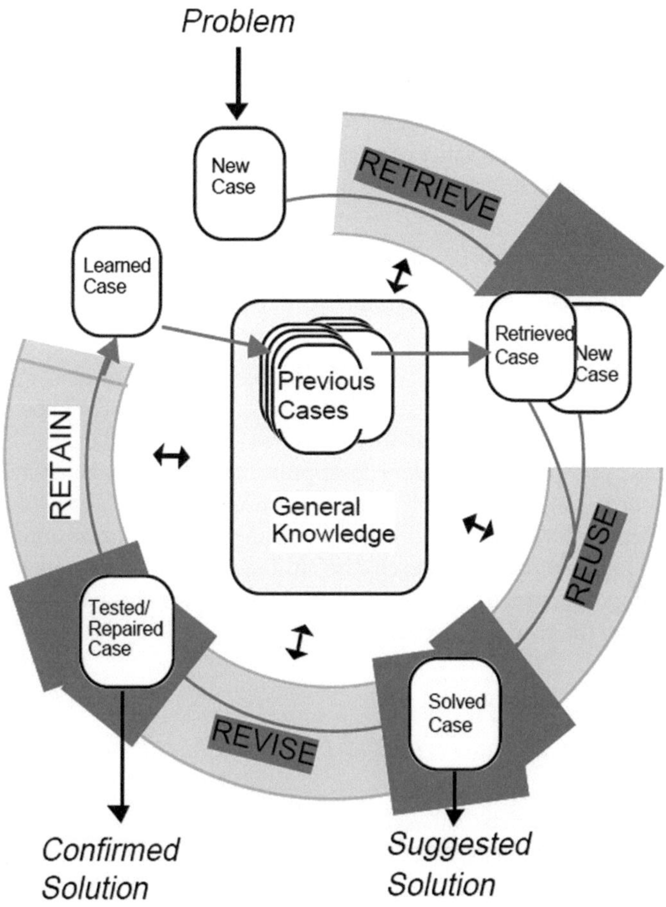

Fig. 1 The CBR cycle [1]

Case-Based Reasoning broadly falls under three categories [6] depending on how we represent the case structure. Textual CBR has cases represented as free-form text. Conversational CBR contains questions and answers in addition to the basic case details. These questions take part in the dialog phase with the user during the case retrieval. Finally, structural CBR has cases represented as attributes or features. This paper is considering structural CBR for discussion.

2.2 Ontology

Ontology is the specification of conceptualization in the domain of interest. Ontology models the domain using concepts (also called as classes) and relations (also called as attributes or properties) connecting the concepts. Building the ontology knowledge models can be manual or automated via ontology learning (OL) [7]. Domain experts define the classes and the attributes in web ontology language (OWL) [8] using popular editors like Protégé [9]. Knowledge modeling using ontology [10] makes it machine readable with reasoning capabilities. CBR system enhanced with domain ontology improves the semantic reasoning capabilities significantly.

2.3 *Ontology-Based Case-Based Reasoning*

A CBR system based on ontology would have cases defined as instances (also called as individuals) of concepts. We search and retrieve cases from the knowledge base for the given user query using similarity measures [11]. Figure 2 depicts the general case retrieval cycle given the problem description. The similarity computation of the problem described in the user query against the cases in the knowledge base is based on global similarity which in turn is the sum of the weighted local similarities. We compute the local similarities for each case attribute describing a case and its contribution to the global similarity is based on the assigned weights per the significance of the feature in the case description.

2.4 *Challenges in Retrieval and Proposed Methodology*

As discussed earlier, the retrieval techniques use global and local similarity measures. However, the effect of constraints in the query on the case characterization has not been studied. This plays a significant role in the retrieval accuracy because based on

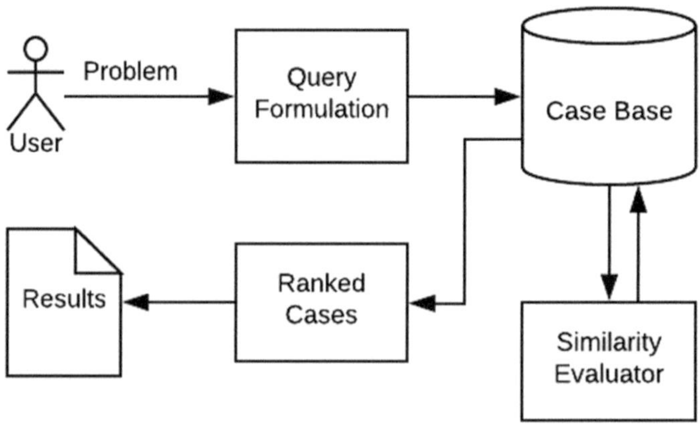

Fig. 2 Case retrieval cycle

the constraints the weight of the case attributes and the local similarity might need tuning. For example, in the Travel domain [12], a query might specify a preference to a hill station. So even though other cases might be retrieved with higher similarity, we will give preference to cases where the location is a hill station. Another example could be a query with no constraints on the budget. This call for a dynamic assignment of weights and local similarity function to the case attributes instead of the default static values depending upon the supplementary conditions in the query. Figure 3 explains the proposed architecture of the Case retrieval cycle.

In the query formation stage based on the user input, the ontology could be consulted to dynamically assign weights and similarity function for the features described in the query. With the computed weight, the ontology search happens to compute the similarity of the cases in the knowledge base. The results are then ranked based on the global similarity. The system presents the search results to the user after adaptions of the retrieved cases if necessary.

3 Case Study

3.1 Design Components

To illustrate the proposed methodology, the camera data set containing over 1000 data instances were taken from Kaggle. We modeled the domain ontology using Protégé [13] a popular ontology modeler tool. Figure 4 summarizes the generated model.

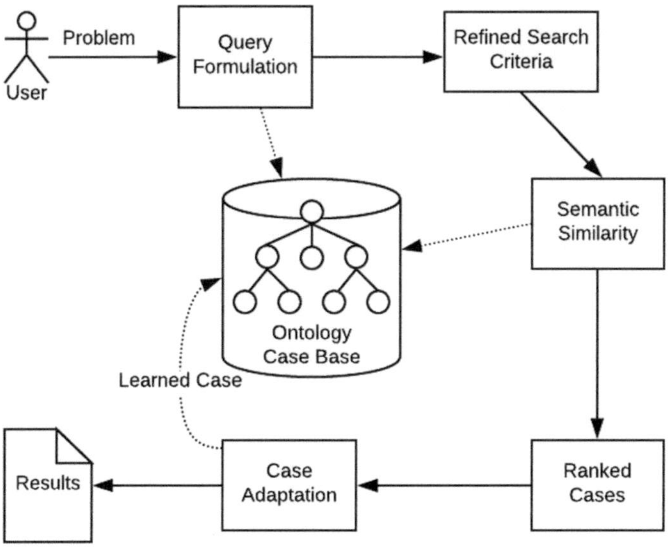

Fig. 3 Case retrieval cycle with dynamic weight assignment

Fig. 4 Domain ontology for camera data set

Apache Jena [14] is the triple store database used for this illustration. We load the triple store with the data set as per the defined ontology. The entities CASE_BASE and CAMERA_CASE are the main Concepts. We store the cases as individuals of the CAMERA_CASE concept with the attributes of the case defined as its data properties. Table 1 tabulates the list of properties along with their domain, range, and weights. The domain for all the properties is cam:CAMERA_CASE. The table includes the range for each attribute along with the default weights. It references the following namespaces.

PREFIX cam "http://www.semanticweb.org/ontologies/2018/cameras"
PREFIX xsd "http://www.w3.org/2001/XMLSchema"

Using the ontology connector classes provided by jCOLIBRI library [15], we extract the case data from the triple store and load the case base. With this setup, the system will be ready to handle user queries.

3.2 System Architecture

Figure 5 explains the overall system architecture in the case study. We load the Apache Jena triple store database with the modeled domain ontology for the camera domain.

Table 1 Data properties for camera domain ontology

No	Data property	Range	Weight
1	HAS_DATA_MODEL	xsd:string	1.0
2	HAS_DATA_RELEASE_DATE	xsd:int	1.0
3	HAS_DATA_MAX_RESOLUTION	xsd:float	1.0
4	HAS_DATA_MIN_RESOLUTION	xsd:float	1.0
5	HAS_DATA_PIXEL	xsd:float	1.0
6	HAS_DATA_ZOOM_WIDE	xsd:float	1.0
7	HAS_DATA_ZOOM_TELE	xsd:float	1.0
8	HAS_DATA_NORMAL_FOCUS_RANGE	xsd:float	1.0
9	HAS_DATA_MACRO_FOCUS_RANGE	xsd:float	1.0
10	HAS_DATA_STORAGE	xsd:float	1.0
11	HAS_DATA_WEIGHT	xsd:float	1.0
12	HAS_DATA_DIMENSIONS	xsd:float	1.0

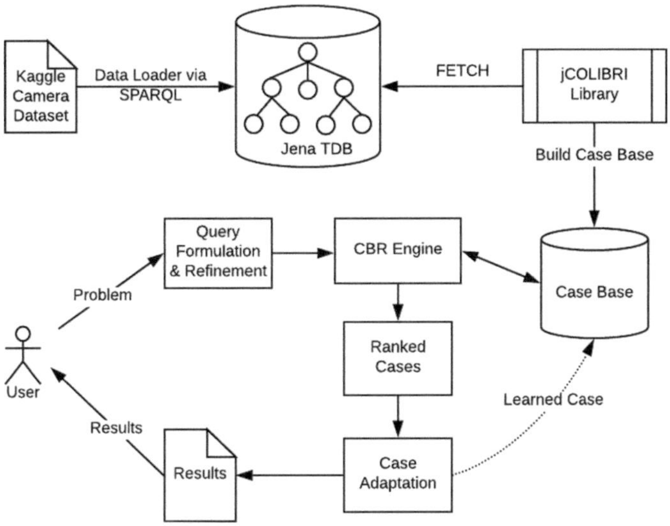

Fig. 5 Case study system architecture

Each row in the camera data set is now added as individuals of the case concept defined. These individuals are case instances in the Case-Base domain. jCOLIBRI is a popular Case-Based reasoning library supporting a range of case representation and its retrieval.

Using the ontology connector APIs, we query the triple store to extract the case data and build the case base. The case structure is a Java Bean in jCOLIBRI. Defining the case base in ontology opens up the possibility of using ontology similarity functions. User queries are now refined to dynamically determine the weights and similarity function before performing the k-nearest neighbor search against the case base. We rank the results based on the defined global similarity function and presented to the user.

3.3 *Interpretation of Results*

We compute the baseline similarity data with preset similarity functions for the various attributes. Local similarity function "MaxString" was used for the model name, "Equal" function for release year and "Threshold" function for all the other attributes of the camera data set. The "MaxString" similarity function returns a value between 0 and 1 which is the ratio of the longest common substring length and the maximum length of the strings being compared. The "Equal" similarity function returns 1 when we have an exact match, otherwise 0. For attributes using the "Threshold" function, we fix the threshold value individually based on the range of the values for that attribute. This similarity function returns 1 if the absolute difference of the values compared is less than or equal to the threshold value, otherwise 0. The global similarity function is the weighted average with the weights equally distributed for all the attributes. Given a query where the user is particularly interested in camera pixel value in addition to the other search attributes a system supporting dynamic assignment of feature weights and similarity function yields better results as shown in Table 2.

We can see that in the default methodology, the results included camera instances that were not equal to 4.0, but still due to matching "name" and "year" attributes those cases got a higher similarity. In the second methodology, the results are more relevant as it picks up only camera instances having the expected pixel value. Here the system dynamically increased the weight of the pixel attribute and also switched the local similarity function to "Equal" since the "pixel" attribute was of particular interest to the user. For brevity, we include only a few camera case attributes in Table 2 in the "Ranked Search Results" column.

Table 2 Similarity comparison for query results

Query attributes	Methodology	Ranked search results	Similarity values
Name = Kodak, Year = 2005, Pixel = 4.0	Static weights	[id = CASE442, model = Kodak Z700, releaseDate = 2005, pixel = 4.0, weight = 270.0, …]	1.0
		[id = CASE448, model = Kodak Z760, releaseDate = 2005, pixel = 6.0, weight = 259.0, …]	0.933
		[id = CASE445, model = Kodak Z730, releaseDate = 2005, pixel = 5.0, weight = 270.0,…]	0.933
	Dynamic weights	[id = CASE442, model = Kodak Z700, releaseDate = 2005, pixel = 4.0, weight = 270.0,…]	1.0
		[id = CASE350, model = Kodak C330, releaseDate = 2005, pixel = 4.0, weight = 180.0,…]	0.9
		[id = CASE349, model = Kodak C310, releaseDate = 2005, pixel = 4.0, weight = 154.0,…]	0.9

4 Conclusion and Future Work

Ontology-Based Case-Based Reasoning is a memory-based problem-solving technique with semantic reasoning capabilities. When it comes to retrieval efficiency for user queries, the system is only as effective as the query formulation step of the case retrieval process. If due importance is not given to this, then this might lead to incorrect results although the data might be captured in an efficient domain ontology. Hence in the current work, we illustrate how we can improve the system performance by dynamically determining the weight and similarity function of the case features based on user input paying due attention to the attributes that are of interest to the user. This will improve the global similarity computations of the cases and alter the ranking of the relevant cases and thereby presenting more accurate results.

As future work, we plan to integrate with NLP packages like Wordnet [16] for understanding the semantics of the user queries to compute the weights dynamically. Also, we will evaluate the effect of formal axioms in the domain ontology in improving the accuracy of the ranked results.

Acknowledgements This work was conducted using the Protégé resource, which is supported by grant GM10331601 from the National Institute of General Medical Sciences of the United States National Institutes of Health.

References

1. Aamodt, A., Plaza, E.: Case-based reasoning: foundational issues, methodological variations, and system approaches. Artif. Intell. Commun. **7**(1), 39–59 (1994)
2. Gruber, T.R.: A translation approach to portable ontology specifications. Knowl. Acquis. **5**(2), 199–221 (1993)
3. Wyner, A.: An ontology in OWL for legal case-based reasoning. Artif. Intell. Law **16**(4), 361–387 (2008)
4. McSherry, D.: Conversational case-based reasoning in medical decision making. Artif. Intell. Med. **52**(2), 59–66 (2011)
5. Kaggle,: 1000 cameras data set, released under CC BY-SA 3.0. https://perso.telecom-paristech.fr/eagan/class/igr204/datasets. Accessed 16 Dec 2017
6. Bergmann, R., Kolodner, J.L., Plaza, E.: Representation in case-based reasoning. Knowl. Eng. Rev. **20**(03), 209–213 (2005)
7. Bergmann, R.: On the use of taxonomies for representing case features and local similarity measures. In: Proceedings of the Sixth German Workshop on CBR, pp. 23–32 (1998)
8. Bergmann, R., Schaaf, M.: Structural case-based reasoning and ontology-based knowledge management: a perfect match? J. Univ. Comput. Sci. **9**(7), 608–626 (2003)
9. Sanchez, D., Moreno, A.: Creating ontologies from web documents. In: Recent Advances in Artificial Intelligence Research and Development, vol. 113, pp. 11–18. IOS Press, Amsterdam (2004)
10. Noy, N., McGuinness, D.: Ontology development 101: a guide to creating your first ontology. Technical report, Stanford University (2000)
11. Allemang, D., Hendler, J.A.: Semantic web for the working ontologist: effective modeling in RDFS and OWL. Morgan Kaufmann, San Francisco (2008)

12. Protégé (Stanford University) Homepage. http://protege.stanford.edu. Accessed 16 Dec 2017
13. Lemnaru, C., Dobrin, M., Florea, M., Potolea, R.: Designing a travel recommendation system using case-based reasoning and domain ontology. In: Proceedings of the 8th IEEE International Conference on Intelligent Computer Communication and Processing, pp. 23–30 (2012)
14. McBride, B.: Jena: implementing the RDF model and syntax specification. Technical report (2001)
15. Bello-Tomás, J.J., González-Calero, P.A., Díaz-Agudo, B.: JColibri: an object-oriented framework for building CBR systems. In: Funk, P., González Calero, P.A. (eds.) ECCBR 2004. LNCS (LNAI), vol. 3155. Springer, Heidelberg (2004)
16. Wordnet Homepage. https://wordnet.princeton.edu/. Accessed 16 Dec 2017

A Survey on Representation for Itemsets in Association Rule Mining

Caryntthia Kharkongor and Bhabesh Nath

Abstract Mining frequent itemset is one of the main challenges in association rule mining. The efficiency of frequent itemsets depends on the computation time and the data structure used for storing the itemsets. The data structure greatly influences the space requirement. Most of the algorithms work well for a sparse dataset. However, if the dataset is large, it becomes difficult for computation, which eventually increases the execution time. This will affect the scalability of the algorithm. With a compact and concise representation of the itemsets, the itemsets can fit in the memory and hence, do not require any I/O operations. The data structures that are mostly used are array, tree, and trie. In this paper, we present a comparison of the different data structures that are used by the mining algorithms.

Keywords Prefix trees · Trie · FP tree · Vertical layout · Frequent itemset · Association rule mining

1 Introduction

Data mining has been identified as one of the promising areas in database research. One such application of data mining is association rule mining [1, 9]. Association rule mining finds the frequent itemsets that play an essential role. These frequent itemsets generated from mining algorithms are required for the production of rules [16]. The algorithm scans the database to find out the itemsets. Many algorithms that are used for mining frequent itemsets are Apriori based or tree based. The algorithms that are tree based are FP Tree [9], CP-Tree [17], CAN tree [10] while Apriori based are FUP [7], MAAP [20], FUP2 [6] algorithms. Apriori involves the generation of many candidate itemsets that require scanning of the database. The tree-based algorithm usually scans the whole database only once and uses the intermediate structure to hold the data. This intermediate structure is difficult to handle and monitor when

C. Kharkongor (✉) · B. Nath
Tezpur Univerity, Tezpur 784028, Assam, India
e-mail: caryn@tezu.ernet.in; carynethia@gmail.com

B. Pati et al. (eds.), *Advanced Computing and Intelligent Engineering*,
Advances in Intelligent Systems and Computing 1082,
https://doi.org/10.1007/978-981-15-1081-6_14

Table 1 Horizontal layout

TID	Items
1	A, B, C
2	B, C
3	D

Table 2 Vertical layout

A	B	C	D
1	3	4	9
2	4	5	9
8	7	1	3

the data needs updating, deletion, merging. Moreover, time is wasted in searching and browsing the entire tree especially when the database is very large. Most of these algorithms have considered the horizontal layout as shown in Table 1. In such layout, the database is arranged as a set of rows where each row represents customer transactions. Other than horizontal layout, there's a vertical layout shown in Table 2 where each item corresponds to a set of values representing the transaction in which it occurs [1, 9].

2 Association Rule Mining Methods

In the era of big data, extremely voluminous and complex datasets are produced. Using analytics techniques, these datasets may reveal trends, association patterns. The main challenges to achieve this goal are data storage and analysis. Storing such large data and processing become impractical as it does not only require time but also resources. Furthermore, if the size of the large database cannot fit in main memory, it becomes infeasible as it involves huge data storage and retrieval of data from secondary memory. Therefore, the representation of itemsets becomes a key role in mining of frequent itemsets. The size of the current main memory ranges from gigabytes to terabytes, so the large database can become memory resident. Analyzing and processing such large datasets will then improve the time and space complexity. In this paper, we provide a brief description of such algorithms that provide concise and compress representation of the itemsets.

- Badon in his paper [4] proposes a data structure for Apriori using a trie. A trie is a hash tree with root at depth 0 and a pointer that links the node. The end character is represented by * which is a special letter. The trie is implemented with arrays and vectors, which require less memory. Advantages of using a trie are as follows:

1. The generation of candidate itemsets and the rules productions are faster.
2. The negative border is generated immediately.

The support count of the item is calculated by scanning the database. The nodes are sorted in an ordered set. There are two indices: one for the items and another for the edges of the node. The first element is assigned to the index. If the two indices are equal then the procedure is called recursively. The steps are continued until the end of the transaction. For traversing the tree, a linear search is used to first compare the item with the searched item. If it is greater, then move to the next node and if it is smaller, then the edge is not present. This searching technique continues until they are equal.
Candidate generation is faster if the nodes are sorted because it is easier to perform union operation and finding the other siblings. Searching technique can be improved if a hash table is used. A certain threshold known as leaf_max_size is chosen so that the hash table need not be altered all the time. Only those nodes whose number of parents edge is greater than the threshold is added to the trie.
Another technique called *apriori-brave* is introduced to keep track of the memory usage and need during candidate generation. After (k+1) candidate itemset are generated, it checks whether it is more than the maximum memory. If it does, then the (k+2) candidate itemsets will not be generated.
While storing the transaction, the reduced transaction is stored that contains only the frequent items in order to save memory. It is stored in a tree and the transactions which are not frequent are deleted [4].

- Pietracapricia introduces PatriciaMine [12] which is memory based and utilizes a trie for the representation of itemset. The advantage is that this representation is space-efficient without using array and standard trie. However, using a trie has some disadvantages. Disadvantage of using trie is during compression while merging common prefixes especially when dealing with large databases. In the worst case, the number of items can be N which is the size of the original dataset. The Patricia trie was introduced that contains the nodes $v_0, v_1, \ldots v_n$ with all v_i having the same count except v_k which contain exactly one child node. This node v_k is conjugated into a single node which inherits its count value.
A transaction is inserted in a trie by starting from the root downward with an associated path. The path is associated with the prefix, which is previously inserted transaction. The other remaining suffix is added as the child node visited. The tree is traversed downward and each node corresponds to a hash table that contains pointers of the children v. The number of pointers in the hash table is considered as the function of the number of children. After the tree is build, the space allocated by hash table is freed as the trie is traversed upward.
In this implementation, the tree is globally traversed from the leaf to the node. This strategy is called as the item guided traversal. Using this strategy, each node is traversed only once. The traversing starts from each node and if the node is visited, then it is marked as visited. The process is continued until no visited node is left unvisited. A comparison between standard trie, array, and Patricia trie is shown in Fig. 1 [12].
- Mingjim et al. [15] presents an approach that compresses and maps the transaction *id* into an interval list using a tree. The support of an itemset is counted by

Dataset	Transactions	AvgTS	min_sup %	Array	Trie	Patricia
Chess	3,196	35.53	20	467,060	678,560	250,992
Connect-4	67,557	31.79	60	8,861,312	69,060	55,212
Mushroom	8,124	22.90	1	776,864	532,720	380,004
Pumsb	49,046	33.48	60	6,765,568	711,800	349,180
Pumsb*	49,046	37.26	20	7,506,220	5,399,120	2,177,044
T10.I4.D100k.N1k.L2k	100,000	10.10	0.002	4,440,908	14,294,760	**5,129,212**
T40.I10.D100k.N1k.L2k	100,000	39.54	0.25	16,217,064	71,134,380	**16,935,176**
T30.I16.D400k.N1k.L2k	397,487	29.30	0.5	48,175,824	163,079,980	41,023,616
POS	515,597	6.51	0.01	15,497,908	32,395,740	13,993,508
WebView1	59,601	2.48	0.054	831,156	1,110,960	618,292
WebView2	77,512	4.62	0.004	1,742,516	4,547,380	**1,998,316**

Fig. 1 A comparison of the space requirement by standard trie, array, and Patricia trie showing that the Patricia trie consumes less space requirement

intersecting these lists. A prefix tree is used to generate candidate itemsets in lexicographic order. Each node in the prefix tree handles collection of itemsets with their support. The root has 1 frequent itemset and goes downward with node d storing *d* frequent itemsets.

The construction of the tree is similar to FP tree. The tree gives a compact representation of the itemsets. Each node corresponds to the *id* and a counter that keeps track of the number of transactions. The steps used in construction of a tree are as follows:

- The database is scan and finds the itemsets that are sorted in descending order.
- Again, the database is scan. The frequent itemsets are inserted in a transaction tree starting with the 1-frequent itemsets as the root. If the root contains child node, the count of this node is incremented by 1.
- After the tree is built, all transactions are represented with interval list along with the item. The support count of the itemset is counted by intersecting the interval lists.
- A lexicographic tree is constructed by storing only the minimum information, if the expansion of the tree is not successful, the search backtracks. When the search continues, the itemsets with support are output. An example of the transaction database is shown in Table 3 and the corresponding tree in Fig. 2 [15].

– Takeaki Uno et al. [18] combines the three data structures: array, bitmap, and prefix tree. The bitmap is represented using matrix such that the presence of the element is marked as 1. The prefix tree is a tree that stores sequences or strings representing the path from the leaf to the root. A common prefix of 2 itemsets can be stored using a common prefix which saves memory. The main structure is the list of array.

A constant c is chosen along with integer bit array, bit (T) and integer array, *ary* (T) of size T^P. The element bit (T) is 1 if T^s contains the element. The *ary* (T) contains the items in increasing order. A complete prefix tree consists of vertices where each vertex is associated with transaction T, bit (T) = i. For each item, there is bucket that keeps track of the number of occurrences. The transaction is input

Table 3 A transaction database

TID	Items	Ordered frequent items
1	2, 1, 5, 3, 19, 20	1, 2, 3
2	2, 6, 3	2, 3
3	1, 7, 8	1
4	3, 1, 9, 10	1, 3
5	2, 1, 11, 3, 17, 18	1, 2, 3
6	2, 4, 12	2, 4
7	1, 13, 14	1
8	2, 15, 4, 16	2, 4

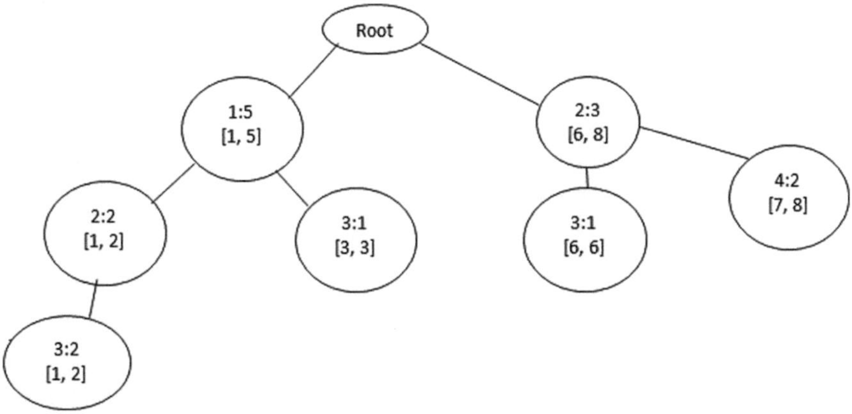

Fig. 2 Transaction tree corresponding to Table 2

only once so we can reuse the prefix trees and buckets. The items in the higher frequencies contain larger indices. Each vertex has two fields: weight, *w(i)* and highest bit, *h*(*i*). The weight contains the number of transaction and *h* is the highest bit *i* equal to 1.

The complete tree is initialized with the empty bucket and weight of vertex as 0. After initialization, the weights are added to the vertex bit (T). Each vertex *i* is then inserted to the *n*-*h*(*i*)th bucket. The prefix tree can be reused again by reserving the buckets and weights during the recursive calls. This saves space and time used for constructing the prefix tree. A complete prefix tree shown in Fig. 3 for 3 elements [18].

- Gosta G and Zhu J proposed a variation of FP tree with an array structure. Using the array structure, the FP tree method will speed up. The main contribution of this paper is reducing traversing time in FP using an additional structure. The proposed FP tree has three fields: base, header, and array. Allocation and deallocation of memory for construction of FP tree takes time. After the memory is deallocated

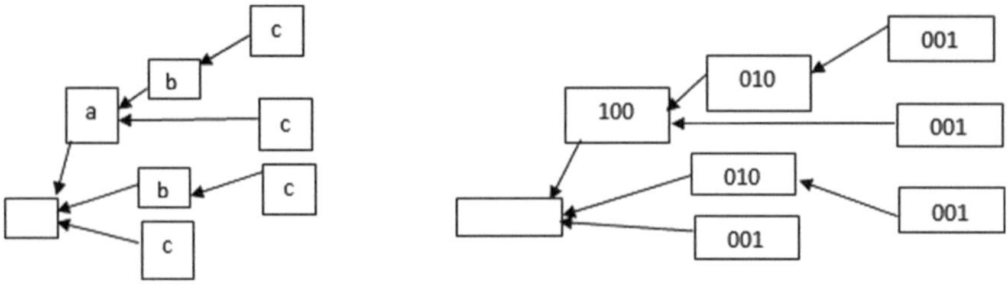

Fig. 3 Complete prefix tree for three items (a, b, c) and bit pattern associated with each cell

at the end of the transaction, the memory is allocated to the tree during creation of the tree. After the generation of frequent items, the memory is discarded [8]. For keeping track of maximal itemsets, an FPmax introduced a structure called a maximal frequent itemset tree (MFI tree). An itemset is inserted into in MFI tree only if its subset is not present in the tree. Each FP tree is associated with each MFI tree. MFI tree contains item name, header node, node link, and level. The link is used for the purpose of subset testing. Nodes with the same item are combined in a FP tree. Subset testing is done in decreasing order of frequencies.
For each node n, the itemsets $i, i_1, \ldots i_n$ are tested for subset. If the level is $l < k$, the test stops since there are no matching ancestors. This is FPmax* which is depth-first algorithm and returns only the maximal frequent itemsets. To mine frequent closed itemsets, FPclose verifies whether itemset is closed or not. A CFI tree is used for the purpose and it considers only count of an itemset [8].

- Schmidt shows that ECLAT performs better than other algorithms in dense datasets. ECLAT starts with the incidence matrix $C\Phi$ and an empty prefix. The incidence matrix contains only those items which are frequent. The items are computed to find 1-frequent itemset and will go on recursively until there are no resulting, C_x which is empty. However, before constructing the incidence matrix, it is necessary to prune the infrequent items, rearrange the items in a specific order and sort the transaction in that mentioned order.
ECLAT rearranges the items by increasing frequency. The improvement of ECLAT is using diffsets. Intersection helps in reducing the size of the incidence matrix by storing only the transactions *id* that are matched. Feature called omission of equipsupport extension is also added. This is when the item x is extended and the prefix contains the same support. The extension can be added without altering the support. Euipsupport can be stored in a separate list, which helps in filtering out the items. Another feature known as interleaving incidence matrix computation is appended. This process involves finding intersection of those transactions which have supported greater than minimum support and filters out those transactions which are not otherwise [14].
- Batezs racz et al. [13] proposed a data structure that is compact based on FP-growth tree. The algorithm has been implemented to suit four different sized FP tree: very large, dense, sparse, and single node. The main data structure used is a trie. Each node in the tree contains a pointer and the counter. These nodes are stored using

Table 4 Checking of frequent itemsets with minimum support 2

	i_1	i_2	i_3	i_4
o_1	0	1	1	1
o_2	0	1	0	1
o_3	0	1	1	0

the array and the pointer indicated the indices for the array. There is no header table and the items are sorted in ascending order.

During allocating and deallocating of memory, a number of overheads are incurred. The only solution is to reuse the memory. First, the array allocation can be used again in the next recursive call. As the recursion continues, the size of array decreases. Therefore, we can insert these arrays on a stack containing decreasing block size. While entering into a new recursion level, the top of the stack is checked. If the block is large, then we pop from the stack else a new memory is allocated and continue with the unmodified stack. In order to clean up the array after usage, the array is inserted with 0 before reusing again. This saves space storage. To present a compact representation for the tree, the infrequent items and the nodes with 0 conditional frequent items are eliminated. However, it does not rearrange or combine the items for further recursion [13].

– Thanh-trung et al. [11] explained in his paper about storing the itemsets. It introduces a structure called constructive set that generates closed set according to a threshold support. The constructive sets are produced from a group of patterns. These group patterns represent the transaction sequences that are related to transaction set, itemset, and their frequencies. The transaction set is stored in the form of bit where 1 represents the presence and 0 represents the absence of the item as shown in Table 4. The transaction set is represented in the form of matrix n × m transaction set T(O,I,R) where O is the transaction objects O_1, O_2, O_N, I is the transaction items i_1, i_2, i_N, and R is the binary relation on O, I. Suppose there are two bits m pattern a_1, a_2, a_m and b_1, b_2, b_m. The composition of and b is depicted by AND operation such that

a & b = c

$c_k = a_k \times b_k$ for all k belongs to {1, 2, 3, …, m}.

A bit pattern with frequency is represented with (.) as the delimination. O_2 with frequency 1 in 0 is written as 110111.1. If a = 11110110 and b = 11110010, c = $a \times b$ = 11110010 which is equal to b.

Therefore, a is covered by b. The number of occurrences in the pattern a is the frequency of a denoted by f_a. If the bit pattern a occurs once, then $f_a = 1$. If the bit values are not defined, * is used for identification of the aggregation. A subset says A is a subset of O, if A is closed, that is, if its subset is identical. For determining closure property, ↑ ↓ operators are used. ↑ ↓ determine the closure O ⊆ if O ϵ ρ (0) → O ↑ ↓ ϵ ρ (0) [17].

$I_0 = \{ i_2, i_4 \} = f_{I0} = 2$

$I_1 = \{ i_2 \} = f_{I1} = 3$

$I_2 = \{ i_2, i_3 \} = f_{I3} = 2$

1	1	1	1	1	1	1	1	1	1	1	1	1	1	1	1
1	0	0	1	1	1	1	1	1	1	1	1	1	0	0	1
1	0	0	1	0	0	0	1	0	0	0	1	0	0	0	0
1	1	1	1	1	1	1	1	0	0	0	0	0	0	0	0
Pattern #1				Pattern #2				Pattern #3				Pattern #4			

Fig. 4 Representation of pattern in 4 × 4 matrix

Table 5 The data pattern for 9 and 0

Pattern #	Features	Label
1	1, 2, 3, 4, 8, 9, 12, 13, 14, 15, 16	0
2	1, 2, 3, 4, 5, 6, 7, 8, 12, 13, 14, 15, 16	9
3	1, 2, 3, 4, 5, 6, 7, 8, 12	9
4	1, 2, 3, 4, 5, 8	0

When depicted as group pattern, the set of closed frequent itemsets = **1*.2,*1 **.3, *11*.2 and maximal frequent itemsets M = *1**.3. Therefore, M⊆C⊆F after closed itemsets and frequencies are computed. Constructive set is produced from the maximal frequent group patterns and each of these patterns is called constructive pattern. From the table, the constructive set = **1*.2, *1*.3, *11*.2, *11*.1, *1*1.1 [11].

- Ananthanarayana et al. [2] paper have proposed a novel data structure called pattern count tree (PC tree) that scans the database only once. This data structure stores the transaction using a tree where each node has four fields: feature, count, child pointer, and sibling pointer. Feature denotes the non-zero value of the pattern. Count represents the number of patterns. Child pointer specifies the pointer to the following path and sibling pointer points to the node in the subsequent path. Each pattern is represented using a matrix shown in Fig. 4. The presence of a feature is indicated by 1 and the absence as 0. The transaction id is represented as the pattern # and the transaction database represents the field feature. Each pattern has a label corresponding to it shown in Table 5. The PC tree is represented using a binary tree. Each node has two fields: the name of the node and the count value [2].
- Mohammad et al. [19] introduces a novel solution by storing only the difference of tid between class member and the prefix itemsets. The difference of tid is stored in diffset. The difference is established starting from the root node to the child node, i.e., leaf. To check whether the itemset is frequent or not, difference operation is used. Suppose for an itemset of size 2 say, AB. d(AB) = t(A) − t(B) = 13. To find out whether it is frequent or not, we compute $\sigma(A) - |d(AD)| = 4 - 2 = 2$ is not frequent. In this example, the tid database has 23 entries whereas difference has only 7 entries. This pattern is shown in Fig. 5 [19].
- Baralis et al. [3] utilize two data structures, i.e., hybrid tree and array. It is called the hybrid tree as it uses both the data structures. The array is used to provide support

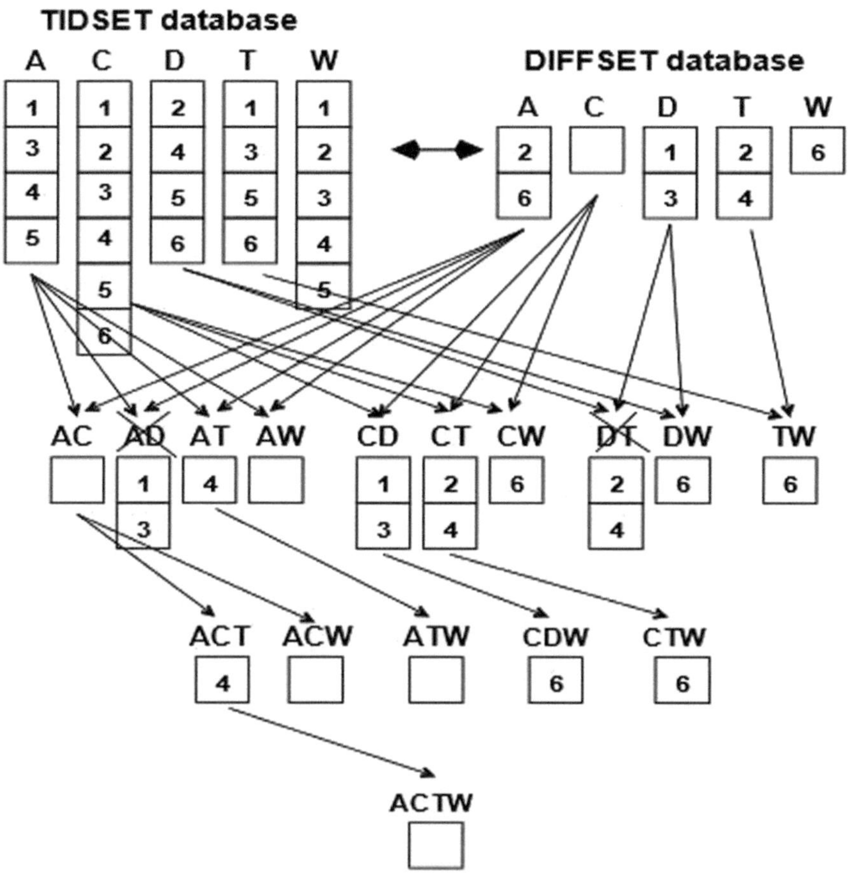

Fig. 5 Pattern counting using diffset

count derived from the transactional database shown in Table 6. Each item in the database is represented by a cell that contains the item id and support. These cells are stored in a descending order. A pattern is represented using a 4×4 matrix. The HY-tree has two node structures: upper and lower nodes. The upper nodes are stored at the higher part of the tree. It is a branched tree that has more than 1 support count. Lower nodes are stored at the lower part of the tree and have only one single node and no brother. Nodes at the lower part are represented as an array with each node stored in a cell. These cells contain the id of item [3].

– Chen et al. [5] introduces bitmap that utilizes bitmap-based representation. These itemsets are represented in n bitmap format and direct support is counted by scanning the database once. Suppose the database D has three items {1, 2, 3}, the itemsets are{ }, {1}, {2}, {1, 2}, {3}, {2, 3}, {1, 3}, {1, 2, 3}. The bitmap representation for the database D are 000, 001, 010, 011, 010, 101, 111, 110. Using BISC, memory requirement for storing direct support is $O(2^n)$ and time complexity is

Table 6 The item array

E	5
A	4
F	4
B	3
C	3
D	3

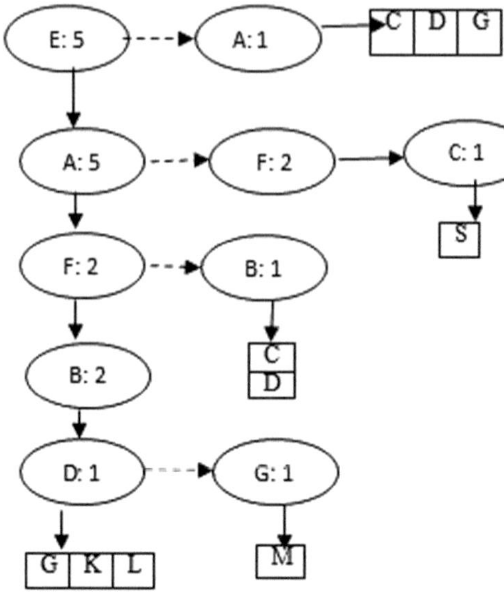

Fig. 6 Representation of pattern in 4 × 4 matrix

$O(2^{n-1})$. BISC can handle 32 bit integer database which means only 32 frequent itemsets. In BISC2, it utilizes prefix/suffix of the bitmap itemset where prefix is associated with the higher ids and suffix with the lower ids. The itemsets are sorted in descending order in such a way that prefix will contain less frequent items. Using BISC2, the space complexity will be $2^f + 2^{s-f}$ direct support where f is for prefix and s is suffix itemsets (which is less than BISC1). Suppose n = 28 and s = 14, f = 14, we need to store only $2^{14} + 2^{14} = 32768$ direct supports as opposed to 2^{28}(268 million) in BISC [5] (see Figs. 6 and 7).

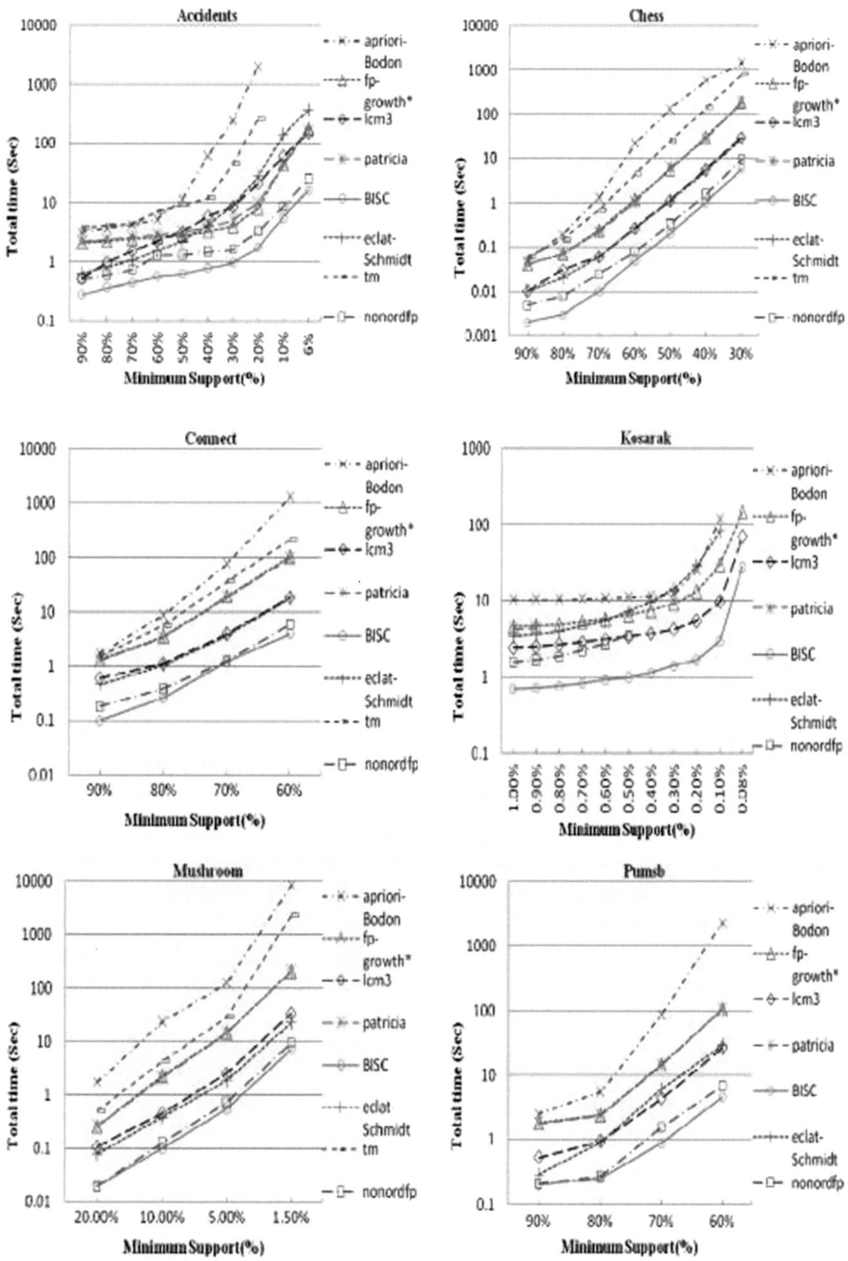

Fig. 7 Total time taken by the mining algorithms

3 Datasets Used

The behavior of the algorithm varies from one dataset to another. Some algorithms show fair performance for sparse dataset and some dataset show poor performance for dense dataset. The publicly available datasets are downloaded from the repository (http://mi.cs.helsinki./). The datasets used are synthetic datasets and real datasets. The synthetic datasets are T10I4D100K and T40I10D100K and real datasets include chess, kosarak, accidents, connect, mushroom, webdocs, retail, pumsb star, and pumsb [5].

3.1 Comparison of the Mining Algorithms

The performance of the algorithms differs with the ranging size of the datasets. The total time and memory consumption by the algorithms is shown in Figs. 8, 9, 10, 11 [21] (see Table 7).

4 Discussion

The mining algorithm is compared on the basis of the memory requirement. The algorithms have their own cons and pros in mining of frequent items. Most of the algorithms have utilized tree, trie, and array for representing the itemsets. With big data revolutionizing in different fields and growing with an exponential rate, mining such voluminous data becomes infeasible and impractical. If the dataset is large, then processing of the algorithms will consume time. This affects the complexity of the algorithm. Most of the mining algorithms show that with increase in the dataset, the efficiency of the algorithms decreases. The representation of itemsets reflects the main reason for memory requirement. The mining algorithms should mine datasets in such a way that it requires less space and time execution.

5 Conclusion

The performance of the algorithm depends on the space requirement. With large datasets nowadays, the mining should make sure that the itemsets should fit in the memory for faster execution. This paper provides a brief description of the mining algorithms. As seen from the comparison, the data structure affects the efficiency of the mining algorithms. With big data evolving, the mining algorithms should be able to handle the large dataset without deteriorating its efficiency.

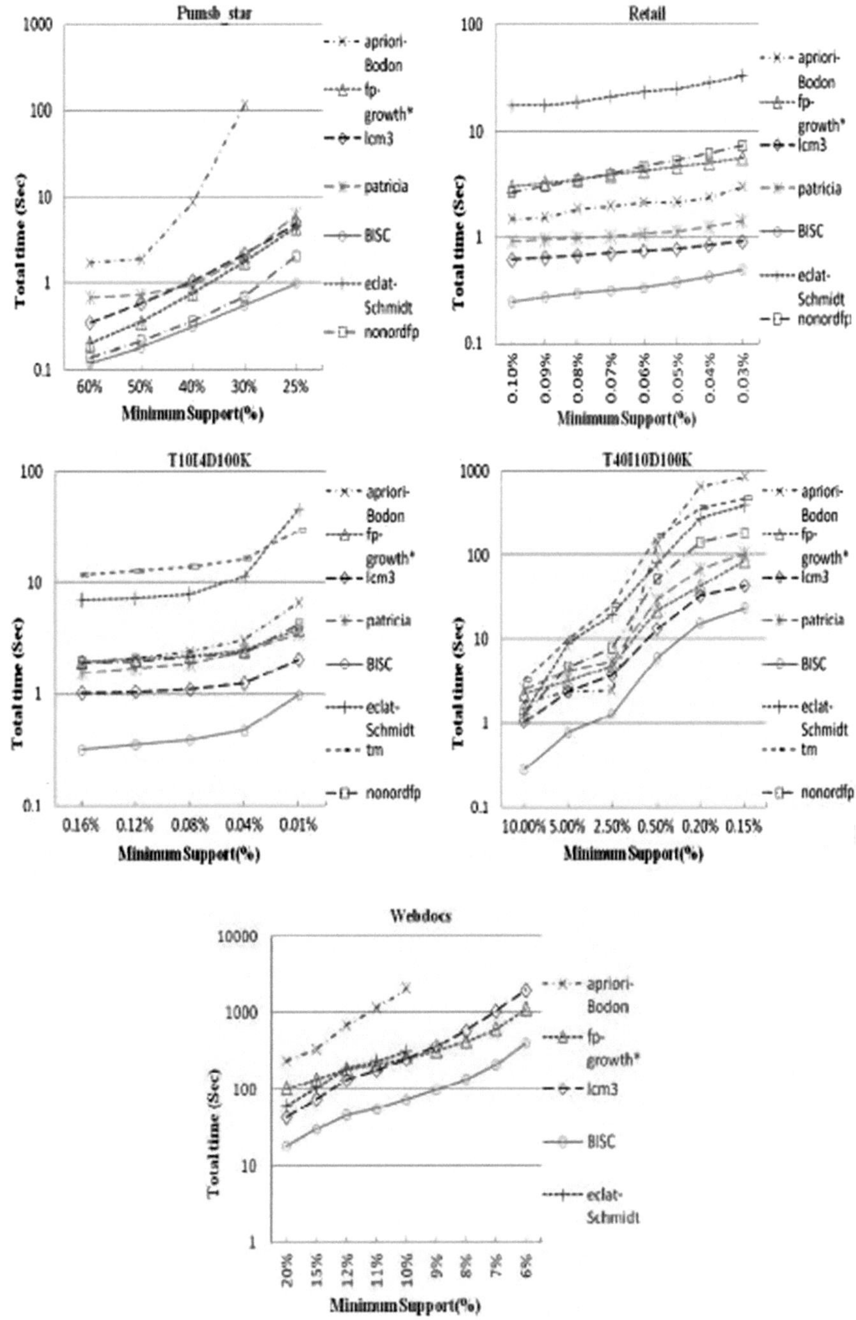

Fig. 8 Continue

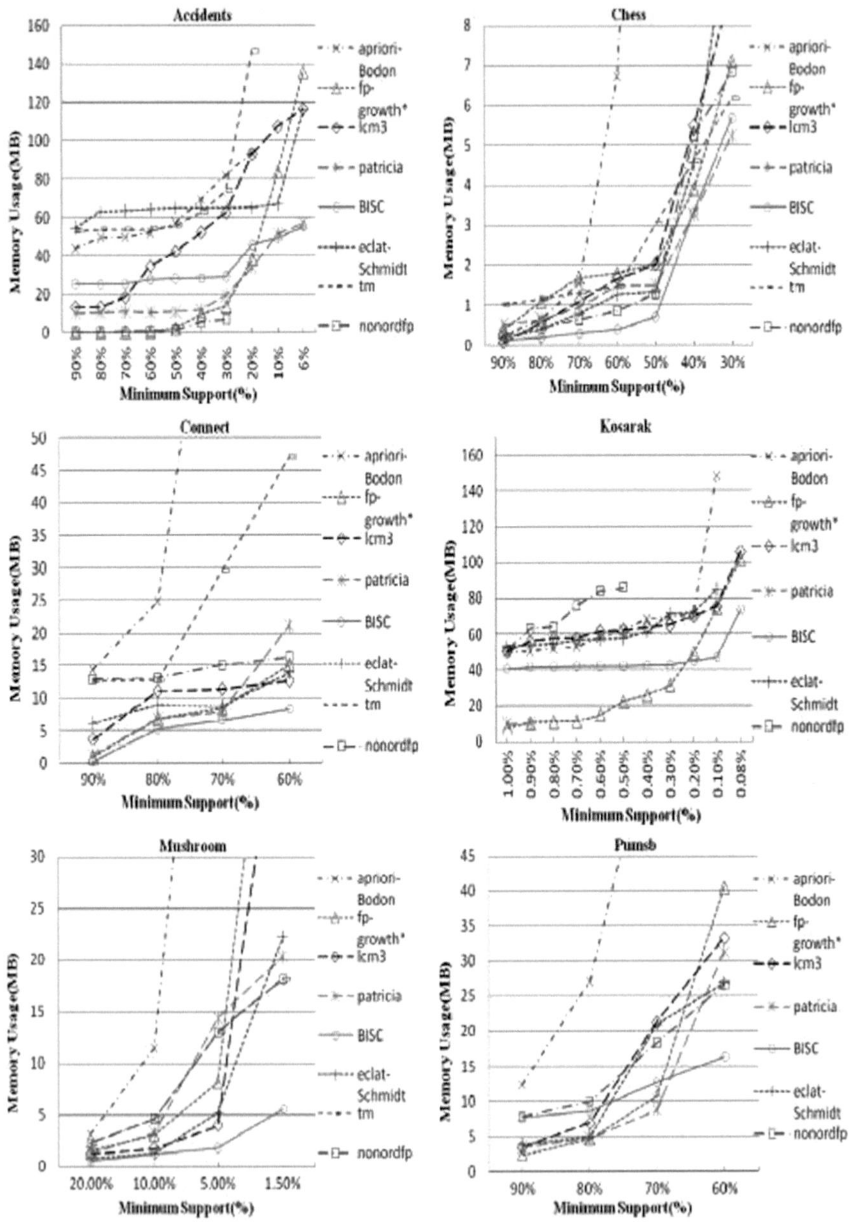

Fig. 9 Memory consumed by the different algorithms

Table 7 The table shows the pros and cons of the algorithms and the different data structures

Paper	Data structure	Advantages	Disadvantages
[11]	Constructive set	Storing the transaction bit id which speeds up the computation	While adding a new transaction, the algorithm uses nested loop, k for checking the status of bit that increases the complexity
[2]	Pattern count tree	Represent the pattern in terms of 0 and 1. Compactness of the algorithm id increases because of the matching of prefix pattern	the node has four fields which consumes space for maintaining it
[19]	Prefix tree	Stores the difference in tids called diffset which reduces the consumption of space. The space requirement is reduced for storing the tid array	For long pattern sequences, the diffset length increases. Keeping track of the large diffset length becomes difficult
[3]	Hybrid tree: array and prefix tree	It reduces the I/O cost and memory during data loading	If no support threshold is specified, HY-tree may not fit the main memory
[5]	Bitmap	Acquire the support based on direct support by scanning database once. Database projection is decrease	Storing the direct support in memory is impractical for large n
[4]	Trie	Uses vector and array makes fast implementation of functions. Store the reduced transactions that allocate memory only to frequent itemsets	It cannot utilize k frequent itemset to prune (k+1) candidate itemsets
[14]	Trie	Replacing item covers by diffsets in ECLAT	Not all the features added can perform equally for both sparse and dense database
[18]	Complete prefix tree	Gives fair performance by combining the data structure. Union and intersection save memory by representing by bit of 1 and 0	Generating extended candidate sets are inefficient which is decided by constant c. For large c, it becomes impractical
[15]	Prefix tree	Runs faster for higher value of support and d-ECLAT. There is no header table	It runs slower for lower value of support
[8]	Frequent pattern tree	Uses array that improve the FP tree performance. The transversal time is reduced	For larger value of n, the FP may not fit in main memory
[12]	Patricia trie	Limit the number of projection of the database	Tree overhead increases. In the merging process, sometimes, n is equal to the size of the original database
[13]	Trie	Array is reuse again that reduce the memory. Storing the frequent itemsets in the memory one per cell	It performs poor in sparse database

References

1. Agrawal, R., Mannila, H., Srikant, R., Toivonen, H., Verkamo, A.I., et al.: Fast discovery of association rules. Adv. Knowl. Discov. Data Min. **12**(1), 307–328 (1996)
2. Ananthanarayana, V., Murty, M.N., Subramanian, D.: Tree structure for efficient data mining using rough sets. Pattern Recognit. Lett. **24**(6), 851–862 (2003)
3. Baralis, E., Cerquitelli, T., Chiusano, S.: A persistent hy-tree to efficiently support itemset mining on large datasets. In: Proceedings of the 2010 ACM Symposium on Applied Computing, pp. 1060–1064. ACM (2010)
4. Bodon, F.: A fast apriori implementation, rpi cs department technical report tr 03-14 (2003)
5. Chen, J., Xiao, K.: Bisc: a bitmap itemset support counting approach for efficient frequent itemset mining. ACM Trans. Knowl. Discov. Data (TKDD) **4**(3), 12 (2010)
6. Cheung, D.W., Lee, S.D., Kao, B.: A general incremental technique for maintaining discovered association rules. In: Database Systems For Advanced Applications' 97, pp. 185–194. World Scientific (1997)
7. Ezeife, C.I., Su, Y.: Mining incremental association rules with generalized fp-tree. In: Conference of the Canadian Society for Computational Studies of Intelligence, pp. 147–160. Springer (2002)
8. Grahne, G., Zhu, J.: Efficiently using prefix-trees in mining frequent itemsets. In: FIMI, vol. 90 (2003)
9. Han, J., Pei, J., Yin, Y.: Mining frequent patterns without candidate generation. In: ACM Sigmod Record, vol. 29, pp. 1–12. ACM (2000)
10. Leung, C.S., Khan, Q.I., Hoque, T.: Cantree: a tree structure for efficient incremental mining of frequent patterns. In: Fifth IEEE International Conference on Data Mining, p. 8. IEEE (2005)
11. Nguyen, T.T.: Mining incrementally closed item sets with constructive pattern set. Exp. Syst. Appl. **100**, 41–67 (2018)
12. Pietracaprina, A.: Mining frequent itemsets using patricia tries (2003)
13. Rácz, B.: nonordfp: an fp-growth variation without rebuilding the fp-tree. In: FIMI (2004)
14. Schmidt-Thieme, L.: Algorithmic features of eclat. In: FIMI (2004)
15. Song, M., Rajasekaran, S.: A transaction mapping algorithm for frequent itemsets mining. IEEE Trans. Knowl. Data Eng. **18**(4), 472–481 (2006)
16. Srikant, R., Agrawal, R.: Mining sequential patterns: generalizations and performance improvements. In: International Conference on Extending Database Technology, pp. 1–17. Springer (1996)
17. Tanbeer, S.K., Ahmed, C.F., Jeong, B.S., Lee, Y.K.: Efficient single-pass frequent pattern mining using a prefix-tree. Inf. Sci. **179**(5), 559–583 (2009)
18. Uno, T., Kiyomi, M., Arimura, H.: Lcm ver. 3: collaboration of array, bitmap and prefix tree for frequent itemset mining. In: Proceedings of the 1st International Workshop on Open Source Data Mining: Frequent Pattern Mining Implementations, pp. 77–86. ACM (2005)
19. Zaki, M.J., Gouda, K.: Fast vertical mining using diffsets. In: Proceedings of the Ninth ACM SIGKDD International Conference on Knowledge Discovery and Data Mining, pp. 326–335. ACM (2003)
20. Zhou, Z., Ezeife, C.: A low-scan incremental association rule maintenance method based on the apriori property. In: Conference of the Canadian Society for Computational Studies of Intelligence, pp. 26–35. Springer (2001)

Efficient Clustering Using Nonnegative Matrix Factorization for Gene Expression Dataset

Pooja Kherwa, Poonam Bansal, Sukhvinder Singh and Tanishaq Gupta

Abstract Gene expression dataset consists of a complex association of gene patterns consisting of tens or hundreds samples. Finding relevant biological information for different tasks from this complex data is really a tedious job. Text mining approaches like classification and clustering are used in the literature to discover relevant aspects of dataset for many biological applications. Gene expression also contains irrelevant data known as noise. In this paper for efficient clustering results, a very powerful dimension reduction technique is presented as preprocessing step to improve clustering results and also cluster the gene expression samples into relevant classes. In this study, the concept of nonnegative matrix factorization and non-smooth nonnegative matrix factorization, which is an extended algorithm of the basic-NNMF algorithm is used for sparser matrix factorization, and the factorization differences are observed. Later on, the performance and the accuracy of K-means, NNMF, and NS-NNMF are compared, and NS-NNMF has shown highest accuracy.

Keywords Dimension reduction · Gene expression · Nonnegative matrix factorization · Clustering

---

P. Kherwa (✉) · P. Bansal · S. Singh · T. Gupta
Maharaja Surajmal Institute of Technology, C-4 Janak Puri, New Delhi 110058, India
e-mail: poona281280@gmail.com

P. Bansal
e-mail: pbansal81@gmail.com

S. Singh
e-mail: sukhvisukh@gmail.com

T. Gupta
e-mail: tanishkgupta@gmail.com

B. Pati et al. (eds.), *Advanced Computing and Intelligent Engineering*,
Advances in Intelligent Systems and Computing 1082,
https://doi.org/10.1007/978-981-15-1081-6_15

1 Introduction

With the rapid development of technology in biomedical sciences, microarray technology makes it possible to understand the expression level of large number of genes and proteins very easily [1]. The main challenge of this type of biological data is to discover and capture valuable knowledge to understand biological process and human diseases mechanism [2]. Gene expression datasset is a very complex and big dataset. Various methods have been developed in last two decades to analyze gene expression data, for various applications in biology and bioinformatics including—for detecting cancer cell and another disease like Alzheimer's disease symptoms in gene dataset [1, 3, 4]. Analyzing complex dataset like Gene is a big challenge for researchers due to its high dimensionality and complex structure and noise. Mining such multidimensional complex data structure poses huge challenges as well as opportunities for researchers to develop an automatic system with high precision and accuracy. Despite the difficulty, the clustering and classification methods are applied to gene expression dataset for many biological application domains [5–8].

The gene dataset can be represented in a form of n*m matrix A, where n represents gene under m number of samples. In this dataset ,column denotes gene signatures (GS) of n genes during m experiments, where each row of matrix A signifies gene profiles (GP) for each gene under all m conditions [9]. This property of microarray can be used to cluster similar gene patterns.

Many dimension reduction techniques exist in literature including SVD and PCA and ICA. Single value decomposition (SVD) is the most widely used approach for dimension reduction in bioinformatics [10]. SVD sorts the genes according to the descending values of eigengenes [11], it gives global view to the gene expression, for further clustering to find similar cellular states of genes. A robust SVD variant [10] has been proposed to cope with outliers or missing values often found in gene expression.

Nonnegative matrix factorizations start its journey by an experiment performed by a Finnish group of researchers under the name positive matrix factorization and after that, its various variants are explored in various applications due to its simplicity [12].

This basic aim of nonnegative matrix factorization is to reduce the dimensional complexity of the data to a point where clustering can be computed faster and computational errors can be reduced. Microarray Gene Pattern datasets using nonnegative matrix factorization method by reducing the Dimensional Complexity of the huge datasets, smoothening the overall factorization, and filtering the noises present in the dataset can be useful for Microarray Gene Pattern datasets. This approach (NNMF) will allow the researchers to analyze the expressions of thousands of genes under different samples at the same time.

The paper deployed a special version of NNMF known as non-smooth nonnegative matrix factorization to introduce sparseness in the factorization result and hence reinforcing the gene patterns. The differences in sparseness due to basic-NMF and

NS-NMF are then observed. The datasets and the factorized matrices are then represented in a pictorial way known as heat maps, this makes easy to analyze a large dataset and its consequent factorization. Lastly, the performance and accuracy of NNMF, NS-NNMF, and K-means have been discussed and it will be proved that NS-NMF has the highest accuracy.

In this paper, Sect. 2 presents some existing works in NNMF using microarray dataset; in Sect. 3, nonnegative matrix factorization methodology is discussed; in Sect. 4, the experimental setup is explained; in Sect. 5, results and evaluation are explained;and in Sect. 6, the paper is concluded with future direction for clustering of gene expression datasets.

2 Related Work

The advancement of technology in the last two decades makes it possible to study complex biological system on multiple levels. At the same time, due to the complex structure, diversity of units, scales, and types the format of multidimensional biological data confront huge challenges.

The microarray dataset consists of thousands of genes on every single chip and disease samples on these genes are much smaller than that of genes. Initial work on biological dataset focused on the application of statistical techniques for identifying pattern and clustering the patterns.

In microarray dataset, many machine learning approaches like classification and clustering have been applied for cancer identification using molecular gene expression dataset [13, 14]. A comparative experiment on filtered set of genes for cancer classification using discrimination methods [15, 16] was done and abnormalities in cell behavior are highlighted with valid proofs. A relevant work also used independent component analysis (ICA) [17] to select specific genes to identify tumors. Classification-based analysis for the classification of Alzheimer's disease using feature selection methods like chi-squared, andgain ration is proposed by Joshi et al. [18], A kernel-based principal component analysis also used dimension reduction and then, linear discriminant analysis (LDA) is used for classification of microarray dataset.

Single value Decomposition and Principal Component Analysis methods have also contributed to biological research including genetic profiling in leprosy [19], analysis of human fibroblast data [20], breast tumor classification [21], and the identification of tissue-specific gene expression pattern.

Till now, limited work has been done that can analyze multidimensional genomic datasets directly, regardless of the specific data types [22]. NNMF has the power to analyze high-dimensional biological dataset including discrete and continuous.

3 Matrix Factorization Methods

Matrix Factorization methods can be categorized as

3.1 Nonnegative matrix factorization (NNMF)
3.2 Non-smooth nonnegative matrix factorization (NSNNMF)

3.1 Nonnegative Matrix Factorization

Nonnegative matrix factorization is a technique to split a matrix $\boldsymbol{V}$ of size $\boldsymbol{m} \times \boldsymbol{n}$ into two matrices viz. $\boldsymbol{W}$ and $\boldsymbol{H}$ such that $\boldsymbol{W}$ is of size $\boldsymbol{m} \times \boldsymbol{k}$ and $\boldsymbol{H}$ is of size $\boldsymbol{k} \times \boldsymbol{n}$, where k is known as the rank of factorization. The factorization problem is a NP-hard problem [23, 24]; therefore, the factorization result is computed approximately.

So, the matrix $\boldsymbol{V}$ is factorized into two matrices $\boldsymbol{W}$ and $\boldsymbol{H}$ such that

$$\boldsymbol{V} \approx \boldsymbol{W} \times \boldsymbol{H} \tag{3.1}$$

Here, W represents the basis factors and H, the decomposition coefficients. It is known as nonnegative matrix factorization because the result of factorization is nonnegative, i.e., all the values obtained in the factorized matrices are positive. Due to this property, this concept has applications in various fields such as text mining or gene clustering [25, 26] where traditional singular value decomposition cannot be applied since matrix representations itself cannot contain negative values.

The main advantage of using this technique will be to reduce the dimensional complexity of our original matrix. The NNMF algorithm reduces a large matrix into two matrices of lower dimensional complexity than the original one, so significantly to point where computations and analysis on factorized matrices become faster.

$\boldsymbol{V}$ is the original matrix, factorized into two matrices $\boldsymbol{W}$ and $\boldsymbol{H}$, where $\boldsymbol{W}$ has the basis factors and $\boldsymbol{H}$ contains the encoding vectors and a linear multiplication can be used to compute column vectors of $\boldsymbol{V}$ matrix such that

$$\boldsymbol{v}_i = \boldsymbol{W} \times \boldsymbol{h}_i \tag{3.2}$$

There are numerous techniques to compute the $\boldsymbol{W}$ and $\boldsymbol{H}$ matrices, the most widely used one being multiplicative update rule [27] which is as explained ahead. Although many new techniques have also been developed for the matrix factorizations.

First, we initialize $\boldsymbol{W}$ and $\boldsymbol{H}$, to which there are several methods of initialization such as random initialization or Nonnegative Double Singular Value Decomposition method [28, 29]. Then, the $\boldsymbol{W}$ and $\boldsymbol{H}$ matrices are updated as follows, with n being the index of the iteration. This is repeated until the convergence is achieved.

$$H_{[i,j]}^{n+1} \leftarrow H_{[i,j]}^{n} \frac{\left((W^n)^T V\right)_{[i,j]}}{\left((W^n)^T W^n H^n\right)_{[i,j]}} \tag{3.3}$$

and,

$$W_{[i,j]}^{n+1} \leftarrow W_{[i,j]}^{n} \frac{\left(V(H^{n+1})^T\right)_{[i,j]}}{\left(WH^{n+1}(H^{n+1})^T\right)_{[i,j]}} \tag{3.4}$$

However, there are also many other ways to compute the factorization with new methods being developed.

3.2 Non-smooth Nonnegative Matrix Factorization

In non-smooth nonnegative matrix factorization, the matrix V is factorized into three matrices instead of two matrices. In original NMF, the factorization is as follows:

$$V \approx W \times H, \tag{3.5}$$

But in non-smooth nonnegative matrix factorization, the result of factorization is as follows:

$$V \cong W \times H \times S \tag{3.6}$$

where
V is our original matrix of gene expression data
W is $n \times k$ matrix of basis vectors
H forms the linear combination of basis vectors of size $k \times n$
S is a matrix of size $k \times k$ known as smoothening matrix and S is calculated as

$$S = (1 - \theta)I + \frac{\theta}{q} 11^T \tag{3.7}$$

where I is an identity matrix, 11^T is a $k \times k$. matrix having each and every item as 1, and θ smoothening factor where $0 \leq \theta \leq 1$.

The NS-NMF algorithm makes the factorized matrices sparse and it significantly reinforces the gene experiments that sustain and represent factors [9] A sparse matrix means that only a few values are successfully used to represent typical data vectors [30]. This is why we use non-smooth nonnegative matrix factorization with sparseness constraints and also due to the fact that the factors are significantly sustained and represented.

4 Experimental Setup

Dataset Description: Two datasets were obtained from Human Brain Atlas. Alzheimer's disease microarray dataset having expressions of 40 genes and Autism microarray dataset having 60 gene expressions [31].

Preprocessing: The datasets obtained from the human brain atlas were already clean, with a few null columns that were automatically and easily removed by Microsoft Excel.

Parameter Setting: The Alzheimer's disease dataset has 40 rows and 841 columns (after preprocessing) which make up the $\boldsymbol{V}$ matrix of $\boldsymbol{m} = \mathbf{40}$ and $\boldsymbol{n} = \mathbf{841}$ for Alzheimer's disease dataset, on which NS-NMF factorization is applied with the following parameters.

Rank of factorization, $\boldsymbol{k}$ is set as **12** and initialization method of $\boldsymbol{W}$ and $\boldsymbol{H}$ matrix is set as "Nonnegative double singular value decomposition". The factorization is run for 100 iterations for $\boldsymbol{k} = \mathbf{12}$ and $\boldsymbol{\theta} = \mathbf{1}$ to introduce max sparseness.

The Autism dataset has 60 rows and 841 columns (after preprocessing) which make up the $\boldsymbol{V}$ matrix of $\boldsymbol{m} = \mathbf{60}$ and $\boldsymbol{n} = \mathbf{841}$ for Autism dataset, on which NS-NMF factorization is applied with the following parameters. Rank of factorization, $\boldsymbol{k}$ is set as **18** and initialization method of $\boldsymbol{W}$ and $\boldsymbol{H}$ matrix is set as "Nonnegative double singular value decomposition". The factorization is run for 100 iterations for $\boldsymbol{k} = \mathbf{18}$ and $\boldsymbol{\theta} = \mathbf{1}$ to introduce max **sparsenesss**.

Similar value of $\boldsymbol{K}$ is passed to the corresponding basic-NMF factorization algorithms with 100 iterations in order to compare the accuracy between basic-NMF and NS-NMF. Similarly, **K-means** clustering is applied to both original samples (V matrix) with the same $\boldsymbol{K} = \mathbf{12}$; for Alzheimer's disease dataset and $\boldsymbol{K} = \mathbf{18}$; for Autism dataset. This is done to compare the performance of all three algorithms.

5 Results and Evaluation

In this experiment, first of all, both the datasets—The Alzheimer's disease dataset and The Autism dataset are factorized with nonnegative matrix factorization (NMF) and non-smooth nonnegative matrix factorization (NS-NMF) methods. The complete dataset is our V matrix and it is factorized into two matrixes known as W and H.

When $\boldsymbol{W}$ and $\boldsymbol{H}$ matrices were successfully factorizing, the $\boldsymbol{W}$ matrix contains the basis factors meta-genes and $\boldsymbol{H}$ matrix contains the encoding vectors known as meta-experiments as shown in Figs. 1, 2, 3, and 4.

Upon successful convergence, with specified iteration as in parameter setting section, the decomposition of $\boldsymbol{V}$ into $\boldsymbol{W}$ and $\boldsymbol{H}$ is done. The automatic clustering information is contained in the $\boldsymbol{H}$ matrix. **K-means** clustering is applied to the original sample, i.e., the $\boldsymbol{V}$ matrix because we need to compare the accuracy and performance of a dimensionality reduction technique like NMF with that of a traditional

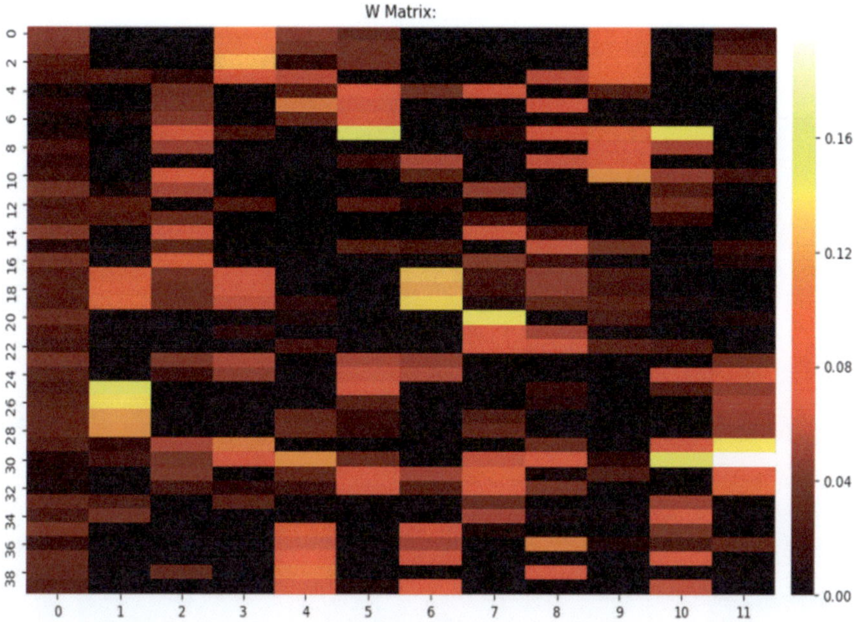

Fig. 1 Decomposed *W* matrix by NS-NMF of Alzheimer's disease dataset

clustering algorithm such as **K-means**. These are the results of the outputs of NMF and NS-NMF shown as heat maps.

All three algorithms, i.e., K-means, NMF, and NS-NMF were applied to both Alzheimer's disease and Autism's datasets and the results are evaluated. K-means clustering result returned as one-dimensional array A in which the entry A[i] belongs to the cluster number i. The genes will be clustered in this way with the help of the *W* matrix: Gene **i** is placed in cluster number **j** if **W[i][j]** is the largest entry in the row vector of the *W* matrix [18]. This way the cluster number of different genes was assigned and the clustering was done in both datasets. In that Metafile, the clusters were already assigned, so it became very easy for us to compare the accuracy of the results we got from the three algorithms viz. K-means, NMF, and NS-NMF.

The comparison of the original cluster versus the output of our algorithms was done this way. Clustered genes were already sorted and were assigned a **gene-id** and a **gene-name.** Obviously, the **gene-id** and the **gene-name** are the same for similar genes under the same cluster. This was compared with the cluster numbers we got for the decomposed W matrix in the same arrangement order of what was in the original Metafile. Since the cluster number may be different but the logical cluster in the sense is same, in order to compare the original cluster and the clusters that we got from our algorithm, we assigned a **pseudo-index** in place of the **gene-id** in such a way that the most frequently occurring cluster number from a list of cluster numbers corresponding to the original cluster was assigned as a pseudo-index in order to easily compare the results.

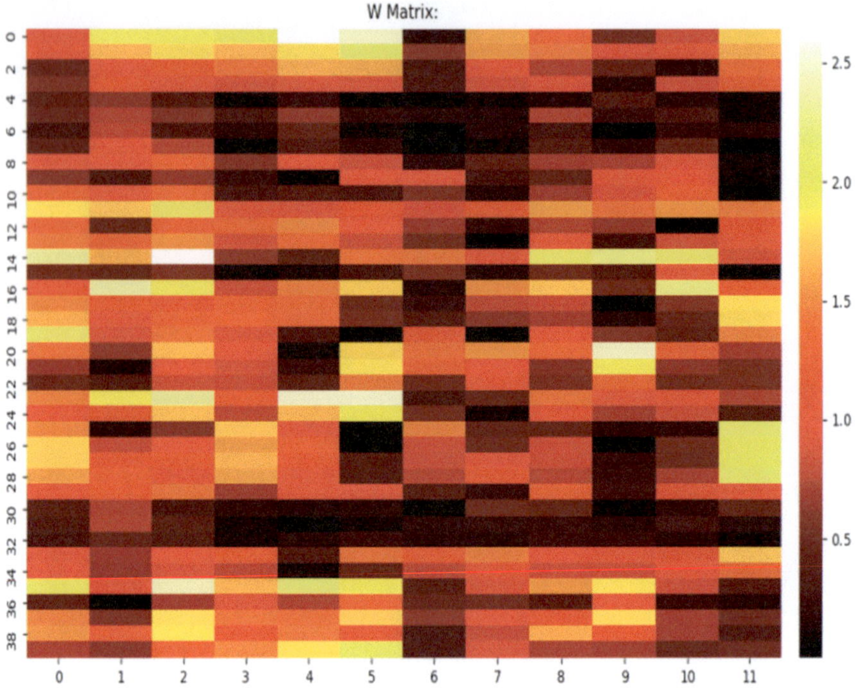

Fig. 2 Decomposed *W* matrix by basic-NMF of Alzheimer's disease dataset

Please do note that this step was done purely for our own convenience in spreadsheet software (MS-Excel) because we needed to compare the original columns with our results automatically and since it was done by software, chances of error are also reduced. In short, we just changed the gene-id number according to the cluster numbers that we got because logically, the cluster is same, but the number referring to it can be anything Then, we found out that how many genes were placed in the correct cluster or not, and we obtained the count of correctly placed genes for each algorithm in both datasets. The correct number of clustered genes divided by original number of genes multiplied by 100 gives us the percentage of correctly clustered genes and this is our accuracy. The results were obtained as follows:

Figure 5 shows the accuracies of different algorithms on the distinct datasets, and clearly, the NS-NMF got the highest accuracy.

The results were as follows.

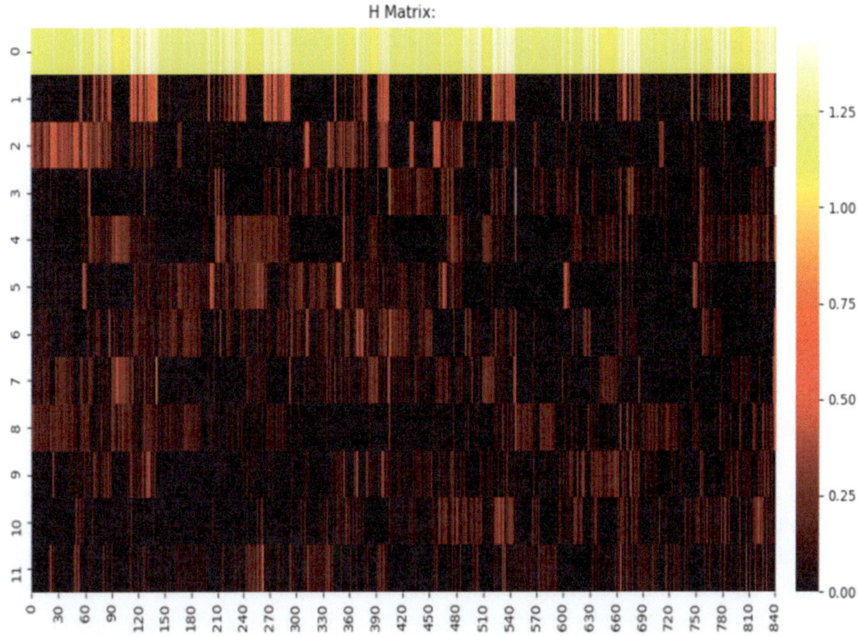

Fig. 3 Decomposed *H* matrix by NS-NMF of Alzheimer's disease dataset

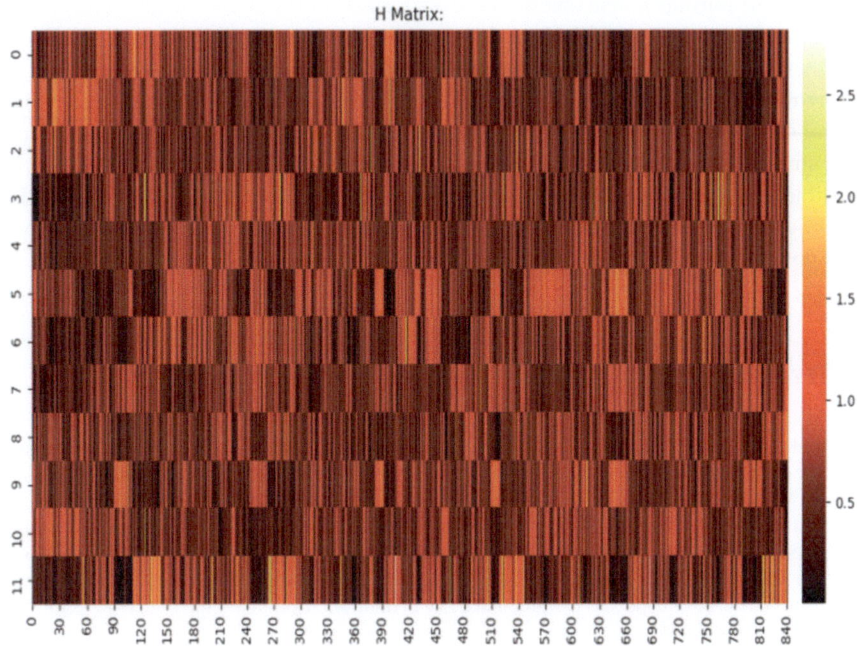

Fig. 4 Decomposed *H* matrix by basic-NMF of Alzheimer's disease dataset

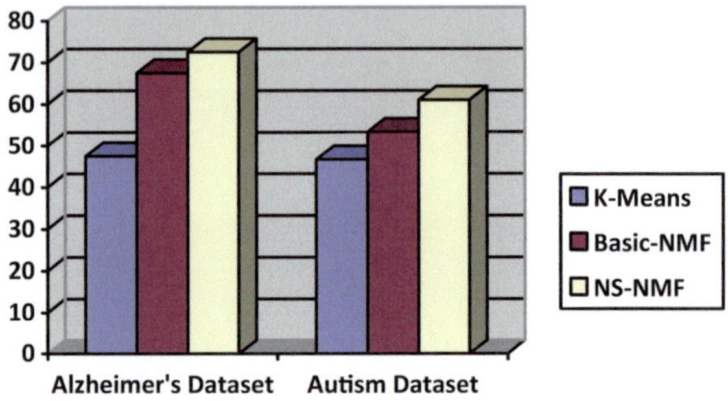

Fig. 5 Chart showing different accuracies of different algorithms on different datasets

6 Conclusions

In this work, it has been concluded that Nonnegative Matrix Factorization can be applied to various complex datasets, such as in this case, we applied on gene expression data for analysis, provided as long as the dataset is in 2D matrix form and does not contain any negative values. Nonnegative Matrix Factorization is a powerful technique to reduce dimensional complexity of the data by splitting or factorizing a larger matrix into two smaller matrices. It has an inherent clustering property and it finds its applications in various fields. For clustering of gene expressions, a special version of NNMF known as non-smooth nonnegative matrix factorization is deployed to introduce sparseness in the factorization result and hence, reinforcing the gene patterns. The differences in sparseness due to basic-NMF and NS-NMF are then observed. The datasets and the factorized matrices are then represented in a pictorial way known as heat maps, to make it easy to analyze a large dataset and its consequent factorization. We have proved that NS-NMF has the highest accuracy rate among Basic-NMF and traditional K-Means clustering approach. In future, it is feasible to embed fuzzy clustering techniques [32] and least square NMF (ls-NMF) to assign genes to multiple groups. Bayesian Decomposition [20] is also emerging as a powerful technique for recognizing useful patterns in microarray dataset.

References

1. Del Buono, N., Esposito, F., Fumarola, F., Boccarelli, A., Coluccia, M.: Breast cancer's microarray data: pattern discovery using nonnegative matrix factorizations. In: International Workshop on Machine Learning, Optimization and Big Data, Springer Champ 281–292 (2016)
2. Moschetta, M., Basile, A., Ferrucci, A., Frassanito, M., Rao, L., Ria, R., Solimando, A., Giuliani, N., Boccarelli, A., Fumarola, F., Coluccia, M., Rossini, B., Ruggieri, S., Nico, B., Maiorano, E., Ribatti, D., Roccaro, A., Vacca, A.: Novel targeting of phospho-cMET overcomes drug resistance and induces antitumor activity in multipllemyeloma. Clin. Cancer Res. **19**(26), 4371–4382 (2013)
3. Jain, M., Dua, P., Lukiw, W.J.: Data adaptive rule-based classification system for Alzheimer classification. J. Comput. Sci. Syst. Biol. **6**, 291–297 (2013)

4. Zheng, C.H., Huang, D.S., Zhang, L., Kong, X.Z.: Tumour clustering using nonnegative matrix factorization with gene selection. IEEE Trans. Inf. Technol. Biomed. **13**(4), 599–607 (2009)
5. Golub, T.R., Slonim, D.K., Tamayo, P., Huard, C., Gaasenbeek, M., Mesirov, J.P., Coller, H., Loh, M.L., Downing, J.R., Caligiuri, M.A., Bloomfield, C.D., Lander, E.S.: Molecular classification of cancer: class discovery and class prediction by gene expression monitoring. Science **286**, 531–537 (1999)
6. Brunet, J.P., Tamayo, P., Golun, T.R., Mesirov, J.P.: Metagenes and molecular pattern discovery using matrix factorization. Proc. Nat. Acad. Sci. USA. **101**(1), 4164–416 (2004)
7. Bryan, K., Cunningham, P., Bolshakova, N.: Application of simulated annealing to the biclustering of gene expression data. IEEE Trans. Inf. Technol. Biomed. **10**(3), 519–525 (2006)
8. Pomeroy, S.L., Tamayo, P., Gaasenbeek, M., Sturla, L.M., Angelo, M., Mclaughlin, M.E., Kim, J.Y., Goumnerova, L.C., Black, P.M., Lau, C., Allen, J.C., Zagzag, D., Olson, J.M., Curran, T., Wetmore, C., Biegel, J.A., Poggio, T., Mukherjee, S., Rifkin, R., Califano, A., Stolovitzky, G., Louis, D.N., Mesirov, J.P., Lander, E.S., Golub, T.R.: Prediction of central nervous system embryonal tumour outcome based on gene expression. Nature **415**, 436–442 (2002)
9. Wei Kong, X.M.: Exploring matrix factorization techniques for significant genes identification of Alzheimer's disease microarray gene expression data. BMC Bioinformat. (2011). https://doi.org/10.1186/1471-2105-12-S
10. Liu, L., et al.: Robust singular value decomposition of microarray data. Proc. Nat. Acad. Sci. USA **100**, 13167–13172 (2003)
11. Alter, O., et al.: Singular value decomposition for genome-wide expression data processing and modeling. Proc. Nat. Acad. Sci. USA **97**, 10101–10106 (2000)
12. Lee, D.D., Seung, H.S.: Learning the parts of the objects by non-negative matrix factorization. Nature **401**, 788–791 (1999)
13. Brunet, J.P., Tamayo, P., Golun, T.R., Mesirov, J.P.: Metagenes and molecular pattern discovery using matrix factorization. Proc. Nat. Acad. Sci. USA **101**(12), 4164–4169 (2004)
14. Pan, W.: A comparative review of statistical methods for discovering differently expressed genes in replicated microarray experiments. Bioinformatics **18**, 546–554 (2002)
15. Dudoit, S., Fridlyand, J., Speed, T.P.: Comparison of discrimination methods for the classification of tumor using gene expression data. J. Am. Stat. Assoc. **97**, 77–87 (2002)
16. Daniela, G.C., Giuliano, G., Marilena, P., Cinzia, V.: Variable selection in cell classification problems: a strategy based on independent component analysis. In: Studies in Classification, Data Analysis, and Knowledge Organization, Part I. pp. 21–29. Springer, Berlin/Heidelberg, Germany (2006)
17. Lopez, M., Ramirez, J., Salas-Gonzalez, D., Alvarez, I., Segovia, F.: Neuro image classification for the Alzheimer's Disease Diagnosis using Kernal PCA and support vector machines. In: Nuclear Science Symposium Conference Record (NSS/MIC) (2009)
18. Futschik, M.E., Kasabov, N.K.: Fuzzy clustering of gene expression data. In: IEEE International Conference on Fuzzy Systems, pp. 414–419. IEEE, Honolulu, HI (2002)
19. Moloshok, T.D., Klevecz, R.R., Grant, J.D., Manion, F.J., Speier, W.F., Ochs, M.F.: Application of Bayesian decomposition for analysing microarray data. Bioinformatics **18**(4), 566–575 (2002)
20. Lee, D.D., Seung, H.S.: Learning the parts of objects by non-negative matrix factorization. Nature **401**, 788–791 (1999)
21. Misra, J., Schmitt, W., Hwang, D., Hsiao, L.L., Gullans, S., Stephanopoulos, G.: Interactive exploration of microarray gene expression patterns in a reduced dimensional space. Genome Res. **12**(7), 1112–1120 (2002)
22. Ghosh, D.: Singular value decomposition regression models for classification of tumors from microarray experiments. Pac. Symp. Biocomput. **7**, 18–29 (2002). [PubMed: 11928474]
23. Boutsidis, C., Gallopoulos, E.: SVD based Initialization: A head start for nonnegative matrix factorization. Pattern Recogn. **41**(4), 1350–1362 (2008)
24. Field, D.J.: What is the goal of sensory coding? Neural Comput. **6**(4), 559–601 (1994)
25. Taslaman, L., Nilsson, B.: A Framework for regularized non-negative matrix factorization, with application to the analysis of gene expression data. Plos One 7–11 (2012)

26. Daniel, D. Lee, H.S.: Algorithms for non-negative matrix factorization. In: Proceedings of the 13th International Conference on Neural Information Processing Systems, 535–541 (2000)
27. Kim, M.H., Seo, H.J., Joung, J.G., Kim, J.H.: Comprehensive evaluation of matrix factorization methods for the analysis of DNA microarray gene expression data. BMC Bioinformat. **12**(13), S8 (2011)
28. Vavasis, S.A.: On the complexity of nonnegative matrix factorization. SIAM J. Optimiz. **20**(3), 1364–1377 (2009)
29. http://www.brain-map.org/ (Online)
30. Li, W., Zhang, S., Liu, C.C., Zhou, X.J.: Identifying multi-layer gene regulatory modules from multi-dimensional genomic data. Bioinformatics **28**(19), 2458–2466 (2012)
31. Bleharski, J.R., Li, H., Meinken, C., Graeber, T.G., Ochoa, M.T., Yamamura, M., Burdick, A., Sarno, E.N., Wagner, M., Rollinghoff, M., Rea, T.H., Colonna, M., Stenger, S., Bloom, B.R., Eisenberg, D., Modlin, R.L.: Use of genetic profiling in leprosy to discriminate clinical forms of the disease. Science **301**(5639), 1527–1530 (2003) [PubMed: 12970564]
32. Holter, N.S., Mitra, M., Maritan, A., Cieplak, M., Banavar, J.R., Fedoroff, N.V.: Fundamental underlying gene expression profiles: simplicity from complexity. Proc. Natl. Acad. Sci. USA **97**(15) 8409–8414 (2000)

Design of Random Forest Algorithm Based Model for Tachycardia Detection

Saumendra Kumar Mohapatra, Tripti Swarnkar and Mihir Narayan Mohanty

Abstract ECG signals are need to be analyzed accurately for better diagnosis. Different parameters of ECG signals provide information regarding the heart disease. In this paper, an attempt has been made to detect tachycardia, a class of arrhythmia. With the help of random forest algorithm, the updated technique has been utilized for the cardiac signal classification and detection. Thirteen attributes are considered as the input to the model. The technique is multiple decision trees based with each tree size is considered as 150. As compared to earlier methods the proposed method found better classification accuracy.

Keywords Cardiac disease · Machine learning · Ensemble learning · Random forest

1 Introduction

Machine learning based classification techniques can be used for the decision-making in health care domain which includes disease diagnosis and prediction, psychoanalysis, survival analysis, and hospital management. In recent years, many countries in the world are also adopting an automated diagnosis system to detect a disease at an early stage. This automatic system can detect disease by analyzing the data

---

S. K. Mohapatra
Department of Computer Science, ITER, Siksha 'O' Anusandhan (Deemed to Be University), Bhubaneswar, India
e-mail: saumendramohapatra@soa.ac.in

T. Swarnkar
Department of Computer Application, ITER, Siksha 'O' Anusandhan (Deemed to Be University), Bhubaneswar, India
e-mail: triptiswarnakar@soa.ac.in

M. N. Mohanty (✉)
Department of Electronics and Communication Engineering, ITER, Siksha 'O' Anusandhan (Deemed to Be University), Bhubaneswar, India
e-mail: mihirmohanty@soa.ac.in

B. Pati et al. (eds.), *Advanced Computing and Intelligent Engineering*,
Advances in Intelligent Systems and Computing 1082,
https://doi.org/10.1007/978-981-15-1081-6_16

collected from different medical test [1]. Machine learning based diagnosis system is a supportive tool in the clinical environment where the machine is trained with different features and class of disease. The class of a disease refers to the decision of a physician after analyzing several features of a disease. The accuracy of the overall system is one of the essential points in the process of disease diagnosis.

In recent years many countries in the world adopting automated diagnosis system to detect a disease at an early stage. This system can detect disease by analyzing the data collected from the different medical test. This automatic diagnosis job can be performed by a machine using a different machine-learning algorithm. Machine learning based diagnosis system is a supportive tool in the clinical environment. The accuracy of the overall system is one of the essential points in the process of disease diagnosis [2].

According to the World Health Organization (WHO), millions of death occurs because of the heart attack. A similar report by the American Heart Association (AHA) is that around 70 million people everywhere throughout the world were diagnosed with the cardiac problem [3, 4]. It happens due to the high blood pressure, cholesterol, obesity, smoking, etc. Tachycardia is one of cardiac disease which happens due to the increase of heart rate. With the proper diagnosis of heart disease at an early stage can reduce the death rate. It is one of the complicated tasks to develop an automatic diagnosis system which can detect the disease accurately at an early stage. Different techniques have been applied by the researcher for getting a better accuracy result, some of them are cited.

In [5], authors have introduced a data resampling approach for arrhythmia detection. They have taken random forests classifier for their classification purpose. This random forest classifier was also used by other researchers for the classification of different disease. The distance of this classifier was also modified using the non-parametric method [6, 7].

Neural network based cardiac arrhythmia classifiers were also proposed in some studies. Block-based neural network classifier for arrhythmia classification was presented in [8]. They have trained their neural network with PSO algorithm and get a high accuracy rate. Authors in [9] used MLP for arrhythmia classification. They have taken wavelet and Fourier features from the ECG for input to the neural network. A feed-forward neural network based classifier was designed in [10] to classify arrhythmia ECG signals. Wavelet and PCA features were taken as the input for the neural network classifier in their study. They have optimized the neural network with the MD-PSO process. This optimization technique can also be applied to any kind of neural network to decrease the error rate. Bayesian neural network based cardiac arrhythmia detection system was planned by the authors in [11]. They have designed the neural network by using logistic regression and backpropagation algorithm. Forgetting the better result authors in [12] have proposed deep learning based arrhythmia classifier which was giving a better result as compare to other types of neural network. A voting feature interval based classifier was proposed in [13] for classifying arrhythmia from 12 lead ECG. They have distinguished different types of arrhythmia by applying this algorithm. In [14] authors have used KNN classifier with kernel difference weight to get 70% accuracy. They have applied 10-fold cross-validation in the arrhythmia data. Serial fusion of support vector machine and logistic regression were applied in arrhythmia data in [15]. Their proposed fusion method gives an adaptable tradeoff between the error and rejection rates. In [16], fuzzy weighted preprocessing with AIRS was applied for the diagnosis of arrhythmia. The cardiac arrhythmia was diagnosed by a fully automatic manner using fuzzy weighted preprocessing with AIRS. It is expected that more accurate results can be obtained by further exploration of data. Markov Blanket algorithm was used by the authors in [17] for the classification of cardiac arrhythmia and they have obtained around 66% accuracy. RBFN, RNN, and Feed-forward neural network were used in [18] to classify healthy and arrhythmia classes using 12-lead UCI-ECG database. Different types of arrhythmia classes were not considered in their work. Multi layer perceptron was compared with the other three types of classifier (KNN, SVM, and Naive Bayes) in [19] and MLP was giving a better result as compare to others.

Congestive heart failure, diabetes, and other cardiac approach were also detected using this ensemble classifier. Some authors have also compared this method with other methods like KNN, SVM, decision tree, neural network, etc. [20–22]. Random Forest algorithm was viewed as a better classifier because of its incredible execution. To discover extra methodology for raising the variety of trees is an exciting task. This technique was also presented in a big data approach. Out of bag error was introduced in for the analysis of big data problem [23, 24]. A technique for predicting the disease risk from the pathological data was introduced in [25]. They have compared the performance different method for predicting the risk of eight persistent diseases. In [26], the ECG signals were classified using the Random Forest algorithm. The classifier model was designed with the DWT features of the ECG signal. By taking this DWT feature of ECG signal a random forest model for human verification was introduced in [27].

The accuracy can be increased with a more accurate machine learning algorithm. In this work, the random forest model is considered for the detection of heart disease. Whatever is left of the paper is sorted out as following way: proposed strategy is portrayed in Sect. 2. Result and the conclusion and further works are contained by Sects. 3 and 4.

2 Proposed Methodology

A three-stage detection model is developed as shown in Fig. 1. The data may be collected from the patients directly or similar data can be collected from different databases. The UCI Cleveland heart disease data is used in this work. This is one of the most used databases in the area of bioinformatics research where different types of data sets are freely available.

The dataset contains 303 patient's samples and each sample is characterized by 13 attributes. The details of the Cleveland database are explained in Table 1.

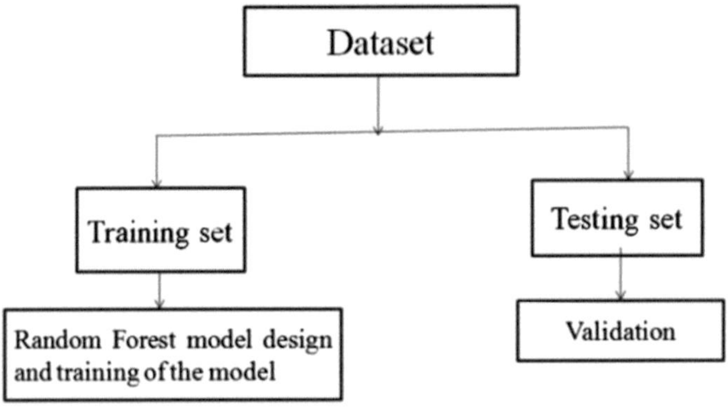

Fig. 1 Proposed methodology

Table 1 Cleveland heart disease dataset

Attributes	Explanation
Age	Patient age
Sex	1 represents male and 0 represents female patients
Cp	Different types of pain in chest(1 = typical angina, 2 = atypical angina, 3 = non-angial pain, 4 = asymptomatic)
Trestbps	Blood pressure of the patient at the admission time to the hospital
Chol	Cholesterol of the patient in mg/dl
FBS	Sugar in blood at fasting time
Restecg	ECG report (For normal it is 0, patients having ST-T wave abnormality it is 1 and left ventricular is 2
Thalch	Pulse rate
Exang	Exercise encouraged angina (for Yes it is 1 and for No it is 0)
Oldpeak	ST depression induced by exercise ST segment
Slope	Peak exercise slope (up sloping = 1, flat = 2, down sloping = 3
Ca	Amount of main vessels (0–3) colored by fluoroscopy
Thal	3 shows the normal, fixed defect is denoted as 6 and the reversible defect is denoted as 7

2.1 Classification Algorithm

For classification of tachycardia random forest model is considered in this work. It is the current method used in many problems. However, the application of biomedical signal with this model is the novelty of the work. Basically, this technique is Decision Tree-based and is a combination of the number of Decision Tree classifiers. It works on the ensemble learning method where a number of learners are trained to solve a

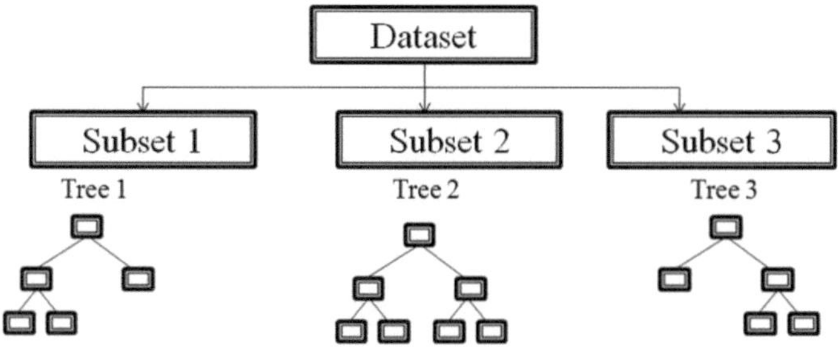

Fig. 2 Structure of the random forest classifier

particular problem. This method tries to construct a set of assumption and combine them to use as in [28, 29]. The formulation of this problem is explained as follows.

Let θ_m in is a random vector which is free from earlier random vectors. The tree is developed by the training data and it generates a classifier $h(y, \theta_m)$, here y is the input vector. The vote for the most accepted class happens when a big amount of tree is generated. This classifier consists of a group of treelike classifiers $\{h(y, \theta_m), m = 1, \ldots.\}$ where $\{\theta_m\}$ is the autonomous identically circulated arbitrary vectors and every tree radiates a vote for the most accepted class on input y.

A group of classifiers $h1$(y), $h2$(y), ..., $hM(y)$ is given and the training data drawn at random from the circulation of the random vector X, Y. The margin function can be defined as:

$$mrg(Y, X) = avg_m I(h_m(Y) = X) - \max_{k \neq X} avg_m I(h_m(Y) = k) \quad (1)$$

Here $I(.)$ is the indicator function. The margin evaluates the degree to which the standard amount of votes at Y, X for the proper class exceed the average vote for other class. Generalization error can be evaluated as:

$$PE^* = P_{Y,X}(mrg(Y, X) < 0) \quad (2)$$

Here the subscripts Y, X points out that the possibility is more than the Y, X space. The structure of the RF model is presented in Fig. 2.

3 Result and Discussion

For estimating the execution of any machine learning method, distinctive methodologies are utilized. The first approach is to separate the entire dataset into two training set and testing set. Both these two sets should be selected separately from each other.

The classifier performance can be measured by calculating sensitivity and specificity. Sensitivity is applied to specify the classifier performance for recognizing the positive samples and it can be defined by

$$\text{Sensit.} = \frac{TrP}{TrP + FlN} * 100 \tag{3}$$

TrP is the amount of true positive samples and *FlN* is the amount of false negative samples in the data. Sensitivity characterizes the number of patients having a cardiac problem. Specificity is used to calculate the classifier performance for recognizing the patients without having cardiac disease and it is calculated by

$$\text{Spec.} = \frac{TrN}{TrN + FlP} * 100 \tag{4}$$

Here *TrN* is the number of true negative samples and *FlP* is the number of false positive samples in the data set. The overall accurateness of the model is calculated by

$$\text{Acc.} = \frac{TrP + TrN}{N} * 100 \tag{5}$$

Here in our work random forest classifier is planned to classify heart disease from the clinical data collected from different people. The original data it is divided into training and testing set. 223 data is in the training and 80 data in the test set. The output is represented as 0 and 1 where 0 represents the healthy patients and 1 represents the affected patients. Then the random forest model is built with the training set with 500 tree size. After successfully building the model it is trained with the training data. Confusion matrix of the training performance is shown in Tables 2 and 3 shows the overall training performance of the model.

Table 2 Confusion matrix of the training set

Reference		
Predicted	0	1
0	123	0
1	0	100

Table 3 Training result

TrP	100
TrN	123
FlP	0
FlN	0
Sensitivity	100%
Specificity	100%

Table 4 Confusion matrix of the test set

Reference		
Predicted	0	1
0	57	7
1	0	13

Table 5 Validation result

TrP	13
TrN	57
FlP	7
FlN	0
Sensitivity	100%
Specificity	89%

Table 6 Comparison of the proposed work with earlier work

Author	Result (%)
Wu et.al. [4]	76
Sagkrasoulis and Montana [7]	77
Genuer et.al. [24]	73
Proposed method	**87**

After successfully completion of training of the model its performance is tested using the test set data. The confusion matrix of the testing step is presented in Tables 4 and 5 shows the overall testing result of the model.

From the above table, it is found that the classification accuracy is 87% which is quite good as compare to previous works shown in Table 6. Figure 3 gives the error rate of the Random Forest model for classification of heart disease.

4 Conclusion

Cardiac problem is a widespread disease and is vital to detect accurately. Though numerous researchers have worked on it and different techniques were developed to test the same. In this work, Random Forest classifier is used for the classification purpose and from this, we have achieved around 87% classification result which quite good as compare to other works. Further accuracy can be improved by applying different classifiers and can be modified this technique.

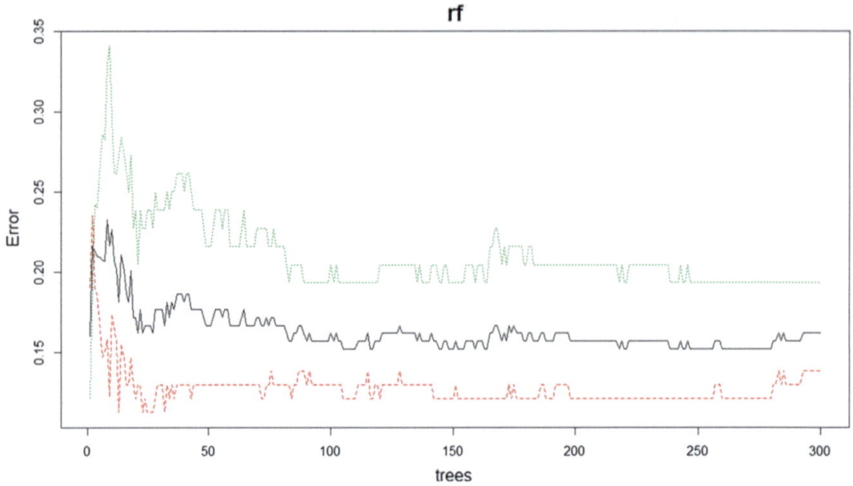

Fig. 3 The error rate of the proposed random forest algorithm

References

1. Hayashi, J., Kunieda, T., Cole, J., Soga, R., Hatanaka, Y., Lu, M., Fujita, H.: A development of computer-aided diagnosis system using fundus images. In: Proceedings Seventh International Conference on Virtual Systems and Multimedia, pp. 429–438. IEEE (2001)
2. Larranaga, P., Calvo, B., Santana, R., Bielza, C., Galdiano, J., Inza, I., Robles, V.: Machine learning in bioinformatics. Brief. Bioinform. **7**(1), 86–112 (2006)
3. Elhaj, F.A., Salim, N., Harris, A.R., Swee, T.T., Ahmed, T.: Arrhythmia recognition and classification using combined linear and nonlinear features of ECG signals. Comput. Methods Progr. Biomed. **127**, 52–63 (2016)
4. Wu, Z., Ding, X., Zhang, G., Xu, X., Wang, X., Tao, Y., Ju, C.: A novel features learning method for ECG arrhythmias using deep belief networks. In: 6th International Conference on Digital Home (ICDH), pp. 192–196. IEEE (2016)
5. Özçift, A.: Random forests ensemble classifier trained with data resampling strategy to improve cardiac arrhythmia diagnosis. Comput. Biol. Med. **41**(5), 265–271 (2011)
6. Mohapatra, S.K., Mohanty, M.N.: Analysis of resampling method for arrhythmia classification using random forest classifier with selected features. In 2018 2nd International Conference on Data Science and Business Analytics (ICDSBA), pp. 495–499. IEEE (2018, September)
7. Sagkrasoulis, D., Montana, G.: Random forest regression for manifold-valued responses. Pattern Recognit. Lett. **101**, 6–13 (2018)
8. Shadmand, S., Mashoufi, B.: A new personalized ECG signal classification algorithm using block-based neural network and particle swarm optimization. Biomed. Signal Process. Control **25**, 12–23 (2016)
9. Mohapatra, S.K., Palo, H.K., Mohanty, M.N.: Detection of arrhythmia using neural network. Ann. Comput. Sci. Inform. Syst. **14**, 97–100 (2017)
10. Ince, T., Kiranyaz, S., Gabbouj, M.: A generic and robust system for automated patient-specific classification of ECG signals. IEEE Trans. Biomed. Eng. **56**(5), 1415–1426 (2009)
11. Gao, D., Madden, M., Chambers, D., Lyons, G.: Bayesian ANN classifier for ECG arrhythmia diagnostic system: a comparison study. In: 2005 Proceedings IEEE International Joint Conference on Neural Networks IJCNN'05, vol. 4, pp. 2383–2388. IEEE, July 2005

12. Shensheng Xu, S., Mak, M.W., Cheung, C.C.: Deep neural networks versus support vector machines for ECG arrhythmia classification. In: 2017 IEEE International Conference on Multimedia and Expo Workshops (ICMEW), pp. 127–132. IEEE, July 2017
13. Guvenir, H.A., Acar, B., Demiroz, G., Cekin, A.: Supervised machine learning algorithm for arrhythmia analysis. In: Computers in Cardiology, pp. 433–436 (1997)
14. Zuo, W.M., Lu, W.G., Wang, K.Q., Zhang, H.: Diagnosis of cardiac arrhythmia using kernel difference weighted KNN classifier. In: Computers in Cardiology, pp. 253–256. IEEE, Sept 2008
15. Uyar, A., Gurgen, F.: Arrhythmia classification using serial fusion of support vector machines and logistic regression. In: 4th IEEE Workshop on Intelligent Data Acquisition and Advanced Computing Systems: Technology and Applications. IDAACS 2007, pp. 560–565. IEEE, Sept 2007
16. Polat, K., Şahan, S., Güneş, S.: A new method to medical diagnosis: Artificial immune recognition system (AIRS) with fuzzy weighted pre-processing and application to ECG arrhythmia. Expert Syst. Appl. **31**(2), 264–269 (2006)
17. Aliferis, C.F., Tsamardinos, I., Statnikov, A.: HITON: a novel Markov Blanket algorithm for optimal variable selection. In: AMIA Annual Symposium Proceedings, vol. 2003, p. 21. American Medical Informatics Association (2003)
18. Pandey, S.K., Janghel, R.R.: ECG arrhythmia classification using artificial neural networks. In: Proceedings of 2nd International Conference on Communication, Computing and Networking, pp. 645–652. Springer, Singapore (2019)
19. Mustaqeem, A., Anwar, S.M., Majid, M., Khan, A.R.: Wrapper method for feature selection to classify cardiac arrhythmia. In: 2017 39th Annual International Conference of the IEEE on Engineering in Medicine and Biology Society (EMBC), pp. 3656–3659. IEEE, July 2017
20. Ozcift, A., Gulten, A.: Classifier ensemble construction with rotation forest to improve medical diagnosis performance of machine learning algorithms. Comput. Methods Progr. Biomed. **104**(3), 443–451 (2011)
21. Masetic, Z., Subasi, A.: Congestive heart failure detection using random forest classifier. Comput. Methods Progr. Biomed. **130**, 54–64 (2011)
22. Joshi, A., Monnier, C., Betke, M., Sclaroff, S.: Comparing random forest approaches to segmenting and classifying gestures. Image Vis. Comput. **58**, 86–95 (2017)
23. Abellán, J., Mantas, C.J., Castellano, J.G., Moral-García, S.: Increasing diversity in random forest learning algorithm via imprecise probabilities. Expert Syst. Appl. **97**, 228–243 (2018)
24. Genuer, R., Poggi, J.M., Tuleau-Malot, C., Villa-Vialaneix, N.: Random forests for big data. Big Data Res. **9**, 28–46 (2017)
25. Khalilia, M., Chakraborty, S., Popescu, M.: Predicting disease risks from highly imbalanced data using random forest. BMC Med. Inform. Decis. Mak. **11**(1), 51 (2011)
26. Emanet, N.: ECG beat classification by using discrete wavelet transform and Random Forest algorithm. In: Fifth International Conference on Soft Computing, Computing with Words and Perceptions in System Analysis, Decision, and Control, pp. 1–4. IEEE, Sept 2009
27. Belgacem, N., Nait-Ali, A., Fournier, R., Bereksi-Reguig, F.: ECG based human authentication using wavelets and random forests. Int. J. Cryptogr. Inform. Secur. (IJCIS) **2**(2), 1–11 (2012)
28. Breiman, L.: Random forests. Mach. Learn. **45**(1), 5–32 (2001)
29. Jones, Z., Linder, F.: Exploratory data analysis using random forests. In: Prepared for the 73rd Annual MPSA Conference, Apr 2015

Detection of Spam in YouTube Comments Using Different Classifiers

Rama Krushna Das, Sweta Shree Dash, Kaberi Das and Manisha Panda

Abstract Nowadays thousands of videos are uploaded to YouTube each minute and people start giving likes and comments immediately. Some popular and viral videos get millions of comments, some are healthy with good complements and some are Spams with unrelated, abusive, sometimes provided with a URL for commercial advertisement or redirection to other sites. In this paper, we have taken YouTube comment datasets of five famous singers and detecting Spam comments using some Artificial Neural Network based Classifiers and some Normal Classifiers. The proposed technique compares the derived results of the classifiers and suggests the best classifiers for detecting Spam comments.

Keywords Spam detection · Spam · Ham · WEKA tool · Accuracy · Specificity · Positive predicted value · Negative predicted value · Matthews correlation coefficient

1 Introduction

Spam nowadays is very common in our day-to-day life. Some people ignore Spams, while a few of the new internet users are not aware of it. Detecting Spam and not getting affected by it is the main aim. Spam refers to the sending of some unnecessary messages haphazardly to a large number of users. Spam is of different types, some

R. K. Das (✉)
National Informatics Centre, Berhampur, India
e-mail: ramdash@yahoo.com

S. S. Dash · K. Das
Institute of Technical Education and Research, Bhubaneswar, India
e-mail: sweta.soa@gmail.com

K. Das
e-mail: kaberidas@soa.ac.in

M. Panda
Berhampur University, Berhampur, India
e-mail: manishapanda2013sai@gmail.com

B. Pati et al. (eds.), *Advanced Computing and Intelligent Engineering*,
Advances in Intelligent Systems and Computing 1082,
https://doi.org/10.1007/978-981-15-1081-6_17

may be harmful, some may not be harmful. The main types are (a) E-mail frauds, like WannaCry ransomware through mail attachment, (b) Phishing Scams, stealing important information such as User ID, Password, Bank Account details, etc., (c) Virus, small programmes capable of data corruption or system corruption, etc. and (d) Chain Messages, those messages which convince the recipient to forward them to a large number of users with false promises. It is seen that expert hackers try to exploit the errors done by human beings, rather than finding system flaws, as a result, users are the target group in social media threats. In the paper in Ref. [1], the authors focus on two major aspects, i.e. URL and Online Social Network (OSN) for spreading Spams. In some popular YouTube videos, the likes and commands are in millions and billions. It is very difficult to segregate the Spams hidden in the comments manually from the huge data. In this paper, the authors propose to use some Artificial Neural Network (ANN) based Classifiers and some Normal Classifiers to classify the Spams in the comments. The results are compared to find out the best classifier for this purpose.

2 Literature Review

Most of the researchers are working on comment mining. The paper in Ref. [2] shows the spread of Spams through malicious websites and the authors proposed an automatic URL classification system, using statistical methods. In [3], the authors have taken a shot at YouTube comment information, keeping in mind the end goal to scrutinise through the Spam comments. Performance measure of various state-of-art text characterization classifier, including Naive Bayes was compared with Bagging (a group classifier). It has been clearly observed that collective classifier gives a better outcome in the majority of the cases. In [4], the authors proposed an alteration that redresses for the discourse inclination while applying tf-idf to remarks on online posts and uploads. Anyway, the fundamental commitment of this paper is in uncovering the potential abuse of tf-idf when connected to the dependent text. In [5], authors gave a top to bottom examination of user remarks in two noticeable social Web destinations, to be specific YouTube and Yahoo! News, going for accomplishing a superior comprehension of network criticism on the social Web. They concentrated on understanding the clients and substance in online networks, and in addition talking about suggestions for outlining collective frameworks that advance and empower support and improve the general user experience. In [6], proposed a Spam identification system in view of continuous subsequence mining. The system supplements current methods for Spam recognition including those in view of machine learning. They have adjusted the PRISM classifier to get another better version of that classifier called mcPRISM, which can mine incessant groupings from Spam comments productively. In [7], the authors took the Amazon dataset, matched the posted rating with the ascertained evaluations of each survey, by delivering a supposition score with the assistance of an in-house word reference. At last, they graphically investigated and demonstrated the diverse highlights of the item which adds to its notoriety or downgrade.

3 Research Methodology

In this research paper, the methodology part provides an overview of the research related works. We found our data sets from UCI machine learning repository [8]. We collected the raw dataset of YouTube Spam Collection, which consists of five different datasets of most popular Actors, such as Psy, Ketty Perry, Eminem, LMFAO, Shakira and their top video, which were in the 10 most viewed list at the collection time [9]. The datasets consisted of comments. We are using WEKA [10] data mining tool which accepts only clean data for further evaluation. Here, we have used R-Programming code to clean the extra spaces, special characters, stemming of words, punctuations, etc. We then found some more special characters which we doubt to be emoji. We could not clean those emoji's using codes, so we cleaned them manually. Then WEKA data-mining tool was able to evaluate the datasets. In the data sets the comments were classified into two classes, i.e. Spam and Ham depending on their types.

4 Experimental Setup

The experimental setup was carried out using a Personal Computer with AMD A8-7410 APU with AMD Radeon R5 Graphics 2.20 GHz Processor, 4 GB RAM, Windows 10 (64 bit) Operating System and WEKA [10] data mining tool. We have collected the YouTube Spam detection dataset from the UCI machine learning repository. Each dataset has five attributes, namely COMMENTID, AUTHOR, DATE, CONTENT, and CLASS. Each class is classified into Spam and Ham, i.e. 1 and 0, respectively.

4.1 Pre-processing of Data

We cleaned the data keeping in mind different parameters such as stop word removal, special character removal, stemming, and removal of numbers from the comment attribute so that the data can be compatible with WEKA data mining tool. We used R programming for stop word removal, special character removal, stemming, and removal of numbers from the comment attribute. Then we found occurrences of some more special characters such as (Ã, ðŸ, Â, my friend Sam loves this song ŠðŸŠðŸŠðŸŠðŸŠðŸŠðŸŠï) in our comments. We cleaned these characters manually.

4.2 *Use of Classifiers*

We selected randomly some ANN Classifiers, like HierarchalLvq, *MultipassLvq, MultipassSom, Olvq1, Olvq3, and Som*. We choose randomly some Normal classifiers, like K-Star, DecisionTable, RandomForest, RandomTree, J48, and JRip. We used WEKA data mining tool to compute the results of the datasets using each classifier.

4.3 *Computation of Results*

We computed the results keeping in mind some parameters such as Accuracy, Specificity, Positive Predictive Value (PPV), Negative Predictive Value (NPV), and Matthews Correlation Coefficient (MCC).

4.3.1 Accuracy

The accuracy of a measurement is how close it is to the true value. The formulae to calculate Accuracy is given below:

$$\text{Accuracy} = \frac{TP + TN}{TP + TN + FP + FN}$$

4.3.2 Specificity

Specificity measures the proportion of actual negatives that are correctly identified. The Formulae to calculate Specificity is given below:

$$\text{Specificity} = \frac{TN}{TN + FP}$$

4.3.3 Positive Predicted Value

Positive Predicted Value (PPV) can be defined as the number of true positives by the sum of positive predicted values of the matrix. PPV is used to indicate the probability that in case of a positive test, the result is positive. The formulae to calculate PPV is given below:

$$\text{PPV} = \frac{\text{TP}}{\text{TP} + \text{FP}}$$

4.3.4 Negative Predicted Value

Negative Predicted Value (NPV) [11] can be defined as the number of true negatives by the sum of the predicted negative values of the matrix. NPV is used to indicate the probability that in case of a negative test, the result is negative. The formulae to calculate NPV is given below:

$$\text{NPV} = \frac{\text{TN}}{\text{TN} + \text{FN}}$$

4.3.5 Matthews Correlation Coefficient

Matthews Correlation Coefficient (MCC) [12] is used in Machine Learning as a measure of the quantity of binary classification. The MCC is, in essence, a correlation coefficient between the observed and predicted binary classifications; it returns a value between -1 and $+1$. A coefficient of $+1$ represents a perfect prediction, 0 no better than random prediction and -1 indicates total disagreement between prediction and observation. While there is no perfect way of describing the confusion matrix of true and false positives and negatives by a single number, the Matthews correlation coefficient is generally regarded as being one of the best such measures. It can be calculated with the formulae given below:

$$\text{MCC} = \frac{(TP * TN) - (FP * FN)}{\sqrt{(TP + FP)(TP + FN)(TN + FP)(TN + FN)}}$$

4.4 Results

4.4.1 Accuracy

Table 1 shows result of accuracy for Artificial Neural Network based algorithms. It shows the highest result in case of HierarchalLvq in LMFAO dataset and lowest result in case of MultipassSom in Eminem dataset.

Table 2 shows result for accuracy for Normal Classifiers. It shows the highest result in case of K-Star for Shakira dataset and lowest result in case of J48 and JRip for Psy and Ketty Perry Dataset.

4.4.2 Specificity

Table 3 shows Specificity for Artificial Neural Network Classifiers. It shows the highest result in case of HierarchalLvq for Eminem dataset and lowest result in case of Som Algorithm for Eminem dataset.

Table 4 shows specificity for Normal Classifiers. It shows the highest result in case of K-Star, DecisionTable, RandomForest, RandomTree and JRip for Shakira dataset and lowest result in case of DecisionTable, J48, and JRip for Shakira dataset.

Table 1 Result of accuracy for artificial neural network classifiers

Algorithms	Psy	Ketty Perry	Eminem	LMFAO	Shakira
HierarchalLvq	62	58.57142857	81.20805369	92.69406393	85.13514
MultipassLvq	63.71428571	57.71428571	84.56375839	91.09589041	82.43243
MultipassSom	65.51724138	59.8265896	54.17607223	90.16018307	84.38356
Olvq1	63.14285714	58	68.90380313	89.04109589	81.35135
Olvq3	60.85714286	56	77.18120805	88.12785388	80.54054
Som	66.28242075	59.24855491	51.12612613	88.99082569	84.74114

Table 2 Result of accuracy for normal classifiers

Algorithms	Psy	Ketty Perry	Eminem	LMFAO	Shakira
K-Star	49.71428571	52	57.49441	58.67579909	70.540541
DecisionTable	49.42857143	49.71428571	51.23042506	53.88127854	61.621622
RandomForest	51.14285714	52	56.59955257	56.62100457	68.648649
RandonTree	49.42857143	50.28571429	55.03355705	53.88127854	60.27027
J48	48.57142857	48.57142857	54.58612975	53.88127854	52.972973
JRip	48.57142857	48.57142857	54.58613	53.88127854	55.405405

Table 3 Result of specificity for artificial neural network classifiers

Algorithms	Psy	Ketty Perry	Eminem	LMFAO	Shakira
HierarchalLvq	62.65060241	58.72093023	97.05882353	94.73684211	85.2071
MultipassLvq	64.63414634	60.62992126	100	93.39207048	85.62092
MultipassSom	63.7755102	63.28125	58.22222222	91.37931034	85.71429
Olvq1	63.69047619	60.9375	92.68292683	91.59292035	83.87097
Olvq3	60.85714286	56	77.18120805	88.12785388	80.54054
Som	64.61538462	64.1025641	55.17241379	89.74358974	85.88957

Table 4 Result of specificity for normal classifiers

Algorithms	Psy	Ketty Perry	Eminem	LMFAO	Shakira
K-Star	49.72067039	52	56.22119816	56.62650602	100
DecisionTable	48.57142857	49.71098266	53.08056872	54.08653846	100
RandomForest	51.36986301	52.4137931	55.70776256	55.4245283	100
RandonTree	49.43820225	50.28248588	54.83146067	53.88127854	100
J48	48.57142857	48.57142857	54.58612975	53.88127854	#DIV/0!
JRip	48.57142857	48.57142857	54.58612975	53.88127854	100

4.4.3 Positive Predicted Value

Table 5 shows positive predicted value for Artificial Neural Network Classifiers. It shows the highest result in case of MultipassLvq for Eminem dataset and lowest result in case of Som for Eminem dataset.

Table 6 shows positive predicted value for Normal Classifiers. It shows the highest result in case of K-Star, DecisionTable, RandomForest, RandomTree, J48 and JRip for Shakira dataset and lowest result in case of J48, and JRip for Eminem dataset and RandomTree, J48 and JRip for LMFAO dataset.

Table 5 Result of positive predicted value for artificial neural network classifiers

Algorithms	Psy	Ketty Perry	Eminem	LMFAO	Shakira
HierarchalLvq	64.57142857	59.42857143	97.53694581	94.05940594	87.2449
MultipassLvq	66.85714286	71.42857143	100	92.57425743	88.77551
MultipassSom	59.1954023	72.83236994	53.69458128	90.0990099	88.0829
Olvql	65.14285714	71.42857143	95.56650246	90.59405941	87.2449
Olvq3	58.85714286	57.71428571	96.55172414	89.10891089	84.69388
Som	60.11560694	75.58139535	48.76847291	88.11881188	88.14433

Table 6 Result of positive predicted value for normal classifiers

Algorithms	Psy	Ketty Perry	Eminem	LMFAO	Shakira
K-Star	48.57142857	52	6.403940887	10.89108911	100
DecisionTable	79.42857143	50.28571429	2.463054187	5.445544554	100
RandomForest	59.42857143	60.57142857	4.433497537	6.435643564	100
RandonTree	48.57142857	49.71428571	0.985221675	0	100
J48	48.57142857	48.57142857	0	0	100
JRip	48.57142857	48.57142857	0	0	100

4.4.4 Negative Predicted Value

Table 7 shows the negative predicted value for Artificial Neural Network Classifiers. It shows the highest result in case of Som for Eminem dataset and lowest result in case of Som for Ketty Perry dataset.

Table 8 shows negative predicted value for Normal Classifiers. It shows the highest result in case of J-48, JRip for Eminem dataset and RandomTree, J48 and JRip for LMFAO dataset and lowest result in case of J48 Shakira dataset.

Table 7 Result of negative predicted value for artificial neural network classifiers

Algorithms	Psy	Ketty Perry	Eminem	LMFAO	Shakira
HierarchalLvq	47.92626728	49.26829268	45.45454545	53.20197044	45.71429
MultipassLvq	47.53363229	38.11881188	46.2962963	53.13283208	42.95082
MultipassSom	54.8245614	39.13043478	54.58333333	53.8071066	44.80519
Olvql	48.41628959	38.42364532	37.01298701	53.07692308	43.18937
Olvq3	51.64319249	48.46938776	43.1884058	53.36787565	44.8505
Som	54.7826087	36.58536585	56.3876652	54.12371134	45.01608

Table 8 Result of negative predicted value for normal classifiers

Algorithms	Psy	Ketty Perry	Eminem	LMFAO	Shakira
K-Star	51.14942529	50	94.94163424	91.43968872	24.904215
DecisionTable	19.65317919	49.42528736	97.81659389	95.33898305	14.035088
RandomForest	41.89944134	41.75824176	96.44268775	94.75806452	22.834646
RandonTree	50.86705202	50.56818182	99.18699187	100	12.107623
J48	50	50	100	100	0
JRip	50	50	100	100	4.3902439

4.4.5 Matthews Correlation Coefficient

Table 9 shows Matthews Correlation Coefficient for ANN Classifiers. It shows the highest result in case of HierarchalLvq for LMFAO dataset and lowest result in case of Som for Eminem dataset.

Table 10 shows Matthews Correlation Coefficient for Normal Classifiers. It shows highest result in case of K-Star for Shakira dataset and lowest result in case of DecisionTable for Ketty Perry dataset.

5 Discussion

From the above experiment, we came across different results of the same parameters, such as accuracy, Specificity, Positive Predicted Value, Negative Predicted Value, and Matthews Correlation Coefficient for different classifiers, i.e. Artificial Neural Network and Normal Classifiers for the same datasets, i.e. Psy, Ketty Perry, Eminem, LMFAO, and Shakira. Given below are the graphical representations of all the parameters used.

Table 9 Result of Matthews correlation coefficient for artificial neural network classifiers

Algorithms	Psy	Ketty Perry	Eminem	LMFAO	Shakira
HierarchalLvq	0.240318019	0.171453766	0.668280784	0.853987288	0.701425
MultipassLvq	0.274829179	0.160438825	0.731623553	0.822109866	0.649282
MultipassSom	0.312855598	0.203538086	0.082500214	0.80235751	0.686806
Olvql	0.263067681	0.166102649	0.471447249	0.781095532	0.626775
Olvq3	0.21731678	0.120070593	0.601848283	0.762309241	0.609215
Som	0.327819384	0.19747751	0.018755099	0.778624016	0.693904

Table 10 Result of Matthews correlation coefficient for normal classifiers

Algorithms	Psy	Ketty Perry	Eminem	LMFAO	Shakira
K-Star	−0.005715779	0.04	0.18974647	0.233927292	0.4899595
DecisionTable	−0.014285714	−0.005714659	− 0.124236491	0.017905313	0.3265653
RandomForest	0.023177601	0.040601035	0.157156046	0.170371738	0.4576043
RandomTree	−0.011430251	0.005714659	0.073499077	#DIV/0!	0.297775
J48	−0.028571429	−0.028571429	#DIV/0!	#DIV/0!	#DIV/0!
JRip	−0.028571429	−0.028571429	#DIV/0!	#DIV/0!	0.1675796

5.1 Accuracy

From Fig. 1, we can analyse that the result of Accuracy is better in case of ANN Classifiers than the Normal Classifiers. In ANN Classifiers, HierarchalLvq gives the best result of 92.69406% in LMFAO dataset. In case of Normal Classifiers, K-Star gives the best result of 70.54054% in Shakira dataset. So we can conclude that Artificial Neural Network Classifiers gives a better result than Normal Classifiers in case of Accuracy.

5.2 Specificity

From Fig. 2, we can analyse that the results of Specificity are better in case of ANN Classifiers than the Normal Classifiers. In Artificial Neural Network Classifiers MultipassLvq gives the best result of 100% in Eminem datasets, and almost all the

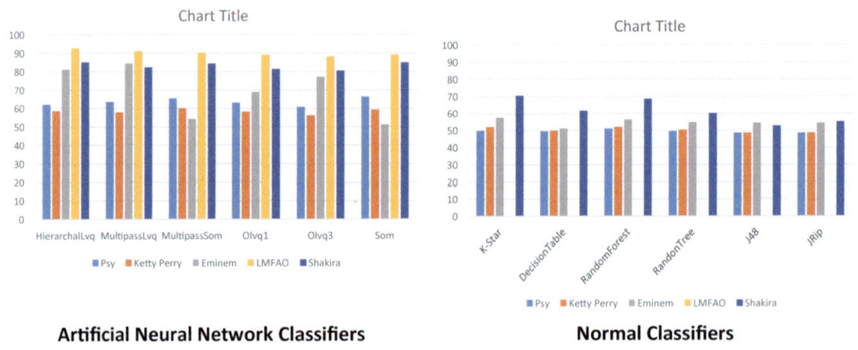

Fig. 1 Classification of accuracy for **a** Psy, **b** Ketty Perry, **c** Eminem, **d** LMFAO, **e** Shakira

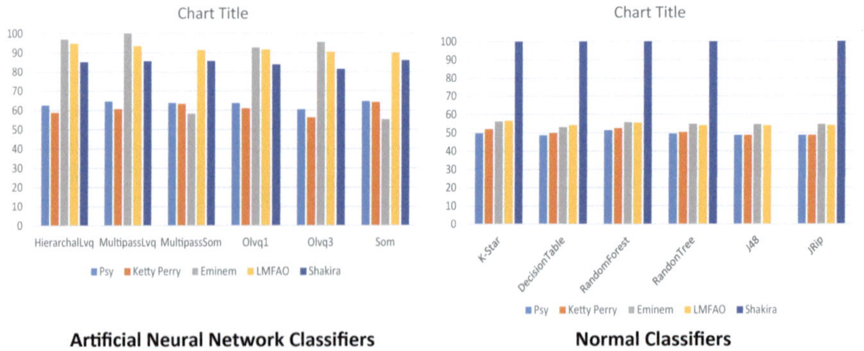

Fig. 2 Classification of specificity for **a** Psy, **b** Ketty Perry, **c** Eminem, **d** LMFAO, **e** Shakira

other classifiers have shown more than 50–90% result in case of ANN Classifiers. In case of Normal Classifiers, K-Star, DecisionTable, RandomForest, RandomTree, and JRip shows 100% result in case of Shakira dataset only. For all the other normal classifiers and datasets it shows 46–56% result only. So we can conclude that Artificial Neural Network Classifiers gives a better result than Normal Classifiers in case of Specificity.

5.3 *Positive Predicted Value*

From Fig. 3, we can analyse that the results of Positive Predicted Value are better in case of Artificial Neural Network Classifiers than the Normal Classifiers. In the case of Artificial Neural Network Classifiers MultipassLvq gives 100% result in Eminem dataset, and almost all the other classifiers give above 60% result. But in the case of Normal Classifiers, Shakira dataset gives 100% result in all the classifiers, but the result of other datasets varies from 0 to 80%. So we can conclude that Artificial Neural Network Classifiers gives better result than Normal Classifiers in the case of Positive Predicted Value.

5.4 *Negative Predicted Value*

From Fig. 4, we can analyse that the result of Negative Predicted Value is better in the case of Normal Classifiers than the Artificial Neural Network Classifiers. In case of Artificial Neural Network classifiers, the result is less than 60%, in the case of all the classifiers and datasets. But in case of Normal Classifiers RandomTree, J48, and JRip give 100% results for Eminem and LMFAO, and other datasets also give result

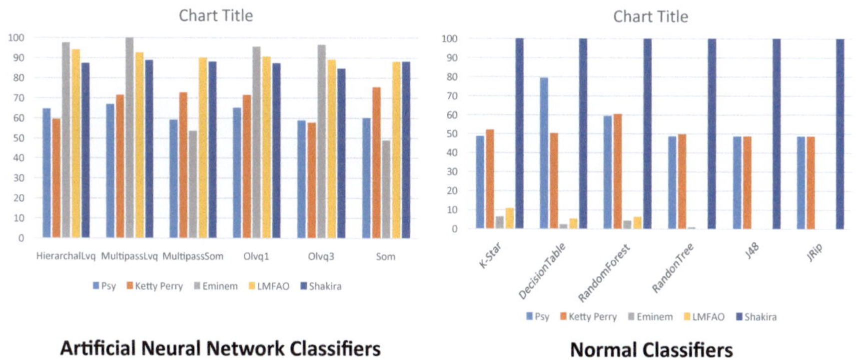

Fig. 3 Classification of positive predicted value for **a** Psy, **b** Ketty Perry, **c** Eminem, **d** LMFAO, **e** Shakira

from 4 to 97%. So we can conclude that Normal Classifiers gives better result than Artificial Neural Network Classifiers in the case of Negative Predicted Value.

5.5 *Matthews Correlation Coefficient*

From Fig. 5, we can analyse that the Artificial Neural Network Classifiers gives better result than Normal Classifiers. All the values lie above 0, and below 1 which gives Random Prediction, and the highest prediction is given by HierarchalLvq for LMFAO dataset, i.e. 0.85. In the case of normal classifiers, no result is above 0.5, i.e. Random Prediction, and also lies below 0, which indicates disagreement. In case of Normal Classifiers, K-Star gives highest prediction for Shakira dataset, i.e. 0.48. So we can conclude that Artificial Neural Network Classifiers gives better result than Normal Classifiers in the case of Matthews Correlation Coefficient.

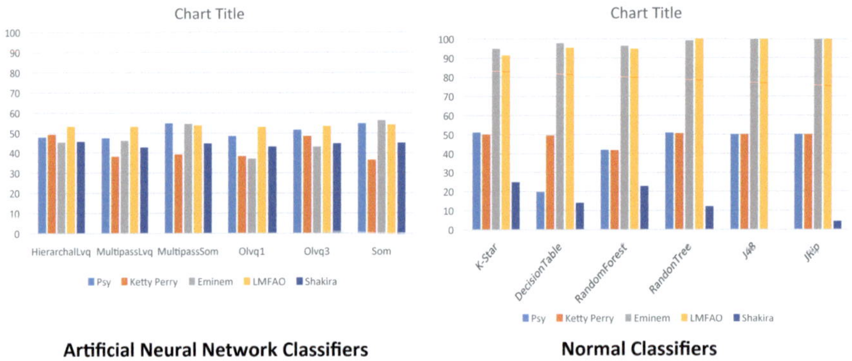

Fig. 4 Classification of negative predicted value for **a** Psy, **b** Ketty Perry, **c** Eminem, **d** LMFAO, **e** Shakira

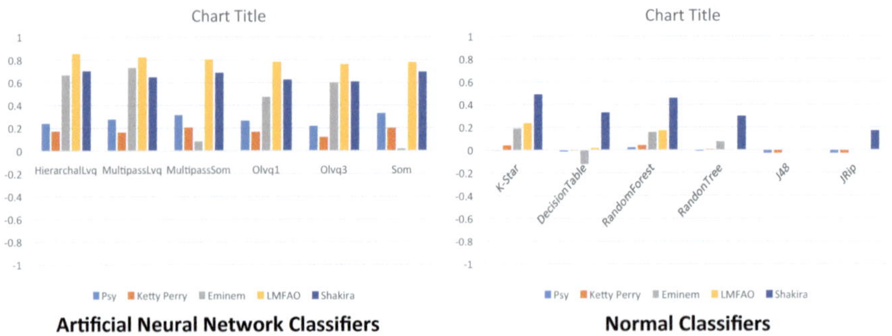

Fig. 5 Classification of Matthews correlation coefficient for **a** Psy, **b** Ketty Perry, **c** Eminem, **d** LMFAO, **e** Shakira

6 Conclusion and Future Work

Spams are major threats these days. In this paper, we have tried to differentiate between Spam and Ham comments depending on their types. From the above Results and Discussions, we can conclude that Artificial Neural Network Classifiers are better than Normal Classifiers because their Accuracy is high as compared to Normal Classifiers. Their Specificity is also high as compared to Normal Classifiers. Their Positive Predicted Value is also high as compared to Normal Classifiers. Their Negative Predicted Value is Low as compared to Normal Classifiers, and their Matthews Correlation Coefficient is also high as compared to Normal Classifiers. From the values, we can conclude that HierarchalLvq is the best Artificial Neural Network Classifier because it has the highest accuracy as well as the highest Matthews Correlation Coefficient for LMFAO dataset. Thus, we can conclude that Artificial Neural Network Classifiers give better results in case of Spam detection. Further, we will try to implement Text Mining and classify them as Spam and Ham. We will further add another class known as Abtutional (Abusive+Unconstitutional) to the dataset. This class will mainly consist of the unconstitutional words and abusive languages which are used in the comment section.

References

1. Alghamdi, B., Watson, J., Xu, Y.: Toward detecting malicious links in online social networks through user behavior. In: 2016 IEEE/WIC/ACM International Conference on Web Intelligence Workshops
2. Ma, J., Saul, L.K., Savage, S., Voelker, G.M.: Beyond blacklists: learning to detect malicious web sites from suspicious URLs. In: KDD'09, June 28–July 1, 2009, Paris, France. Copyright 2009 ACM
3. Zaman, Z., Sharmin, S.: Spam detection in social media employing machine learning tool for text mining. In: 2017 13th International Conference on Signal-Image Technology and Internet-Based Systems, pp. 137–142. https://doi.org/10.1109/sitis.2017. IEEE (2017)
4. Yahav, I., Shehory, O., Schwartz, D.: Comments mining with TF-IDF: the inherent bias and its removal. J. Latex Class Files **14**(8). IEEE (2018)
5. Siersdorfer, S., Chelaru, S., Pedro, J.S., Altingovde, I.S., Nejdl, W.: Analyzing and mining comments and comment ratings on the social web. In: ACM Trans. Web 8, 3, Article 17 (June 2014), 39 pages. http://dx.doi.org/10.1145/2628441
6. Kant, R., Sengamedu, S.H., Kumar, K.: Comment spam detection by sequence mining. In: WSDM'12, February 8–12, 2012, Seattle, Washington, USA, pp 183–192. 2012 ACM 978-1-4503-0747-5/12/02
7. Chauhan, S.K., Goel, A., Goel P., Chauhan, A., Gurve, M.K.: Research on product review analysis and spam review detection. In: 2017 4th International Conference on Signal Processing and Integrated Networks (SPIN)
8. https://archive.ics.uci.edu/ml/datasets/YouTube+Spam+Collection. Accessed 02 Feb 2018
9. Weka.: Data Mining Software in Java. http://www.cs.waikato.ac.nz/~ml/weka/. Accessed 02 Feb 2018
10. https://en.wikipedia.org/wiki/Positive_and_negative_predictive_values. Accessed 02 Feb 2018

11. The Matthews correlation coefficient hosted at https://en.wikipedia.org/wiki/Matthews_correlation_coefficient. Accessed 02 Feb 2018
12. https://archive.ics.uci.edu/ml/index.php. Accessed 02 Feb 2018

Deep Learning Architectures for Named Entity Recognition: A Survey

Anu Thomas and S. Sangeetha

Abstract Extracting named entities from natural language text is an important task in natural language processing, with applications in sentiment analysis, information retrieval, and answer selection in question answering. For identifying named entities, many methods have been developed ranging from knowledge-based methods to supervised machine learning methods. In recent years, deep learning models have achieved cutting-edge results in language processing tasks, particularly in Named Entity Recognition (NER). NER aims to locate and categorize proper names in natural language text into predefined classes such as people, organization, location names, etc. In this paper, we provide a comprehensive survey on existing deep learning architectures used in the CoNLL-2003 NER task. We first introduce the basic deep learning models employed in the NER task followed by reviewing the prominent deep learning NER architectures. Furthermore, the present study throws light upon top factors impacting NER performance which includes the design choices of deep-learning-based NER architecture, and the significance of incorporating character-level information, additional lexical features, external training data, etc. while training the NER model.

Keywords Information extraction · Named entity recognition · Deep learning models · Deep active learning · CoNLL 2003 shared task

---

A. Thomas (✉) · S. Sangeetha
Department of Computer Applications, National Institute of Technology, Tiruchirappalli, Tamil Nadu, India
e-mail: anugraha707@gmail.com

S. Sangeetha
e-mail: sangeetha@nitt.edu

B. Pati et al. (eds.), *Advanced Computing and Intelligent Engineering*, Advances in Intelligent Systems and Computing 1082,
https://doi.org/10.1007/978-981-15-1081-6_18

1 Introduction

In this digital era, a great amount of information that would be used to layman, researchers, and other professionals comes in textual form. The unstructured textual content is rich with information, but finding what is relevant is always a challenging task. This resulted in the emergence of Information Extraction (IE) technique that supports the automatic extraction of relevant information from the natural language text. IE generally deals with the extraction of named entities, relations between named entities, and the events in which the entities are involved. The sub-task of IE that locates and classifies proper names in natural language text into person, location, organization names, etc. is known as Named Entity Recognition (NER). The accuracy of the NER task affects the performance of further IE tasks such as extraction of relations and events [10].

NER plays a major role in a wide variety of domains. Nowadays, news agencies generate huge amounts of online content on a daily basis. Categorizing these contents correctly is very significant to get the most use of each news article. NER can simplify the task of content categorization by automatically scanning the entire articles and then extracting what are the important people, locations, and organizations discussed in them. Categorization of articles based on relevant tags enables smooth content search.

NER enables the efficient handling of customer complaints in an electronic store. The complaints are categorized on the basis of location names and product names mentioned in them and are then dispatched to the relevant department within the organization.

Currently, social media have become a popular platform for information sharing of live updates. Location detection from social media texts has a major role in security applications. Extracting the location information from social media posts could be useful in emergency cases such as fire or traffic accidents. Extracting the user's location based on their entire social media postings can also help since not all users reveal their location in their social media profile. Using the messages annotated with location information (from the users who declared their location) as the training data, a classifier can predict the location of any user. User location can be of interest to security applications also, in order to predict the possible location of a user in case many disturbing messages are posted by him.

Thus, in news, customer care, and social media domains, NER plays a leading part in providing answers to many real-world questions, such as

- Which are the company names mentioned in a news article?
- Which product names are mentioned in complaints or reviews?
- Does a text contain the name of a person? Does the text also provide his current location?

In addition to these, NER has a wide range of applications in tasks, such as Information Retrieval (IR), question answering, sentiment analysis, etc. NER plays a significant role in IR to accurately find the documents related to a particular person or company [15], and in opinion mining to link opinions to particular entities [1]. These practical applications have provided solid motive for research in NER.

There exists numerous techniques for NER, made up of supervised and supervised methods. Under supervised settings, many systems were developed using supervised machine learning classifiers (such as conditional random fields, maximum entropy models, support vector machines, etc.) and taking combination of lexical, syntactic, or semantic features. Unsupervised approaches include techniques that utilize handmade rules and gazetteers.

In recent times, deep learning has revolutionized many application domains, ranging from image processing to Natural Language Processing (NLP). Deep-learning-based solutions to NER task have also turned out to be very common recently and are showing improved performance. Two most common deep learning models for NER are Convolutional Neural Network (CNN) and Recurrent Neural Network (RNN). According to [24], CNN is good at modeling sub-word information and RNN at modeling units (words) in a sequence. However, literature shows that combining CNN and RNN architectures at various levels of NER task yields cutting-edge results.

This aim of this survey is to give answers to the following questions:

- How various deep learning models are combined for an effective NER task?
- What is the importance of character-level information in deep-learning-based NER task?

This work is the first of its kind to cover the current practices in the NER domain by surveying the leading deep neural network architectures for the NER task.

The structure of the paper is as follows: In Sect. 2, the concept of deep learning and the basic deep learning models employed in NER task are discussed; next, in Sect. 3, we review the prominent deep learning NER architectures. Section 4 discusses the key findings of the survey followed by the conclusion.

2 Deep Learning

Deep learning, a refined form of machine learning, uses multilayered artificial neural networks to learn tasks. These neural networks process raw input data through many layers of nonlinear transformations in order to calculate a target output. The layers that are near to the input layer learn simple features, whereas the higher layers learn complex features resulting from the lower layers.

The idea of neural network is derived from the biological brain structure. The information processing units in the neural network called neurons learn to perform various tasks by modifying the weights between two neurons, similar to the working of an original brain. A feedforward neural network consists of an input layer, multiple hidden layers, and an output layer. The neurons in the hidden layers process the output from the preceding layer and deliver output to the following layer.

Deep learning models for NER are mainly categorized into two: CNN [12] and RNN [6]. Even though RNN is capable of handling long-distance dependencies, sometimes, they fail because of the vanishing gradient problems [2, 19]. This led to the development of more sophisticated types of RNN: Long Short-Term Memory (LSTM) [8], Bidirectional long short-term memory (BI-LSTM) [7], and Gated Recurrent Unit (GRU) [4].

2.1 Convolutional Neural Network (CNN)

CNN [12] is a special type of deep, feedforward artificial neural network mainly applied in visual imagery applications. CNN is composed of multiple convolutional layers to mimic the function of cells in the human visual cortex.

CNN consists of an input layer, output layer, and multiple hidden layers. The hidden layers are composed of multiple convolutional layers and pooling layers, followed by fully connected layers as well as normalization layers. The filters in the convolution layer scan each region in the input data and end up in an activation map or feature map. The pooling layer that follows the convolutional layer operates on each feature map independently thus reducing the spatial size of representation. There can be multiple such convolutional and pooling layers in CNN architecture. Finally, CNN ends up with a fully connected layer trailed by a softmax layer with output classes for classification tasks [26]. It is widely accepted that CNN works perfectly for image problems and outperformed most of the other methods in image classification tasks. These convolutional and pooling layers make CNN capable of extracting character level features for NER tasks [3, 14, 21] and are discussed in Sect. 3.

2.2 Recurrent Neural Network (RNN)

RNN [6] is a type of neural network in which the connections between the neurons form a directed cycle and is good at handling sequential data such as text data. In RNN, the hidden state at each time step depends on the value of the prior hidden state and the input vector at present time step. The model predicts the current output based on long-distance features using the connection between the previous hidden state and the present hidden state. This feature of RNN produced promising results in various NLP applications [16, 17].

2.3 Long Short-Term Memory Network (LSTM)

To overcome the limitation of standard RNN caused by the vanishing (exploding) gradient problem, researchers have developed a distinct type of RNN called LSTM [8] that is able to manage long-term dependencies. In LSTM, memory cells replace the hidden layer updates. A memory cell consists of an input gate, a forget gate, a neuron with a self-recurrent connection, and an output gate. The self-recurrent connection ensures the state of a memory cell to be constant from one time step to another. The input gate decides the influence of the incoming signal on the state of the memory cell, whereas the output gate resolves the influence of memory cell's state on other neurons. At the end, the forget gate allows each cell to retain or forget its last state, by controlling the memory cell's self-recurrent connection.

2.4 Bidirectional LSTM Networks (BI-LSTM)

The performance of the sequence tagging task can be improved if there is access to both past and future input contexts for a given time. Since the LSTM's hidden state receives information only from the past, an elegant solution is to utilize a bidirectional LSTM network [7]. One LSTM network scans the input sequence toward right while the other runs in reverse order.

2.5 Gated Recurrent Units (GRU)

In Gated Recurrent Unit [4], both "forget" and "input" gates are combined together to form a single update gate. In addition to this, the cell state is merged with the hidden state to make some variations. According to Yang et al. [25], GRUs are good at representing character and word-level information in NER tasks.

3 Deep Learning Architectures for NER

Under deep learning setting, NER is solved as a sequence tagging problem. In sequence tagging, each token (element) in the input sequence (sentence) is predicted with a label considering the labels of the neighboring elements. In NER, the goal is to predict the named entity tag for each element in the input sentence. Furthermore, NER falls under the category of multi-class classification problems where each token is classified into one among multiple tags. Deep learning models such as CNNs, RNNs, or its variation like LSTMs are promising choices for solving multi-class classification problems where CNN is capable of extracting sub-word features and RNN at modeling elements in the sequence.

In NER task, neural networks in deep learning provide a hierarchical architecture, in which the lower layers discover sub-word features, the middle layers represent word-specific features, and the higher layer uses information coming from the lower layers to identify named entities.

Table 1 CoNLL 2003 dataset statistics

	#Sent	#Token	#PER	#LOC	#ORG
Train	14987	204567	6600	7140	6321
Dev	3466	51578	1641	1434	1154
Test	3684	46666	1617	1668	1661

Below, we review various deep learning architectures that showed cutting-edge performance in the Conference on Computational Natural Language Learning (CoNLL)-2003 NER task dataset [23]. This dataset contains four different kinds of named entities: PERSON, ORGANIZATION, LOCATION, and MISC. Table 1 shows the corpora statistics, where Sent, Token, PER, LOC, and ORG denote the total number of sentences, tokens, person, location, and organization names, respectively.

Below is the overview of the state-of-the-art neural network architectures for NER.

The use of Deep Neural Networks (DNN) for NER was pioneered by CNN-CRF model [5] in which CNN is applied to capture word-level information from the input word embeddings. Then, a Conditional Random Field (CRF) layer is attached on its top, as tag decoder(). By applying a fixed-sized contextual window, the model achieved 89.59% F1 score in NER task using both word embeddings [5] and external training data.

Huang et al. [9] proposed a range of RNN-based models for NER task. These models include LSTM network, BI-LSTM network, LSTM network connected to a CRF layer (LSTM-CRF), and BI-LSTM network linked to a CRF layer (BI-LSTM-CRF). In BI-LSTM-CRF, word embeddings and additional word features (spelling and context features) were fed into the BI-LSTM network, for producing the word-level representation. This word-level representation was then passed to the output CRF layer to predict output tags. The BI-LSTM-CRF model is less dependent on word embeddings as compared to CNN-CRF model [5] and had achieved an F1 score of 90.10% using both Senna embeddings (the same as in [5]) and gazetteer features.

Chiu and Nichols [3] successfully employed CNNs to extract character features for each word from character embedding and character-type features. For each word, these character vectors are concatenated with the word embeddings and additional word features. Then, this concatenated input is supplied to multiple layers of LSTM units that are connected to each other in sequence. At each time stamp, both linear layer and log-softmax layer individually calculate the log probabilities (vectors) for each tag category by decoding the output of each forward and backward layers. Finally, these two vectors are summed to produce the final output. Apart from these, a list of known named entities from DBpedia is also used as an external knowledge source. The model used publicly available word embeddings trained by Collobert et al. [5]. The model trained on both training and validation datasets of CoNLL-2003 dataset [23] achieved an F1 score of 91.62 ± 0.33.

Lample et al. [11] proposed an architecture similar to [3], but instead used BI-LSTM layer to extract the character-level features. The character embeddings for each character in a word are passed to a forward and backward LSTM in direct and reverse order. The character-level information is concatenated with pretrained word embeddings, created using skip-n-gram [13]. The fine-tuning of word embeddings during the training phase resulted in 90.94% F1 score.

Ma and Hovy [14] proposed a neural network architecture consisting of BI-LSTM, CNN, and CRF that could automatically extract word as well as character-level features by using GloVe 100-dimensional word embeddings [20]. The model achieved a leading performance of 91.21% F1 score without using any character-type features as in [3].

Yang et al. [25] proposed a hierarchical NER model by employing deep GRUs on both character and word levels, followed by a CRF layer to make the output tag. The GRUs captured the morphological information at the character-level and n-gram patterns, semantics on the word level using the word embeddings trained by [5]. The fine-tuning of word embeddings during the training phase, enabled the model to achieve 91.20% F1 score.

Strubell et al. [22] utilized an Iterated Dilated Convolutional Neural Network (ID-CNN) and CRF hybrid model for NER. In comparison to traditional CNNs, ID-CNNs are good at large context and structured prediction. The existing forefront models [3, 11, 14] fail fully to make use of the parallelism opportunities of a GPU. While processing the entire document, ID-CNNs allows the fixed-length convolutions to run in parallel across the input, and this leads to a faster sequence tagging process. The model achieved a better F1 score of 90.54 ± 0.18 using ID-CNN and CRF as token encoder and tag decoder, respectively.

Shen et al. [21] introduced deep active learning algorithms for the NER task by proposing a lightweight architecture "CNN-CNN-LSTM model", consisting of convolutional character and word encoders and an LSTM tag decoder. In this work, they demonstrated that the amount of labeled data can be drastically reduced with deep active learning. The model is initialized with latent word embeddings with word2vec training [18] and these embeddings were fine-tuned during training. Compared to the BI-LSTM-CNN-CRF model [3], CNN-CNN-LSTM provided significant improvement in training time and achieved comparatively high performance with an F1 score of 90.69 ± 0.19. The model showed cutting-edge performance in NER using just 25% of the original training data.

In the above section, we reviewed the leading-edge works on neural-network-based NER. The next section discusses the key findings related to the design choices of DNN-based NER architecture as well as the factors influencing its performance.

4 Discussion

Based on the above survey, we arrived at some findings regarding the design choices of DNN-based NER architecture and the top factors influencing its performance.

Design Choice

1. Character-level encoder: for extracting character-level features corresponding to each word in the input.
2. Word-level encoder: extracts context features for every single word in the input.
3. Tag decoder: induces a probability distribution over any sequences of output tags.

Based on the above decomposition, Table 2 shows the design choices of the state-of-the-art DNN-based NER architecture. From Table 2, we infer that for the character-based representations, the CNN approach by [14] and the BI-LSTM approach by [11] performed on par. Between these two, the CNN approach should be preferred due to higher computational efficiency. The models with LSTM or GRU word-level encodings gave a slight (but not significant) boost over CNN word-level encoders in terms of F1 score. However, according to [21], CNN word-level encoders are considerably faster than RNN variants, which is critical for iterative training in the active learning scheme. At the tag decoder level, LSTM showed a better performance than CRF, when used with CNN encoders.

Key factors

As shown in Table 3, the type of word embeddings used, use of additional lexical features, external training data, etc. also have an impact on the NER performance. The selection of the pretrained word embeddings has a large effect on the total performance of the system. According to [3], the embeddings created from an in-domain text perform better compared to the published embeddings, like Standford's GloVe embeddings and Google's word2vec embeddings.

Table 2 Design choices of the DNN-based NER architecture

Work	Character-level encoder	Word-level encoder	Tag decoder	F1 score
[5]	None	CNN	CRF	89.59
[9]	None	BI-LSTM	CRF	90.10
[3]	CNN	BI-LSTM	Softmax	91.62 ± 0.33
[11]	BI-LSTM	BI-LSTM	CRF	90.94
[14]	CNN	BI-LSTM	CRF	91.21
[25]	GRU	GRU	CRF	91.20
[22]	None	Dilated CNN	CRF	90.54 ± 0.18
[21]	CNN	CNN	LSTM	90.69 ± 0.19

Table 3 Key factors influencing the DNN-based NER task

Work	Word embeddings	Use of additional lexical features	Use of external training data	Tuning of word embeddings
[5]	Senna	No	Yes	No
[9]	Senna	Yes	No	No
[3]	Senna	No	Yes	No
[11]	Word2vec	No	No	Yes
[14]	Glove	No	No	No
[25]	Senna	No	No	Yes
[22]	None	No	No	No
[21]	Word2vec	No	No	No

Importance of character-level information

Literature shows that all models except [5] used character-level information (such as prefix or suffix of a word) either by employing heavy feature engineering [22] or by a character-encoder [3, 11, 14, 21, 25]. Based on the performance of the above-mentioned models, we conclude that character-level information plays a significant role in the NER task.

Role of deep active learning

Scarcity of labeled data for NER tasks renders many approaches like deep learning unusable. When deep learning is combined with active learning the amount of training data can be drastically reduced, still showing better performance [21]. They carried out incremental training with each batch of new labels, i.e., before querying for labels in a new round; they mixed newly annotated samples with the older ones and updated the network weights for a trivial number of epochs. This incremental active learning significantly reduced the computational requirements of active learning methods.

5 Conclusion

Applying deep learning techniques to named entity recognition task has turned out to be a common research area recently. This paper surveys the various frontline deep-learning-based NER architectures that have achieved cutting-edge results on CoNLL-2003 NER dataset. The survey shows that for representing character and word-level features, the CNN models are preferred to RNN variants in terms of computational efficiency. At the tag decoder level, the choice between LSTM and CRF directly depends on the model used at the word-encoding level. Furthermore, the word embeddings trained on in-domain text as well as the character-level information also have an impact on the performance of the system. Apart from these factors, internal parameters of the network called hyperparameters such as the optimizer, drop-out mechanism also influence cutting-edge performance. Another interesting finding is that the introduction of active learning in deep neural network architectures eliminated the need for large-sized training datasets. In future, there will be more prolific research in deep-learning-based NER with the advancements in deep learning techniques.

References

1. Batra, S., Rao, D.: Entity based sentiment analysis on twitter. Science **9**(4), 1–12 (2010)
2. Bengio, Y., Simard, P., Frasconi, P.: Learning long-term dependencies with gradient descent is difficult. IEEE Trans. Neural Netw. **5**(2), 157–166 (1994)
3. Chiu, J.P., Nichols, E.: Named entity recognition with bidirectional lstm-cnns. Trans. Assoc. Comput. Linguist. **4**, 357–370 (2016)
4. Cho, K., Van Merriënboer, B., Bahdanau, D., Bengio, Y.: On the properties of neural machine translation: Encoder-decoder approaches. arXiv:1409.1259 (2014)
5. Collobert, R., Weston, J., Bottou, L., Karlen, M., Kavukcuoglu, K., Kuksa, P.: Natural language processing (almost) from scratch. J. Mach. Learn. Res. **12**, 2493–2537 (2011)
6. Elman, J.L.: Finding structure in time. Cogn. Sci. **14**(2), 179–211 (1990)
7. Graves, A., Schmidhuber, J.: Framewise phoneme classification with bidirectional lstm and other neural network architectures. Neural Netw. **18**(5–6), 602–610 (2005)
8. Hochreiter, S., Schmidhuber, J.: Long short-term memory. Neural Comput. **9**(8), 1735–1780 (1997)
9. Huang, Z., Xu, W., Yu, K.: Bidirectional lstm-crf models for sequence tagging. arXiv:1508.01991 (2015)
10. Jiang, J.: Information extraction from text. In: Mining text data, pp. 11–41. Springer (2012)
11. Lample, G., Ballesteros, M., Kawakami, K., Subramanian, S., Dyer, C.: Neural architectures for named entity recognition. In: Proceedings NAACL-HLT (2016)
12. LeCun, Y., Boser, B., Denker, J.S., Henderson, D., Howard, R.E., Hubbard, W., Jackel, L.D.: Backpropagation applied to handwritten zip code recognition. Neural Comput. **1**(4), 541–551 (1989)
13. Ling, W., Tsvetkov, Y., Amir, S., Fermandez, R., Dyer, C., Black, A.W., Trancoso, I., Lin, C.C.: Not all contexts are created equal: Better word representations with variable attention. In: Proceedings of the 2015 Conference on Empirical Methods in Natural Language Processing. pp. 1367–1372 (2015)
14. Ma, X., Hovy, E.: End-to-end sequence labeling via bi-directional lstm-cnns-crf. In: Proceedings of the 54th Annual Meeting of the Association for Computational Linguistics (Volume 1: Long Papers). pp. 1064–1074 (2016)
15. Mihalcea, R., Moldovan, D.: Document indexing using named entities. Stud. Inf. Control **10**(1), 21–28 (2001)
16. Mikolov, T., Deoras, A., Povey, D., Burget, L., Černockỳ, J.: Strategies for training large scale neural network language models. In: 2011 IEEE Workshop on Automatic Speech Recognition and Understanding (ASRU), pp. 196–201 (2011)
17. Mikolov, T., Karafiát, M., Burget, L., Černockỳ, J., Khudanpur, S.: Recurrent neural network based language model. In: Eleventh Annual Conference of the International Speech Communication Association (2010)
18. Mikolov, T., Sutskever, I., Chen, K., Corrado, G.S., Dean, J.: Distributed representations of words and phrases and their compositionality. In: Advances in neural information processing systems. pp. 3111–3119 (2013)
19. Pascanu, R., Mikolov, T., Bengio, Y.: On the difficulty of training recurrent neural networks. In: International Conference on Machine Learning. pp. 1310–1318 (2013)
20. Pennington, J., Socher, R., Manning, C.: Glove: Global vectors for word representation. In: Proceedings of the 2014 Conference on Empirical Methods in Natural Language Processing (EMNLP). pp. 1532–1543 (2014)

21. Shen, Y., Yun, H., Lipton, Z.C., Kronrod, Y., Anandkumar, A.: Deep active learning for named entity recognition. arXiv:1707.05928 (2017)
22. Strubell, E., Verga, P., Belanger, D., McCallum, A.: Fast and accurate sequence labeling with iterated dilated convolutions. arXiv:1702.02098 **1** (2017)
23. Tjong Kim Sang, E.F., De Meulder, F.: Introduction to the conll-2003 shared task: Language-independent named entity recognition. In: Proceedings of the Seventh Conference on Natural Language Learning at HLT-NAACL 2003, vol. 4. pp. 142–147 (2003)
24. Wenpeng, Y., Katharina, K., Mo, Y., Hinrich, S.: Comparative study of cnn and rnn for natural language processing. arXiv: 1702.01923 (2017)
25. Yang, Z., Salakhutdinov, R., Cohen, W.: Multi-task cross-lingual sequence tagging from scratch. arXiv:1603.06270 (2016)
26. Zhang, L., Wang, S., Liu, B.: Deep learning for sentiment analysis: A survey. Wiley Interdisciplinary Reviews: Data Mining and Knowledge Discovery p. e1253 (2018)

Effect of Familiarity on Recognition of Pleasant and Unpleasant Emotional States Induced by Hindi Music Videos

Syed Naser Daimi, Soumil Jain and Goutam Saha

Abstract Valence is an important dimension representing the hedonic value of emotion labeled as positive or negative. Inducing these emotions and making an understanding from brain responses have huge practical significance. Music is a powerful tool to induce emotions, and the induced emotions are generally getting influenced by factors external to music such as familiarity. This work presents a novel study on the effect of familiarity on recognition of pleasant (positive) and unpleasant (negative) emotional states induced by Hindi music videos. For this, we recorded 32-channel EEG from six healthy subjects while they watched Hindi music videos and self-reported ratings of felt emotions on valence and familiarity scale. We used a machine learning framework for emotion classification from power spectral and functional connectivity features. The framework consists of SVD-QRcp and F-ratio based feature selection and an SVM classifier. The classification was performed under three cases of familiarity, namely, familiar, unfamiliar, and regardless of familiarity of the music videos. We found that for the familiar case, the classification performance was higher than unfamiliar and regardless of familiarity cases for all considered features. The best performing features were from the individual electrodes and these features were from the frontal and left parietal regions which indicate the lateralized processing of valence. In addition to classification, we analyzed the feature and electrode usage for all the cases of familiarity. It was found that the features from theta, alpha, and gamma band covering the frontal and parietal brain regions were dominantly involved.

Keywords Familiarity · Valence · Power spectral density · Functional connectivity · Support vector machine

S. N. Daimi (✉) · S. Jain · G. Saha
Department of Electronics and Electrical Communication Engineering,
Indian Institute of Technology Kharagpur, Kharagpur 721302, India
e-mail: sndaimi123@gmail.com

S. Jain
e-mail: jainsoumil2@gmail.com

G. Saha
e-mail: gsaha@ece.iitkgp.ernet.in

B. Pati et al. (eds.), *Advanced Computing and Intelligent Engineering*,
Advances in Intelligent Systems and Computing 1082,
https://doi.org/10.1007/978-981-15-1081-6_19

1 Introduction

Music is universally enjoyed, known to ease stress, used to express and modify emotions, and also in coping with emotions [1]. Investigating and analyzing the human emotions from brain signals is necessary for Human-computer interaction or brain-computer interface research [2]. Music induces strong emotions and widely used as stimuli [3]. However, most of the research on emotions primarily focused on listeners' sensitivity to emotions in the music of their own culture, however, the emotions are communicated through a combination of universal and cultural cues [4]. This cues can be internal (acoustic features of the music) and external (cultural background, age, gender, the familiarity of the music, music preference, etc.) to the music which influences the shaping of the induced emotions. For example, listening to familiar music activates the episodic memory which is one of the six mechanisms by which emotions are induced as proposed by [1].

Most of the research investigating the relationship between the external factors to music and emotions focus on familiarity and the majority of these studies are psychological [5–7] and some neuropsychological [8–10]. In general, the preference studies focus on individual's familiarity to music, and it is well established that the familiarity is an influential factor in preference [6]. However, with exposure to the music, that is familiarity, unpleasant music was liked even less, whereas pleasant music was liked even more [11]. Moreover, it was observed in [5] that the repetition of western classical music increases the pleasant subjective ratings specifically, for the music which induces pleasant affect initially tends to evoke increased pleasantness. The reason might be that as the familiarity with the cords grows which successively increases the consonance experience and accordingly the pleasantness [12]. Also, it was observed that the unfamiliar music induces high arousal potentials and with exposure, it decrease [13].

The neuropsychological studies, for example, [8] analyzed fMRI data of 14 subjects and observed that limbic, paralimbic, and reward circuits of the brain were activated for familiar music. The event-related potentials of fronto-central electrode region were found to be correlated with the familiar melodies [14]. Another study [10] found that the familiarity should be considered while selecting the stimuli for emotion recognition. However, all these research usually involves western music and few Indian music [4, 15–17]. Most of the studies that involve Indian music are psychological [4, 15, 16] and very few neuropsychological [17, 18]. None of these studies investigate the effect of familiarity on emotion recognition induced by Indian music. In this work, we investigate the influence of familiarity on emotion recognition induced by Hindi music videos. Here, we considered emotion recognition on valence scale as it is one of the important dimensions of modeling the emotions and, analyzing its brain responses has wide practical implications. For this, we recorded 32-channel EEG of six healthy participants while they watched 4–5 minutes long Hindi music videos. After completion of watching each video, participants have provided the assessment of their felt emotions on valence and familiarity scales. The valence represents the manner one feels a situation and varies from pleasant or positive to unpleasant or negative [19]. On the other hand, the familiarity scale indicated the degree of familiarity with the music.

Next, we performed emotion recognition using the standard machine learning framework that involves preprocessing, feature extraction, feature selection, and classification. In this study, we investigated the emotion recognition performance using conventional spectral features and functional connectivity features, SVD, QRcp and F-ratio based feature selection, and SVM classifier.

2 Methods and Materials

2.1 Dataset

The experimental data were collected from six healthy adult participants (of IIT Kharagpur, India), consisting of three males and three females between 21 and 34 years. The written consent of each participant was obtained before the beginning of the experiment. To induce the emotions, Hindi music videos were selected in four sets representing happy, elated, excited (Set I), relaxed, serene, calm (Set II), sad, depressed, fatigue (Set III), and upset, stressed, tense (Set IV). The videos were selected with the help of four musicians with up to 5 years of experience in Indian music. With the agreement between the musicians, 15, 9, 13 and 20 videos were selected representing the four sets respectively. The Hindi music videos used as stimuli are listed in Appendix A. The following experimental protocol was followed: each participant was instructed to select at least four songs (two familiar and two unfamiliar) from each set. The chosen video was presented after 5 s of screen fixation. The EEG was continuously recorded using 32-channel EnoBio[1] system with 500 Hz sampling frequency. The EEG cap was placed in accordance with the 10–20 international system and reference at the ear-lobe electrode. At the end of each video participants reported their felt emotions on arousal, valence, dominance, liking on continuous 9-point rating scale estimated using the Self-Assessment Manikin (SAM) [20] and familiarity on a discrete scale ranging from 1 ('never heard before the experiment') to 5 ('knew the song very well'). The EEG signals were preprocessed using a bandpass filter (8th order Butterworth IIR) in the frequency range 4–45 Hz, and ICA (Independent Component Analysis) was used to remove eye blink artifacts.

2.2 Feature Extraction

Spectral features The Welch's method [21] was used to estimate the power spectral density (PSD) of the EEG signals. First, the time series x is segmented into N

[1] https://www.neuroelectrics.com/products/enobio/enobio-32/.

segments, and each of them is windowed using a Hamming window. Next, the discrete Fourier transform (DFT) was computed. Finally, the DFT magnitudes were averaged over all segments to get the estimate of PSD as shown below.

$$\hat{S}_{XX}(k) = \frac{1}{N}\sum_{n=1}^{N}|X_n(k)|^2 \tag{1}$$

where $X_n(k)$ is the DFT of the signals x corresponding to the nth segment after windowing at the kth frequency bin. In this study, we computed spectral features as the relative PSD values (see Eq. 2 for each channel in five EEG bands theta (4–8 Hz), slow alpha (8–10 Hz), alpha (8–12 Hz), beta (12–30 Hz), and gamma (30–45 Hz). The resulting feature dimension is $32 * 5 = 160$.

$$\hat{P}_i = \frac{\sum_{k=k_1}^{k_2}\hat{S}_{XX}(k)}{\sum_{k=1}^{N_{FFT}}\hat{S}_{XX}(k)} \quad \text{for} \quad i = 1, 2, \ldots, 5 \tag{2}$$

where $\hat{P}_i$ and $k_1, k_1 + 1, \ldots, k_2$ denote the relative power and the indices corresponding to the ith band and kth sample respectively.

Coherence The coherence is a correlation of a pair of time-series in the frequency domain and provides the information of these signals operating at the same frequency. It is widely used functional connectivity in EEG signals [22–24]. The coherence of two signals $x(t)$ and $y(t)$ is given by the ratio of cross-spectral density and the square root of the product of individual spectral densities [25] as

$$C(f) = \frac{S_{XY}(f)}{\sqrt{S_{XX}(f)S_{YY}(f)}} \tag{3}$$

where $C(f)$ is the coherence function at frequency f, $S_{XY}(f)$, and $S_{XX}(f)$ and $S_{YY}(f)$ are the cross-spectral and auto-spectral density of $x(t)$ and $y(t)$ respectively. Coherence takes values between 0 and 1, where 0 indicates no linear relationship between X and Y while 1 indicates the perfect linear relationship. Here, we used Welch estimates of auto- and cross-PSD to estimate coherence in each of the four frequency bands according to [26] as shown below:

$$C_{XY}(\bar{k}) = \frac{\Sigma_{k=k_1}^{k_2}\hat{S}_{XY}(k)}{\sqrt{\Sigma_{k=k_1}^{k_2}\hat{S}_{XX}(k)\Sigma_{k=k_1}^{k_2}\hat{S}_{YY}(k)}} \tag{4}$$

where $\hat{S}_{XY}(k)$ and $\hat{S}_{XY}(k)$ and $\hat{S}_{XY}(k)$ are the estimate of cross-spectral density and auto-spectral density of $x(t)$ and $y(t)$ respectively. The frequency indices $k_1, k_1 + 1, \ldots, k_2$ correspond to the considered frequency band.

2.3 Feature Selection

The computed features contain different amount of information; few may be redundant while few may be relevant for separating the classes [27]. In this study, we used singular value decomposition (SVD), QR factorization with column pivoting (QRcp) and F-ratio based feature selection method which is successively used earlier in speaker recognition system [28], in cardiac diseases [29] and emotion classification [30]. In this method, the SVD determines the number of features to be selected, while the QRcp and F-ratio ranks them based on the orthogonality and discriminative aspect respectively. The feature selection steps are as follows.

1. We compute SVD, QRcp and F-ratio and determine the number of singular values n for which the cumulative energy was 99% of the total energy.
2. We determine QRcp_Rank and F-ratio_Rank of the features.
3. Subsequently, we combined both the ranks as

$$(\text{Combined Score})_i = \frac{(\text{QRcp}_{\text{Rank}} + F - \text{ratio}_{\text{Rank}})}{2}$$

   If there was a tie between any two features, the feature having better QRcp ranking was given priority in selection.
4. The classification performance was evaluated using the n number of top-ranked features. Further improvement in performance was sought by investigating the $(n - 10)$ to $(n + 10)$ neighborhood to get the best input features.

2.4 Classification

In this study, we used SVM classifier which is a two-class linear classifier that maps the input data as a point in space and determines the decision plane such that the gap between the data points of two classes is as wide as possible. Non-linear classification can be performed with the help of kernel functions. Here, the performance of valence classification was evaluated in a k-fold cross-validation and SVM classifier (linear kernel). In each step of cross-validation, $(k - 1)$ folds were used for feature selection and training classifier, and one-fold for testing. Here, we used $k = 10$.

3 Results and Discussions

3.1 Single Trial Classification

The primary aim of this study was to investigate the influence of familiarity on valence, that represents positive or pleasant and negative or unpleasant emotional states. For this purpose, we grouped the valence trials into three groups, depending upon familiarity ratings, familiar (rating covering 3–5), unfamiliar (1–2), and regardless of familiarity (all songs), further in each group, we separated the valence ratings into two classes - pleasant and unpleasant. The threshold to separate these valence classes was placed at the midpoint on the 9 point scale. So we have three groups (familiar, unfamiliar, and regardless of familiarity) and two classes (pleasant and unpleasant) in each group.

For each EEG channel, we computed the spectral and functional connectivity features on trial of 3 mins duration. Spectral features used were the relative PSD values computed by dividing the power of each frequency band by total energy of that EEG channel. We also computed the hemispheric asymmetry features by subtracting a feature in the right-hemisphere electrode from the symmetric-pair electrode in the left-hemisphere. In this way, we got 160 (all channels) and 70 (hemispheric asymmetry) relative PSD features. The connectivity features using coherence were computed for each pair of EEG electrodes in four frequency bands theta, alpha, beta, and gamma. The total number of connectivity features was $496 \times 4 = 1984$.

Next, we perform a two-class classification between pleasant and unpleasant trials of each group. There was a class imbalance as the felt emotions for a song changes from person-to-person; therefore, we used a simpler method, random downsampling to balance the classes, where the same number of trials as in the minority class were randomly selected from majority class. We evaluated the classification performance in each category using 10-fold cross-validation. The complete procedure was repeated 20 times and mean classification accuracy was computed. The performance was evaluated for both spectral, and coherence features—relative PSD of 32 EEG electrode (rPSD32), hemispheric asymmetry of relative PSD of 14 symmetric EEG electrode pairs (rPSD14), feature-level concatenation of rPSD14 and rPSD32 (rPSD), and coherence (Cohr) features. The classification performance for all features is shown in Table 1.

For classification with two classes, the chance level is usually set at 50%, but in practice, this level could be substantially higher due to limited sample size [31]. Assuming the classification errors following a binomial distribution [31], with our sample size, the empirical chance level was calculated to be 71.42%, 63.33%, and 62.22% for familiar, unfamiliar, and regardless of familiarity cases, respectively. Hence, the classification performance across groups based on familiarity was compared with the empirical chance level and Fig. 1 shows the performance above the calculated chance level. It was observed that for familiar music videos the classification performance of pleasant/unpleasant was best as compared to unfamiliar and regardless to familiarity. This trend, that is, the classification performance for familiar was best, followed by unfamiliar and regardless of familiarity, for rPSD32 and rPSD features. The familiar music induces discriminative brain responses for pleasant and unpleasant suggesting the suitability of familiar music to induce emotions and to achieve better valence classification. Further, it was observed that the rPSD32 features, from 32 EEG electrodes, performed better as compared to all other features.

Table 1 The average classification accuracy of pleasant and unpleasant in three different cases of familiarity for different features

	Accuracy (%)	Sensitivity (%)	Specificity (%)
	Familiar		
rPSD14	79.41	82.00	77.50
rPSD32	84.75	87.00	82.50
rPSD	82.91	84.50	82.50
Cohr	80.83	82.00	80.00
	Unfamiliar		
rPSD14	64.16	60.16	68.50
rPSD32	67.32	67.41	67.08
rPSD	67.66	69.75	65.33
Cohr	63.60	63.91	63.58
	Regardless of familiarity		
rPSD14	64.34	63.26	65.46
rPSD32	65.38	64.56	66.20
rPSD	66.22	65.73	66.80
Cohr	66.36	67.83	65.20

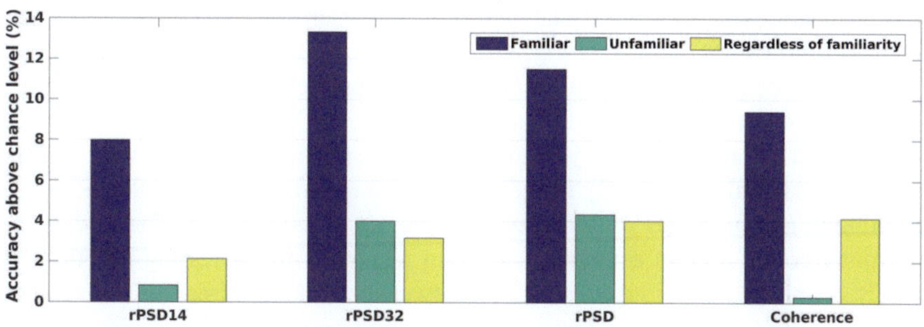

Fig. 1 The classification accuracies (above the chance level) of pleasant and unpleasant in three different cases of familiarity for different features

3.2 *Feature and Electrode Usage in Pleasant and Unpleasant Classification*

Here, we investigated the nature of selected features for each category—specifically, the EEG bands and electrodes they represent. The analysis was performed on the set of features, which were selected by the feature selection method at each step of cross-validation. For this analysis, we considered best performing PSD features (rPSD) and coherence (Cohr) features. First, a feature occurrence histogram was created from these selected feature sets. Each bin of the histogram represents a feature which is further normalized by dividing the occurrence of a feature by 200 (10-fold cross-validation × 20-time repetition). Next, the features which achieved

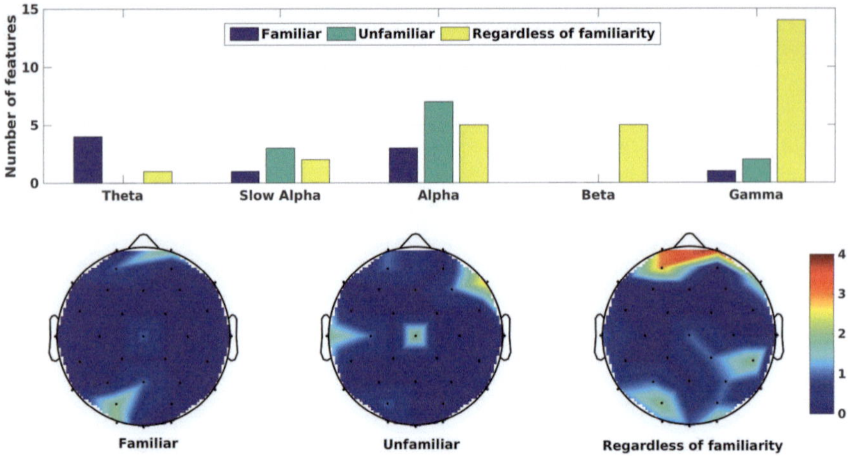

Fig. 2 The bar-charts shows the feature (rPSD32) usage and scalpmaps the EEG electrode usage. The color bar represent the number of occurrence of electrode in the selected features

greater than or equal to 50% of maximum achieved occurrence level were further grouped in accordance with the band and electrode they belong. The proportion of selected features in each band and region is shown in Fig. 2. For coherence, the selected features were shown in Fig. 3.

It was observed that for pleasant/unpleasant classification the selected spectral features were dominantly from low and mid-frequency bands for the familiar case whereas mid and high-frequency for unfamiliar and regardless of familiarity. In [10] it was observed that a slightly higher gamma-power over the parietal lobe while listening to an unfamiliar song representing an act of recollection. The role of the prefrontal cortex and its asymmetrical brain responses in processing emotions is widely established. Previously, in [32] the power spectral features in high-frequency bands over frontal and parietal brain electrode region were discriminative for emotion classification. These results are similar to earlier studies showing that brain responses in mid to high-frequency bands are actively involved in processing different emotions [33, 34].

The brain regions involved were frontal, central, and left-parietal for familiar, left-temporal, central, and right-frontal for unfamiliar, and frontal and parietal for regardless of familiarity. The selected features show hemispheric asymmetry this supports the earlier findings that the brain hemispheres are associated with different information processing. For example, the outcome of the information processed in the left hemisphere of the brain is generally more positive than information processed in the right hemisphere [35]. This is true while processing emotions positive or pleasant emotions in the right hemisphere and negative or unpleasant emotions in the left [36–38].

Further, for coherence features, we observe the interaction between the right and left hemisphere in theta and alpha frequency bands for the familiar case. We also observe the interactions between the mid-line electrode (Cz) and the left or right

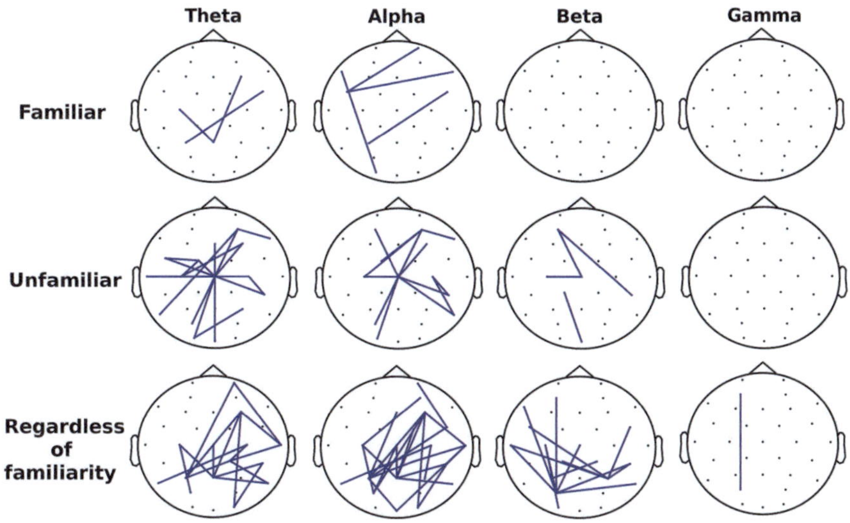

Fig. 3 Shows the selected coherence features for different frequency bands and cases of familiarity

hemisphere electrodes for the unfamiliar case. Whereas, for regardless of familiarity case the connectivity interactions were within and across the left or right hemisphere. In [39] it was observed that an increase in functional connectivity in the higher frequency bands during an experience of recollection compared to that during an experience of familiarity. And delta and gamma band functional connectivity was observed during the performance of autobiographical memory tasks [40].

4 Conclusion

In this study, we investigated the effect of familiarity on the pleasant and unpleasant emotion classification induced by Hindi music videos. For this, we analyzed the classifier performance with three power spectral (rPSD14, rPSD32, and rPSD) and one functional connectivity (cohr) feature. The relative PSD from 32 electrodes (rPSD32) features have performed better as compared to other features. The brain responses of induced valence were more discriminative for the familiar case than unfamiliar and regardless of familiarity case. Next, the feature and electrode usage reveals that involvement of low and mid-frequency bands features of frontal and parietal electrode regions distinguishing pleasant/unpleasant emotion. This study discloses the importance of familiarity while classifying pleasant/unpleasant emotions. Despite the promising results in this paper, the mechanisms underlying the effect of familiarity remain unclear and need to be explored further. The present study can be extended by including more participants, a large number of music stimuli, and investigating more advanced signal processing methods and machine learning tools.

Acknowledgements All procedures performed in this study involving human participants were in accordance with the ethical standards of the institutional and/or national research committee and with the 1964 Helsinki declaration and its later amendments or comparable ethical standards.

5 Appendix A

Table 2 List of Hindi music videos used as stimuli

<table>
<tr><th>SET I</th><th>SET III</th></tr>
<tr><td>Mehbooba mehbooba - Sholay</td><td>Faasle | Coke Studio Season 10</td></tr>
<tr><td>Aankhon mein teri - Om Shanti Om</td><td>Khoon chala - Rang De Basanti</td></tr>
<tr><td>Bandaya ho - Khuda Ke Liye</td><td>Mahi ve - Khuda Ke Liye</td></tr>
<tr><td>Galwakdi | Tarsem Jassar</td><td>Lambi judai - Jannat</td></tr>
<tr><td>I hate u like love u - Delhi Belly</td><td>Sau dard hai - Jaanemann</td></tr>
<tr><td>Matargasthi - Tamasha</td><td>Tu hi re - Bombay</td></tr>
<tr><td>Naalayak - Bawra</td><td>Waqt ne kiya kya - Kaagaz Ke Phool</td></tr>
<tr><td>Nashe si chad gayi - Befikre</td><td>Ya rabba - Salaam-E-Ishq</td></tr>
<tr><td>Phir se ud chala - Rockstar</td><td>Us rah par | Coke Studio Season 10</td></tr>
<tr><td>Illahi - Yeh Jawaani Hai Deewani</td><td></td></tr>
<tr><td>Pehla pehla pyaar - Hum Aap Ke Hain Kon</td><td>SET IV</td></tr>
<tr><td>Ajj din chadheya - Love Aaj Kal</td><td>Paisay da nasha | Bohemia</td></tr>
<tr><td>Enna sona - OK Jaanu</td><td>Alvida - Life In A Metro</td></tr>
<tr><td>Maahi ve - Highway</td><td>Aur ho - Rockstar</td></tr>
<tr><td>Main agar kahoon - Om Shanti Om</td><td>Musafir - Sweetiee Weds NRI</td></tr>
<tr><td></td><td>Allah ho - Kudha Ke Liye</td></tr>
<tr><td>SET II</td><td>In circles - AsWeKeepSearching</td></tr>
<tr><td>Zara zara - Rehnaa Hai Tere Dil Mein</td><td>Jee le zara - Talaash</td></tr>
<tr><td>Phir le Aya dil - Barfi</td><td>Kajar bin kare - karsh</td></tr>
<tr><td>khuda ke liye - Khuda Ke Liye</td><td>Koi fariyad - Tum Bin</td></tr>
<tr><td>Lab pe aati hai dua | Ahmed Hussain</td><td>Lo aa gayi unki yaad - Do Badan</td></tr>
<tr><td>Jab samne tum aa jaate ho | Jagjit Singh</td><td>Tanhayee - Dil Chahta Hai</td></tr>
<tr><td>Kal ho na ho - Kal Ho Naa Ho</td><td>The local train - Bandey</td></tr>
<tr><td>Kuch is tarah - Doorie | Atif Aslam</td><td>To kia hua | Bilal khan</td></tr>
<tr><td>Kun faya kun - Rockstar</td><td>Tu yaad naa aaye | Himesh Reshammiya</td></tr>
<tr><td>Lagi tumse mannn ki lagan - Paap</td><td>Tum ho - Rockstar</td></tr>
<tr><td></td><td>Tune jo naa kaha - New York</td></tr>
<tr><td>SET III</td><td>The tattava | AsWeKeepSearching</td></tr>
<tr><td>Jag soona soona lage - Om Shanti Om</td><td>Main wo chaand hoon - Teraa Surror</td></tr>
<tr><td>Ja ja ja bewafa - Aar Paar</td><td>Mann jaage - Bittoo Boss</td></tr>
<tr><td>Aaoge jab tum - Jab we met</td><td>At long last | AsWeKeepSearching</td></tr>
<tr><td>Tadap tadap ke - Hum Dil De Chuke Sanam</td><td></td></tr>
</table>

References

1. Juslin, P.N., Västfjäll, D.: Emotional responses to music: The need to consider underlying mechanisms. Behavioral and brain sciences **31**(05), 559–575 (2008)
2. Calvo, R.A., D'Mello, S., Gratch, J., Kappas, A.: The Oxford handbook of affective computing. Oxford University Press, USA (2014)
3. Juslin, P.N., Sloboda, J.: Handbook of music and emotion: Theory, research, applications. Oxford University Press (2011)
4. Balkwill, L.L., Thompson, W.F.: A cross-cultural investigation of the perception of emotion in music: Psychophysical and cultural cues. Music perception: an interdisciplinary journal **17**(1), 43–64 (1999)
5. Bornstein, R.F.: Exposure and affect: overview and meta-analysis of research, 1968–1987. Psychological bulletin **106**(2), 265 (1989)
6. Schubert, E.: The influence of emotion, locus of emotion and familiarity upon preference in music. Psychology of Music **35**(3), 499–515 (2007)
7. Schellenberg, E.G., Peretz, I., Vieillard, S.: Liking for happy-and sad-sounding music: Effects of exposure. Cognition & Emotion **22**(2), 218–237 (2008)
8. Pereira, C.S., Teixeira, J., Figueiredo, P., Xavier, J., Castro, S.L., Brattico, E.: Music and emotions in the brain: familiarity matters. PloS one **6**(11), e27241 (2011)
9. Hadjidimitriou, S.K., Hadjileontiadis, L.J.: EEG-based classification of music appraisal responses using time-frequency analysis and familiarity ratings. IEEE Transactions on Affective Computing **4**(2), 161–172 (2013)
10. Thammasan, N., Moriyama, K., Fukui, K.-I., Numao, M.: Familiarity effects in EEG-based emotion recognition. Brain informatics **4**(1), 39–50 (2017)
11. Witvliet, C.V., Vrana, S.R.: Play it again sam: Repeated exposure to emotionally evocative music polarises liking and smiling responses, and influences other affective reports, facial emg, and heart rate. Cognition and Emotion **21**(1), 3–25 (2007)
12. McLachlan, N., Marco, D., Light, M., Wilson, S.: Consonance and pitch. Journal of Experimental Psychology: General **142**(4), 1142 (2013)
13. Stang, D.J.: Methodological factors in mere exposure research. Psychological Bulletin **81**(12), 1014 (1974)
14. Daltrozzo, J., Tillmann, B., Platel, H., Schön, D.: Temporal aspects of the feeling of familiarity for music and the emergence of conceptual processing. Journal of Cognitive Neuroscience **22**(8), 1754–1769 (2010)
15. Gregory, A.H., Varney, N.: Cross-cultural comparisons in the affective response to music. Psychology of Music **24**(1), 47–52 (1996)
16. Mathur, A., Vijayakumar, S.H., Chakrabarti, B., Singh, N.C.: Emotional responses to hindustani raga music: the role of musical structure. Frontiers in psychology **6**, 513 (2015)
17. Banerjee, A., Sanyal, S., Patranabis, A., Banerjee, K., Guhathakurta, T., Sengupta, R., Ghosh, D., Ghose, P.: Study on brain dynamics by non linear analysis of music induced EEG signals. Physica A: Statistical Mechanics and its Applications **444**, 110–120 (2016)
18. Balasubramanian, G., Kanagasabai, A., Mohan, J., Seshadri, N.G.: Music induced emotion using wavelet packet decomposition–An EEG study. Biomedical Signal Processing and Control **42**, 115–128 (2018)
19. Mehrabian, A.: Pleasure-arousal-dominance: A general framework for describing and measuring individual differences in temperament. Current Psychology **14**(4), 261–292 (1996)
20. P. Lang and M. M. Bradley, "The international affective picture system (iaps) in the study of emotion and attention," *Handbook of emotion elicitation and assessment*, vol. 29, 2007
21. Welch, P.: The use of fast fourier transform for the estimation of power spectra: a method based on time averaging over short, modified periodograms. IEEE Transactions on audio and electroacoustics **15**(2), 70–73 (1967)
22. Quiroga, R.Q., Kraskov, A., Kreuz, T., Grassberger, P.: Performance of different synchronization measures in real data: a case study on electroencephalographic signals. Physical Review E **65**(4), 041903 (2002)

23. Sherman, D.L., Patel, C.B., Zhang, N., Rossell, L.A., Tsai, Y.C., Thakor, N.V., Mirski, M.A.: Sinusoidal modeling of ictal activity along a thalamus-to-cortex seizure pathway i: new coherence approaches. Annals of biomedical engineering **32**(9), 1252–1264 (2004)
24. Srinivasan, R., Winter, W.R., Ding, J., Nunez, P.L.: EEG and MEG coherence: measures of functional connectivity at distinct spatial scales of neocortical dynamics. Journal of neuroscience methods **166**(1), 41–52 (2007)
25. Goodman, N.R.: On the joint estimation of the spectra, cospectrum and quadrature spectrum of a two-dimensional stationary gaussian process. NEW YORK UNIV NY ENGINEERING STATISTICS LAB, Tech. Rep. (1957)
26. Nolte, G., Bai, O., Wheaton, L., Mari, Z., Vorbach, S., Hallett, M.: Identifying true brain interaction from EEG data using the imaginary part of coherency. Clinical neurophysiology **115**(10), 2292–2307 (2004)
27. I. Guyon and A. Elisseeff, "An introduction to variable and feature selection," *Journal of machine learning research*, vol. 3, no. Mar, pp. 1157–1182, 2003
28. Chakroborty, S., Saha, G.: Feature selection using singular value decomposition and QR factorization with column pivoting for text-independent speaker identification. Speech Communication **52**(9), 693–709 (Feb. 2010)
29. Ari, S., Saha, G.: In search of an SVD and QRcp based optimization technique of ANN for automatic classification of abnormal heart sounds. International Journal of Biomedical Sciences **2**(1), 1013–1016 (2007)
30. Daimi, S.N., Saha, G.: Classification of emotions induced by music videos and correlation with participants' rating. Expert Systems with Applications **41**(13), 6057–6065 (2014)
31. Combrisson, E., Jerbi, K.: Exceeding chance level by chance: The caveat of theoretical chance levels in brain signal classification and statistical assessment of decoding accuracy. Journal of neuroscience methods **250**, 126–136 (2015)
32. Lin, Y.-P., Wang, C.-H., Jung, T.-P., Wu, T.-L., Jeng, S.-K., Duann, J.-R., Chen, J.-H.: EEG-based emotion recognition in music listening. IEEE Transactions on Biomedical Engineering **57**(7), 1798–1806 (2010)
33. Boksem, M.A., Smidts, A.: Brain responses to movie trailers predict individual preferences for movies and their population-wide commercial success. Journal of Marketing Research **52**(4), 482–492 (2015)
34. Hu, X., Yu, J., Song, M., Yu, C., Wang, F., Sun, P., Wang, D., Zhang, D.: EEG correlates of Ten Positive Emotions. Frontiers in Human Neuroscience **11**, (2017)
35. Anand, P., Holbrook, M.B., Stephens, D.: The formation of affective judgments: The cognitive-affective model versus the independence hypothesis. Journal of Consumer Research **15**(3), 386–391 (1988)
36. Wheeler, R.E., Davidson, R.J., Tomarken, A.J.: Frontal brain asymmetry and emotional reactivity: A biological substrate of affective style. Psychophysiology **30**(1), 82–89 (1993)
37. Tomarken, A.J., Davidson, R.J., Henriques, J.B.: Resting frontal brain asymmetry predicts affective responses to films. Journal of Personality and Social Psychology **59**(4), 791 (1990)
38. Altenmüller, E., Schürmann, K., Lim, V.K., Parlitz, D.: Hits to the left, flops to the right: different emotions during listening to music are reflected in cortical lateralisation patterns. Neuropsychologia **40**(13), 2242–2256 (2002)
39. Burgess, A.P., Ali, L.: Functional connectivity of gamma EEG activity is modulated at low frequency during conscious recollection. International Journal of Psychophysiology **46**(2), 91–100 (2002)
40. C. Imperatori, R. Brunetti, B. Farina, A. M. Speranza, A. Losurdo, E. Testani, A. Contardi, and G. Della Marca, "Modification of EEG power spectra and EEG connectivity in autobiographical memory: a sLORETA study," *Cognitive processing*, vol. 15, no. 3, pp. 351–361, 2014

The Case Study of Brain Tumor Data Analysis Using Stata and R

Prisilla Jayanthi, Murali Krishna Iyyanki, Prakruthi Vadakattu and Padmaja Megham

Abstract The analysis and prediction of brain tumor in statistical approach provide a challenging and quantitative evaluation of the medical reports, giving more valuable information for the neuropathologist. The rise of concern about human health risk for brain tumors has made the study more enthusiastic. The analysis of study with the several combinations from the datasets shows that the F-value in ANOVA is identical as F-statistics in the regression using R-GUI, hence the null hypothesis will be accepted. The adjusted R-squared values are calculated and found to be nearly the same using Stata and R-GUI software. The data analysis in all tables using Stata shows that when the p-value is less than the significance level ($p < 0.0005$), that is, probability $> F = 0.0000$ and probability $> F = 0.0004$, then the regression model is said to be a good fit of the data. The linear regression facilitates tumor analysis by allowing the estimation of extent and accurate investigation by calculating the p-value.

Keywords Analysis · Categorize · Dataset · Linear regression

P. Jayanthi (✉)
Assistant Professor, KG Reddy College of Engineering and Technology, Hyderabad, India
e-mail: prisillaj28@gmail.com

M. K. Iyyanki
R&D JNTUH, Hyderabad, India
e-mail: iyyanki@gmail.com

P. Vadakattu
BioMedical Engineering, MIT, Manipal University, Manipal, India

P. Megham
Mahatma Gandhi Institute of Technology, Hyderabad, India

B. Pati et al. (eds.), *Advanced Computing and Intelligent Engineering*,
Advances in Intelligent Systems and Computing 1082,
https://doi.org/10.1007/978-981-15-1081-6_20

1 Introduction

For the data analysis of any disease, regression analysis is the widely used statistical method. It helps in recognition and categorization of relationships between various features. In this study, Stata and R-programming are explored to bring out the analysis of tumors using linear regression. The data are categorized based on the type of brain tumor, namely glioma, meningioma, parietal, astrocytoma, lymphoma, medulloblastoma and recurrent. For this study, 100 brain tumor reports were collected from Gandhi Hospital, Secunderabad, India and the datasets glioma, meningioma and parietal are categorized and analyzed.

2 Literature Survey

Hardell et al. (2013) analyzed that the usage of mobiles and cordless phones are linked with brain tumors based on the outputs obtained from the datasets. The calculations of odds ratio (OR) and confidence interval (CI) using unconditional logistics regression analysis were done using Stata SE 10.1. The data analyzed from various articles made accessible to the same team of authors is not a replicated data. The heterogeneity test model for overall >= 10 years and >= 1640 h was preferred. Cox proportional hazards model was preferred to evaluate hazard ratios and confidence intervals. The study suggests that incidence facts to be dismissed in the investigating epidemiology and radio frequency electronic magnetic fields are bioactive and was found to be a potential cause for health impacts [1].

Montgomery et al. (2013) analyzed the minority subset from each cohort with a primary brain tumor and the patient's mortality. The study utilized a unit of Swedish patients with multiple sclerosis and the matched general population unit. Cox regression model was used to estimate the mortality risk within 5 years. Few adjustments for sex, geographical region and socioeconomic index were made while analyzing linear regression. The analysis was carried out using SPSS statistical software and the results proved that survival with multiple sclerosis is dreadful [2].

Patil (2016) predicted the levels of addiction to the WhatsApp group by an individual. The study highlights the usage of R statistical software, which is used to extract and work on a particular datasets. The analysis was carried out on gender and age groups. The result showed that the age group between 20 and 30 years was found to be most addicted to WhatsApp [3].

3 Statistical Analysis

One of the statistical techniques used is regression analysis when the continuous variables are response and predictor variables; the response variable/dependent (y) and predictor variable/independent variable (x). The quintessence of regression analysis is to assess parameters values and their standardized errors through sample data. The relation between the dependent variable and independent variable is described using a simple linear model given as $y = a + bx$, a and b are the parameters and the y-intercept is a and b stand for the slope; when $x = 0$ it is the value of y. One of the statistical procedures is the linear regression for predicting the value of a response variable from a predictor variable when the variables are described in a linear model [4].

3.1 About Stata

In mid-1980s, Stata got started in California, with initially named as DiAL/later Stata by Bill Gould and Sean Becketti in January 1984. In January 1985, regression package with data management in Stata 1.0 version was released. Stata is used popularly in the science for manipulating and summarizing data and conducting statistical analyses. Stata is efficient in importing/exporting files, and executing other commands for data manipulation.

3.2 R Programming

R is famous for its programming features and in-built environment for graphical and statistical computing used by statisticians for data analysis using programming. It was developed at University of Auckland, New Zealand by Ihaka Ross and Gentleman Robert.

R is a command-line interpreter, supports features like procedural programming with functions and object-oriented programming with standard functions. R is an open-source project supported by the community developing it with a free software license. Basically, R is used for solving data-oriented problems using machine learning.

4 Experimental Exploration

Ensuring the integrity of research, accurate data collection is essential. The knowledge of the actual source of data, methods of extraction and purpose of data is significant, and if the data extracted is historic data, it provides the key to understand data over time. The purpose of collecting data helps in predicting the disease. The regression analysis comes with line equation that fits through cluster of points with the minimal amount of deviations from the line.

4.1 Exploring Through R

The data analysis is carried out on R-GUI version 3.5.0 supported by The R Foundation for Statistical Computing on Windows 7 professional. The main dataset is segregated based on the categories. Tables 1, 2 and 3 describe the types of brain tumor categorized as glioma, meningioma and parietal, respectively.

Figures 1 and 2 represent the boxplot graph for glioma and parietal dataset where the attributes lie between 0 and 1 as the severity of attributes are represented with low—0, high—1 and very high—2. Hence the box for attribute is plotted between 0 and 1 value. The box of age with minimum value is 36, $Q1 = 44$, median is 57, $Q3 = 60$ and maximum value is 80.

1. Let's find thc linear regression between age and sex of the datasets using R-GUI.

```
reg < −lm(age ~ sex, data = P1)
summary (reg)
```

From the datasets one can calculate variance shown in Tables 5, 7 and 9. F-value helps to determine the p-value. F-statistics judges on multiple coefficients taken together at the same time, gives power to judge whether that relationship is statistically significant or not, and thus, helps to decide whether to accept or reject the null hypothesis. The output of regression of various comparisons is shown in Tables 4, 6 and 8.

```
reg < −lm(age ~ sex, data = P1)
with(P1, plot(sex, age))
anova(reg)
abline(reg)
```

The abline() supports by specifying the arguments a and b, a line can be drawn that fits the mathematical equation $y = a + b * x$. The output of abline(reg) for age and sex attributes is shown in Fig. 3.

Table 1 Glioma dataset

S. no.	Sex	Type—Brain tumor	Age	Head ache	Vomit	Seizure	Alter behavior	Decreased vision	Difficult in speech	Brain tumor
1	0	Glioma	71	1	2	1	0	3	0	1
2	0	Glioma	60	1	1	0	0	2	1	1
3	1	Glioma	65	3	0	1	3	1	0	1

Table 2 Meningioma dataset

S. no.	Sex	Type—Brain tumor	Age	Head ache	Vomit	Seizure	Alter behavior	Decreased vision	Difficulty in speech	Brain tumor
44	1	Meningioma	35	0	0	0	0	0	0	0
45	0	Meningioma	48	0	3	0	1	0	0	1
46	1	Meningioma	40	1	0	1	0	0	1	1

Table 3 Parietal dataset

S. no.	Sex	Type—Brain tumor	Age	Head ache	Vomit	Seizure	Alter behavior	Decreased vision	Difficulty in speech	Brain tumor
80	0	Parietal	60	0	0	2	0	0	0	1
81	0	Parietal	38	0	0	1	0	0	0	0
82	1	Parietal	46	0	0	1	0	0	0	1

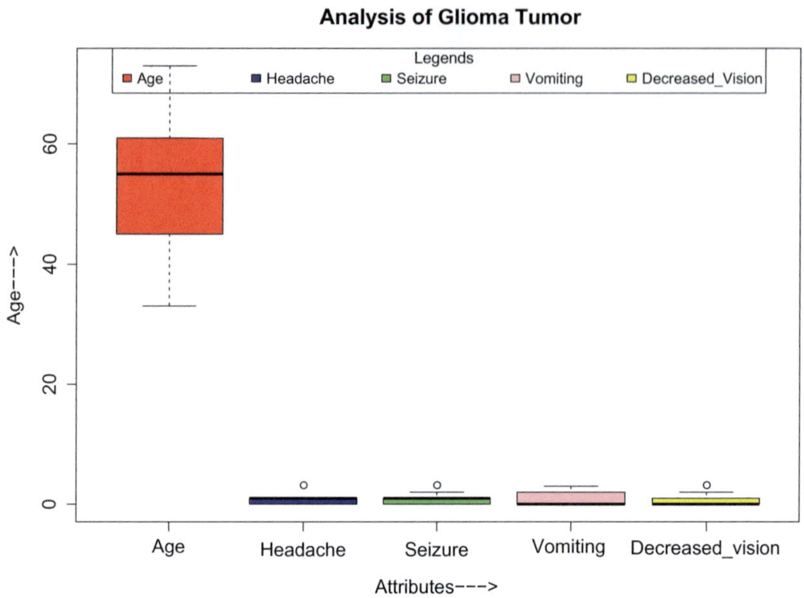

Fig. 1 R-GUI boxplot graph for glioma dataset

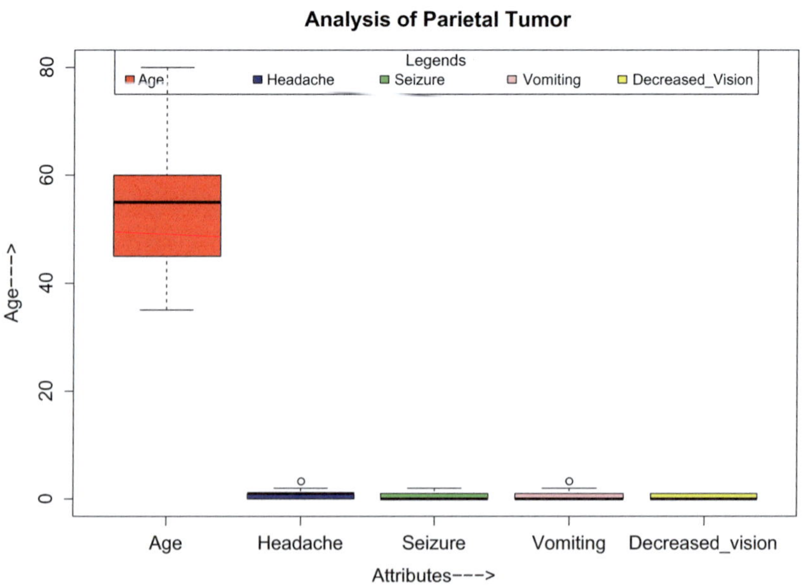

Fig. 2 R-GUI boxplot graph for parietal dataset

Table 4 Regression implemented on the attributes—age and sex

Residuals				
Min	1*Q*	Median	3*Q*	Max
−19.667	−8.667	3.333	5.333	19.333
Coeff. df: 1 to 403				
	Estimate	Std. Err	t value	*p*(>|t|)
Intercept	55.6667	0.6714	82.91	<2e−16
Sex	−5.0000	1.1629	−4.30	2.15e−05
RSE	MR-squared	AR-squared	*F*-statistic	*p* value
11.03	0.04386	0.04149	18.49	2.149e−05

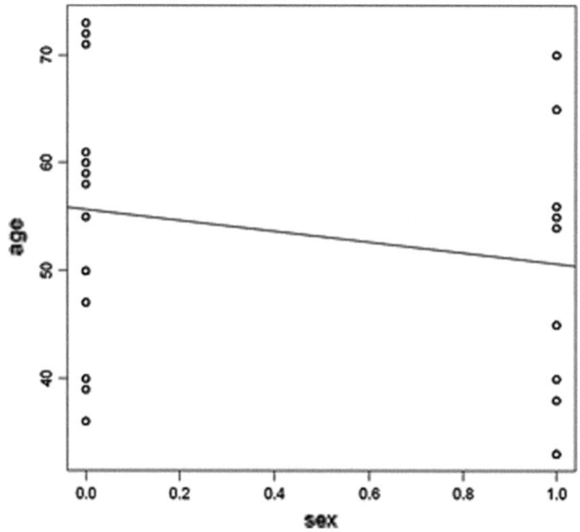

Fig. 3 abline() graph for age and sex using R-GUI

2. Linear regression between the attributes vomiting and decreased vision

lm(formula = vomiting ~ decreased-vision, data = P1)
anova(b)

3. Linear regression between the attributes seizure and sex

a < −lm(seizure ~ sex, data = P1)
with(P1, plot(sex, seizure))
abline(a)

lm(formula = seizure ~ sex, data = P1)
anova (a)

From the Tables 4 and 5 in a test implies that F-statistic = 18.49 is similar as F-value = 18.486 and the Tables 6 and 7 specifies F-statistic = 19.26 which is equivalent to F-value = 19.256, and the Tables 8 and 9 postulates F-statistic = 0.5191, which is same as F-value = 0.5191 using R-GUI, so the null hypothesis is accepted. The output of abline(b) graph is displayed in Fig. 4.

Table 5 Age and sex variance

Variance of age and sex (response variable)					
	df	SS	MS	*F*-value	*p* (> F)
Sex	1	2250	2250.00	18.486	2.149e−05
Residuals	403	49,050	121.71		

Table 6 Regression of vomiting and decreased vision

Min	1*Q*	Middle	3*Q*	Max
−1.0521	−0.7638	−0.4756	0.6596	2.5244
Coeff. df 1 to 403				
	Estimate	Std. Err	t value	*p*(>|t|)
Intercept	0.47558	0.06824	6.970	1.31e−11
Decreased vision	0.28826	0.06569	4.388	1.46e−05
RSE	MR-squared	AR-squared	*F*-statistic	*p*-value
1.067	0.0456	0.04323	19.26	1.462e−05

Table 7 Variance of vomiting and decreased vision

Response: vomiting					
	df	SS	MS	*F*-value	*p* (>F)
Decreased vision	1	21.90	21.9045	19.256	1.462e−05
Residuals	403	458.43	1.1375		

Table 8 Regression of seizure and sex

Low	1*Q*	Middle value	3*Q*	High
−0.8667	−0.8667	0.1333	0.2074	2.2074
Coeff. df 1 to 403				
	Estimate	Std. err	*t* value	*p*(>|t|)
(Intercept)	0.86667	0.05936	14.60	<2e−16
Sex	−0.07407	0.10281	−0.72	0.472
RSE	MR-squared	AR-squared	*F*-statistic	*p*-value
0.9754	0.001286	−0.001192	0.5191	0.4717

Table 9 Variance of seizure and sex

Response: seizure					
	df	SS	MS	F-value	p (>F)
Sex	1	0.49	0.49383	0.5191	0.4717
Residuals	403	383.39	0.95135		

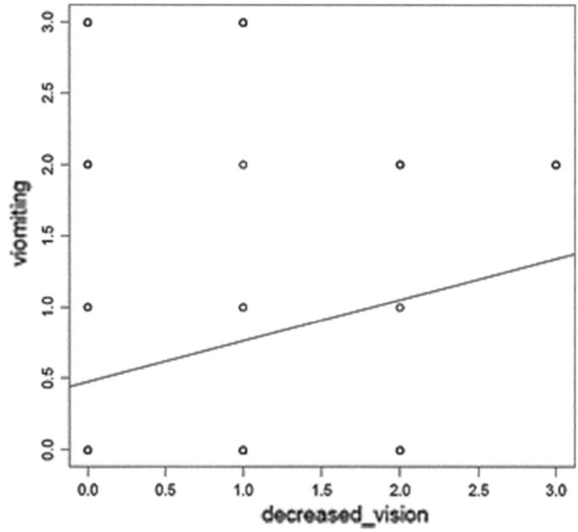

Fig. 4 Vomiting and decreased vision abline() graph

4.2 Using Stata

The data analysis is carried out on a licensed Stata IC 12.1 on Windows 7 professional 64-bits. Stata IC 12.1 has the competency of running types of regression models linear and nonlinear, parametric and nonparametric, and so on. This paper emphasizes on linear regression in comparison with R programming and essences on glioma dataset. The computation of Stata is faster in comparison to R-GUI on the same given dataset. Figure 5 shows the boxplot graph created using Stata IC for the analysis of glioma dataset.

Tables 10, 11, 12 and 13 show regression between the various attributes. The data analysis from all the tables using Stata shows in three cases where the lower the p-value than the significance level ($p < 0.0005$), that is, probability > $F = 0.0000$ and probability > $F = 0.0004$, the regression model is a good fit of the data.

The adjusted R-squared = 0.0415 is the same for age and sex in Tables 4 and 13. Tables 6 and 11 indicate the adjusted R-squared = 0.04323 of the attributes vomiting and decreased vision, and the Tables 8 and 12 indicate the adjusted R-squared = − 0.0012 and 0.001286 for the attributes seizure and sex are nearly equal using the Stata and R software.

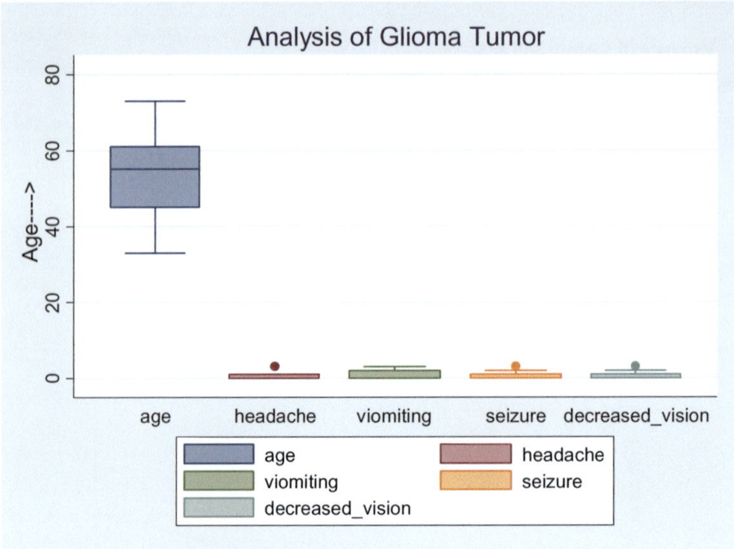

Fig. 5 Glioma dataset boxplot graph using Stata

Table 10 Regression of headache and age using Stata

Source	SS	df	MS	
Model	12.0734698	1	12.0734698	Number of observations = 405 Prob > F = 0.0004
Residual	375.704308	403	0.93226875	$F(1, 403) = 12.95$ R-Squared = 0.0311
Total	387.777778	404	0.95984598	Adj. R-Squared = 0.0287
Headache	Coeff.	Std. err	t	95% conf. interval
Age	0.0153411	0.004263	3.60	0.0069607 – 0.0237216
Constant	0.0975049	0.2351468	0.41	−0.3647627 – 0.5597724

Table 11 Regress vomiting and decreased vision

Source	SS	df	MS	
Model	21.9044592	1	21.9044592	Number of observations = 405 Prob > F = 0.0000
Residual	458.426405	403	1.1375345	$F(1, 403) = 19.26$ R-Squared = 0.0456
Total	480.330864	404	1.18893778	Adj. R-Squared = 0.0432
Vomiting	Coeff.	Std. err	t	95% conf. interval
Decreased vision	0.2002634	0.0656909	4.39	0.1591237 – 0.4174031
Constant	0.4755807	0.0682368	6.97	0.3414362 – 0.6097253

Table 12 Regress seizure and sex

	SS	df	MS	
Model	0.49382716	1	0.49382716	Number of observations = 405 Prob > F = 0.4717
Residual	383.392593	403	121.712159	$F(1, 403) = 0.52$ R-Squared = 0.0013
Total	383.88642	404	126.980198	Adj. R-Squared = −0.0012
Seizure	Coeff.	Std. err	t	95% conf. interval
Sex	−0.0740741	0.102813	−0.72	−0.2761909 – 0.1280427
Constant	0.866667	0.0593591	14.60	0.7499745 – 0.9833589

Table 13 Regress age and sex

Source	SS	df	MS	
Model	2250	1	2250	Number of observations = 405 Prob > F = 0.0000
Residual	49,050	403	121.712159	$F(1, 403) = 18.49$ R-Squared = 0.0439
Total	51,300	404	126.980198	Adj. R-Squared = 0.0415
Age	Coeff.	Std. err	t	95% conf. interval
Sex	−5	1.162909	−4.30	−7.286125 – −2.713875
Constant	0.4755807	0.6714058	82.91	54.34677 – 56.98656

5 Conclusion

For data scientist, the R programming software is an eye-opener, whereas Stata continues to be the heart-winning software for the statistician and for data management. It has a great degree of flexibility. Running more complex models, along with multilevel models are found in Stata IC. One can find that there is not much difference in the values calculated with Stata IC 12.1 and R. The analysis shows that the smaller the p-value than the significance level ($p < 0.0005$), the regression model is an ideal fit and the variable is a statistically significant part of the model. As the result of the F-value is similar as F-statistics, hence the null-hypothesis is accepted. Stata is faster in computation, efficient, and the new version of Stata IC 15.0 has few machine learning algorithms implemented. The analysis of odds ratio (OR) and confidence interval (CI) calculations using unconditional logistics regression can be performed using Stata. This feature is not available in R; it helps the prediction of brain tumor in statistics more elegant, but R provides a definite platform for machine learning algorithms.

Acknowledgements The authors would like to thank the ethical committee of Gandhi Hospital, Secunderabad for providing us the required resources for conducting the analysis successfully.

References

1. Hardell, Lennart, et al.: Use of mobile phone and cordless phones is associated with increased risk for glioma and acoustic neuroma. Pathophysiology **20**, 85–110 (2013)
2. Montgomery, S., et al.: Mortality following a brain tumour diagnosis in patients with multiple sclerosis. BMJ Open 2013-003622
3. Patil, S.: Whatsapp group data analysis with R. Int. J. Comput. Appl. **154**(4) (2016). ISSN 0975-8887
4. Shazmeen, S.F., et al.: Regression analysis and statistical approach on socio-economic data. Int. J. Adv. Comput. Res. **3**(no. 3(11)) (2013). ISSN (print): 2249-7277 ISSN (online): 2277-7970

Predictive Data Analytics for Breast Cancer Prognosis

Ritu Chauhan and Neeraj Kumar

Abstract The leading healthcare explosion in datasets has created furious interest among the researchers and scientists to determine automated technology for clinical future implications. Hence, data mining in the past decade has resolved and discovered hidden information and knowledge from large-scale databases. Data mining tends to be widely anticipated technology which is utilized in varied application domain for retrieval of patterns for future decision-making. In the past several research studies are based on the prevalence and prognosis of breast cancer in females, but our proposed discuss the interception of disease for both the genders. Hence, we integrate the cancer data with predictive data analytics to determine the incidence and mortality rate of breast cancer both in males and females from the year 1990 to 2014.

Keywords Predictive data analytics · Data mining · Breast cancer · Spectral analysis

1 Introduction

Cancer is one the leading cause of deaths in developing and developed nations. Hence, data generated by these nations consist of automated healthcare technology which has vastly increased in the last decade. This has forced healthcare practitioners and researchers around the globe to develop new innovative technology for future clinical decision-making. Subsequently, machine learning (ML) technology has significantly widened the era of data analytics to search for hidden and unknown patterns from large-scale databases [1]. Moreover, different ML techniques are applied to develop new strategies for future prediction of disease. Substantially, ML techniques are extensively applied with complex cancer datasets to determine the effective cause

R. Chauhan (✉) · N. Kumar
Amity Institute of Biotechnology, AUUP Sec-125, Noida, India
e-mail: rituchauha@gmail.com

N. Kumar
e-mail: nkumar8@amity.edu

B. Pati et al. (eds.), *Advanced Computing and Intelligent Engineering*,
Advances in Intelligent Systems and Computing 1082,
https://doi.org/10.1007/978-981-15-1081-6_21

of relationship among varied relative attributes. However, ML techniques use data mining as an algorithmic technique to predict models for future knowledge discovery [2, 3].

Data mining techniques are widely anticipated techniques to discover knowledge from big databases. Hence, data analytics from big databases is a trivial process due to its complex nature. So, manual or tradition tools tend to be unsuccessful for handling and retrieving of patterns for clinical decision-making. However, retrieval of knowledge from big databases should be an automatic process for real-time decision-making. Knowledge discovery in databases or KDD tends to be an efficient tool for discovery of hidden knowledge and patterns from big databases. In the past, several studies have utilized KDD as an imperative process to detect information knowledgably from big databases for varied application domains [4–6].

KDD process involves varied steps from pre-processing of data, which include data cleaning, data selection, data transformation, data mining algorithms to determine the relevant technique for retrieval of information from big databases. The application of data mining applied to clinical databases can retrieve validated results which can improve the overall susceptibility, prognosis and diagnosis of disease [7–10]. The outcome of the predicted result can improve the overall accuracy of cancer prevention.

However, clinical decision-making has been an imperative tool where the focus of researchers and scientists around the globe is to detect patterns which can benefit end users. At the foremost, there exist continuous constraints which adhere with the process of advent nature among data where the dimensionality, dynamic and voluminous plays a major role for analysis of data. In the current scenario viable technology of big data has made decision-making implicative for researchers. Hence, big data is the new ambiguity technology where the focus relies on data. Data source can be in any form of structured and unstructured but real-world data is heterogeneous in nature and are available from static to dynamic environment.

In the current scenario, most of the studies are relied on perpetual static data which has been accumulated from past years and the emphasis of the studies relies on detecting patterns associated with correlated patterns. However, medical data mining can be related to past exponential learning to up grow the new trends of diseases which can be correlated with the past and future decision-making that can be made for patient diagnostic system.

The proposed work in the study integrates the cancer data with predictive data analytics to determine the incidence and mortality rate of breast cancer both in males and females from the year 1990 to 2014. The comparative work will generalize the predictive data analytics with SPSS for future prognosis of disease. The paper is organized as per guidelines: Sect. 2 discusses the related works; the methodology and the results are discussed in Sect. 3 and 4, respectively, and conclusion is the last section.

2 Related Works

The vast literature studies have shown the technological advances in data mining and statistical methods for knowledge discovery among large-scale databases. As we know, the larger databases are more complex nature of databases. Hence, the traditional tools are not enough to handle and retrieve effective and efficient patterns from such databases. To deal with such interfaces several studies are discussed in the past to handle high data volumes. A recent research has shown the missing values among the data can retrieve inconsistent output for future prediction modelling. Hence, the noise was classified as class and attribute noise for the determination of represented data model [11].

The new intervention technique in data mining to handle large data with automatic selection process is genetic algorithms. These algorithms process significantly and classify data using varied strategies, which include modification among data, mutation and natural selection of data. The genetic algorithms are anticipated in data mining techniques to generate and formulate general hypotheses to discuss and mutually interrelate between variables in closer dependency for association rules [12].

The electronic health record (EHR) is a widely anticipated term which may consist of records, such as demographic features, physical attributes, clinical notes, vital signs, laboratory data and other clinical implications [13]. These databases can help healthcare providers and researchers to detect the ongoing clinical care and future prognosis of disease.

Recently, a healthcare framework was deployed and designed to determine the classifier best suited for healthcare application domain. The three deterministic classifier approaches were synthesized to determine the most appropriate classifier for clinical datasets. Further, varied accuracy techniques were used to determine the best accurate model suited for classification techniques [4].

Herewith, the overall objective of data mining is to extract meaningful hidden information from large-scale raw data for knowledge discovery purpose. In the past varied data mining techniques are involved in healthcare framework, which include neural network, association rules, Bayesian models, fuzzy rule system, tree induction, decision tress and genetic algorithms where the sole aim was to determine knowledgeable patterns which can benefit end-users for future decision-making [14, 15].

In healthcare databases the major concern is the predictive results which need to be accurate, as it is inversely related to survival rate. In the current scenario, work of data mining with big data using healthcare databases is widely anticipated using novel techniques for knowledge discovery. A similar approach was studied to handle big data challenges using varied pre-processing techniques for knowledge discovery. Further, the implemented techniques were applied on application of healthcare databases to determine the novel technique for detection of patterns [16]. However, the contest is to deal with the dimensionality of data, which can arbitrarily change the overall orientation of results if the features are not correctly selected.

A similar approach was identified to identify the appropriate features from large-scale databases using varied classifiers on pancreatic cancer databases. The overall study was effective to determine the efficiency of feature selection algorithms for medical diagnosis [17]. The prediction data model is been validated for varied diagnostic healthcare domains. An analogous study was conducted to characterize the skin disorder.

The objective of the research was to design a predicative data model for skin orders by using combinatorial approach of supervised learning technique in which varied techniques were integrated, that is, classification tree with neural network and classification methods for knowledge discovery. The overall focus of this work was influenced on six major skin disorders. For evaluation of data, five major experiments were conducted. The results were accomplished where neural network technique had synthesized 92.62% accuracy while predicting skin disorders [18].

3 Mining of Data in Healthcare Application

The diagnostic system in healthcare integrates itself from varied treatments, which is accountable for future prevention of disease. In the current scenario, day-to-day lifestyle behaviour of users is affecting the overall global burden of disease. The challenge among the researchers around the globe is to reduce the healthcare costs among the end users for healthy livelihood. Therefore, the growth of Information, Communication and Technology (ICT) has widely increased the data resources with utmost speed. The contest is to develop an automation technique which can handle the voluminous nature of data. Hence, in the past decade data mining has been widely anticipated into extraction of knowledge from large healthcare data for future medical diagnosis. The healthcare practitioners are widely utilizing the data mining as an efficient tool for early diagnostic treatment for future medical prognosis of disease. The various diagnostic application areas where data mining can be utilized as an effective tool are discussed below:

1. Electronic Health Records (EHR)
   The electronic data records are increasing at an immense speed which is exceeding terabyte every day. The resources of EHR include patient socioeconomic details, treatment applied, hospitalization stay related to cost, Mediclaim, and other details while going with treatment. The electronic healthcare records consist of most important details of patients. If effectively analysed it can lead to early prognosis and diagnosis of disease. Herewith, data mining has the capability to deal with real time which is complex and voluminous in nature. The data mining technique compromises predictive and descriptive data analytical techniques and can be utilized in correspondence to the need of the user orientation. If the class is not required to be pre-defined then supervised learning techniques are utilized further. If the class is to be defined, then the unsupervised learning techniques are widely utilized [17, 19, 20].

2. Imaging
   The imaging data which include PET CT, CT scan, MRI, ECHO and other devices have far exceeded and benefited patient diagnostic system for early prognosis of disease. In the context for the same, the novel data mining techniques have figured to analyse the overall size of tumour and discussing about the features included with them. The data mining is used as predictive data analytical tool for early detection of diseases.
3. Computational databases
   The genomics and proteomics databases is gaining momentum from the past decade where the diseases are correlated with their structures. However, data mining are widely applied to measure the prognosis of disease using computational databases for knowledge extraction [18, 21, 22].

4 Methodology

In the current study, we have approached data mining technique to discover time series analysis of data to determine the knowledgeable patterns for future prognosis of disease. We have selected the cancer raw data from the Scotland registry and pre-processed it to discover definite patterns from them. As we know, there are varied internal and external features or attributes that can lead to cancer. But the existing datasets are complex in nature, so relating each attribute and identifying patterns from data is trivial for future prediction of disease. In our comparative study the breast cancer dataset is assessed for both the genders, male and female, to analyse the incidence and mortality rate from 1990 till 2014 among different age groups. The incidence rate is measured as number of new cases emerged for a particular disease in a given time period. It is also stratified as incidence density rate or person time incidence rate. The study of incidence rate provides information about the cause or etiologic of a disease and its effects for a given time period. It also helps researchers to determine the risk factors which may be concluded for a particular disease or other medical condition. Further, incidence can be discussed or measured as a rate or proportion. When it is measured as a proportion, it determines the risk of an occurrence of a disease in a given period of time. And when it is measured as a rate, it determines the number of new cases which may occur in population over a time. Thus, in order to calculate the incidence, three factors must be discussed. First, the number of new cases, and second the population at risk and the time period.

Similarly, mortality rate is discussed as a measure of the total number of deaths that occurs due to any specific causes, such as disease, injury and any other medical conditions in a particular population for specific period of time. It is usually expressed in the units of deaths for 1,000 individuals per year in terms of survival rate. Thus, the survival rate is equivalent to 1 minus the cumulative death rate.

The investigated study determines the risk involved for each age group within the time period using SPSS (version 16.0) toolkit. The tool provides the statistical and data mining interface to predict the knowledge for future prognosis of disease. The data is pre-processed and irrelevant data values are removed to maintain consistency among the datasets. However, the results retrieved determine the hidden pattern for future investigation studies in healthcare application domain.

5 Experimental Results

In the approach study the breast cancer datasets was statistically processed to determine relevant information from Scotland registry from 1990 to 2014 [23]. The data comprises incidence and mortality rate for patients suffering from breast cancer among male and female to determine the descriptive analysis. The data was pre-processed using SPSS where raw data was converted into field data to discover the hidden knowledge and patterns for future medical diagnosis. Further, field data was initialized for experimental data analytics.

The comparative study of descriptive statistics was conducted to determine the predominance among varied age groups for male and female. Figure 1 indicates the time series spectral analysis for incidence among male patients from year 1990 till 2014.

Figure 1 represents the incidence rate varying from year 1990 to 2014 with a time interval of 5 years. The blue line indicates the incidence rate of year 1990 having a peak of highest incidence at the age of 69, whereas the peak value in 1995 move towards the age group of 85–89 shown by a green line. In the year 2005, the incidence rate decreases from a mean of 6 in 1995 to 5 in 2005 and the peak value also moved to the age group of 69–74. Then a drastic increase in the incidence rate has been seen in the year 2010, shown by a yellow line with a mean of 9 in the age group of 59–64.

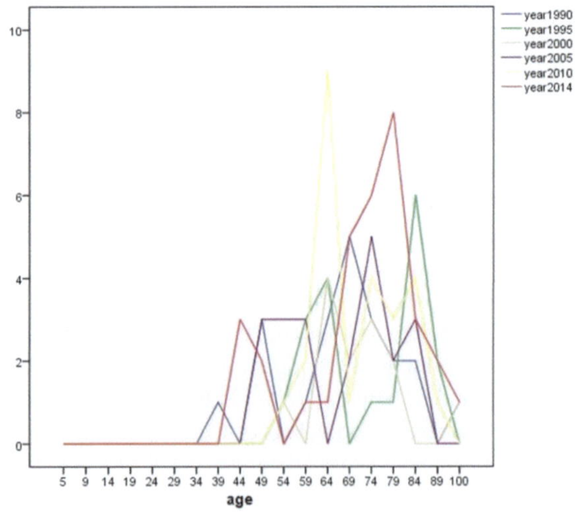

Fig. 1 Spectral analysis for male incidence rate from year 1990 to 2014

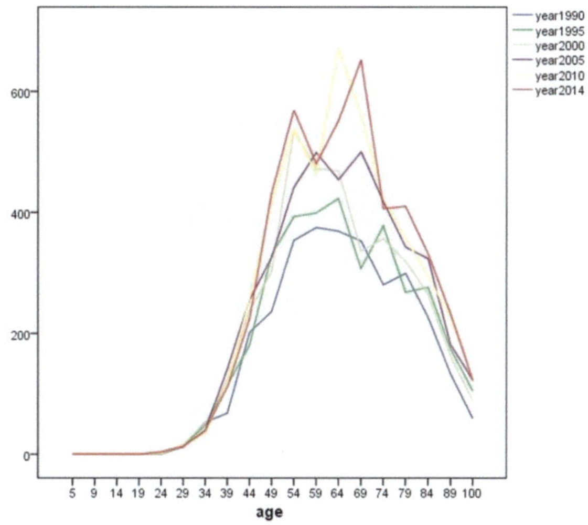

Fig. 2 Spectral time series analysis of incidence among female

However, in the year 2014, the incidence rate in 2014 is decreased with a mean of 1.5 compared to the year 2010 having a peak value in the age of 79. Further, it was observed that in 2014 the onset of cancer is occurring in the age group below 5 years.

In Fig. 2 a time series analysis is represented for female incidence rate from the year 1990 till 2014. The analysis represents a spectral increase from 1990 till 2014. It was observed that in the year 1990, the onset of cancer was stratified from the age group of 19–24 in females while having a highest incidence rate in the age group of 54–59 years. While in the year 2014, it was clearly indicated that the major shift of onset among breast cancer is now occurring from a very low age group that is below 5 and having a very high incidence rate in the age of around 70.

In Fig. 2 the change in incidence rate of female breast cancer from 1990 to 2014 with a gap variation of five years, that is, 1990, 1995, 2000, 2005, 2010, and 2014 is represented. The observed value for incidence rate signifies with an explosive variation from year 1990 to 2014 with a highest peak value of incidence varying from a mean of 300 to 700.

In the year 1990, shown by a blue line, we can see that the incidence rate is as low as compared to other years having a highest peak of incidence in the age group of 54–59. The incidence is increasing drastically from year 1990 to 2014 with a peak value moving towards the elder age. The highest breast cancer incidence is seen in the year 2010 with an incidence mean of 700, while in 2014, there is a decrease in the breast cancer incidence mean from 700 to 650 as compared to the year 2010. We also observed that as compared to other years, 2014 has the highest onset of cancer patients from a very low age group that is below 5.

In Fig. 3 we have represented the mortality rate of male patients in the year 1990 and 2014. The results indicate the steep increase in the mortality rate of breast cancer in males in the age group of 69–74. However, in the year 2014, the mortality rate is

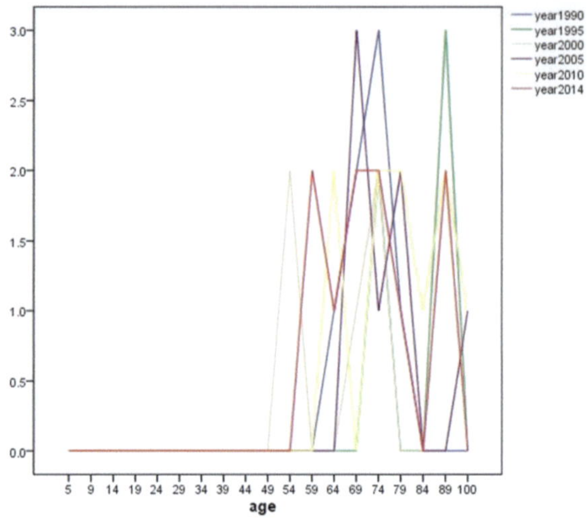

Fig. 3 A spectral analysis of mortality rate of breast cancer in males

varying from the age of 54 to 100, that is, a decrease in the mortality rate is also seen in the age of 64 and 84.

In Fig. 3 we observe that the mortality rate has decreased from year 1990 to 2014. In 1990, the highest mortality rate of breast cancer in male is observed to be in the age of 74, which is then increased towards age group of 89 in the year 1995. The mortality rate in the year 2014 is varying from age 59 to 100 with a mean of around 1.8. However, the death rate is also observed to be higher in the lower age group for year 2014.

However, Fig. 4 indicates the breast cancer mortality rate in females from 1990 to 2014.

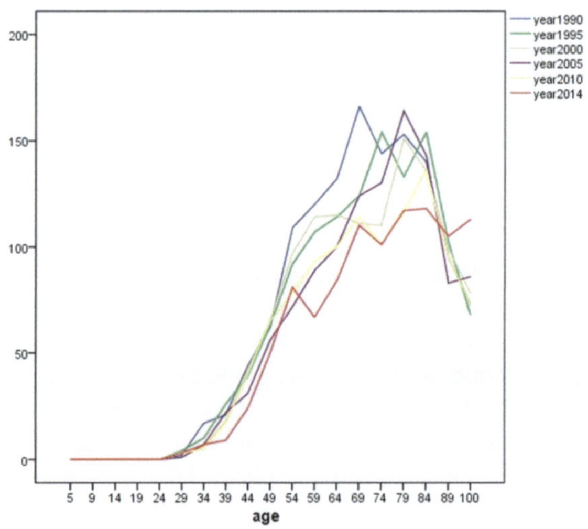

Fig. 4 Spectral analysis in mortality rate of breast cancer in females

In Fig. 3 the retrieved time serial spectral analysis represents the decrease in mortality rate from year 1990 to 2014. There is continuous decrease in mortality rate in subsequent years due to advancement in technology. However, the highest mortality rate is observed in the age group 79–84 in 2014. This may occur due to low immunity factor and other vice versa factors.

6 Conclusion and Discussion

Data mining techniques are providing challenges for researches and scientists to determine the best model for clinical implications of disease. Hence, it has played a major role to classify and determine varied causes which can lead to prognosis of cancer among the society. Subsequently, cancer is among the leading cause of deaths in both males and females. Our proposed study uses breast cancer incidence and mortality rate for both genders from 1990 till 2014 to develop and predict analytics for future prognosis of disease. A complete time series analysis is performed to determine the influence among the prevailing society in variation with several age groups. The study can effectively determine the age group at major risk with time analysis.

References

1. Bellaachia, A., Guven, E.: Predicting breast cancer survivability using data mining techniques. In Ninth Workshop on Mining Scientific and Engineering Datasets in conjunction with the Sixth SIAM International Conference on Data Mining (2006)
2. Huang, Y., McCullagh, P.J., Black, N., Harper, R.: Evaluation of outcome prediction for a clinical diabetes database. In: KELSI, pp. 181–190 (2004)
3. Yingjie, L., Yisheng, Z., Yuhong, X.: The nonlinear dynamical analysis of the EEG in schizophrenia with temporal and spatial embedding dimension. J. Med. Eng. Technol. **25**, 79–83 (2001)
4. Kaur, H., Tao, X. (eds.): ICTs and the millennium development goals: A United Nations perspective. Springer, New York (2014)
5. Chauhan, R., Kaur, H.: Big data analytics for ICT monitoring and development. In: Catalyzing Development Through ICT Adoption, pp. 25–36. Springer (2017). ISBN 978-3-31956522-4
6. Breiman, L.: RFtools—For predicting and under-standing data. Technical Report. http://oz.berkeley.edu/users/breiman/RandomForests (2004)
7. Breiman, L., Friedman, J., Stone, C.J., Olshen, R.A.: Classification and regression tree. Chapman & Hall, London (1993)
8. Holsheimer, M., Siebes, A.: Data mining: the search for knowledge in databases. Technical report CS-R9406, CWI 1994, January
9. Hua, W., Qicheng, J., Xuegang, H.U.: Application of data mining to medicine. Anhui Med. Pharm. J. **12**, 746–748 (2008)
10. Massoud, T., Lamy, J.B., Philippe, L.T.: Using data mining techniques to explore physicians' therapeutic decisions when clinical guidelines do not provide recommendations: methods and example for type 2 diabetes. BMC Med. Inform. Decis. Mak. **6**, 1471–2288 (2009)

11. Zhu, X., Khoshgoftaar, T., Davidson, I., Zhang, S.: Special issue on mining low-quality data. Knowl. Inf. Syst. **11**, 131–136 (2007)
12. Ngan, P.S., Wong, M.L., Lam, W., Leung, K.S., Cheng, J.C.: Medical data mining using evolutionary computation. Artif. Intell. Med. **16**, 73– 96 (1999)
13. Maglogiannis, I.: Introducing intelligence in electronic healthcare systems: state of the art and future trends. Artif. Intell. LNAI **5640**, 71–90 (2009)
14. Kaur, H., Chauhan, R., Wasan, S.K.: A Bayesian network model for probability estimation. In: Khosrow-Pour, M. (ed.) Encyclopedia of Information Science and Technology, 3rd edn, pp. 1551–1558. Accessed 10 Dec 2014. https://doi.org/10.4018/978-1-46665888-ch148 (2015)
15. Chauhan, R., Kaur, H.: SPAM: an effective and efficient spatial algorithm for mining grid data. In: Geo-intelligence and Visualization Through Big Data Trends. IGI Global (2015), pp. 245–263. Web. 9. https://doi.org/10.4018/978-1-4666-8465-2.ch010
16. Chauhan, R., Kaur, H.: Big data application in medical domain. In: Computational Intelligence for Big Data Analysis: Frontier Advances and Applications, Adaptation, Learning, and Optimization, vol. 19, pp. 165–179. Springer International Publishing, Switzerland (2015)
17. Chauhan, R., Kaur, H., Sharma, S.: A feature based approach for medical databases. In: AICTC '16 Proceedings of International Conference on Advances in Information Communication Technology and Computing, Article 94
18. Chang, L., Chen, C.H.: Applying decision tree and neural network to increase quality of dermatologic diagnosis. Expert Syst. Appl. (Elsevier) **36,** 4035–4041 (2009)
19. Kaur, H., Chauhan, R.: In-silico study of computational modelling and GLP-1 receptor inverse agonist compounds on a cancer cell line inhibitory bioassay dataset. Int. J. Comput. Biol. Drug Des. IJCBDD 072116 **8**(3), pp. 293–309 (2015)
20. Chauhan, R., Kaur, H.: A feature based reduction technique on large scale databases. Int. J. Data Anal. Tech. Strateg. **9**(3), 207–221 (2017)
21. Chauhan, R., Kaur, H., Alam, M.A.: Data clustering method for discovering clusters in spatial cancer databases. Int. J. Comput. Appl. 10(6), 9–14 (2010). ISBN/ ISSN 975-8887
22. Chauhan, R., Kaur, H., Chang, V.: Advancement and applicability of classifiers for variant exponential model to optimize the accuracy for deep learning. J. Ambient Intell. Humaniz. Comput. (impact Factor: 1.6 SCI) https://doi.org/10.1007/s12652-017-0561-x (2017)
23. Scottish Cancer Registration: https://www.isdscotland.org/Health-Topics/Cancer/Cancer-Statistics/

A Semantic Approach of Building Dynamic Learner Profile Model Using WordNet

T. Sheeba and Reshmy Krishnan

Abstract The learners' interest forms the essential characteristics of the learner profile in various applications, such as information retrieval, classification, and recommender systems. This paper proposes a method to improve learner interest extraction from the frequently used documents of the learner by exploring the concept of WordNet. Initially, the web log files of each learner are obtained from the learning management system, and then the frequently visited documents of each learner are downloaded and processed to identify domain-related words. The learner's interest is then extracted initially using the standard vector space model and then improved using the semantic-based representation of WordNet. The WordNet identifies a set of semantic concepts related to the document words. To select the appropriate meaning of a word from a set of concepts, "Word Sense Disambiguation (WSD)" semantic similarity algorithm is used. The experiments were performed in NetBeans IDE using Java language and WordNet 2.1. The effect of the proposed method is examined with classification experiments, and the result proved that the use of WordNet concepts in learner interest retrieval shows better classification performance than compared to the existing method of term representation, thereby obtaining a classification accuracy of 89%.

Keywords Dynamic learner profile · WordNet · Learner interest · Semantic representation · Word sense disambiguation

T. Sheeba (✉)
Department of Computer Science & Engineering, Karpagam University, Coimbatore, India
e-mail: tsheebat2002@yahoo.co.in

R. Krishnan
Department of Computing, Muscat College, Ruwi, Oman
e-mail: reshmy_krishnan@yahoo.co.in

B. Pati et al. (eds.), *Advanced Computing and Intelligent Engineering*,
Advances in Intelligent Systems and Computing 1082,
https://doi.org/10.1007/978-981-15-1081-6_22

1 Introduction

A learner profile plays a vital role in identifying a learner's view of the access and retrieving of relevant information. It provides a more appropriate and better learning environment by enabling learners to use the proper facilities which best suits their requirements, which in turn enhances the usage of learning content. Learner's interest forms the essential features of the learner profile. The most common method used for learner interest extraction is keyword based, which extracts interest from the search documents of the learner while using the learning system. Keywords are the subset of words that represent information about the content of the document. The keyword-based extraction method is used to get the essential keywords from documents. It is a process used to take out essential words of a whole document purpose. This process requires to be performed systematically and with least or no human intervention. There are various methods used for keyword extraction: statistical, linguistic, machine learning and hybrid approaches. The main drawback of these methods is the lack of adequate training data or the specific knowledge and skills required in that field.

To overcome the drawback of the existing methods, the proposed system explores the concept of WordNet to enhance the learner interest representation, which provides semantic representation and improves the overall performance of learner interest representation of the system.

2 Literature Review

One of the most critical issues faced in a learning system is to extract keywords and concept findings from learning content used by learners to find the learning interest. A novel model [1] is presented to improve keyword extraction from learning objects using data mining techniques along with the WordNet dictionary. The WordNet is used to remove the unrelated concepts from the keywords extracted using the standard TF-IDF method. The essential keywords are selected by the highest similarity score obtained for the remaining concepts in the learning object. A novel approach [2] is proposed to improve and increase keyword extraction accuracy in learning objects. The decision tree algorithm which has better accuracy is used for feature selection using the WordNet dictionary. For each word, WordNet semantics is used to compare and eliminate the words having a similar meaning. In the final step, keywords having the highest similarity are selected as output keywords. The documents [3] from different journals are collected, and then keywords are obtained from these documents using the standard TF-IDF method. Then WordNet is applied to these keywords to find the similarity between the selected words, and the words having the highest similarity are taken as keywords. These keywords are stored in the database to evaluate using standard classifiers, such as decision tree, K-nearest neighbor (KNN) and naive Bayes algorithms, using ten-fold cross-validation. The final result shows that the accuracy obtained for text classification using decision tree algorithm is better compared to other algorithms. Web access logs [4] of each learner are analyzed to extract learners' interested terms using the traditional vector space model (VSM) method. The extracted e-learners' interests are used for the clustering and prediction of learners using statistical K-means clustering method. This work recommends for ontology-based user profiles as future implementation, to maintain a sophisticated learner interest profile representations.

Many research papers are using the concept of WordNet to extract learner interest. Most of these works use the concept of WordNet to find the highest similarity word as the learner's interest and to remove the unrelated concepts. The proposed approach makes use of WordNet in a more efficient way by using semantic similarity algorithm which helps to improve the representation of learner interest. Also, the proposed approach makes use of learner interest extraction in the construction of learner profile which is not found in the previous works. The identified concepts obtained from the documents using WordNet would mostly reflect the real interest of learners to be used by the learning system to retrieve learning content that suit their interest. It would enhance the effectiveness of learner interest acquisition semantically and is profoundly necessary for a learning system.

3 Methodology

3.1 Proposed Approach

Extracting learner search interest is an essential component in the learner profile. In the proposed system, learner interest is processed from the frequently visited documents of the learner by using web log analysis. The frequency of visited documents is determined by the number of times the learner views the document for each type such as "notes", "PowerPoint" and "hyperlinks" provided in the learning management system (LMS). Figure 1 shows the proposed architecture for extracting learner interest from web log files.

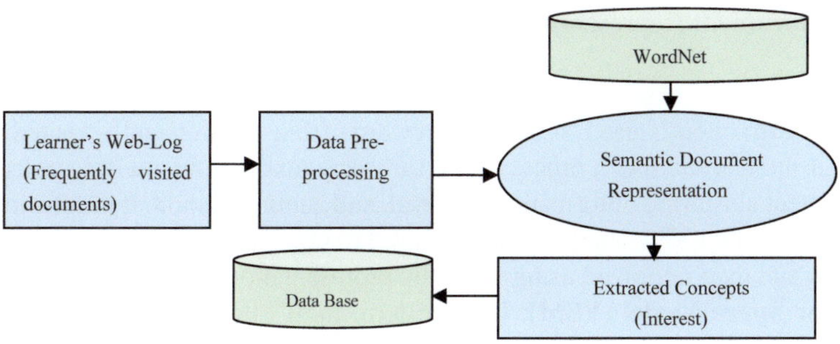

Fig. 1 Architecture of extracting learner interest

3.1.1 Data Collection

Initially, web log files are collected from Moodle LMS for the registered courses. Each web log contains records of IP address, full student name, actions and the information of action done during the interaction. The name identifies the learner in these web logs. From the web logs, the frequently visited documents of each learner are obtained from the "information" field and then send to the pre-processing step to extract the learner interest. The document types included in the courses are in the form of text, PowerPoint, and hyperlinks, and so on.

3.1.2 Document Pre-processing

This stage is used to represent each document selected by the learner as a feature vector, that is, to separate the text in documents into individual words. The extracted documents are initially downloaded from Moodle LMS, and then each document is processed using stop words elimination and stemming methods to identify domain-related words representing the document.

Stop Words Elimination: Stop words are words commonly used in natural language which do not specify any special meaning in an information retrieval (IR) system. They are extremely common words which have little value in helping documents to match with the keyword of user's need. The common words in a text such as prepositions, pronouns, are stop words that do not contribute any special meaning to the documents. Some of the examples of stop words are: a, in, the, an, with, and so on. The stop words elimination method is used to remove stop words from the frequently used documents of each learner which would help to prevent some unnecessary words from being selected as keywords.

Stemming: Stemming techniques are applied to transform the words in texts into their stems. This technique includes a great deal of knowledge in language dependent linguistics. For example, the words, documentation, documents, documented and documenting can be stemmed to the word "document". The proposed system uses standard porter stemming algorithm [5] to find the root word in the document.

3.1.3 Document Representation

The learner interest extraction is based on the primary method of keyword extraction. Keyword extraction is used to extract keywords from the frequently visited documents of the learner. It is a process used for many text mining analysis tasks, such as document clustering, information retrieval, and summarization. It is done initially using the standard VSM method which is used to represent each document by a bag of words and then improved using semantic representation using WordNet.

Vector Space Model (VSM). It is the term most often used for weighting the indexed terms to enhance the retrieval of documents. It is numerical statistics that reflect the importance of a word in a document of the corpus. The importance increases proportionally with the frequency of the same word occurs in the document. This method prevents the user from selecting the words which are often used in specific subjects and the words distributed over most of the documents in the same manner. It is successfully applied for filtering stop words in several applications, including classification and text summarization.

Semantic Representation Using WordNet. The main drawback of VSM is that it does not consider the semantic relatedness of keywords in the documents. To improve the performance of the traditional VSM, the concept of WordNet has been introduced which identifies a set of semantic concepts related to the document words represented by VSM. It is mainly used to obtain a semantic representation of the terms used in the documents.

WordNet [6]. It is a large database of English lexical items which is freely available online. It is widely used in natural language processing and artificial intelligence. It is one of the richest thesauruses that contain more than 150,000 different unique words organized into a hierarchical structure of more than 115,000 synsets. It is compatible with many different dictionaries and other semantic sources such as DBpedia. It supports the conceptual and lexical semantic relationship that structures the concepts and links. It was designed to establish the connection between the lexical items of WordNet into four types of English parts of speech (POS): noun, verb, adjective and adverb. The basic unit of WordNet is synset, which describes synonyms sets that share a common meaning of a specific word. A synset is identified with three parts, which include the word itself, explanation and synonyms. Each synset represents a specific sense of that word. A word may have different meanings. Sense gives the specific meaning of a given word under one type of POS. Each sense of a word may appear in more than one synset. If a word and one type of POS have more than one sense, then WordNet organizes them in the order of the most to least frequently used (Semcor). Synsets connect word and its corresponding sense using explicit semantic relations. Some of these relationships include hypernym, hyponym for nouns and troponym for verbs. The different kinds of semantic relations used are hyponymy vs. hypernymy (is a), holonymy vs. melonymy (is part of), and so on. It can be used as a lexical ontology in which concepts or classes represent the word sense nodes based on the formalism.

The primary purpose is to obtain three different semantic relations such as synonyms, hypernyms and hyponyms from the WordNet database for the document keywords selected using VSM method. WordNet gives an ordered set of different synsets for each keyword in the document. The ordering reflects how common a term reflects the concepts in "standard" English language. The more common term meanings are listed before less common ones.

To find the correct synonyms from the list of synsets, a word sense disambiguation (WSD) technique is used for more precise identification of the proper synonym of the target word when there are multiple meanings. This technique uses a semantic similarity method based on Lin's method to find the exact meaning of the word. The algorithm first checks all the senses of the target word and calculates the similarity between them based on the domain area. The terms having highest similarity are taken as the correct sense, and it is used as a learner interest.

The relation between concepts is essential as it plays a crucial role in capturing the main concepts in the documents. To show the relationships, hypernyms and hyponyms between concepts are considered. These relations are then obtained from the WordNet database for the selected concept. Hypernyms of concepts represent concepts that refer to a broad category of actions and hyponyms represents the concept that refers to more specific meaning applicable to it.

3.2 *Proposed Algorithm*

The algorithm of the learner interest extraction process is shown in Fig. 2.

4 Results

The web log files stored in Moodle LMS are used for the learner interest extraction experiments. These files contain the date and time of access, full student name, IP address, actions and the information of action done during the interaction. Fig. 3 shows a sample of web log file created in Moodle LMS while the students are using the system. From the "information" field of these web log files, the "frequently visited documents" visited more than twice for around 300 learners are downloaded for five different online courses "networking", "network operating system", "data structures", "multimedia" and "introduction to programming" created in the computer science domain of Moodle LMS. The document types are in the form of text, PowerPoint and hyperlinks.

The extracted documents are then pre-processed using stop word elimination and stemming methods. The stop words excluded from the documents are "a", "are", "about", "as", "because", "by", "from", "I", "our", "yours" and so on. Special characters excluded are "!", "$", "*", ":", ";", "+" and so on. In addition to that, numbers and extra spaces are also excluded from the documents. Then a stemming method used to stem the words used in the documents. The porter stemmer algorithm is used in the proposed work, which is the frequently used algorithm in English. The next stage is the learning phase, where the commonly used text representation VSM model is used to represent the document by a bag of words. The final stage relates to the classification stage, where the bags of words are represented using semantic concepts and relations.

The evaluation of the algorithm in Fig. 2 is done with a series of classification experiments on around 20 frequently visited documents from every five pre-defined classes such as networking, network server operating system, data structures, multimedia and introduction to programming. The classification is based on the percentage split of 70% documents for training and 30% documents for testing. The first method

Algorithm **Input: Weblog files of each learner** **Output: Concepts representing learner interest**
1. Access the 'information' field of weblog files of each learner.
2. Download the document visited more than twice.
3. Preprocess documents a. Remove stop words, extra spaces, punctuation symbols, numbers etc. b. Apply standard Porter Stemming Algorithm to transform words into their stem.
4. Apply keyword extraction method using the standard vector space model to find the importance of a word is to a document. a. Construct a document term matrix b. Compute normalized term frequency (TF) TF(w) = (No of times word w appears in a document) / (Total no of words in the document) c. Calculate Inverse Document Frequency (IDF) IDF(w)=log_e(Total no of documents/No of documents with word w in it) d. Calculate TF-IDF weight of words in the selected document W W (w) = TF (w) * IDF (w). e. Select top high-weighted words which are above the threshold value (n% terms from each document according to tf-idf value).
5. Look up WordNet to find the semantic relationship of target keywords a. For each selected words in the document do b. Obtain the synset Si: Si= {S1i, S2i, ….., Sni}, in the hierarchy, such that concept Wi has |Si |=n senses from WordNet. c. Mark Si d. Apply Word Sense Disambiguation (WSD) algorithm to the list of synsets in order to find the correct sense based on the semantic similarity method of Lin's method. i. For each synsets in the hierarchy do ii. Find similarity using the formula: $sim_{lin} = 2*IC(lcs(c1,c2))IC(c1)+IC(c2)$ where lcs(c1, c2) is the lowest common subsumer of concepts c1and c2, and IC returns the information content of the concept. iii. The words are ordered based on the similarity score. iv. The words with the highest similarity score are taken as the final sense. v. Insert the sense into the database.
6. Look up WordNet to find the Hypernyms and Hyponyms: Obtain the hypernym and hyponyms for the selected sense in the previous step and insert them into the database. Hypernyms of concepts represent broad concepts up to a certain level of generality $Hf_e = \sum b_\epsilon H\ (c,r)\ Cf_b$ Hyponym of concepts represent specific concepts from the level of generality. $Hf_o = \sum b_\epsilon H\ (c,r)\ Cf_b$ where H(c,r) is the set of concepts C_H

Fig. 2 Algorithm of extracting learner interest

Time	IP address	User full name	Action	Information
Tue 3 March 2015, 8:23 AM	192.168.200.2	Ayman Al-hashmei	course view section	Outcome 3
Tue 3 March 2015, 8:22 AM	192.168.200.2	Hilal Alqurri	course view	Network Technology and Data Communications (New)
Tue 3 March 2015, 8:07 AM	192.168.200.2	Hilal Alqurri	resource view	L3-PTO Switched Services
Tue 3 March 2015, 8:07 AM	192.168.200.2	Hilal Alqurri	course view section	Outcome 3
Tue 3 March 2015, 8:07 AM	192.168.200.2	Hilal Alqurri	assign view	View own submission status page.
Tue 3 March 2015, 8:04 AM	192.168.200.2	Rashad Albalushi	assign submit	Submission status: Submitted for grading. The number of file(s) : 1 file(s).
Tue 3 March 2015, 8:04 AM	192.168.200.2	Rashad Albalushi	assign submit	Submission status: Submitted for grading. The number of file(s) : 1 file(s).
Tue 3 March 2015, 8:03 AM	192.168.200.2	Rashad Albalushi	assign submit	Submission status: Submitted for grading. The number of file(s) : 1 file(s).
Tue 3 March 2015, 8:02 AM	192.168.200.2	Rashad Albalushi	assign view submit assignment form	View own submit assignment page.
Tue 3 March 2015, 8:02 AM	192.168.200.2	Rashad Albalushi	assign view	View own submission status page.
Tue 3 March 2015, 8:02 AM	192.168.200.2	Hilal Alqurri	course view section	Outcome 3
Tue 3 March 2015, 8:01 AM	192.168.200.2	Ayman Al-hashmei	course view section	Outcome 3

Fig. 3 Sample web log file

extracts the highest weightage keywords from documents of all classes and the second method is used to expand these keywords with synonyms. For both the methods, the following quantities have been computed using Eqs. (1) and (2):

$$\text{Precision} = \frac{\text{Number of relevant document retrieved}}{\text{Total number of documents retrieved}} \tag{1}$$

$$\text{Recall} = \frac{\text{Number of relevant document retrieved}}{\text{Total number of relevant documents}} \tag{2}$$

F-Measure is derived from the precision and recall values. This measure is used to assess the overall performance of the approach on given datasets. It produces a high result only when precision and recall values are both balanced, and thus it is very significant.

$$\text{F-Measure}(\%) = (2 \times \text{Recall} \times \text{Precision})/(\text{Recall} + \text{Precision}) \tag{3}$$

The classification experiment aims to prove that the semantic-based representation using WordNet is better than the term-based representation (TF-IDF) method to express learners' interest. The experimental results shown in Table 1 compare the

Table 1 Experimental results

Topics	Term-based representation (TF-IDF)			Semantic-based representation using WordNet		
	Precision	Recall	F-Measure	Precision	Recall	F-Measure
Networking	0.749	0.544	0.63	0.75	0.703	0.726
Network operating system	0.754	0.583	0.658	0.786	0.725	0.754
Data structures	0.745	0.58	0.652	0.777	0.755	0.766
Multimedia	0.723	0.571	0.638	0.743	0.746	0.744
Introduction to programming	0.691	0.547	0.611	0.732	0.76	0.746

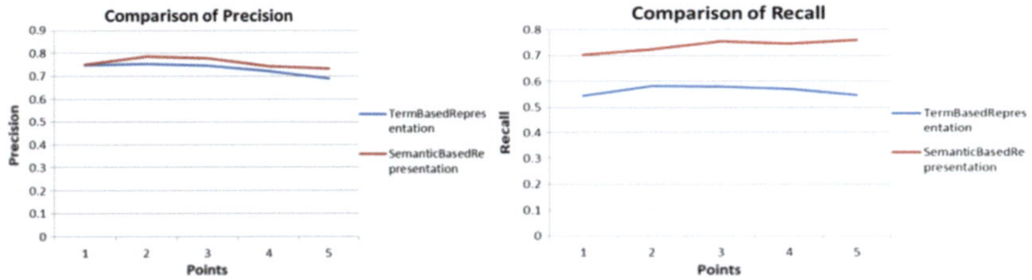

Fig. 4 Comparison of precision and recall of TF-IDF vs WordNet

recall, precision and F-Measure values obtained for both the traditional term-based and semantic-based document representation methods.

Based on the values obtained in Table 1, the graph drawn for the precision and recall values of both methods is shown in Fig. 4. The graph indicates that both the precision and recall values are increased for semantic-based representation compared to the term-based representation.

Evaluation is also done by calculating the overall classification accuracy obtained for both the methods using the formula in Eq. (4):

$$\text{Classification Accuracy} = \frac{\text{No of documents correctly classified}}{\text{Total No of documents}} \tag{4}$$

The graph in Fig. 5 shows that with a training sample of 16–96 documents, the performance of classification obtained using semantic-based representation using WordNet is much better than that obtained using term-based representation. The overall classification accuracy obtained is 89%.

The reason is that in term-based representation, the words per topic are very small, whereas in WordNet-based representation, the words per topic are expanded in a weighted manner. As a result, with a term-based representation for a document, a document has given a low similarity measure with documents from the same topic since they have only a few in common words. With WordNet representation, a document has given a high similarity measure with documents from the same topic, since the words have expanded using WordNet. Hence, the results obtained in the

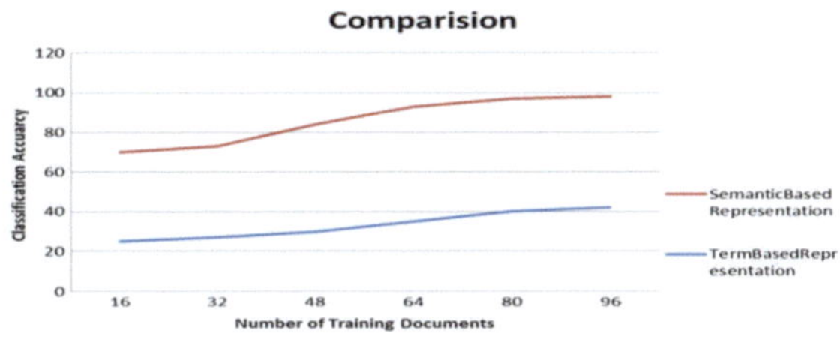

Fig. 5 Classification accuracy

experiments suggest that the integration of WordNet has improved the classification results compared with the traditional term-based representation method.

5 Conclusion

The proposed system is used to extract learners' interest using semantic-based representation using WordNet. The use of WordNet in learner interest retrieval shows better classification performance than compared to the existing method of term representation. The overall classification accuracy obtained is 89%, which proves that the semantic document representation would help to achieve a more powerful identification of relevant terms of the learner. The educators could use the extracted interests for the prediction and recommendation of learning content to the learners.

References

1. Ahmad, A., Kardan., Farahmandnia, F., Omidvar, A.: A novel approach for keyword extraction in learning objects using text mining and WordNet. Glob. J. Inf. Technol. **3**(1), 1–6 (2013)
2. Thakyhkar, A., Chauhan, S.: A novel approach for keyword extraction in learning objects using text mining. IJARIIE. **2**(3), 2395–4396 (2016)
3. Menaka, S., Radha, N.: Text classification using keyword extraction technique. Int. J. Adv. Res. Comput. Sci. Softw. Eng. **3**(12), 734–740 (2013)
4. Fouad, K.M.: Adaptive E-learning system based on semantic web and fuzzy clustering. Int. J. Comput. Sci. Inf. Secur. **8**(9), 308–315 (2010)
5. Porter, M.: An algorithm for suffix stripping. Program Electron. Libr. Inf. Syst. **14** (3), pp. 130–137 (1980)
6. Biasiotti, M., Francesconi, E., Palmirani, M., Sartor, G., Vitali, F.: Legal informatics and management of legislative documents. Global Centre for ICT in Parliament, Working Paper, vol. 2, pp. 62–76 (2008)

Prioritizing Public Grievance Redressal Using Text Mining and Sentimental Analysis

Rama Krushna Das, Manisha Panda and Sweta Shree Dash

Abstract After successful implementation of online grievance monitoring systems by different government agencies, the grievance submission by common citizen has increased many folds. As the number exponentially increases, it becomes difficult for the government authorities to redress the grievances timely, efficiently, and effectively. In this paper, the authors are proposing different text mining and sentimental analysis techniques, on the content of the grievance, to prioritize the grievances submitted to the Chief Minister (CM) grievance cell, Odisha Province. Using these techniques, the grievances are prioritized as high priority, medium priority, and low priority. It helps the concerned government authorities to redress the top priority grievances within a stipulated time period, in comparison to medium and low priority grievances. This helps the needy and common citizen to get timely public services and government support, and their faith and confidence increases on the government machinery.

Keywords Text mining · Sentiment analysis · Lexicon · Afinn · Bing · Grievance · Grievance redressal · High priority · Medium priority · Low priority

R. K. Das (✉)
National Informatics Centre, Berhampur 760004, India
e-mail: ramdash@yahoo.com

M. Panda
Berhampur University, Berhampur 760001, India
e-mail: manishapanda2013sai@gmail.com

S. S. Dash
Institute of Technical Education and Research, Bhubaneswar 751030, India
e-mail: sweta.soa@gmail.com

B. Pati et al. (eds.), *Advanced Computing and Intelligent Engineering*,
Advances in Intelligent Systems and Computing 1082,
https://doi.org/10.1007/978-981-15-1081-6_23

1 Introduction

The measure of information is expanding at exponential rates step by step. All kinds of establishments, associations, and business enterprises are putting away their information electronically. A huge proportion of content is spilling over the web as cutting edge libraries, files, and other printed information, for instance, online diaries, electronic life framework, and messages. It is the most difficult assignment to choose legitimate models and examples to isolate productive data from this broad volume of data. Conventional information mining tools are unable to deal with this textual information since it necessitates more time and push to extricate data. The issue in text mining is finding of valuable information from the unstructured or semi-organized text or document, and is expanding its attention and challenge. Text mining and information mining (data mining) are more or less comparative, aside from data mining which works on organized information, text mining deals with semi-organized and unstructured information [1]. "Data mining is in charge of extraction of certain, obscure, and potential information, whereas text mining is in charge of unequivocal extraction of expressed information in the given content". Then again potential data extraction is basic to both. In this paper, we have put forward a new framework for text mining based on the combination of information extraction (IE) and knowledge discovery from databases (KDD) using data mining. The use of R programming is made for this process. Here we have proposed the use of text mining and sentiment analysis in the sample grievances dataset having 10,050 grievances. The use of text mining first preprocesses the grievances and then assigns a score based on the number of positive and negative words. Sentiment analysis is applied here to find out how much negativity is present in a particular grievance. The application recognizes the grievance which needs quick attention and high priority. Based on negativity and the score, we can judge which grievance need much attention and can prioritize them basing on same. We can create a sorted list depending on the grievances based on the priority of their solving from the above method. This method of solving grievances may take less time and the huge number of grievances can easily be looked into.

2 Literature Review

There are many papers incorporating the text mining and sentimental analysis. Some of the paper works are briefly discussed in this section. In the paper [2], the authors have presented an approach for extricating learning from content reports containing depictions of information in a specialized space. Data extraction in their methodology depends on the clarification of the substance with learning parts (a thought starting in information designing), which we do manually for semantic parts found in framework semantics. The structure executed for this object depends on profound NLP and active learning. Tests have shown a vigorous learning execution, and the results and explanations were of high caliber. In the paper [3], the authors have utilized the text mining methods to examine the fascinating and pertinent data adequately and proficiently from expansive measure of unstructured information. This gives a concise review of text mining strategies that benefit to enhance the text mining process. Particular examples and successions are connected keeping in mind the end goal to remove valuable data by dispensing with unessential points of interest for prescient investigation. In the paper [4], the authors proposed an information portrayal and its logarithmic activities to coordinate ontologies with OLAP frameworks to dissect a colossal arrangement of literary records. By utilizing the strategy, two sorts of data (organized and unstructured data) can commonly upgrade data revelation and examination capacity. In the paper [5], the author outlined a novel and strong machine learning framework for opinion mining and extraction. The model normally coordinates numerous phonetic highlights into programmed learning. The framework can anticipate new potential item and assessment elements in view of the examples it has realized, which is to a great degree valuable in text and web mining because of the many-sided quality and adaptability of natural language.

3 Text Mining and Its Applications

Text mining is otherwise called text analytics. "Text mining is the process of examining large collections of written resources to generate new information, and to transform the unstructured text into structured data for use in further analysis". Text mining can be defined, in other words, as a knowledge-intensive process, as under this the user relates with a group of document. The analysis is done with the help of the tools associated with it [1]. The use of text mining is mainly used for the identification of facts, relationships, and assertions that would have remained hidden because of the vast data [6]. After processing these proofs are removed and turned into structured data for the examination, perception, and combination with organized information in databases or stockrooms. Additionally, refining with the machine learning tools is performed [7]. There are several applications of text mining [8]. Some of them are:

1. Risk management
2. Knowledge management
3. Cybercrime prevention
4. Customer care service
5. Fraud detection through claims investigation
6. Contextual Advertising
7. Business intelligence
8. Content enrichment
9. Spam filtering
10. Social media data analysis.

4 Steps Associated in Text Mining

The steps associated with the text mining processes are discussed underneath:

Step 1: Information Retrieval
The very first step in the data mining process is information retrieval. This step includes the assistance of a web index to discover the accumulation of text, otherwise called corpus of writings, which may require some change [9]. These writings ought to be united in a specific format which will be useful for the clear understanding by the clients. Generally, XML is the standard for text mining.

Step 2: Natural Language Processing
This step empowers the system to play out a linguistic examination of a sentence to import the substance/content. It also dismembers the content in structures.

Step 3: Information Extraction
This is the second stage where the ultimate objective is to recognize the centrality of a particular content. A metadata is made for the substance in this stage and added to the database. It also incorporates adding names or regions to the substance. This movement lets the web searcher for getting the information and finding the associations among the compositions using their metadata.

Step 4: Data Mining
The final stage included here is the data mining utilizing different tools. This progression is used to find out the likenesses among the data those having a similar significance which will be generally hard to discover [9]. Text mining is nothing but a tool supporting the examination procedure and testing of the questions.

5 Odisha State Grievance Redress System

Odisha is a province in the eastern part of India, where the submission of grievances by common citizen to government is increasing steadily day-by-day. The Odisha Chief Minister's Office gets an extensive number of online grievances identified by a number, with segregated topics. Such grievances are sent by the concerned authorities to the appropriate departments of the state government [10]. There are numerous numbers of grievances that are submitted daily. Owing to their vast numbers, some are proceeded further and solved, while some of them are neglected due to many reasons. These neglected grievances may also include some serious ones which may put adverse impact on our society. So, all the grievances should be solved based on their severity rate. e-Abhijoga is a special online grievance redressal portal of Odisha government. Its goal is to encourage online accessibility of the grievance redressal mechanism, to common citizens of the state, thereby providing him/her facility to lodge the grievances at any time and from anywhere. It has the facility to track the status and send updates to the common citizens. e-Abhijoga additionally incorporates paperless concept for handling of grievances. The portal automatically generates a uniquely identified registration number and a grievance ID [11].

6 Grievance Redressal

Grievance redressal is a management and administration-related process used ordinarily in India. While the articulation "grievance redressal" essentially covers the acceptance and handling of complaints from common citizens, a broad definition defines it as the moves made effectively and accurately on any issue raised by the common citizen for better profit and benefits [12]. There are many sorts of grievances, such as individual, group, and policy.

6.1 Individual Grievance

In this one, an individual mourns about the organization/administration action that has ignored their rights under the total comprehension. Cases consolidate educating, minimize goading, wretched gathering, or refuse of earned additional minutes. This helps in examining and helping the part with the grievance. If an individual declines to lament, numerous agreements allow the association lament for the benefit of the nearby. This protects the understanding and guarantees the benefits of various workers.

6.2 Group Grievance

A group grievance mourns that an organization/administration movement has hurt a social event of individuals likewise. For example, when a business failing to pay a legitimately compulsory development premium, a gathering complaint demonstrates solidarity and collects control in a work gathering. In the event that it includes cash, make a point to incorporate all individuals influenced and that they are recorded on the grievance. You may even wish to incorporate dialect in the cure segment of the grievance to "make all members affected, whole in every way".

6.3 Policy or Union Grievance

With a policy or union grievance, the whole associated workers cry that an organization/administration movement insults the agreement. It normally oversees contract interpretation, not an individual complaint.

7 Scope and Theme of the Paper

Most of the serious and urgent grievances are left out unsolved or delayed. This paper mainly focuses on prioritizing of the grievances based on the positive and negative words present in the sentence. Each grievance is given a fixed score based upon which they are sorted. The negative indexed grievances need urgent solution and they are categorized as "high priority". The grievances with 0 indexed value is considered as "medium priority", whereas the positive indexed grievances are categorized as "low priority".

Basing on Positive and Negative Words:

Use of lexicon Bing
Positive percentage = (Total number of positive words in a complaint/Total number of words in that grievance) * 100
Negative percentage = (Total no. of negative words in a complaint/Total no. of words in that grievance) * 100
Need instant solution, if (Positive percentage − Negative percentage) < 0
Maybe solved later, if (Positive percentage − Negative percentage) > 0
Should be looked into, if Positive percentage = Negative percentage.

Basing on Cumulative Score:

Use of lexicon Afinn
Cumulative score = (Total sum of the sentiment scores present in that sentence).

The total score is calculated using the above method and then they are sorted in ascending order. The grievances are solved according to their priority and their scores. The grievances having high negative values are to be solved first and then the lower ones. So the high priority, then the medium priority, and then the low priority grievances need to be solved. By the use of this smart technique of solving the grievances, needed grievances will be solved by getting high attention. The extraction of needed grievances to be solved from the vast number of grievances is very tough. Employing this technique can solve the required grievances fast.

8 Experiment

The experiment is carried out using the R Studio [13]. The version of R Studio used for our experiment is Version 1.0.44—© 2009–2016 RStudio, Inc. Each and every code is written in R programming language and is run under the same version. Several codes relating all the steps are written and run manually. The dataset used here is the sample of grievance dataset that is received from the e-Abhijoga [14]. The dataset consists of 10,050 grievances.

The two lexicons used in our paper are the "Afinn" and "Bing". Afinn lexicon allocates words with a score that keeps running between −5 and 5, with negative scores showing negative estimation and positive scores demonstrating positive feeling. The Bing lexicon arranges words in a paired manner into positive and negative classes. It categorizes each sentimental word of the grievance into positive and negative words, whereas the Afinn lexicon assigns a constant score to the sentimental words in the grievance. Basing on these two lexicons, we have calculated the scores of all grievances and then categorized then into three categories. When the lexicon "Bing" fails to assign any score to the grievance, the "Afinn" lexicon is used. By this, we have segregated the high, medium, and low priority grievances.

Stepwise algorithm for using the lexicons and finding out the total cumulative score of each grievance:

1. Needed packages are installed such as tidyverse, tidytext, glue, stringr, reshape2, ngram.
2. After installation these packages are called and attached from the library.
3. The path of the working directory is set.
4. The sample of grievance dataset is read from that directory.
5. A new table is created with seven columns such as "Grievance ID", "Positive", "Negative", "Total Words", "Positive Percentage", "Negative Percentage", "Total Percentage".
6. A loop is created for reading each grievances.
7. Each grievance is extracted from the dataset. They are cleaned and preprocessed in order to use the text mining techniques to make them suitable for sentimental analysis.
8. Only the sentimental words are taken out of the whole grievance. After that using the "Bing" lexicon, we categorized them into positive and negative grievances.
9. If there is no positive and negative words in that particular grievance, the "Afinn" lexicon is used to calculate the score.
10. After the positive and negative words are found, then the total number of sentimental words in that grievance is calculated.

11. The positive and negative percentage of each grievance is calculated as such:

$$\text{Positive }\% = \frac{\text{Positive Words}}{\text{Total No. of Words}} * 100$$

$$\text{Negative }\% = \frac{\text{Negative Words}}{\text{Total No. of Words}} * 100$$

12. From that the total percentage is found out:

$$\text{Total }\% = \text{Positive }\% - \text{Negative }\%$$

13. Then the loop continues until the scores of all the grievances are calculated.
14. After finishing the table is stored as a .CSV file.

Stepwise algorithm for finding out the status of the grievance based on the score calculated:

1. A grievance is known by its specific unique ID. Each grievance is recognized by its grievance ID.
2. The table stored from the above algorithm is imported here for further processing.
3. Each grievance ID and its score of total percentage is screened based on which they are categorized.
4. By screening each grievance score, they are assigned a status of priority such as:

$$\text{Score} < 0, \text{High Priority}$$
$$\text{Score} = 0, \text{Medium Priority}$$
$$\text{Score} > 0, \text{Low Priority}$$

5. After all the grievance IDs are categorized, they are saved onto a .CSV file in the working directory.

Stepwise algorithm for getting the comparison word cloud of high priority, medium priority, and low priority grievance ids:

1. Needed libraries are added such as wordcloud, tm, etc.
2. The .CSV file having the status of the grievance IDs is imported to R.
3. Only the first 300 grievances are considered for this purpose as the whole datasets grievance IDs cannot be accommodated in the wordcloud.
4. The term document matrix was calculated.
5. The term document matrix was converted into the normal matrix.
6. Using the function comparison.cloud(), the comparison wordcloud was plotted.

Stepwise algorithm for getting comparison wordcloud of all the positive and negative words present in the grievances:

1. The needed libraries were installed and attached.
2. The .CSV file containing the whole grievance dataset was taken.
3. The specific column containing the grievance contents was converted to text.
4. Full grievance text was extracted from the dataset. They are then cleaned and preprocessed in order to use the text mining techniques to make them suitable for sentimental analysis.
5. Only the sentimental words are taken out of the whole grievance. After that using the "Bing" lexicon, we categorize into positive and negative grievances.
6. Then a term document matrix was created based on the negative and positive grievances.
7. The term document matrix was converted to the normal matrix.
8. Using the function comparison.cloud(), the comparison wordcloud was plotted.

Stepwise algorithm for getting the bar plot representation of whole grievance dataset categorizing the high, medium, and low grievances:

1. The needed libraries was installed and attached such as ggplot2.
2. The .CSV file having the grievance status that was done in the previous code was imported.
3. Using the status column the bar plot was made using the function barplot().

9 Discussion

The codes were run through the mentioned R Studio version and the graphs shown in Fig. 3 are extracted from the results. For the pictorial representation in wordcloud the full grievance dataset was not able to accommodate itself. So here we have taken only the first 300 grievances for the representation purpose. Form the whole dataset containing 10,050 number of instances, out of that 6435 number of grievances fall into "high priority" category, 3039 number of grievances fall into the "medium priority" category, and 576 number of grievances fall into the "low priority" category. Table 1 shows the number of grievances that fall into different category from the whole dataset.

Table 1 Representation of the category and their number of grievances

Row labels	Count of grievance ID
High priority	6435
Low priority	576
Medium priority	3039

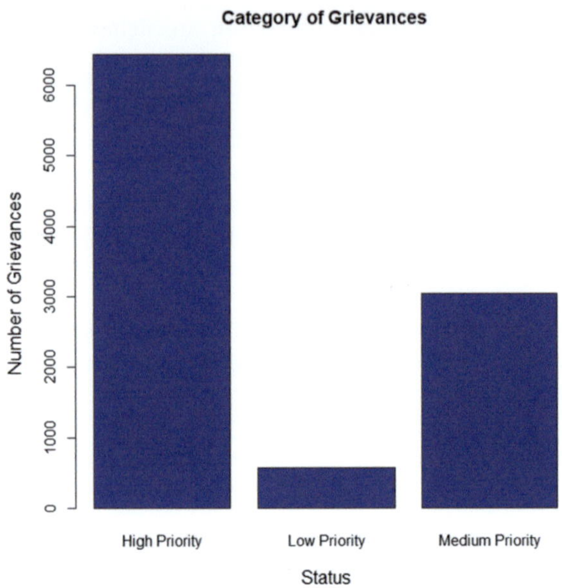

Fig. 1 Bar plot representation of high, low, and medium priority grievances

In Fig. 1, the bar plot representation of Table 1 is shown. Here the number of high priority, medium priority, and low priority grievances are represented graphically. X-axis shows the category of grievances under which it falls, whereas the Y-axis represents the number of grievances.

In Fig. 2, the comparison cloud is plotted for the first 300 grievances. The grievance IDs are shown in the cloud for representing a particular grievance. From first 300 grievances, it is seen that 156 grievances fall under the category of "high priority", 125 grievances fall into "medium priority", and 20 grievances fall under the category

Fig. 2 Comparison wordcloud of high, medium, and low priority up to 300 grievance IDs

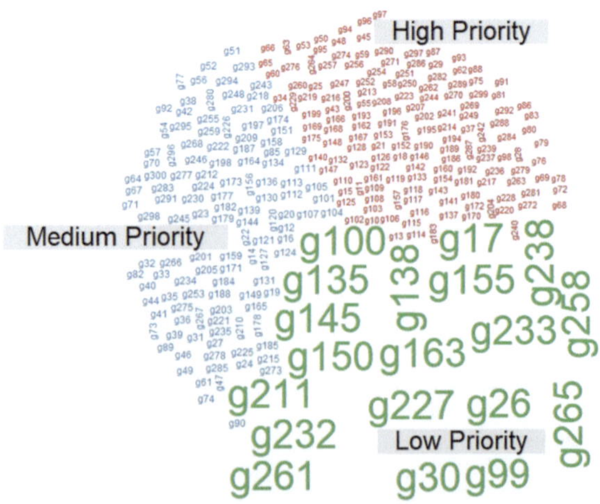

Fig. 3 Comparison wordcloud of positive and negative words in the grievances

of "low priority". The blue region shows the grievance IDs with medium priority; the red region shows the grievance IDs with high priority, and the green region shows the grievance IDs with low priority.

In Fig. 3, the comparison cloud is plotted for the positive and negative words present in the whole grievance dataset. We can see that there are many words present in the whole grievance dataset categorizing into positive and negative words. But here we have taken only those words with a minimum frequency limit of 30. This comparison cloud consists of positive and negative words that have frequency >30. The green region represents the positive words, whereas the red region represents the negative words.

10 Conclusion and Future Work

Prioritizing the grievances narrow downs the number of grievances to give more focus on the top ones by the government authorities. This helps the needy and common citizen to get timely public services and government support, and their faith and confidence increases on the government machinery. The experiment can also be carried out creating a "bag-of-words" for public grievance and finding out the priority based on the number of occurrence of the words in the content of the grievance. Further, the same experiment can be extended for local language grievances (Odia), using together text mining and natural language processing (NLP) techniques on the grievance data. We want to further modify the codes such that we can also apply the sentence-wise sentiment analysis and text mining for prioritizing the public grievances which would be helpful in detecting the tricky, yet diplomatic grievances needing high priority.

References

1. Kumar, S.B., Karthika R.: A survey on text mining process and techniques. Int. J. Adv. Res. Comput. Eng. Technol. (IJARCET) **3**(7) (2014)
2. Mustafaraj, E., Hoof, M., Freisleben, B.: Mining diagnostic text reports by learning to annotate knowledge roles. In: Natural Language Processing and Text Mining. ISBN-10: 1-84628-175-X, ISBN-13: 978-1-84628-175-4
3. Talib, R., Hanif, M.K., Ayesha, S., Fatima, F.: Text mining: techniques, applications and issues. Int. J. Adv. Comput. Sci. Appl. (IJACSA) **7**(11) (2016)
4. Inokuchi, A., Takeda, K.: A method for online analytical processing of text data. In: CIKM'07, 6–8 Nov 2007, Lisboa, Portugal, Copyright 2007 ACM 978-1-59593-803-9/07/0011
5. Jin, W., Ho, H.H., Srihari, R.K.: OpinionMiner: a novel machine learning system for web opinion mining and extraction. In: KDD'09, June 28–July 1, 2009, Paris, France, Copyright 2009 ACM 978-1-60558-495-9/09/06
6. Yogapreethi, N., Maheswari, S.: A review on text mining in data mining. Int. J. Soft Comput. (IJSC). **7**(2/3) (2016)
7. Fatima, E.B., Abdelmajid, E.M.: A new approach to text classification based on naïve Bayes and modified TF-IDF algorithms. In: SCAMS 17, 25–27 Oct 2017, Tangier, Morocco, ACM 978-1-4503-5211-6/17/10
8. Jusoh, S., Alfawareh, H.M.: Techniques, applications and challenging issue in text mining. IJCSI Int. J. Comput. Sci. **9**(6), no. 2 (2012). ISSN (Online) 1694-0814
9. Yuan, C., Yue, Y., Wei, S., Yin, N.: A quality evaluation model for android system based on forum text mining. In: IEEE International Conference on Knowledge Engineering and Applications (2016)
10. http://www.pmindia.gov.in/en/status-of-public-grievances/. Accessed 18 June 2018
11. https://pgportal.gov.in/. Accessed 18 June 2018
12. https://en.wikipedia.org/wiki/Grievance_redressal. Accessed 18 June 2018
13. https://www.rstudio.com/. Accessed 18 June 2018
14. http://www.gaodisha.gov.in/node/727. Accessed 18 June 2018

An Automatic Summarizer for a Low-Resourced Language

Sagarika Pattnaik and Ajit Kumar Nayak

Abstract In the current scenario with the availability of huge volumes of information has given rise to the quench for auto summarizers. Our paper proposes a simple auto summarizer for text document in Odia language, a language that is computationally impoverished. It is a statistical-based extractive text summarizer and does a shallow approach. The summarizer also considers some linguistic features in the process. The sentences for the summary are extracted on the basis of their significant values. The program-generated summary is evaluated against human-generated summaries on the basis of F score values and has got a performance score of 66.92% giving a clear gist of the input text.

Keywords NLP · Text summarization · Extractive · Abstractive · F score

1 Introduction

The imminence of internet has made us available with information that has taken the size of a mammoth. To handle this we entered into natural language processing (NLP). Text summarization is one such task in NLP that retrieves important information from larger texts and serves our need in lesser time and effort [1]. It is the process of making an abridged version of large text, keeping intact its concept and doing it automatically is the task of auto summarizer. There are mainly two approaches to this process: one is extraction-based [2, 3] and the other one is abstraction-based [4]. In extractive approach, the sentences having higher score of significance are extracted in their original form and constitute the summary. But, in abstractive approach there is emphasis on the semantic of the text, and the output summary is a modified condensed form of the input document, keeping intact its

S. Pattnaik (✉)
Department of CSE, ITER, S'O'A University, Bhubaneswar, India
e-mail: sagarika.pari@gmail.com

A. K. Nayak
Department of CS&IT, ITER, S'O'A University, Bhubaneswar, India
e-mail: ajitnayak2000@gmail.com

B. Pati et al. (eds.), *Advanced Computing and Intelligent Engineering*, Advances in Intelligent Systems and Computing 1082,
https://doi.org/10.1007/978-981-15-1081-6_24

meaning. This is a knowledge-rich approach requiring deep linguistic analysis. The proposed summarizer follows the extractive path analyzing the statistical features of the given text to be summarized. Some linguistic features are used in categorizing similar words. The purpose of developing an auto summarizer is many folds but the foremost objective is to get a consistent and unbiased summary. Human-generated summaries require more time and effort and are generally biased. There is lack of consistency. So to get a consistent and less-biased summary, we go for machine-generated summaries or auto abstracts. Auto summarizers have been developed for different languages, particularly for English [5–9] but for Indian languages mainly for Odia language, this field is not much explored [10]. Few attempts have been done related to this task [11, 12], but our method does it in a cost-effective manner using a simple statistical approach considering the significant factors of sentences.

The rest of the segments is arranged as follows: Sect. 2 presents some of the relevant works on the topic. Section 3 provides an overview of Odia language and its characteristics. A discussion on the proposed text summarization model is done in Sect. 4. Sect. 5 explains the experiment and the evaluation process. Finally, in Sect. 6 we conclude our paper with a future direction of further research in this field.

2 Related Works

Shah and Desai [10] have given a brief summary of automatic extractive text summarization techniques adopted for various foreign and Indian languages. Techniques adopted are: sentence scoring method, combination of random indexing and page rank, Bayesian theory-based method, clustering, SVM, stochastic methods using word frequency, TF-IDF, position of phrases in sentences and so on. The general compression ratio is varied within 20–25%. From the study it is found that proper auto text summarizer for Indian languages is still lacking, and no emphasis is given on the ideality of human summaries taken for comparison. Future work includes increasing the number of features for extracting Hindi sentences, using different machine learning techniques to achieve higher accuracy and to test the proposed technique rigorously on large dataset of various domains.

Hans et al. [13] have proposed an automatic text summarizer for English language, implemented with TF-IDF algorithm. They have claimed of obtaining 67% of accuracy in their research work with three data samples which are higher compared to the other online summarizers. The sample used is pure text document. The auto summarizer makes use of existing libraries such as NLTK and Text Blob, so it does not need any machine learning, which is a time-consuming process. But the summarizer is constrained by the fact that it is corpus dependent, that is, with the increase in volume of the corpus its evaluation metric F measure becomes higher. Some improvements that can be applied to this program to produce a more accurate summary are suggested like making the summary biased on the title of the document and increasing the number of experiment with various types of sample documents.

Jagadeesh et al. [14] have proposed a single document extractive-based summarization model. The proposed system receives only text document and is divided into two components, one related to text analysis and the other related to summary generation. They have adopted a shallow approach for the sentence extraction process. The defined feature functions determine the individual sentence scores and their extraction. The sentences are ranked according to their scores, and accordingly, higher ranked sentences are chosen for the summary. They have also given emphasis on the discourse coherence in the summary. They have evaluated their model with human ranking, ranging from 5 (best) to 0 (worst) and have got an average score of 3.25 and coherence value of 4. The limitation of the system is that it has used arbitrary weights by trial and error method. So, they have proposed to use more NLP tools to get precise weights for sentences in the document as a part of the future work. They have also planned to extend the topic for multi-document summarization.

Luhn [15] in his paper "Automatic Creation of Literature Abstracts" suggests that programs for creating auto abstracts must be based on properties of writing perceived by analysis of specific type of literature. His algorithm is based on word frequency and relative position of words within a sentence for calculating significant sentences. His work is considered as the very first work in automatic text summarization. The summarizer is limited at the point when the writings of the author deviate from the average writing pattern. As a result, it may select sentences of inferior significance.

So, after going through some of the related works, it can be concluded that inadequate amount of work has been done for Indian languages. Limited emphasis has been given on anaphora resolution which is an important point in summarization task. As far as Indian language is concerned, its morphological richness brings complexity in the computation process. It is also seen that the size of the corpus is an important constraint for the performance summarizer and most of the language processing works are statistical based.

3 Morphological Features of Odia Language

Odisha is a land of rich cultural heritage and so is its language Odia. It belongs to the Indo-Aryan branch of the Indo European language family. In the year 2014, it has been declared as sixth classical language of India by the Indian constitution. Earlier it was a Scheduled Language under the Eighth Schedule of the Constitution of India. Odia is a morphologically rich language [16], that is, number of word forms per lexeme is more comparative to English language. It also has an agglutinative character. Unlike English language, the structure of the sentences is in the form of *subject object verb (SOV)* and prepositions are rare in Odia and are replaced by postpositions [17]. Unlike Hindi, in Odia language most of the postpositions are attached to the main word (noun/pronoun) similar to Marathi. There are no capital letters to distinguish the words as noun. From the attached suffixes words can be identified as nouns and verbs to some extent and lastly this language is computationally less explored.

4 Proposed Model

The model has been divided into three subsections, that is, preprocessing, word analysis and sentence analysis and extraction for summary generation.

The compression factor is kept at around 50%.

Algorithm 1 describes the overall procedure carried out by the model.

The model first preprocesses the text, which involves tokenization followed by stripping of noun suffixes and verb suffixes from words or tokens containing them. This is done with reference to the maintained suffix list.

For example, the word ପିଲାଙ୍କୁ(pilāṅku) is reduced to ପିଲା(pilā)

ପିଲାଙ୍କୁ(pilāṅku) → ପିଲା(pilā)

This step is done to reduce the complexity of the words as Odia words have long suffixes attached to them.

Then removal of Odia pronouns, postpositions, prepositions, conjunctions, question words and punctuations is carried out by referring to the dictionary that contains this list.

Example

Pronouns: ସେ, ତୁ(se, tu)

Postpositions: ଉପରେ, ତଳେ, ଆଗରେ(upare, taḷe, āgare)

Prepositions: ଶ୍ରୀ, ଶ୍ରୀମତି, ଶ୍ରୀଯୁକ୍ତ(sri, srimati, srijukta)

Conjunctions: ଓ, ତ, ଏବଂ(o, ta, ebaṁ)

Question words: କିଏ, କେ, କି(kie, ke, ki)

Punctuations: , ? !

The remaining words are then arranged in an alphabetic order.

On the basis of similarity percentage among words, consolidation process is carried out as described in Algorithm 1a.

Based on the frequency of word occurrence, significant words are listed, that is, words of a stipulated low frequency (<5) are considered as nonsignificant and the rest considered as significant words. This threshold value is set in consultation with human experts.

The model then determines the significant factor of each sentence, as described in Algorithm 1b.

Sentences above a certain threshold significant factor value are considered for the summary.

The threshold significant factor value for each document is made to vary to maintain the compression factor at around 50%.

Algorithm 1 Text Summarization Algorithm

```
1. Preprocessing the text document:

   (i)  Tokenization
   (ii) Refer to the suffix dictionary and remove noun
        suffixes and verb suffixes from words to which
        they are attached.
```

(iii) Remove the pronouns, postpositions, prepositions, conjunctions and punctuations with reference to the dictionary containing the list.

2. Arrange the remaining words in an alphabetical order.
3. Consolidation of words is done in algorithm 1a.
4. Count the occurrence of similar words derived through algorithm 1a
5. Words of a stipulated low frequency (<5) are deleted from the list.
6. The remaining words are considered as the significant words.
7. Determine the ***significant factor*** of each sentence according to algorithm 1b.
8. Sentences above a certain threshold significant value are considered for summary.

Input: Text document to be summarized

ଦେଶବାସୀଙ୍କ ପ୍ରତି ଏକ ବେପରୁଆ ଉପହାସ ଭଳି ବର୍ତ୍ତମାନ ଦେଶର ଦୁଇ ପ୍ରମୁଖ ରାଜନୈତିକ ଦଳ ମଧ୍ୟରେ ଏକ ଶଠତାପୂର୍ଣ୍ଣ ଉପବାସ ପ୍ରତିଯୋଗିତା ଦେଖିବାକୁ ମିଳିଛି । (1.8)
ପ୍ରଥମ ରାହୁଲ ଗାନ୍ଧୀଙ୍କ ନେତୃତ୍ୱରେ ମୋଦୀ ସରକାରର ତଥାକଥିତ ଦଳିତ ବିରାଧୀ ଆଭିମୁଖ୍ୟ ବିରୁଦ୍ଧରେ କଂଗ୍ରେସ ନେତାମାନେ ଅନଶନରେ ବସିବା ପରେ ଆଜି ଇଟାର ଜବାବ ପଥରରେ ଦେବା ଭଳି ନରେନ୍ଦ୍ର ମୋଦୀଙ୍କ ନେତୃତ୍ୱରେ ତାଙ୍କ ଦଳ ଭାରତୀୟ ଜନତା ପାର୍ଟିର ସମସ୍ତ ସାଂସଦ ଦିନ ତମାମ ଅନଶନରେ ବସୁଛନ୍ତି । (1.6)

desabāsiṅka prati eka beparuā upahāsa bhaḷi barttamāna desara dui pramukha rājanaitika daḷa madhẏare eka saṭhatāpurṇṇa upabāsa pratijogitā dekhibāku miḷichhi. (1.8)
prathama rāhula gāndhiṅka netrutuare modi sarakārara tathākathita daḷita birādhi ābhimukhẏa biruddhare kaṁgresa netāmāne anasanare basibā pare āji iṭāra jabāba patharare debā bhaḷi narendra modiṅka netrutuare tāṅka daḷa bhāratiẏa janatā pārṭira samasta sāṁsada dina tamāma anasanare basuchhanti. (1.6)

Output: Program generated summary

ଦେଶବାସୀଙ୍କ ପ୍ରତି ଏକ ବେପରୁଆ ଉପହାସ ଭଳି ବର୍ତ୍ତମାନ ଦେଶର ଦୁଇ ପ୍ରମୁଖ ରାଜନୈତିକ ଦଳ ମଧ୍ୟରେ ଏକ ଶଠତାପୂର୍ଣ୍ଣ ଉପବାସ ପ୍ରତିଯୋଗିତା ଦେଖିବାକୁ ମିଳିଛି । (1.8)

desabāsiṅka prati eka beparuā upahāsa bhaḷi barttamāna desara dui pramukha rājanaitika daḷa madhẏare eka saṭhatāpurṇṇa upabāsa pratijogitā dekhibāku miḷichhi. (1.8)

The numbers in the bracket denote the significant factor of the sentence.

Sentences above significant factor value >1.7 are chosen for summary in the given example.

Algorithm 1a describes the process of finding similarity among words and consolidating them.

Similarity is found through character matching in each pair of words.

Algorithm 1a Consolidation of Words

```
For each word in the list of sorted words
```

1. `Do character by character matching for each pair of words.`
2. `Calculate percentage matching for each pair.`
3. `Based on the match percentage the similarity of words are tested and determined.`

Example

Input: ଘର, ଘରେ, ଘରୁ(hara, hare, haru)

Output: ((ଘର:3): 3) (hara:3)

The number 3 denotes the frequency of word ଘର(hara).

Algorithm 1b determines the significant factor of each sentence.

The input document is split into individual sentences based on the sentence boundaries. The significant words in each sentence are then identified referring to the significant word list. The linear distance between these significant words is determined on the basis of number of intervening nonsignificant words between them. Word clusters are formed taking significantly related significant words. Words are said to be significantly related if they are found at close vicinity to each other. So a limit is set on the number of intervening nonsignificant words in a cluster. On this criterion, a sentence can have more than one cluster. This limit is varied for values ≤3, 4, 5 and 6 and the ideal limit value ≤3 is considered after testing. The significant factor for a word cluster and finally for the whole sentence is determined by using the Eq. 1.

$$Significant\ factor = \frac{Square\ of\ the\ number\ of\ bracketed\ significant\ words}{Total\ number\ of\ bracketed\ words} \tag{1}$$

Example

1 2 3 4 5 6 7

Sentence1: [ନିଉଟନ୍‌ଙ୍କ ସୁପରିଚିତ ତୃତୀୟ ନିୟମ ଅନୁସାରେ ପ୍ରତି କାର୍ଯ୍ୟର] ଏକ ସମାନ ଓ ବିପରୀତ ପ୍ରତିକ୍ରିୟା ସୃଷ୍ଟି ହୋଇଥାଏ।

The words underlined are significant words.

Number of significant words = 4

Total number of bracketed words = 7

Using Eq. 1 significant factor of the sentence is evaluated, that is, significant factor $= 4^2/7 = 2.28$

Based on a certain threshold value of significant factor, sentences are extracted for the summary.

Algorithm 1b Significant factor of each sentence

```
1.Initialize w_i=0, sw=0, nsw=0, count=0,
  START_CLUSTER=NULL, END_CLUSTER=NULL
2.For each word w_i in sentence S_i
  If w_i[]= Significant word
    Then START_CLUSTER= w_i
3. count=count + START_CLUSTER
4. sw=sw+1
5. w_i=w_i+1
6.If w_i[] = significant word
   Then
   Goto step4
   reset nsw=0
  Else
     If w_i[]= Non significant word
     Then nsw=nsw+1
       If nsw≤3
        Then
        Go to step5
       Else END_CLUSTER= w_i-2
       Count= END_CLUSTER
7. Find Significant Factor as in equation1
```

If there is more than one cluster in a sentence, the significant factor value of the highest one is considered as a measure for the sentence. Here nsw stands for nonsignificant word index and sw stands for significant word index.

Input: ନିଉଟନ୍‌ଙ୍କ ସୁପରିଚିତ ତୃତୀୟ ନିୟମ ଅନୁସାରେ ପ୍ରତି କାର୍ଯ୍ୟର ଏକ ସମାନ ଓ ବିପରୀତ ପ୍ରତିକ୍ରିୟା ସୃଷ୍ଟି ହୋଇଥାଏ

Output: Significant factor = 2.28

5 Experiment and Evaluation

Excerpts of news article from online sources are taken as input for auto text summarization task.

Three human-generated summaries for each document are considered for validation of the resultant output summary. F score is a composite measure that combines precision and recall is taken as the evaluation metric for the resultant summary [18, 19].

$$F\,score = \frac{2pr}{p+r} \tag{2}$$

where p is precision and r is recall

$$p = \frac{No.of\ sentences\ similar\ in\ system\ generated\ summary\ and\ ideal\ summary}{No.of\ sentences\ in\ system\ generated\ summary} \tag{3}$$

$$r = \frac{No.of\ sentences\ similar\ in\ system\ generated\ summary\ and\ ideal\ summary}{No.of\ sentences\ in\ ideal\ summary} \tag{4}$$

Table 1 shows the F score values of different auto summaries generated with respect to three human-generated summaries. Table 2 shows the standard deviations of the human-generated summaries from their mean value and from Table 3 we derive the average F score value of our auto summarizer.

These program-generated summaries are compared with three human-generated summaries and their F score values are calculated. The program-generated summary that gives the highest F score value for the ideal human-generated summary is considered for our result evaluation. The ideal human-generated summary is found out on the basis of standard deviation value. The standard deviation value gives variation of the data from their mean value.

From Table 1 it has been analyzed that the overall average F score of program summary 1 (PS1) considering number of intervening nonsignificant word length distance $\leq$3 with three human-generated summaries is found to be 62.89, with $\leq$4 (PS2) is found to be 55.95, with $\leq$5 (PS3) is found to be 56.32 and for program summary 4 (PS4) with number of intervening nonsignificant word length distance $\leq$6 is found to be 53.222.

So, it can be concluded that the nonsignificant word length distance $\leq$3 gives a better result and hence we considered the program summary (PS1) for our result evaluation.

Based on this standard deviation value, ideal human-generated summary is selected for output evaluation. The human-generated summary that shows least deviation from its mean value is considered for individual documents, so as to strengthen the validity of the output.

The output generated by the program is almost at par with the summary generated by human experts and is giving a meaningful result.

Table 1 Auto summaries with their F score values

Document	Total no. of lines	Program generated summary (PS)			
		Program summary (PS) with no. of intervening NSW	F score w.r.t. HS1 (in %)	F score w.r.t. HS2 (in %)	F score w.r.t. HS3 (in %)
Doc1	32	≤3	54.54	60.60	**64.70**
		≤4	64.51	51.61	56.25
		≤5	60.60	54.54	58.82
		≤6	51.61	45.16	50
Doc2	35	≤3	68.57	61.11	52.94
		≤4	68.57	55.55	41.17
		≤5	**70.27**	57.89	44.44
		≤6	61.11	64.86	51.43
Doc3	37	≤3	66.56	71.79	64.86
		≤4	**77.1**	57.89	47.36
		≤5	71.72	56.4	46.15
		≤6	64.92	50	50
Doc4	40	≤3	**65**	53.65	60
		≤4	55	43.90	40
		≤5	60	43.90	40
		≤6	55	43.90	30
Doc5	64	≤3	66.66	**72.13**	60.31
		≤4	61.53	66.66	52.30
		≤5	62.5	64.51	53.12
		≤6	61.53	63.49	55.38

HS1: Human-generated summary1, PS1: Program summary 1
HS2: Human-generated summary2, PS2: Program summary 2
HS3: Human-generated summary3, PS3: Program summary 3
NSW: Nonsignificant words, PS4: Program summary 4
PS5: Program summary 5

Table 2 Standard deviation values for human-generated summaries

Document	F score1 (HS1 and HS2) in %	F score2 (HS2 and HS3) in %	F score3 (HS1 and HS3) in %	Average	Standard deviation
Doc1	75	72.72	60.60	69.44	7.74
Doc2	70.27	72.72	51.42	64.64	11.48
Doc3	55	73.68	63.15	63.94	9.36
Doc4	55.81	41.86	71.43	56.36	14.79
Doc5	80.64	64.51	78.12	64.42	8.67

Table 3 Resultant ideal auto summary and human summary with F score value

Document	Program summary	Human summary	F score (in %)
Doc1	PS1	HS3	64.70
Doc2	PS1	HS1	70.27
Doc3	PS1	HS1	66.56
Doc4	PS1	HS1	59.55
Doc5	PS1	HS2	72.13

Average F score value = 66.928 ≈ 67%

6 Conclusion

This is a novel work in the text summarization of Odia text. The statistical method employed overcomes the need of tagging. Proposed model is not limited by the unavailability of annotated corpus nor does it require a stemmer, thus making the summarization process simple. The summary generated by the program has kept the document concept intact. On testing our automatically generated summaries with manually generated ones, we got an output result of 66.92%. Though summary generation is a complex task, simple statistical methods employed in our experiment have made the task simpler, giving output relevance for judgment. As far as Odia language is concerned, their morphological feature like the attachment of postpositions to the words and long suffixes make the computation a bit complex, so we have solved this problem by application of linguistic method. Hope this work will be a small contribution to the society as auto abstracts are reliable, consistent and stable. Further enhancement of the work can be done by taking into consideration more linguistic features of the language and solving issues like dangling anaphors.

References

1. Siddiqui, T., Tiwari, U.S.: Natural Language Processing and Information Retrieval. Oxford University Press, pp. 343–358
2. Moratanch, N., Chitrakala, S.: A survey on extractive text summarization. In: International Conference on Computer, Communication and Signal Processing (ICCCSP). IEEE, pp. 1–6 (2017)
3. Gupta, V., Lehal, G.S.: A survey of text summarization extractive techniques. J. Emerg. Technol. Web Intell. **2**(3), 258–268 (2010)
4. Moratanch, N., Chitrakala, S.: A survey on abstractive text summarization. In: International Conference on Circuit, Power and Computing Technologies (ICCPCT). IEEE, pp. 1–7 (2016)
5. Wu, Z., Lei, L., Li, G., Huang, H., Zheng, C., Chen, E., Guandong, Xu: GA topic modeling based approach to novel document automatic summarization. Expert Syst. Appl. **84**, 12–23 (2017)
6. Baralis, E., Cagliero, L., Jabeen, S., Fiori, A., Shah, S.: Multi-Document Summarization Based on The Yago Ontology. Expert Syst. Appl. **40**(17), 6976–6984 (2013)
7. Haque, Md, Majharul, S.P., Zerina, B.: Literature review of automatic single document text summarization using NLP. Int. J. Innov. Appl. Stud. **3**(3), 857–865 (2013)

8. Salton, G., Christopher, B.: Term-weighting approaches in automatic text retrieval. Inf. Process. Manag. **24**(5), 513–523 (1988)
9. Seki, Y.: Sentence extraction by tf/idf and position weighting from Newspaper Articles (2002) cite seer
10. Shah, P., Desai, N.P.: A survey of automatic text summarization techniques for Indian and Foreign languages. In: International Conference on Electrical, Electronics, and Optimization Techniques (ICEEOT). IEEE, pp. 4598–4601 (2016)
11. Desai, N.P., Shah, P.: Automatic text summarization using supervised machine learning technique for Hindi langauge. Int. J. Res. Eng. Technol. (2016)
12. Krishnaprasad, P., Sooryanarayanan, A., Ramanujan, A.: Malayalam text summarization: an extractive approach. In: International Conference on Next Generation Intelligent Systems (ICNGIS). IEEE (2016)
13. Hans, C., Agus, M.P., Suhartono, D.: Single document automatic text summarization using term frequency-inverse document frequency (TF-IDF). ComTech: Comput. Math. Eng. Appl. 7(4), 285–294 (2016)
14. Jagadeesh, J., Prasad, P., Varma, V.: Sentence extraction based single document summarization. Int. Inst. Inf. Technol. Hyderabad, India 5 (2005)
15. Luhn, H.P.: The automatic creation of literature abstracts. IBM J. Res. Dev. **2**(2), 159–165 (1958)
16. Sethi, D.P.: A survey on Odia computational morphology. Int. J. Adv. Res. Comput. Eng. Technol. (IJARCET) 3(3) (2014)
17. Pradhan, K.C., Hota, B.K., Pradhan, B.: Saraswat Byabaharika Odia Byakarana, Styanarayan Book Store, Fifth Edition (2006)
18. Steinberger, J., Karel, J.: Evaluation measures for text summarization. Comput. Inform. **28**(2), 251–275 (2012)
19. Indu, M., Kavitha, K.V.: Review on text summarization evaluation methods. In: International Conference on Research Advances in Integrated Navigation Systems (RAINS). IEEE (2016)

Printed Odia Symbols for Character Recognition: A Database Study

Sanjibani Sudha Pattanayak, Sateesh Kumar Pradhan and Ramesh Chandra Mallik

Abstract Optical Character Recognition (OCR) is a popular tool which helps in converting a character image to a text form. It is also recognized as one of the most popular devices to digitize the printed documents. The main challenge in Odia character recognition is to capture a set of symbols which have been used in the printing technology. There are many letters and symbols with similar structure, and also different printing materials follow different font styles. Again, there are allographs which can be used while writing in Odia. So, it is a challenge to classify the orthographs and allographs accurately. It is a fact that no standard database can be found in Odia for training and testing purposes. The main goal of this paper is to develop a database for complete set of printed Odia character symbols and then recognize using SVM classifier through a graphical user interface and the DWT is specifically used to extract the features of the character image and process them in a scientific manner.

Keywords OCR · Letters · Symbols · Image interpolation · Gaussian blur · DWT · SVM

1 Introduction

Odia is the official language of Odisha. It is specified as one of the schedule languages of the constitution of India. In the year 2014, Odis is declared as one of the languages which holds a status of classical languages. It is the language mostly spoken by the people of Odisha which is a state that lies in the eastern part of India. According to

S. S. Pattanayak · S. K. Pradhan (✉)
Department of Computer Science and Application, Utkal University, Bhubaneswar, India
e-mail: sateesh.cs@utkaluniversity.ac.in

S. S. Pattanayak
e-mail: sanjibani@gmail.com

R. C. Mallik (✉)
P.G. Department of Odia Language and Literature, Utkal University, Bhubaneswar, India
e-mail: ramesmalik@gmail.com

B. Pati et al. (eds.), *Advanced Computing and Intelligent Engineering*,
Advances in Intelligent Systems and Computing 1082,
https://doi.org/10.1007/978-981-15-1081-6_25

the 2011 census, globally more than 36 million people use Odia language in their intra-communication. It is the official language of the state Odisha. It is a high time to justify the importance of language technology, i.e., specific to Optical Character Recognition (OCR) tool for the development Odia language and revitalization of the Odia written documents.

The printed Odia texts normally include 47 atomic symbols, 10 numerals, around 15 modifiers, punctuation marks, and many conjunctions which are combination of two or more consonants. Identifying such a large set of symbols is a real challenge. Besides which, there are different font styles. Separating the modifier from the actual symbol is also a difficult task as the modifier's position is not fixed always. The above-stated requirement and problems motivate to work in this field.

2 Literature Review

There are a few research works on the OCR that have been carried out by the scholars and most of the works on character recognition for Odia language, that are reported till now, are for handwritten Odia numerals. But printed character recognition has not yet got significant amount of attention by the researchers. A few databases for Odia handwritten characters have been developed and discussed, whereas the database for printed Odia letters and symbols has not yet given emphasize. We can find an image database for handwritten Odia numeric symbols in ISI, Kolkata website. This database contains 5970 samples collected from 356 persons [1]. Similarly, NIT Rourkela team has also developed a database for 47 atomic Odia characters and 10 Odia numeric symbols which are handwritten [2]. This contains 17,100 characters collected from 150 different persons. This database is available in NIT, Rourkela website. Another database creation is also reported by Dash [3]. A database of 5000 samples of handwritten Odia numerals and 35,000 samples of handwritten Odia characters which are collected from 500 individuals is available in IITBBS database.

Nayak and Nayak have developed a training process for tesseract engine for printed Odia atomic symbols [4]. Shanti and Duraiswamy have proposed a system for recognition of Tamil characters using SVM. Here, zonal features are extracted from the image which results in an accuracy rate of 82% [5]. Liu and Fujisawa have given interest about different types of classification strategies for character recognition in detail [6]. The kind of learning-based classification methods included here are statistical method, artificial neural network, support vector machine, multiple classifier combination, etc. In Ref. [7], DWT has been used for feature extraction of printed Odia atomic character. Here, Euclidean distance method has been used for classification. A comprehensive review of different feature extraction techniques of image data has been provided in Ref. [8] which includes statistical features, geometrical features, moment-based features, wavelet features, hybrid features, etc. Bhowmick et al. have proposed an SVM-based hierarchical system for recognition of Bangla characters [9]. With 45 character symbols, a comparative study has been performed with multilayer perceptron and RBF. Here, SVM's performance is found to be better in comparison to other two. Jyothi and group [10] have studied on Telugu character recognition with the help of discrete wavelet transform, projection profile, and singular value decomposition feature extraction techniques. The feature sets are passed through K-nearest neighbor and SVM classifier. It is found that SVM and discrete wavelet transform combination is giving better result.

3 Proposed Method

The present paper basically focuses on developing an extensive database of complete Odia symbols and classifying them. To make the image recognition process user-friendly, a graphical user interface has been developed.

4 Tools Used

4.1 DWT

The Discrete Wavelet Transform (DWT) extracts the features of the image by passing it through low-pass filter and high-pass filter. The filters are basically one dimensional in nature, whereas the images are two dimensional. So, first, it would be applied along rows, then along columns. Discrete wavelet transform splits up the input into low-pass level and high-pass level and the bandwidth to half. After down-sampling, one low-frequency sub-band and three high-frequency sub-bands will be generated for the image. The low-frequency sub-band (LL) is known as approximation coefficient, whereas the high-frequency sub-bands (LH, HL, HH) are known as detailed coefficient. LL provides row-wise low and column-wise low frequency. HL provides row-wise high and column-wise low frequency. LH provides row-wise low and column-wise high frequency. HH represents both row-wise and column-wise high frequencies, i.e., it provides diagonal features. After the end of a single level, LL holds half of the original frequency but does not lose any of the original information. That is why we can say that LL is very similar to original image. In our experiment, db1 wavelet is used for two levels of decomposition (Figs. 1 and 2).

4.2 Support Vector Machine

It is a supervised learning tool which can classify two linearly separable classes with the help of a suitable hyperplane. To work with multiclass problem, here one-against-one approach has been adopted. That means several SVMs are gathered together to

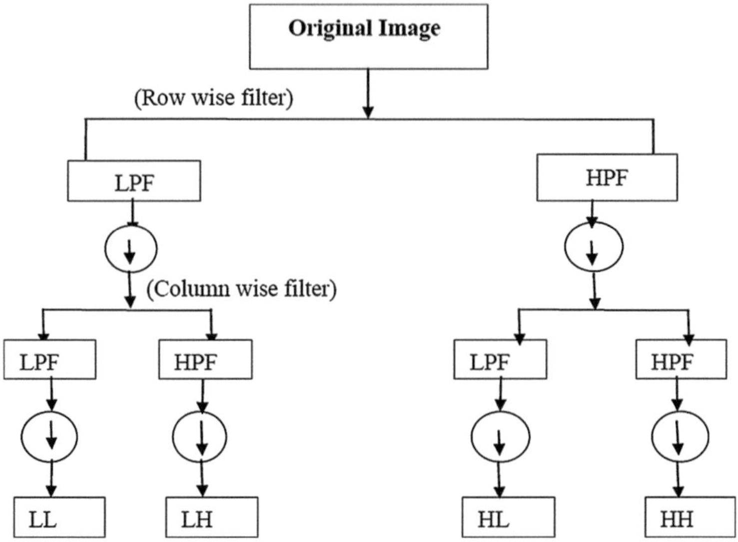

Fig. 1 Processing of a single level of DWT

perform the classification. Internally, libSVM is used to perform all the computations. In comparison to polynomial and RBF kernel, linear kernel is found to perform better.

5 Experiment and Result

For the experiment purpose, Python scipy, numpy are used here which has a rich set of tools for image processing, machine learning, and other mathematical computations. For the preparation of database of printed characters, around 300 Odia character symbols were listed. These character samples are first typed in a text document. Multiple documents are prepared for more than 32 different Odia fonts (e.g., Sarala, Kalinga, Lohita-Oriya, Akruti, etc.). Simple formatting like bold, italic, and normal was also applied for more variation of the same font. Printed copies of these documents are then scanned to get the image form of the same. Then these image forms of documents are line segmented and character segmented to extract the image of individual symbols. Different varieties of the same symbols were accumulated in a single folder. Like this, there are around 300 folders each for a particular symbol. In the database, there are more than 26,000 images of two hundred and thirty-nine symbols of Odia language. A few symbol could not be collected because either it is not included in different fonts or could not be segmented by our underdeveloped character segmentation tool. It will be included soon by our next target. Except few, most of the symbols have approximately 100 varieties. To bring all the image data to

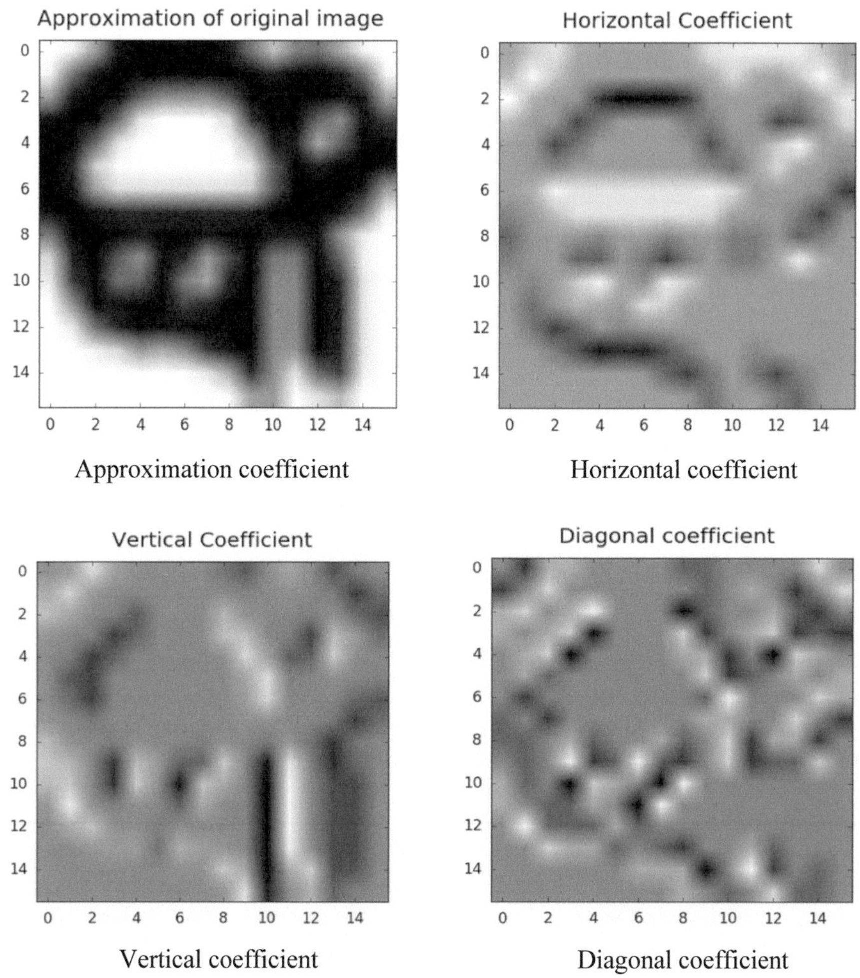

Approximation coefficient

Horizontal coefficient

Vertical coefficient

Diagonal coefficient

Fig. 2 DWT output for image ଙ୍ଗ in order

the same feature space, basic preprocessing has been applied. Bounded box is formed for all the images and the size is normalized to 32 × 32 pixel space. An example of images of Odia symbol with diverse fonts is shown in Fig. 3.

For the recognition process, from the database, 25,327 image data are considered which are free from all sorts of noise. Out of which, 22,794 numbers of images are used for training and remaining 2,533 images are used for testing purpose. This database may be available to other researchers by contacting the authors.

Here, the images which are involved in this process are pretty clean. Noisy images are not considered for this experiment. The noise issue will be focused in the next

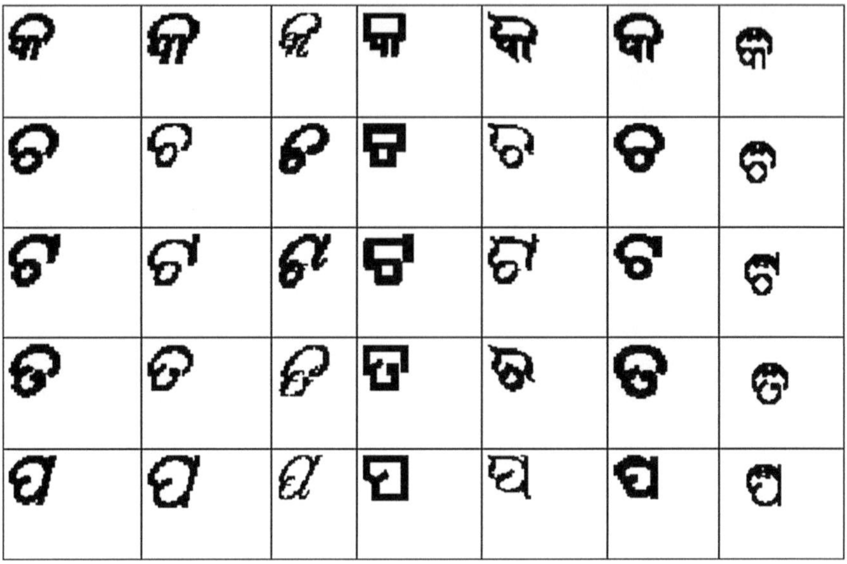

Fig. 3 Odia symbols କ(*ka*) ଚ(*ca*) ତ(*ta*) ଟ(*Ta*) ପ(*pa*)in different fonts

phase of this research process. So, we aim to provide a system which will obtain a high accuracy in Odia character recognition.

A set of characters that has been considered here for the experiment is listed in Table 1. Remaining few characters will become a part of the database soon.

The block diagram shows the steps that we have followed for the character recognition process (Fig. 4).

Preprocessing of images is a very essential phase of character recognition process. Boundary box creation and image size normalization are two important steps of preprocessing carried out for all the images of the database. Image binarization is the next step of preprocessing phase. Otsu's method of binarization is very popular, and hence adopted here.

As images are resized, for pixel adjustment and clarity, image interpolation has been applied and it is found that reflect mode is giving the best result. To get a smooth edge for the image, Gaussian blur technique is applied.

The paper mainly focuses on the performance of DWT feature extraction technique with SVM classifier by using a two-level DWT, and as a result 448 features have been generated for each image. So, for training purpose, a total of 22794×448 samples are accumulated. These data are then passed to an SVM tool for supervised learning.

For recognition of character, a graphical user interface is used to develop with Python tkinter (3.0) tool. A snapshot of the same is shown below. Here, there is a command button (*Browse*) which helps to browse the image of a character that is to be recognized. The image is displayed in the interface. Once the *convert* command

Table 1 List of Odia symbols that are considered for this experiment

ଅ	ଆ	ଇ	ଈ	ଉ	ଊ	ଋ	ଏ	ଐ	ଓ	ଔ	କ	ଖ
ଗ	ଘ	ଙ	ଚ	ଛ	ଜ	ଝ	ଞ	ଟ	ଠ	ଡ	ଢ	ଣ
ତ	ଥ	ଦ	ଧ	ନ	ପ	ଫ	ବ	ଭ	ମ	ଯ	ର	ଳ
ସ	ଶ	ଷ	ହ	କ୍ଷ	ୟ	ଲ	ଡ଼	ଢ଼	କ୍ର	ଖ୍ର	ଗ୍ର	ଘ୍ର
ଝ୍ର	ଜ୍ର	ଟ୍ର	ଡ୍ର	ଢ୍ର	ଦ୍ର	ଧ୍ର	ପ୍ର	ବ୍ର	ଭ୍ର	ମ୍ର	ଯ୍ର	ଶ୍ର
ହ୍ର	କ୍ଲ	ଗ୍ଲ	ଶ୍ଲ	ପ୍ଲ	ମ୍ଲ	ଲ୍ଲ	ହ୍ଲ	କ୍ୱ	ଜ୍ୱ	ଟ୍ୱ	ଣ୍ୱ	ତ୍ୱ
ଥ୍ୱ	ଢ୍ୱ	ଧ୍ୱ	ନ୍ୱ	ଯ୍ୱ	ଶ୍ୱ	ହ୍ୱ	ଗ୍ନ	ଘ୍ନ	ଢ୍ନ	ନ୍ନ	ପ୍ନ	ମ୍ନ
ଶ୍ନ	ଯ୍ନ	ହ୍ନ	କ୍ଷ୍ମ	ଶ୍ମ	ଯ୍ମ	ଙ୍ମ	ଣ୍ମ	ଷ୍ମ	ଷ୍ଣ	ସ୍ମ	ଡ୍ମ	ହ୍ମ
ଷ୍ମ	ର୍କ	ର୍ଖ	ର୍ଗ	ର୍ଘ	ର୍ଚ	ର୍ଛ	ର୍ଜ	ର୍ଝ	ର୍ନ	ର୍ପ	ର୍ଣ	ର୍ଥ
ର୍ବ	ର୍ଭ	ର୍ମ	ର୍ଶ	ର୍ଷ	ର୍ହ	ଙ୍କ	ଙ୍ଖ	ଙ୍ଗ	ଙ୍ଘ	ଞ୍ଚ	ଞ୍ଛ	ଞ୍ଜ
ଞ୍ଝ	ଣ୍ଟ	ଣ୍ଠ	ଣ୍ଡ	ଣ୍ଢ	ତ୍ତ	ତ୍ଥ	ଦ୍ଦ	ଦ୍ଧ	ନ୍ତ	ନ୍ଥ	ନ୍ଦ	ନ୍ଧ
କ୍କ	କ୍ଟ	କ୍ତ	କ୍ଷ	ଗ୍ଦ	ଗ୍ଧ	ଗ୍ଧ	ଗ୍ମ	ଜ୍ଜ	ଜ୍ଝ	ଜ୍ଞ	ଡ୍ଡ	ଡ୍ଢ
ଜ୍ଞ	ଟ୍ଟ	ଡ୍ଡ	ଡ୍ଢ	ଣ୍ଣ	ତ୍ତ	ତ୍ମ	ତ୍ନ	ଦ୍ଭ	ଦ୍ମ	ଦ୍ୱ	ଧ୍ମ	ଧ୍ନ
ଦ୍ଭ	ଧ୍ମ	ନ୍ନ	ପ୍ତ	ପ୍ପ	ପ୍ସ	ବ୍ଦ	ବ୍ଧ	ବ୍ବ	ଲ୍କ	ଲ୍ପ	ଲ୍ଫ	ଳ୍ଳ
ଲ୍ମ	ଲ୍ଲ	ଶ୍ଚ	ଶ୍ଛ	ଷ୍ଟ	ଷ୍ଠ	ଷ୍ପ	ଷ୍ଫ	ସ୍ଥ	ସ୍ଫ	ସ୍କ	ସ୍ଖ	ସ୍ତ
ସ୍ତ	ସ୍ଥ	ସ୍ପ	ସ୍ମ	ସ୍ଥ	ଙ୍କ୍ଷ	ଞ୍ଜ	ଣ୍ଡ୍ର	ଦ୍ଧ୍ର	ହ୍ନ	କ୍ଷ୍ଣ	ହ୍ମ	ହ୍ମ
ମ୍ପ୍ର	ନ୍ତ୍ର	ଙ୍ଗ୍ର	ର୍ଡ	ର୍ଟ	ର୍ଦ	ର୍ଧ	ର୍ଲ	ର୍ଦ୍ର	ର୍ଣ୍ଣ	ଷ୍କ୍ର	ସ୍ତ୍ର	ସ୍ପ୍ର
ସ୍ତ	ର୍ତ୍ତ	ର୍ଦ୍ଧ	ଷ୍ଟ୍ର	ଯ୍ୱ	ର୍ଯ୍ୟ	୦	୧	୨	୩	୪	୫	୬
୭	୮	୯	୰	ଽ								

button is clicked, the image is classified and the result, which is in Unicode format, is redirected to a text document (Figs. 5 and 6).

In this particular context, the accuracy is measured in terms of number of sample correctly identified divided by total number of test samples. A comparative study has been made with varied training and testing sample ratio where 90:10 training and testing samples showed the best result (Table 2).

Here in Table 3, few incorrectly classified cases are demonstrated.

A comparative study of earlier work for printed Odia symbols and proposed work is shown here in Table 4.

Fig. 4 Steps of recognition process

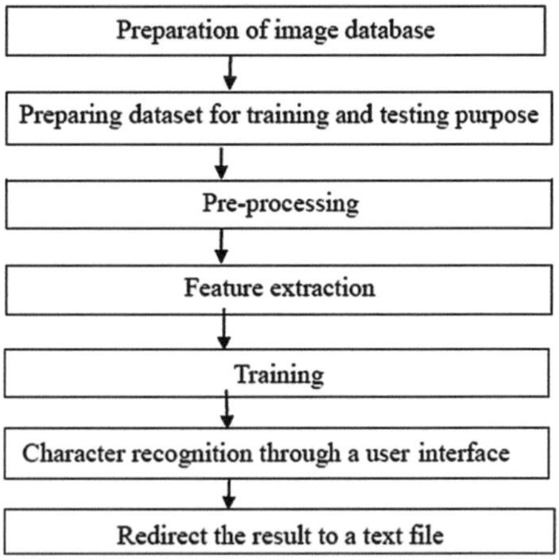

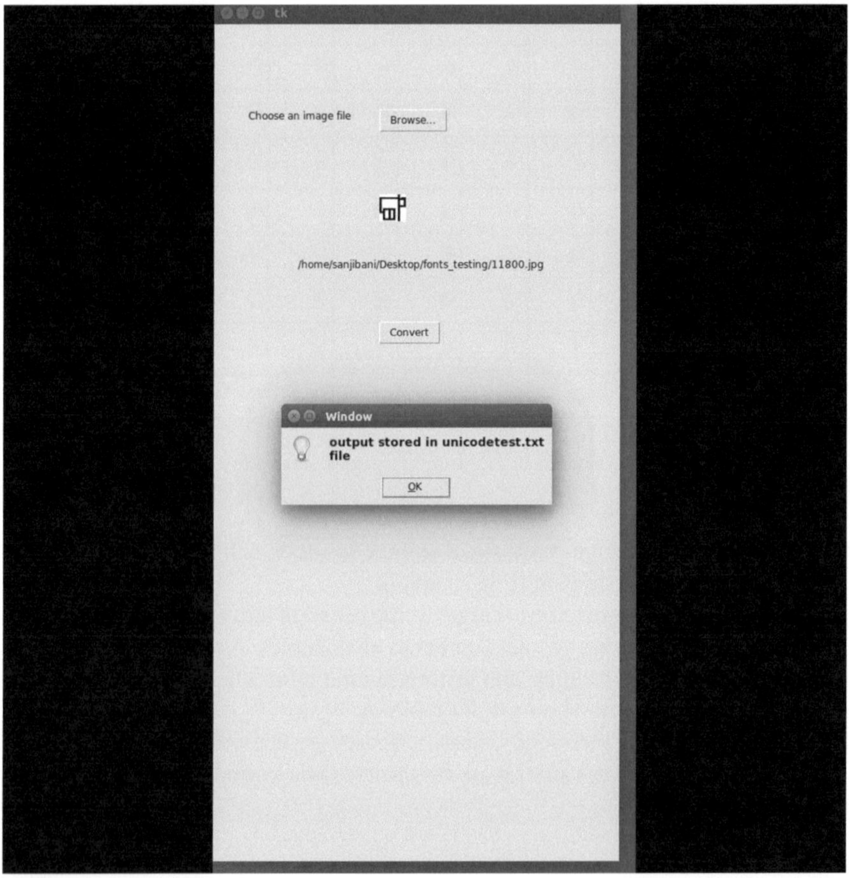

Fig. 5 Screenshot of OCR tool for character

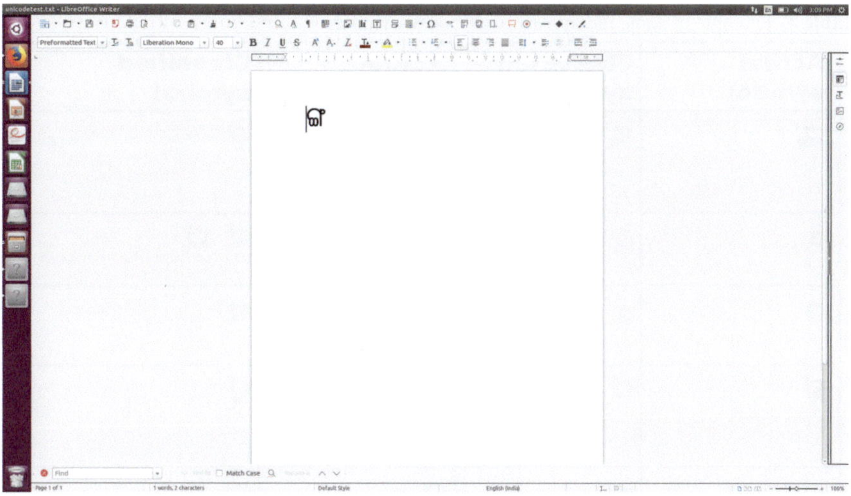

Fig. 6 Unicode format of the recognized image is redirected to a text file

Table 2 Rate of accuracy with varied training and testing ratio

Ratio of number of training and testing samples	Number of training samples	Number of testing samples	Accuracy rate (%)
90:10	22794	2533	99
80:20	20261	5066	98
70:30	17728	7599	98
60:40	15194	10133	94

6 Conclusion and Future Work

The printed Odia Character Recognition (OCR) is a demanding technology which deserves more attention from the researchers. In this paper, it is attempted to create a complete database for printed Odia symbols. A few symbols could not be included here as many characters are not defined in different fonts. Also, many characters could not be segmented properly by the underdeveloped segmentation tool (e.g., ଲ୍ୟ, ଖ୍ୟ, etc.). This database has been studied with the help of DWT feature extraction technique along with SVM classifier. Image interpolation and Gaussian blur techniques are major preprocessing methods which are adopted here. The following works have been planned as a part of future work:

(i) Recognizing the character along with its modifiers.
(ii) Line and character segmentation in a printed Odia page.
(iii) Collecting the remaining printed Odia characters for the growing database.
(iv) Recognizing noisy images.

Table 3 Example of some incorrect recognition of symbols

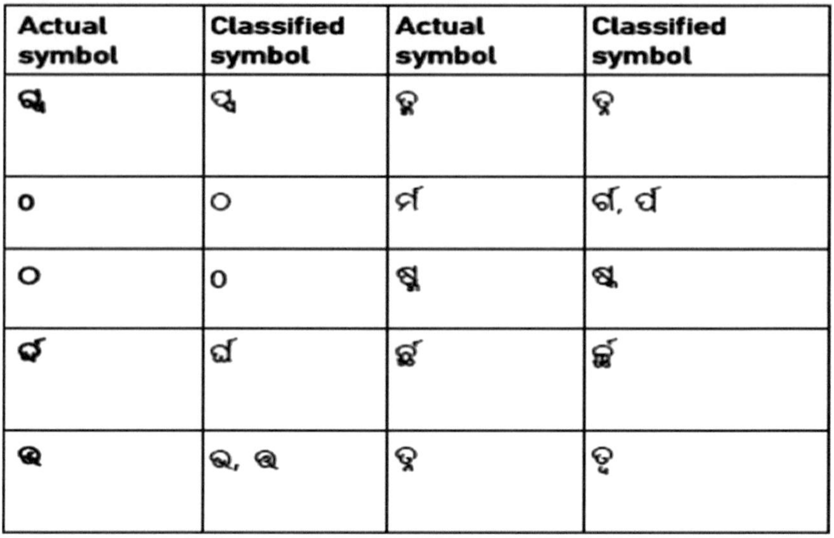

Table 4 Comparative study with other works for printed Odia symbols

Paper reference number	Feature extraction technique used	Classifier used	Number of classes	Number of data samples	Accuracy
[4]	–	Tesseract OCR	Only Odia atomic symbols	Not mentioned	100%
[7]	DWT	Euclidean distance method	47	Not mentioned	90%
[11]	–	NN, K-NN	38	1600 characters	82.33%(NN), 72.27% (k-NN) with normal character features 41.88% (NN), 39.41% (k-NN)
Proposed work	DWT	SVM	239	Training: 22794 Testing: 2533	99%

References

1. Bhattacharya, U., Chaudhuri, B.B.: Databases for research on recognition of handwritten characters of Indian scripts. In: Proceedings of the Eight International Conference on Document Analysis and Recognition (2005)
2. Mohapatra, R.K., Mishra, T.K., Panda, S., Majhi, B.: OHCS: a database for handwritten atomic Odia character recognition. In: NCVPRIPG, IIT, Patna (2015)
3. Dash, K.S., Puhan, N.B., Panda, G.: Odia character recognition: a directional review. Artif. Intell. Rev. **48**(4), 473–497 (2017)
4. Nayak, M., Nayak, A.K.: Odia characters recognition by training tesseract OCR engine. In: International Conference in Distributed Computing & Internet Technology. Int. J. Comput. Appl. 25–30 (2014)
5. Shanthi, N., Duraiswamy, K.: A novel SVM-based handwritten Tamil character recognition system. Pattern Anal. Appl. **13**, 173–180 (2010)
6. Liu, C.L., Fujisawa, H.: Classification and learning methods for character recognition: advances and remaining problems. In: Marinai, S., Fujisawa, H. (eds.) Machine Learning in Document Analysis and Recognition. Studies in Computational Intelligence, vol. 90. Springer (2008)
7. Dash, B., Pradhan, S., Rana, D.: Odia offline character recognition using DWT features. IOSR J. Electron. Commun. Eng. (IOSR-JECE) 31–37 (2016). e-ISSN: 2278-2834
8. Soora, N.R., Deshpande, P.S.: Review of feature extraction techniques for character recognition. IETE J. Res. (2017)
9. Bhowmik, T.K., Ghanty, P., Roy, A., Parui, S.K.: SVM-based hierarchical architectures for handwritten Bangla character recognition. IJDAR **12**, 97–108 (2009)
10. Jyothi, J., Manjusha, K., Anand, M., Soman K.P.: Innovative feature sets for machine learning based Telugu character recognition. Indian J. Sci. Technol. **8**(24) (2015)
11. Pati, P.B., Ramakrishnan, A.G., Aravinda Rao, U.K.: Machine recognition of printed Oriya characters. In: Proceedings of III International Conference on Information Technology ICIT, pp. 227–232, 21–23 Dec 2000

Importance of Data Standardization Methods on Stock Indices Prediction Accuracy

Binita Kumari and Tripti Swarnkar

Abstract Stock market indices prediction has drawn huge attention due to its impact on economic stability. Accurate stock market indices prediction is highly essential to reduce the risk associated with it so as to decide good investment strategies. To acknowledge exact prediction, different strategies have been attempted, amid which the machine learning techniques have pinched consideration and been refined achieving extraordinary results in applying machine learning approaches. In our study, we have adopted Support Vector Machine (SVM) for stock market forecasting due to its capacity to deal with risk. SVM in forecasting requires some preliminary works on the data and one of them is standardization. In this study, we analyze four normalization techniques and their influence on the forecasting results. The investigation demonstrates high affectability of the regularly utilized strategies to input information standardization calculations and shows the requirement for a wary way to deal with the outcomes achieved by them.

Keywords Input data standardization · Support vector machines · Stock market indices

1 Introduction

In the data processing field, a very speedily developing technology is data mining. It has been connected to different disciplines, for example, military, engineering, administration, science, and also the business. Within the money-related space,

B. Kumari (✉)
Department of Computer Science & Engineering, Siksha 'O' Anusandhan Deemed to be University, Bhubaneswar, Odisha, India
e-mail: binitakumari@soa.ac.in

T. Swarnkar
Department of Computer Application, Siksha 'O' Anusandhan Deemed to be University, Bhubaneswar, Odisha, India
e-mail: triptiswarnakar@soa.ac.in

B. Pati et al. (eds.), *Advanced Computing and Intelligent Engineering*,
Advances in Intelligent Systems and Computing 1082,
https://doi.org/10.1007/978-981-15-1081-6_26

data mining may be utilized to help with the expectation of stock costs, financial assessments, etc.

Stock market indices forecast is viewed as a demanding assignment for the prediction process of financial time-series data as the budgetary market is an intricate, developmental, and nonlinear powerful framework [1]. In the most recent decade, numerous investigations have been led in mining financial time-series information, along with traditional statistical methodologies and data mining procedures. In the territory of financial stock market foreseeing (forecasting), numerous investigations were concentrated on the use of support vector machines [2–5]. As of late, the Support Vector Machine (SVM) strategy that was first proposed by Vapnik in 1995 has been utilized as a part of a scope of uses, including stock market forecast in [2–6]. The SVM procedure is broadly viewed as great classifier and authors in [6, 7] show that SVM forecast ways are better than neural network ways. At first developed for taking care of classification issues, SVM systems can be effectively connected to regression problems. A stock forecast process comprises numerous parts like information gathering, creating integrated information, normalizing information, and classification/prediction.

A piece of the stock market indices forecast process is delivering the criterions (parameters) that depict the outcome imparted in diverse units and scales to a typical and equivalent numeric range. This movement, considered standardization, may have basic effect on the estimation's outcome. In this paper, we will take an insight at the impact of data standardization on stock market prediction.

2 Related Work

Anticipating the stock's trends and critical patterns are appallingly eye-catching to the stock exchange's scientists and any individual who wants to settle on the appropriate stock or potentially the best possible time to look for or offer the stocks [8]. Be that as it may, the right expectation is amazingly troublesome in view of the creaky nature and non-static stock expenses. A few full-scale monetary elements like political occasions, organization's approach, general financial conditions, item esteem lists, interest rates and stocks, desire for speculators, and psychological variables affect the stock expenses [9]. Additionally, government arrangement and administrative measures considerably affect the development of the stocks showcase in general. As per the authors in [10], soft computing procedures are generally utilized for stock market issues and are useful devices for foreseeing the nonlinear behavior. Artificial Neural Network (ANN) and SVM have been used by many researchers for stock foreseeing [11]. But even after constructing so many dynamic models, artificial neural network includes few hindrances inside the learning technique which influences the outcome as shown in [7]. Therefore, a few analysts like picking advanced methodologies based on powerful statistical basis like SVM [12]. As of late, the SVM procedure which is a supervised learning methodology has utilizations in classification and regression problems. SVM shows high performance by minimizing the structural risk as shown by authors in [13]. Given the over improvements, after all SVM was presented upheld Vapnik's statistical learning hypothesis, a few investigations have practical experience in the theory and its utilizations. Numerous studies utilize the SVM to foresee the time-series data [3, 13] The SVM is a machine learning system which has been created by Vapnik in 1995 and as a result of its eye-catching choices and superb execution in different issues, it has been used for nonlinear predictions. Tai and Cao in [3] attempt to utilize this sort of neural system to anticipate measurement found the SVM to be better than multilayer neural network system with regard to prediction of monetary time series.

Normalization is an integral part of any method wherever data processing techniques are applied. Thus, the result analysis of applying normalization techniques on different domains has been done recently. A large portion of the research work preprocesses the data while not paying any worry to the data complexity. Inquiries have been raised by authors in [14–16] on the requirement of preprocessing based on the data complexity. A preprocessing system called SMOTE ENN for oversampling the unbalanced datasets has been utilized in [17] so as to evaluate the various interims, wherever the usage of oversampling is helpful for the unbalanced datasets. As discussed by authors in [16, 15], the execution of any classification process is also touched with the companionship of noise inside the dataset.Han and Men [18] try and value the impact of normalization on RNA-seq sickness identification. In another paper, Sukirty [19] have evaluated 14 standard learning approaches for constructing a dynamic selection model so as to choose the best normalization process.

Thus, from the literature, it is clear that the normalization technique chosen for performing any data mining functionality may affect the output accuracy. In our paper, we will have a closer look into the importance of normalization for stock prediction.

3 Methods and Materials

3.1 Datasets Used

In order to verify the influence of input data standardization on forecasting performance, this study chooses the NASDAQ and S & P 500 as experimental datasets. The study chooses the data from 4/1/2010 to 30/4/2013. The gathered information comprises every day high, open, closing, and low costs. They are utilized just as informational indexes. The data has been collected from Yahoo finance (https://in.finance.yahoo.com/).

In this paper, the investigation is to foresee the direction of every day stock value record. A major problem in any stock dataset is that it does not contain any class label for up/down. Thus, we use an attribute Δc which indicates change in closing price as described in [20]. Δc has been used as a class label. "1" and "−1" mean the following day's index is higher or lower than the present day's index, respectively. Forecast miniature is fabricated and the performance is utilized to assess the efficiency.

3.2 *Normalization*

Normalization is a scaling procedure or a mapping strategy or a pre-handling stage, where we scale input information to fall inside a little indicated range. Basically, normalization of the information is required when managing attributes of various units and scales with the end goal to merge for better outcomes. Unless normalized at preprocessing, variables with disparate ranges or varying precision acquire different driving values. Stronger drivers may obfuscate meaningful variables.

On the other hand, if the mining algorithm has a random sampling component, then normalizing for sample size may help ensuring that all sources are treated equally, and that data-availability bias (and its corresponding misrepresentation of the data universe) is reduced. Normalization of input data plays an important role in the stock prediction process.

We have used the following four standardization methods to examine their influence on stock prediction—Euclidean formula, Manhattan formula, Linear formula, and Weitendorf's linear formula. Jüttler–Korth linear standardization was not used since for positive data values, it is similar to linear formula. The standardization formulas for the four methods used in our paper are listed in Table 1, where A_i represents the *i*th element of a given dataset and n is the total number of records.

As indicated by the authors in [8, 10, 21] and literature study, we found that the standardization methods listed in Table 1 are the widely used standardization methods in various domains like medicine, business, finance, etc.

Based on literature survey, we utilize 70% of the data points (closing cost) as the training information. The rest 30% outstanding data points are utilized as the test information. With the end goal to boost the foreseeing capacity of the miniature, we generated a synthesized dataset which is a dataset consisting of general stock data features along with the technical indicators mentioned in Table 3. **It also consists of Δc as mentioned in** [20] **along with the class label (1/−1).**

Table 1 List of normalization techniques used for comparison

Sl. No.	Normalization technique	Formula
1	Euclidean	$A_i = \frac{A_{i,j}}{\sqrt{\sum_{i=1}^{n}(A_i)^2}}$
2	Manhattan	$A_i = \frac{A_i}{\sum_{i=1}^{n}\lvert A_i\rvert}$
3	Linear	$A_i = \frac{A_i}{\max A_i}$
4	Weitendorf's linear	$A_i = \frac{A_i - \min A_i}{\max A_i - \min A_i}$

Table 2 List of some commonly used technical indicators

Technical indicators
20-day bias
Rate of change
Stochastic indicator
Relative index
10-day moving average
Moving average convergence/divergence (MACD)
Commodity channel index (CCI)
Buying/selling willingness indicator
Moving average oscillators (MAO)
Buying/selling momentum indicator
Psychological line
Relative strength index (RSI)
Rate of change (ROC)
Stochastic slow
Disparity 5
Momentum
Disparity 10

3.3 Technical Indicators

The input features which are typically utilized for stock market indices are opening value, closing cost, lowest cost, highest cost, and total volume. It has appeared in numerous articles that the technical indicators are useful for stock forecasting [21–23]. Thus, beneath completely extraordinary conditions, a few imperative technical indicators sketched out in Yongtao Vietnamese money-related unit, 2017 has been taken into thought alongside the daily cost and trading volume of the particular stocks. The technical indicators are determined by implementing an equation to the opening value, the lowest value, the highest cost, and trading volume information. Some of the widely used technical indicators are listed in Table 2.

3.4 Support Vector Machines

As shown by authors in [13, 14], Support Vector Machines (SVMs) are administered learning miniatures that examine information and find out the patterns, utilized for regression analyses and classification. It works by developing hyperplanes in a multidimensional space that isolates instances of various class labels. It can deal with multiple continuous and categorical variables. They are powerful in high-dimensional spaces, notwithstanding when the sum of dimensions is more than the sample numbers. They are memory proficient and flexible.

When applying SVM to monetary prediction, the vital factor that must be thought about is the selection of kernel function. Since the elements of financial time series are powerfully nonlinear, it is naturally considered that the nonlinear kernel functions will deliver higher achievement in comparison to the linear kernel. Several analysts have mentioned the selection of kernel functions [24] *in financial forecasting. In this paper, we have used the Gaussian kernel function due to their flexible nature.*

At the point when the kernel function is picked, two vital parameters (C, γ) *should be settled. Parameter C is the expense of C-SVM and parameter* γ *is the estimation of gamma in kernel function. The estimation of C and* γ *can clearly influence the execution of SVM. In our test, we have picked* $C = 35$ *and* $\gamma = 0.6$ *after trial and error method.*

4 Results and Discussion

The data was collected from Yahoo Finance for two datasets, namely, NASDAQ and S & P 500. In this paper, the test is to foresee the heading of every day stock value record as "1" or "−1" indicating a rise or fall in the closing price.

Along with the opening cost, closing cost, lowest cost, highest cost, the total trading volume, five fitting technical indicators have been treated as starting feature pool. As per the authors in [25, 26], the technical indicators are viable apparatuses to portray the genuine market circumstance in financial time-series forecast. They can be more instructive than utilizing pure prices [26]. In light of the audit of domain specialists and literary works, the chosen five technical indicators are Momentum (MTM), Exponential Moving Average (EMA), Relative Strength Index (RSI), Moving Average Convergence/Divergence (MACD), and Moving Average (MA). In Table 3, the formulae for the technical indicators used in our study are given. The details about the formulae can be referred from [20].

Based on literature study, we utilize 70% of the data points (closing cost) as the training information. The rest 30% outstanding data points are utilized as the test

Table 3 Used technical indicators formulae

Technical indicator	Formulae
MA	$\mathrm{MA(N)} = \frac{1}{N}\sum_{i=1}^{n} A_{i,\mathrm{close}}$
EMA	$\mathrm{EMA(N)} = A_{1,\mathrm{close}}$
MACD	$\mathrm{MACD} = \mathrm{EMA}_{12,i} - \mathrm{EMA}_{26,i}$
RSI	$\mathrm{RSI(N)} = 100 - \frac{100}{1+\mathrm{EMA(N)}_{\mathrm{up}}/\mathrm{EMA(N)}_{\mathrm{down}}}$
MTM	$\mathrm{MTM(i, N)} = A_{i,\mathrm{close}} - A_{i-N,\mathrm{close}}$

Table 4 Accuracy results for NASDAQ between Euclidean + SVM, Manhattan + SVM, Linear + SVM, and Weitendorf's linear + SVM

Method	Prediction accuracy
Euclidean + SVM	87
Manhattan + SVM	89
Linear + SVM	88
Weitendorf's linear + SVM	88

information. With the end goal to improve the forecasting capacity of the model, we generated a synthesized dataset.

The synthesized dataset is needed to be normalized so as to get good prediction results. The normalization technique used for the intake data greatly influences the output of the machine learning methods. We have analyzed four different normalization techniques for each of the two datasets. In our study, the normalization techniques which have been considered are Euclidean, Manhattan, Linear, and Weitendorf's linear.

We adequately check the forecasting performance and impact of standardization methods between Euclidean + SVM, Manhattan + SVM, Linear + SVM, and Weitendorf's linear + SVM with the same set of training dataset and testing dataset of NASDAQ and S & P 500, respectively. The evaluation of the model has been done using Matthews correlation coefficient (MCC) so as to avoid the accuracy bias due to data skew [20]. MCC is a single summary value including all four cells of a 2X2 confusion matrix. Given a confusion matrix (TP, FN, FP, TN), MCC is given by

$$\mathrm{MCC} = \frac{\mathrm{TP} \cdot \mathrm{TN} - \mathrm{FP} \cdot \mathrm{FN}}{\sqrt{(\mathrm{TP} + \mathrm{FP})(\mathrm{TP} + \mathrm{FN})(\mathrm{TN} + \mathrm{FP})(\mathrm{TN} + \mathrm{FN})}}$$

Table 4 lists the accuracy results of NASDAQ between Euclidean + SVM, Manhattan + SVM, Linear + SVM, and Weitendorf's linear + SVM for predicting two class labels, namely, up or down for the test dataset.

From Table 4, we can see that the prediction efficiency of SVM varies when different input data standardization techniques are applied. We can also see that the prediction accuracy of SVM based on Manhattan data standardization is better as compared to Euclidean + SVM, Linear + SVM, and Weitendorf's linear + SVM. Thus, we can say that the prediction accuracy is dependent on the normalization technique implemented for the input data along with other parameters like parameter tuning in the machine learning technique used, etc. As we know, normalization is a scaling procedure to scale input information to fall inside a little indicated range. Thus, when variables with disparate ranges or varying precision acquire different driving values, they may influence the final outcome. Thus, applying same normalization technique on different types of datasets along with the same data mining technique may have different outputs. Similarly, application of different types of normalization techniques on a single dataset may also have different outcomes due to the characteristics of the underlying dataset.

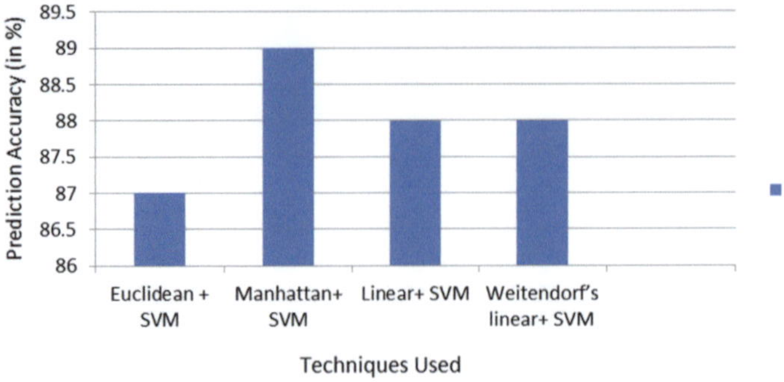

Fig. 1 Comparison results for NASDAQ between Euclidean + SVM, Manhattan + SVM, Linear + SVM, and Weitendorf's linear + SVM

Table 5 Accuracy results for S & P 500 between Euclidean + SVM, Manhattan + SVM, Linear + SVM, and Weitendorf's linear + SVM

Method	Prediction accuracy
Euclidean + SVM	88
Manhattan + SVM	89.1
Linear + SVM	89.8
Weitendorf's linear + SVM	87

Figure 1 shows and compares the results obtained for different techniques in Table 4.

Table 5 lists the accuracy results for S & P 500 between Euclidean + SVM, Manhattan + SVM, Linear + SVM, and Weitendorf's linear + SVM for predicting two class labels, namely, up or down for the test dataset.

From Table 5, we can see that the prediction efficiency of SVM varies when different input data standardization techniques are applied. We can also see that the prediction accuracy of SVM based on linear data standardization is better compared to Euclidean + SVM, Manhattan + SVM, and Weitendorf's linear + SVM. Thus, as seen from Tables 4 and 5, we can say that application of different types of normalization techniques on a single dataset may have different outcomes due to the characteristics of the underlying dataset. Accordingly, we can say that the prediction accuracy is dependent on the normalization technique implemented for the input data along with other parameters. Figure 2 shows and compares the results obtained for different techniques in Table 5.

From our analysis, we find that application of same normalization technique to different datasets may give different levels of results. Thus, the prediction error evaluation results vary from one dataset to another.

Normalization is used to scale input information to fall inside a little indicated range. There may be an influence on the final output when variables with disparate ranges or varying precision acquire different driving values. Thus, application of same normalization technique on different types of datasets using the same data mining

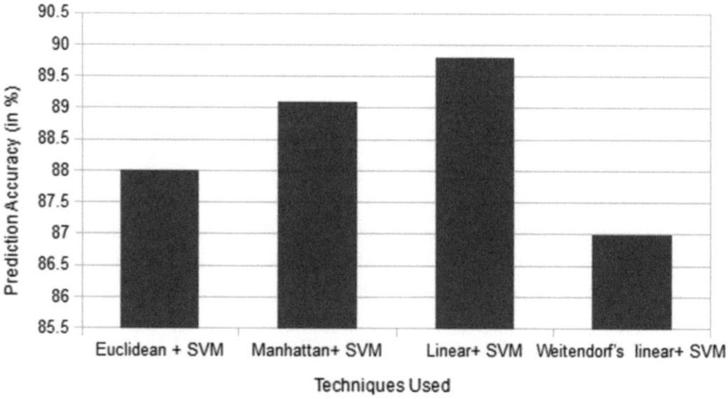

Fig. 2 Comparison results for S & P 500 between Euclidean + SVM, Manhattan + SVM, Linear + SVM, and Weitendorf's linear + SVM

technique may have different outputs. Similarly, application of different types of normalization techniques on a single dataset may also have different outcomes due to the characteristics of the underlying dataset.

Thus, the prediction accuracy results vary from one normalization technique to another. Different normalization techniques may give different prediction accuracy results for the same machine learning algorithm and dataset. Thus, the error accuracy results may also differ for different datasets.

References

1. Abu-Mostafa, Y.S., Atiya, A.F.: Introduction to financial forecasting. Appl. Intell. **6**(3), 205–213 (1996)
2. Huang, W., et al.: Forecasting stock market movement direction with support vector machine. Comput. OR **32**, 2513–2522(2005)
3. Tay, F., Cao, L.: Application of support vector machines in financial time series forecasting. Omega **29**(3), 309–317 (2001)
4. Kim, K.-j.: Financial time series forecasting using support vector machines. Neurocomputing **55**, 307–319 (2003)
5. Cao, L.J., Tay, F.: Support vector machine with adaptive parameters in financial time series forecasting. IEEE Trans. Neural Netw. **14**(3), 1506–1518 (2003)
6. Chen, W.-H., Shih, J.-Y.: Comparison of support-vector machines and back propagation neural networks in forecasting the six major Asian stock markets. Int. J. Electron. Financ. **1**(3), 49–67 (2006)
7. Huang, Z., Chen, H., Hsu, C.J., Chen, W.H., Soushan, W.: Credit rating analysis with support vector machines and neural networks: a market comparative study. Decis. Support Syst. **37**(4), 543–558 (2004)
8. Sahin, U., Ozbayoglu, M.: TN-RSI: Trend-normalized RSI indicator for stock trading systems with evolutionary computation. Procedia Comput. Sci. **36**, 240–245 (2014)
9. Majhi, B., Rout, M., Baghel, V.: On the development and performance evaluation of a multiobjective GA-based RBF adaptive model for the prediction of stock indices. J. King Saud Univ. Comput. Inf. Sci. **32**, 319–331 (2014)

10. Barak, S., Arjmand, A., Ortobelli, S.: Fusion of multiple diverse predictors in stock market. Inf. Fusion **36**, 90–102 (2017)
11. Anbalagan, T., Maheswari, S.U.: Classification and prediction of stock market index based on fuzzy metagraph. Procedia Comput. Sci. **47**, 214–221 (2015)
12. Fernandez-Lozano, C., Canto, C., Gestal, M., et al.: Hybrid model based on genetic algorithms and SVM applied to variable selection within fruit juice classification. Sci. World J. (Article ID 982438), 1797–1805 (2013)
13. García, S., Fernández, A., Luengo, J., Herrera, F.: Advanced nonparametric tests for multiple comparisons in the design of experiments in computational intelligence and data mining: experimental analysis of power. Inf. Sci. **180**(6), 1506–1518 (2010)
14. Garcia, L.P.F., de Carvalho, A.C.P.L.F., Lorena, A.C.: Effect of label noise in the complexity of classification problems. Neurocomputing **160**, 108–119 (2015)
15. Sáez, J.A., Galar, M., Luengo, J., Herrera, F.: Tackling the problem of classification with noisy data using multiple classifier systems: analysis of the performance and robustness. Inf. Sci. **247**, 1–20 (2013)
16. Leigh, W., Modani, N., Hightower, R.: A computational implementation of stock charting: Abrupt volume increase as signal for movement in New York stock exchange composite index. Decis. Support Syst. **37**(4), 515–530 (2004)
17. Xie, B., Passonneau, R.J., Wu, L., Creamer, G.G.: Semantic frames to predict stock price movement. In: The Proceedings of the 51st Annual Meeting of the Association for Computational Linguistics, Bulgaria, pp. 873–883 (2013)
18. Han, H., Men, K.: How does normalization impact RNA-seq disease diagnosis?. J. Biomed. Informat. **85**, 80–92 (2018)
19. Jain, S., Shukla, S., Wadhvani, R.: Dynamic selection of normalization techniques using data complexity measures. Exp. Syst. Appl. **106**, 252–262 (2018)
20. Chen, Y., Hao, Y.: A feature weighted support vector machine and K-nearest neighbor algorithm for stock market indices prediction. Expert Syst. Appl. **80**, 340–355 (2017)
21. Żbikowski, K.: Using volume weighted support vector machines with walk forward testing and feature selection for the purpose of creating stock trading strategy. Expert Syst. Appl. **42**(4), 1797–1805 (2015)
22. Hsu, M.-W., Lessmann, S., Sung, M.-C., Ma, T., Johnson, J.E.V.: Bridging the divide in financial market forecasting: machine learners vs. financial economists. Expert Syst. Appl. **61**, 215–234 (2016)
23. Neely, C.J., Rapach, D.E., Jun, T., Zhou, G.: Forecasting the equity risk premium: the role of technical indicators. Manag. Sci. **60**(7), 1617–1859 (2014)
24. Cao, L.J., Tay, F.E.H.: Support vector machine with adaptive parameters in financial time series forecasting. IEEE Trans. Neural Netw. **14**(6), 1506–1518 (2003)
25. Yeh, T.-L.: Capital structure and cost efficiency in the Taiwanese banking industry. Serv. Ind. J. **31**(2), 237–249 (2011)
26. Nikfarjam, A., Emadzadeh, E., Muthaiyah, S.: Text mining approaches for stock market prediction. In: The 2nd International Conference on Computer and Automation Engineering (ICCAE), Singapore, pp. 1–2 (2010)

Advanced Image and Video Processing

Feature Relevance Analysis and Feature Reduction of UNSW NB-15 Using Neural Networks on MAMLS

Smitha Rajagopal, Katiganere Siddaramappa Hareesha and Poornima Panduranga Kundapur

Abstract Feature relevance is often investigated in classification problems to determine the contribution of each feature, especially when a dataset comprises of numerous features. Feature selection or variable selection aids in creating an accurate predictive model because fewer attributes tend to reduce computational complexity, thereby promising better performance. Machine learning, a preferred approach to intrusion detection, manifests on the appropriate usage of features to improve attack detection rate. A new benchmark dataset, UNSW NB-15, has been used in the study which comprises of five classes of features. This work attempts to demonstrate the relevance of each feature class along with the importance of various combinations of feature classes. During the course of this analysis, 31 possible combinations of features were taken into consideration and their relevance was examined. Empirical results pertaining to feature reduction have shown that an accuracy of 97% could be obtained by using only 23 features. The entire sequence of experimentation was conducted on Microsoft Azure machine learning studio (MAMLS), a scalable machine learning platform. Two-class neural network was used to perform the classification task. Since UNSW NB-15 is a contemporary dataset with modern attack vectors, the research community is still in the process of exploring various facets of this dataset. This article thus intends to offer valuable insights on the significance of features found in UNSW NB-15 dataset.

Keywords UNSW NB-15 · Neural networks · MAMLS · Feature relevance · Feature reduction

S. Rajagopal (✉) · K. S. Hareesha · P. P. Kundapur
Manipal Institute of Technology, Manipal Academy of Higher Education, Manipal, India
e-mail: smitha.mit1012@gmail.com

B. Pati et al. (eds.), *Advanced Computing and Intelligent Engineering*,
Advances in Intelligent Systems and Computing 1082,
https://doi.org/10.1007/978-981-15-1081-6_27

1 Introduction

The entities within a network topology are constantly compromised owing to the proliferation of modern-day attack vectors threatening the confidentiality, integrity and availability of information [1]. The practical problems associated with any intrusion detection system (IDS) normally pertain to the generation of large number of false positives whenever it is exposed to newer attack types [2]. A pre-requisite to effectively train the IDS necessitates the usage of a modern dataset [3]. Traditional datasets like KDD cup 99 or its successor NSL-KDD no longer seem to be the ideal choice due to the absence of modern-day attack scenarios [3].
Drawbacks of KDD cup 99 and NSL-KDD datasets [3, 4]:

- Numerous repetitive records are present in KDD cup 99 due to which machine learning algorithms tend to be biased toward specific attack type.
- The probability distribution of training and testing datasets is different due to which skew or bias is introduced instead of balancing normal and anomalous network instances. NSL-KDD does not include modern-day attack patterns.

In the light of the above discussion, this article considers UNSW NB-15 dataset to evaluate the efficiency of two-class neural networks available as a module on Microsoft Azure machine learning studio (MAMLS) to perform predictive analysis.

UNSW NB-15 dataset was created at the Cyber Range Lab of the Australian Centre for cyber security to spawn both normal activities and attack patterns. IXIA perfect storm and tcpdump tools were used to record 100 GB of network traffic [3, 4]. Tools like Argus and Bro-IDS were used for extracting 49 features of the dataset. The outcome of this simulation process was the generation of four CSV files, which included both normal and anomalous records [3, 4]. Another interesting characteristic of IXIA perfect storm tool is that many modern attack vectors are continuously updated from common vulnerability exposure (CVE) site, which contains entries of publicly known cyber security vulnerabilities. Training and testing datasets have 175,341 and 82,332 records, respectively, categorized into nine different attack types [3, 4].

1.1 Features in UNSW NB-15 Dataset

Argus tool was used to process raw pcap files and create feature sets which resulted in the generation of five classes of features [3, 4]. Flow features are listed in Table 1.

Flow features reveal only the basic information of any network packet related to IPs, port numbers and the protocol used in the transaction. Basic features are more inclined toward time-to-live (TTL), service, packet counts, and so on, as shown in Table 2 [3, 4].

Content features correspond to window sizes of source and destination, sequence numbers, flow packet sizes and so on, as given in Table 3. Time features pertain to

Table 1 Flow features

Sl. no	Feature	Description
1	srcip	IP address of the source
2	sport	Port number of the source
3	dstip	IP address of the destination
4	dsport	Port number of the destination
5	proto	Protocol used in the transaction

Table 2 Basic features

Sl. no	Feature	Description
6	state	State of the associated protocol
7	dur	Duration of transaction
8	sbytes	Bytes from source to destination
9	dbytes	Bytes from destination to source
10	sttl	Time to live of source to destination
11	dttl	Time to live of destination to source
12	sloss	Packets dropped from source
13	dloss	Packets dropped from destination
14	service	Used service, e.g.: http, smtp, ftp, etc.
15	sload	Source bits per second
16	dload	Destination bits per second
17	spkts	Packet count from source to destination
18	dpkts	Packet count from destination to source

Table 3 Content features

Sl. no	Feature	Description
19	swin	Window advertisement of source tcp
20	dwin	Window advertisement of destination tcp
21	stcpb	Sequence number of source tcp
22	dtcpb	Sequence number of destination tcp
23	smeansz	Mean of the flow packet size transmitted by the source
24	dmeansz	Mean of the flow packet size transmitted by the destination
25	trans_depth	Indicates the depth of http request response transaction
26	res_bdy_len	Data transferred from the http service

jitters (delays) of the packets on the network, inter packet arrival time, round trip time, and so on, as shown in Table 4. Additional features focus primarily on the number of records or network instances with same IPs, port numbers, flows in ftp session, same service and so on, as listed in Table 5 [3, 4].

Table 4 Time features

Sl. no	Feature	Description
27	sjit	Jitter at the source
28	djit	Jitter at the destination
29	stime	Record start time
30	ltime	Record last time
31	sintpkt	Inter packet arrival time at the source
32	dintpkt	Inter packet arrival time at the destination
33	tcprtt	The sum of synack and ackdat
34	synack	The time taken between SYN and SYN_ACK packets
35	ackdat	The time taken between SYN_ACK and ACK packets

Table 5 Additional features

Sl. no	Feature	Description
36	is_sm_ips_ports	If source IP equal to destination IP and sport is equal to dport, then value is 1 or else 0
37	ct_state_ttl	Values are assigned to states (6) for source/destination time to live
38	ct_flw_http_mthd	Get and post methods count in http service
39	is_ftp_login	ftp session logged in by user or not
40	ct_ftp_cmd	Count of commands in ftp session
41	ct_srv_src	Count of connections with the same service and source address in 100 connections as per last time
42	ct_srv_dst	Count of connections with the same service and destination address in 100 connections as per last time
43	ct_dst_ltm	Count of connections with the same destination address in 100 connections as per last time
44	ct_src_ltm	Count of connections with the same source address in 100 connections as per last time
45	ct_src_dport_ltm	Count of connections with the same source and destination port in 100 connections as per last time
46	ct_dst_sport_ltm	Count of connections with the same destination and source port in 100 connections as per last time
47	ct_dst_src_ltm	Count of connections with the same source and destination address in 100 connections as per last time

2 Microsoft Azure Machine Learning Studio (MAMLS)

MAMLS enables predictive analytics using machine learning algorithms available as modules [5]. This work involves the application of two-class neural networks to classify the instances of UNSW NB-15 dataset. Since network intrusion detection requires massive data for thorough investigation, a scalable machine learning platform is inevitable to its study. Microsoft claims that the MAMLS encompasses state-of-the-art algorithms convenient for usage by data scientists [6]. Another intriguing aspect of MAMLS is the presence of visual workspace which facilitates the development and validation of predictive models [6, 7]. An extensible requirement could be met by integrating Python or R into the workflows. Partition and sample is a module available on the studio through which tenfold cross-validation was performed [6–8]. Technically, this module facilitates division of the dataset into several subsections of the same size in order to avoid bias of any form through stratified random sampling [7, 8]. The experimentation conducted on MAMLS took into account the dataset with 175,341 samples with size of 30.8 Mb. Although MAMLS offers a split module which divides the data into training and testing sets, partition and sample is a preferred choice as it has stratified sampling capability much more desirable than a simple random split of 80:20 [7, 8].

3 Two-Class Neural Networks

A binary neural network is the chosen classifier for the given task at hand. Given a tagged dataset, the neural network model predicted whether a network instance was anomalous or normal. Often touted as an efficient classifier, most predictive tasks on MAMLS can be accomplished with only few hidden layers [5–8]. Number of hidden nodes used for both feature relevance and feature reduction tasks were 100. The default learning rate is fixed as 0.1 by MAMLS. Since there was no major change reported after varying the learning rate, default value was only kept intact [5–8]. The style of network architecture applied for the creation of the model is a fully connected case which supports one hidden layer. Typically, it has three layers, namely input, hidden and output layers. Input layer relies on the number of features provided to the training process. Number of nodes in the output layer for any two-class classification problem would be two, which obviously suggests that all the inputs are mapped to either of the nodes [5–8].

4 Semantics of Two-Class Neural Networks

Generally, for two-class problems, the neural network architecture uses a cross-entropy loss function which optimizes the log-likelihood of the training data as defined in Eq. 1 [9].

$$E = -\sum_{i=1}^{n} (t_i \log(y_i) + (1 - t_i)\log(1 - y_i)) \tag{1}$$

t_i refers to the target output and y_i refers to the calculated output.

In order to standardize data, the neural network model uses min–max normalization defined in Eq. 2 [10].

$$A = \frac{[X_i - \min(X)]}{[\max(X) - \min(X)]} \tag{2}$$

X_i is the ith data point. min and max represent minimum and maximum values of a particular feature, respectively.

5 Feature Relevance

Feature relevance, at a conceptual level, brings to the forefront all those features which can possibly improve the prediction rate often desirable for classification problems. Since UNSW NB-15 has five classes of features, it is quite important to understand the relevance of each class of feature. It is worthwhile to note that each feature in this dataset holds an intrinsic piece of information about various aspects of a network packet. Thus, the current research problem gave rise to two seemingly pertinent research questions.

RQ 1: What is the relevance of each feature class towards prediction rate?
RQ 2: Does feature reduction improve the classification outcome?

Both questions are answered in Sect. 7.

6 Feature Reduction

MAMLS allows integration of customized python code which can be seamlessly added onto the workflows to introduce an extended functionality. SelectFromModel [11] is a meta estimator found in ScikitLearn library predominantly used to select features based on their scores. Threshold value is used to select significant features while feature scores less than threshold are considered irrelevant by the meta estimator [12, 13]. Twenty-three features were integral to the feature reduction task. Subsequently, two-class neural network was used as the classifier. Various combinations of features greater than and lesser than 23 were also experimented with, but none yielded better results than time, additional, basic and flow (TABF) combination. Table 6 lists 23 features obtained from SelectFromModel.

Table 6 23 features chosen by SelectFromModel

1. spkts	9. tcprtt	17. ct_src_dport_ltm
2. rate	10. synack	18. ct_srv_dst
3. dttl	11. dmean	19. response_body_len
4. sload	12. sbytes	20. is_ftp_login
5. ackdat	13. sloss	21. dload
6. sinpkt	14. trans_depth	22. synack
7. djit	15. ct_srv_src	23. ct_state_ttl
8. dwin	16. ct_ftp_cmd	

7 Results

In this section, 31 combinations of features are explored and their contribution toward classification outcome is explicated in terms of accuracy, precision, recall, F1-score and area under the curve (AUC), as shown in Table 7. Equations 3–6 can be used to calculate the performance metrics [14].

$$\text{Accuracy} = \frac{\text{TP} + \text{TN}}{\text{TP} + \text{TN} + \text{FP} + \text{FN}} \tag{3}$$

$$\text{Precision} = \frac{\text{TP}}{\text{TP} + \text{FP}} \tag{4}$$

$$\text{Recall} = \frac{\text{TP}}{\text{TP} + \text{FN}} \tag{5}$$

$$\text{F1-score} = \frac{2 \times \text{Precision} \times \text{Recall}}{\text{Precision} + \text{Recall}} \tag{6}$$

It can be observed from Table 7 that a combination of **TABF** (altogether 39 features) has resulted in the best prediction outcome but cannot be considered feasible due to the presence of 39 features, thereby offering ample scope for feature reduction. Table 8 summarizes the results obtained by using 23 features for classification task.

Random sampling is preferable to split the data. An 80:20 rule is followed in order to produce a smaller error of estimation as per widely accepted pareto principle [15]. It is more appropriate when datasets are heterogeneous in nature [16], like UNSW NB-15, wherein attack instances (119,341) dominate over normal instances (56,000). Tenfold cross-validation is used [16]. Figure 1 summarizes the critical aspects of the results pertaining to TABF combination. Figure 2 sums up the results when 23 features were considered by the neural network model for classification. Both Figs. 1 and 2 represent the results of evaluation upon executing the workflows on MAMLS. Figures 3 and 4 represent the ROC curve by considering false positives on the X-axis and true positives on the Y-axis pertaining to TABF combination and 23 features respectively.

Table 7 Feature class and their relevance

Feature class	Accuracy	Precision	Recall	F1-score	AUC
All features	0.6	0.59	0.57	0.58	0.66
Basic (B)	0.68	0.69	0.51	0.73	0.65
Flow (F)	0.56	0.50	0.53	0.66	0.59
Content (C)	0.58	0.49	0.52	0.69	0.64
Time (T)	0.44	0.46	0.45	0.51	0.52
Additional(A)	0.5	0.54	0.49	0.56	0.57
BF	0.78	0.82	0.77	0.8	0.83
BC	0.67	0.71	0.65	0.78	0.72
BT	0.54	0.57	0.63	0.66	0.62
BA	0.74	0.59	0.67	0.68	0.7
FC	0.63	0.55	0.6	0.61	0.74
FT	0.49	0.47	0.52	0.58	0.56
FA	0.57	0.51	0.53	0.55	0.66
CT	0.48	0.42	0.46	0.5	0.55
CA	0.52	0.58	0.53	0.57	0.69
TA	0.61	0.56	0.52	0.54	0.63
BFC	0.72	0.78	0.72	0.74	0.79
BCT	0.77	0.73	0.71	0.8	0.82
BTA	0.74	0.76	0.72	0.75	0.8
FCT	0.79	0.77	0.74	0.78	0.81
CTA	0.58	0.55	0.56	0.53	0.65
CAF	0.75	0.76	0.7	0.72	0.84
CBF	0.63	0.69	0.64	0.62	0.67
TFB	0.8	0.76	0.78	0.77	0.85
FTA	0.78	0.79	0.75	0.76	0.82
BAF	0.84	0.83	0.8	0.82	0.86
BFCT	0.69	0.84	0.59	0.66	0.76
FCTA	0.65	0.87	0.58	0.66	0.79
BCTA	0.84	0.81	0.89	0.85	0.88
BAFC	0.87	0.86	0.9	0.95	0.86
TABF	**0.93**	**0.9**	**1.00**	**0.95**	**0.92**

Table 8 Results obtained from feature reduction task with 23 features

Accuracy	Precision	Recall	F1-score	AUC
0.97	0.95	1.00	0.97	0.99

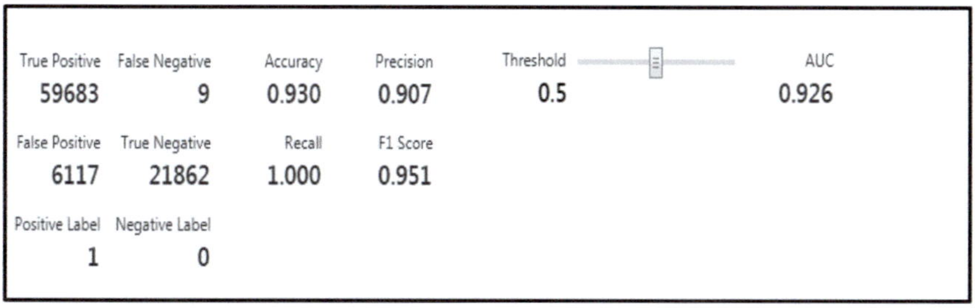

Fig. 1 Summary of results pertaining to TABF combination

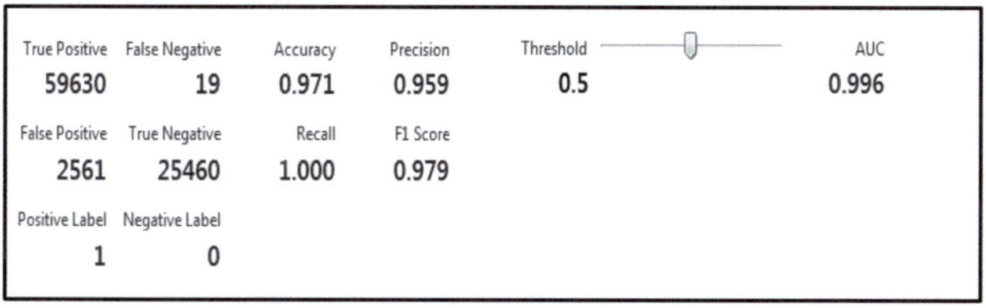

Fig. 2 Summary of results pertaining to feature reduction with 23 features

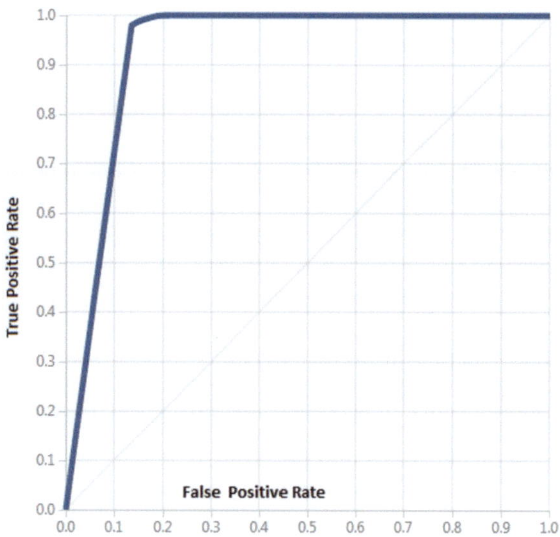

Fig. 3 ROC curve: TABF

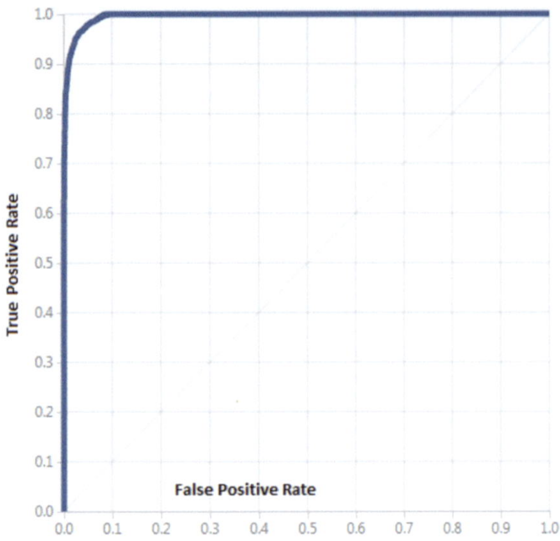

Fig. 4 ROC curve: 23 features

8 Discussion

As discussed in the entire course of the article, feature engineering is a vital component of any classification task. Intrusion detection systems (IDS) invariably deal with numerous features, which if chosen wrongly could hinder the performance of IDS. Going by this assertion, it is extremely important to perform feature analysis to comprehend the performance of IDS. Feature analysis could be of immense help to know the significance of each feature and feature reduction provides ample scope to use fewer features for improving attack detection rate. Feature relevance analysis has not been conducted so far on UNSW NB-15 dataset although a few authors [3, 4, 17, 18] have performed feature reduction to improve the classification outcome.

Two research questions pertaining to feature relevance as well as feature reduction have been formulated and results of both the tasks are presented in Sect. 7.

RQ 1: What is the relevance of each feature class toward prediction rate?

Every feature in a network dataset may not seem to be pertinent to detect intrusions. As an integral part of this work, a detailed study was undertaken to discern the implication of each and every feature without which feature selection becomes absurd. Founders of UNSW NB-15 [3, 4] have emphasized a lot on the mechanism of deriving 47 features but not much has been explained about the applicability of feature classes. This definitely necessitated the feature relevance task explained thoroughly in this article and illustrated in Table 7. To name a few, combinations like BAFC and TABF yielded a good accuracy but cannot be considered as the best subset of features due to the presence of 39 features often considered impractical in the study of machine learning. Feature relevance analysis on UNSW NB-15 has not been conducted so far, which is why finding answer to the first research question becomes even more imperative.

RQ 2: Does feature reduction improve the classification outcome?

Feature relevance alone cannot guarantee optimal results, especially when 39 features are involved. Hence feature reduction becomes inevitable. **RQ 2** has been answered by emphasizing on using only 23 features to achieve favorable results as shown in Table 8. The idea behind conducting feature reduction was to examine whether there exists a feature subset with lesser than 39 features. This paved the way for feature reduction which eventually resulted in better predictions by using only 23 features as retrieved by SelectFromModel. As explained by the authors in [17, 18], in spite of using classifiers like logistic regression, the highest accuracy was reported to be 89.26 but the sequence of experimentation on MAMLS has resulted in an accuracy of 97% notably higher than the existing works. Authors in [17] also used logistic regression to obtain an accuracy of 81.42 with 20 features. It is worthwhile to mention that neural network model on MAMLS has been quite successful in generating optimal predictions with 23 features not yet accomplished so far using UNSW NB-15 dataset which apparently makes this work eminent.

9 Conclusion

This work attempted at exploring the UNSW NB-15 dataset thoroughly by considering its five classes of features. A methodical investigation of this dataset was necessary to comprehend the differences in the prediction rates generated by various combinations of feature classes. Two research questions pertaining to feature relevance and feature reduction were answered in the Discussion section. Feature relevance helped in exploring the features of the dataset, whereas feature reduction contributed towards improving the classification outcome. Since intrusion detection is a promising area of research, use of new benchmark datasets is quite indispensable to its study. Thus an initiative was undertaken to study the dataset and highlight the results in an effective manner. As a case of future study, multiclass classification task can be investigated to determine the relevance of feature classes towards predicting a particular attack type. Subsequently, feature reduction can be applied to achieve better results. Since MAMLS offers various other modules too as detectors, performance evaluation of these detectors could be a possible frontier for data scientists.

References

1. Von Solms, R., Van Niekerk, J.: From information security to cyber security. Comput. Secur. **38**, 97–102 (2013)
2. Garcia-Teodoro, P., et al.: Anomaly-based network intrusion detection: techniques, systems and challenges. Comput. Secur. **28**, 18–28 (2009)
3. Moustafa, N., Slay, J.: UNSW-NB15: a comprehensive data set for network intrusion detection systems (UNSW-NB15 network data set). In: Military Communications and Information Systems Conference (MilCIS), pp. 1–6. IEEE, Canberra (2015)
4. Moustafa, N., Slay, J.: The evaluation of network anomaly detection systems: Statistical analysis of the UNSW-NB15 data set and the comparison with the KDD99 data set. Inf. Secur. J. Glob. Perspect. **25**(1–3), 18–31 (2016)
5. Mund, Sumit: Microsoft azure Machine Learning. Packt Publishing Ltd., U.K. (2015)
6. Barga, R., Fontama, V., Tok, W.H.: Predictive Analytics with Microsoft Azure Machine Learning: Build and Deploy Actionable Solutions in Minutes. Apress (2014)
7. Chappell, D.: Introduction to Azure Machine Learning. Chappell & Associates, San Francisco (2015)
8. Barga R., Fontama V., Tok W.H.: Introducing Microsoft Azure machine learning. In: Predictive Analytics with Microsoft Azure Machine Learning. Apress, Berkeley, CA (2015)
9. Golik, P., Doetsch, P., Ney, H.: Cross-entropy vs. squared error training: a theoretical and experimental comparison. In: Interspeech, vol. 13, pp. 1756–1760 (2013)
10. Patro, S., Sahu, K.K.: Normalization: A Preprocessing Stage. arXiv preprint arXiv:1503.06462 (2015)
11. Hackeling, G.: Mastering Machine Learning with Scikit-Learn. Packt Publishing Ltd. (2014)
12. Garreta, R., Moncecchi, G.: Learning Scikit-Learn: Machine Learning in Python. Packt Publishing Ltd. (2013)
13. Kadiyala, Akhil, Kumar, Ashok: Applications of python to evaluate the performance of decision tree-based boosting algorithms. Environ. Prog. Sustain. Energy **37**(2), 618–623 (2018)
14. Aggarwal, P., Sharma, S.K.: An empirical comparison of classifiers to analyze intrusion detection. In: 2015 Fifth International Conference on Advanced Computing & Communication Technologies (ACCT), pp. 446–450. IEEE, Haryana, India (2015)
15. Suthaharan, S.: Machine learning models and algorithms for big data classification. In: Thinking with Examples for Effective Learning. Springer, New York (2015)
16. Fushiki, T.: Estimation of prediction error by using K-fold cross-validation. Stat. Comput. **21**(2), 137–146 (2011)
17. Khammassi, Chaouki, Krichen, Saoussen: A GA-LR wrapper approach for feature selection in network intrusion detection. Comput. Secur. **70**, 255–277 (2017)
18. Bhamare, D.: Feasibility of supervised machine learning for cloud security. In: International Conference on Information Science and Security (ICISS), pp. 1–5. IEEE, Pattaya, Thailand (2016)

Video Summarization Based on Optical Flow

Dipti Jadhav and Udhav Bhosle

Abstract The explosive growth in digital videos demands a technique that effectively identifies informative parts from the video. Video summarization refers to creating a video summary as a collection of keyframes that depicts key actions and events in the video. The authors propose to generate a video summary based on apparent motion information in the video, that is, optical flow. The proposed algorithm uses optical flow technique to estimate the change in the local flow of pixel intensities to identify the keyframes. The proposed algorithm is tested on two standard databases, such as Open Video Project and YouTube database. The results and the quantitative evaluation validate the effectiveness of the proposed algorithm for generation of a video summary.

Keywords Video summarization · Motion estimation · Optical flow · Keyframes · Quantitative analysis

1 Introduction

The tremendous technological progress over the last few decades has led to digital video becoming ubiquitous. There is a widespread application of videos in areas such as information technology, telecommunications, consumer electronics and entertainment. This has led to huge increase in uploading and downloading of videos on the web. The YouTube statistics proves that users watch billions hours of video daily [1]. Thus there is a need for an efficient technique for abstracting the video contents. This has led to the development of video summarization (VS) technique. Video summarization aims for the development of effective and comprehensive video summary

D. Jadhav (✉)
Department of Computer Engineering/Information Technology, RAIT, Nerul, Navi Mumbai, Maharashtra, India
e-mail: dipti.jadhav@rait.ac.in

U. Bhosle
Department of Electronics and Telecommunication, RGIT, Versova, Mumbai, Maharashtra, India
e-mail: udhav.bhosle@mctrgit.ac.in

B. Pati et al. (eds.), *Advanced Computing and Intelligent Engineering*, Advances in Intelligent Systems and Computing 1082,
https://doi.org/10.1007/978-981-15-1081-6_28

by extracting salient frames of the video. Video summary is of two types. The static video summary generates a storyboard comprising static keyframes. Dynamic video summary consists of keyframes as well as audio abstract from the video [2]. This paper aims at developing a static video summary as it is very effective for various video analysis and retrieval applications.

The most recent work in video summarization is based on techniques such as color estimation, object detection, motion estimation, attention modeling, graph modeling and deep neural networks. Video summarization based on attributes such as color histogram and line profiles is presented in [2]. Video synopsis and indexing method for surveillance videos is presented in [3]. Video summary based on moving object detection and trajectory extraction is given in [4]. An object-driven summary for egocentric videos is done by predicting important people and objects [5]. Hierarchical hidden Markov model (HHMM) is used to generate a video summary [6]. The authors of [7] presented attention models based on audio–visual saliency for building the video summary. Video summarization using web images as a prior is proposed in [8]. In [9] summary of egocentric videos is presented. Kim et al. [10] proposed a video summarization based on estimating visual motion dissimilarities. The authors in [11] proposed a clustering algorithm for video summarization.

Motion is an important key source of information in a video. The motion in the video is due to camera motion as well as object motion. This motion results in spatio-temporal changes that are captured or identified by optical flow. This information can be used for capturing representative frames in a video. Thus we have formulated the video summarization as a dense motion field problem that calculates the optical flow vectors between successive frames for generating an expressive video summary.

The proposed algorithm based on optical flow is presented in Sect. 2. Experimental results and performance analysis is presented in Sections 3 and 4, respectively. Conclusion and future work is presented in Sect. 5.

2 Proposed Methodology

Optical flow helps to identify the amount of pixel movement between adjacent images. Consider a pixel at the centre of the frame as $I(x, y, t)$ in an n × n neighborhoods. Consider it moves by δx, δy in time δt to $I(x + \delta x, y + \delta y, t + \delta t)$. The frames $I(x, y, t)$ and $I(x + \delta x, y + \delta y, t + \delta t)$ are the frames of the same point and we have [12]:

$$I(x, y, t) = I(x + \delta x, y + \delta y, t + \delta t) \tag{1}$$

Equation (1) is true for small local translations provided δx, δy, δt are small. The first-order Taylor series expansion about $I(x, y, t)$ in Eq. (1) gives:

$$I(x + \delta x, y + \delta y, t + \delta t) = I(x, y, t) + \frac{\partial I}{\partial x}\delta x + \frac{\partial I}{\partial y}\delta y + \frac{\partial I}{\partial t}\delta t + \text{H.O.T} \tag{2}$$

where H.O.T. are the higher order terms, which we assume are small and can be ignored. Equations (1) and (2) give:

$$\frac{\partial I}{\partial x}\delta x + \frac{\partial I}{\partial y}\delta y + \frac{\partial I}{\partial t}\delta t = 0 \quad (3)$$

Equation (3) can be rewritten as:

$$\frac{\partial I}{\partial x}V_x + \frac{\partial I}{\partial y}V_y + \frac{\partial I}{\partial t}V_t = 0 \quad (4)$$

In Eq. (4), $V_x = \frac{\delta x}{\delta t}$ and $V_y = \frac{\delta y}{\delta t}$ are the x and y components of image velocity or optical flow. $\frac{\partial I}{\partial x}, \frac{\partial I}{\partial y}, \frac{\partial I}{\partial t}$ are image intensity derivatives at (x, y, t). The image velocity extracted above can be used for extracting the motion flow information in the video sequence. This optical flow information can be used for extracting the keyframes in the video.

2.1 Algorithm: Optical Flow-Based Video Summarization

1. Represent the video by a set of frames I_i such that $V = \sum_{i=1}^{k} I_i$
2. For each frame I_i where $i = 1$ to k.
   a. Extract the image velocity V_x and V_y in x and y direction for every pixel of frame I_i as given in Eq. (4).
   b. Calculate the image velocity of a frame I_i as $I_{Vi} = \sum V_x + \sum V_y$.
3. Calculate the mean of the image velocity of the entire video as: Mean_{IV} = mean of all the image velocity I_{Vi} calculated for all the frames I_i ($i = 1$ to k) in Step 2b.
4. Calculate standard deviation of the image velocity of the entire video as: STD_{IV} = standard deviation of image velocity I_{Vi} calculated for all the frames I_i ($i = 1$ to k) in Step 2b.
5. Calculate threshold as: $\text{Th}_{IV} = \text{Mean}_{IV} + (\alpha * \text{STD}_{IV})$ where α = tuning parameter whose values vary from 0 to 2.
6. For each frame I_i where $i = 1$ to k,
   Calculate the difference D_{IV} in the image velocity I_V between frames I_i and frame I_{i+1}. If the difference D_{IV} between frames I_i and I_{i+1} is less than Th_{IV} then both frames are similar and delete frame I_{i+1}. If the difference D_{IV} is greater than Th_{IV}, means of both the frames are different, so retain frame I_i as keyframe. Next, frame I_{i+1} is the new reference frame and the difference between the image velocity I_V of frames I_{i+1} and I_{i+2} are compared and the process is repeated.
7. Output the video summary.

Table 1 Experimental video dataset

S. no.	Video	Source	Genre	Length of video	Number of frames
1.	v109.avi	YouTube database	Advertisement	00:00:52	1577
2.	v23.mpg	Open Video Project	Documentary	00:00:58	1761
3.	v25.mpg	Open Video Project	Documentary	00:00:59	1794

3 Experimental Results and Discussion Section

3.1 Database and Experimental Settings

The experimental results of the proposed work are obtained using MATLAB R2015a. The proposed algorithm is tested on videos from two different types of benchmark databases. The first database contains videos from the VSUMM [2] database which is a collection of videos from Open Video Project (OVP) [13]. It has videos from various genres like lectures, documentary, historical, advertisement and so on. The run time of the videos varies from 1 to 4 min and is in MPEG-1 format. The total length of the videos is approximately 75 min. VSUMM database is referenced by many classical video summarization methods for validating the results. The second database is random collection of videos from YouTube. It is also provided by [2]. It has a collection of large set of videos from various genres. The durations of the videos vary from 1 to 10 min. The authors label it as YouTube database. Table 1 lists the videos on which the proposed algorithm is tested.

3.2 Experimental Results

(a) **v109 video**:
 See Fig. 1.

Fig. 1 Keyframes extracted from video v109 (YouTube database) by the proposed algorithm

Fig. 2 Keyframes extracted from video v23 (Open Video Project) by the proposed algorithm

Fig. 3 Keyframes extracted from video v25 (Open Video Project) by the proposed algorithm

(b) **v23 video**:
See Fig. 2.

(c) **v25 video**:
See Fig. 3.

The video summary results for two datasets demonstrate the effectiveness of the proposed technique. It is observed that the image velocity information correctly captures all motion changes in the video and extracts correct keyframes.

3.3 Discussion

We know that as the optical flow method directly recovers the image motion at each pixel from spatio-temporal image brightness variation, and thus it accurately extracts the keyframes from the video v23. It is observed that the proposed algorithm calculates dense motion fields but is sensitive to large appearance variations. Thus the proposed algorithm generates a video summary with some redundancy when the videos have large image motion as observed for videos v109 and v25. It is also observed that the proposed algorithm fails to generate satisfactory video summary for videos that have huge PAN and TILT of the camera, or when video gets blur, and there is huge change in illumination. These results encourage the usage of parametric flow methods or feature-based method for video summarization.

4 Performance Analysis

We prove the efficiency and validity of the proposed algorithm by: (1) subjective performance analysis; (2) quantitative performance analysis.

4.1 Subjective Performance Evaluation

The video summary generated by the proposed algorithm is displayed to five testers. The testers are categorized into various groups as students, research scholars and faculty. The testers subjectively evaluated the generated video summaries. The testers graded the video summaries based on three scaling grades: 1 = Bad, 2 = Good and 3 = Acceptable. The testers were first introduced with the videos from two datasets and were then presented with the video summary. The testers were not aware of the video summarization techniques used. Table 2 shows the subjective evaluation.

The results in Table 2 validates that the proposed algorithm generates semantically meaningfully video summary and it can be satisfactorily applied to various genres of videos. It is also observed that the number of keyframes extracted is relevant to the amount of motion in the videos.

4.1.1 Comparison with VSUMM2 [2] and STIMO [14]

The competence of the proposed algorithm is also established by calculating the average subjective performance evaluation (ASPE) score of the videos v23 and v25

Table 2 Subjective performance evaluation score

Video	Frames in the video	Keyframes extracted by the proposed algorithm	Good (%)	Acceptable (%)	Bad
v109	1577	13	95	05	0
v23	1761	18	90	10	0
v25	1794	22	95	05	0

Table 3 Average subjective performance evaluation (ASPE) score of the video summary generated by VSUMM2 [2], STIMO [14] and the proposed algorithm

Video	VSUMM2 [2]	STIMO [14]	Proposed algorithm
v23	1.8	1.1	2.5
v25	1.8	1.5	2

from Open Video Project and comparing it with the ASPE score of several classical video summarization algorithms, such as VSUMM2 [2] and STIMO [14]. The average subjective performance evaluation (ASPE) score is the average of the subjective scores of the testers, which is obtained as given in Sect. 4.1.

As observed in Table 3 the ASPE score of the proposed algorithm outperforms VSUMM2 and STIMO. Thus the proposed technique captures more representative frames than the classical video summarization techniques given in Table 3. The technique is compared against the ground truth video summary provided by [2], VSUMM2 [2], STIMO [14] and VRCVS [11]. As observed in visual comparative analysis in Fig. 4, the frames which have significant change in the motion are extracted as keyframes by the proposed algorithm. Thus it extracts more representative frames from the video. It is even visually observed that improved results are obtained than the recent method such as VRCVS [11].

4.2 Quantitative Performance Evaluation

The proposed algorithm is quantitatively evaluated by comparison of the generated video summary and ground truth or user summary given in [2]. The ground truth video summary is considered as the reference video summary. We propose a surjective function (Onto) S such as:

O = {Set of frames generated from the proposed video summarization technique}
R = {Set of frames from the ground truth summary}

Then the function S is surjective if and only if f (O) = R.

The matching of the frames from set O to set R is done using the robust SURF features. SURF features help to match the frames even if the frames have objects that are occluded or transformed due to motion in the video [15]. This helps for matching frames that are similar in context but may appear to be redundant. The matched value from the above mapping helps in calculating:

$$\text{Precision} = \frac{nMatc}{nVS} \tag{5}$$

$$\text{Recall} = \frac{nMatc}{nGT} \tag{6}$$

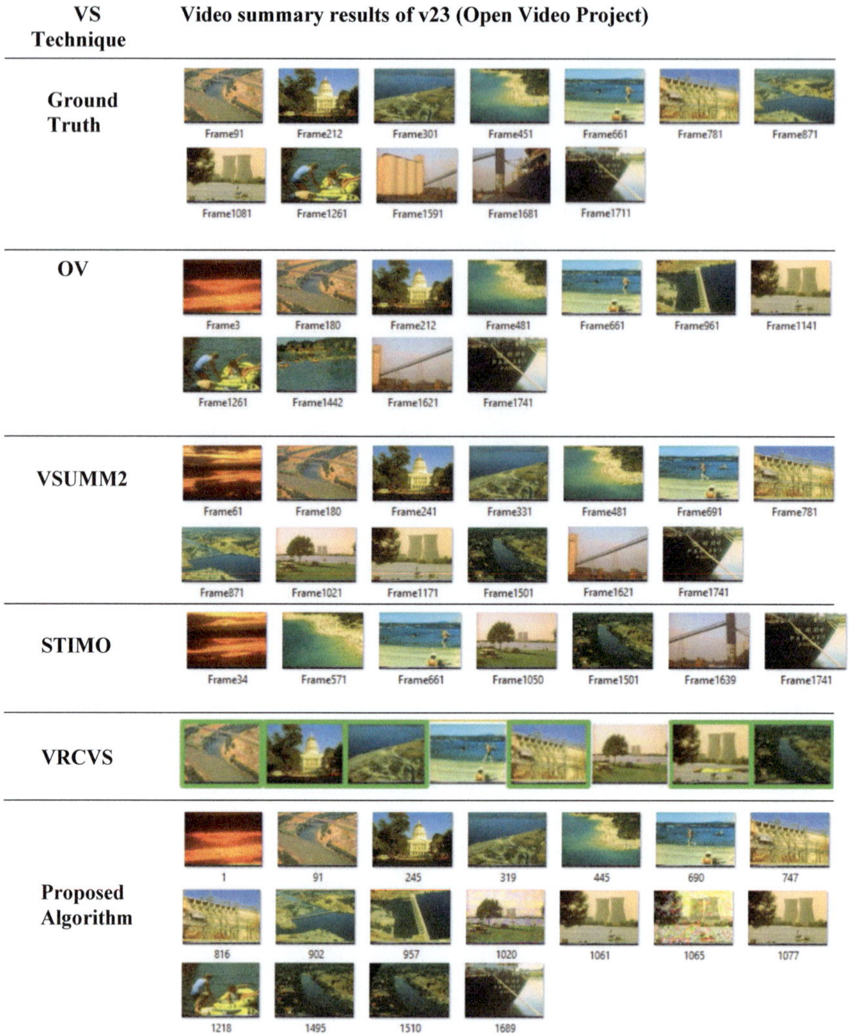

Fig. 4 Visual comparative analysis of the proposed algorithm with OV [13], VSUMM2 [2], STIMO [14], VRCVS [11] for video v23

$$\text{F-Score} = \frac{2 * \text{Precion} * \text{Recall}}{\text{Precision}} \tag{7}$$

where *nMatch* = number of frames matched among ground truth summary and generated summary. *nVS* = frames in the video summary. *nGT* = frames in the ground truth summary [11]. The quantitative evaluation of the proposed algorithm is done by averaging the above mentioned values for videos v25, v23 from Open Video Project. It is demonstrated in Table 4.

Table 4 Quantitative evaluation score

Video summarization technique	Precision	Recall	F-Score
OV	0.7	0.8	1.6
VSUMM2	0.8	0.8	1.6
STIMO	0.7	0.7	1.4
Proposed algorithm	0.8	1	2

As observed in Table 4, the proposed algorithm outperforms other video summarization techniques. It is evident that the proposed algorithm generates video summary which is comparable and in certain videos even outperforms VSUMM and STIMO. The proposed algorithm outperforms most of the classical video summarization methods in both subjective as well as objective evaluation. It is also evident that the proposed algorithm can generate representative video summary based on motion information.

5 Conclusions and Future Scope

The authors have formulated the video summarization technique using spatio-temporal change information extracted from the consecutive video frames. For this purpose we extract the change in image velocity between successive frames and extract representative keyframes in the video. The results and the successful performance analysis prove the effectiveness of the optical flow method for generating a video summary. Future scope includes enhancing the results using feature-based method.

References

1. Youtube Statistics. http://www.youtube.com/t/press/statistics
2. de Avila, S.E.F., da_Luz Jr., A., de Albuquerque Araújo, A., Cord, M.: VSUMM: an approach for automatic video summarization and quantitative evaluation. In: Proceedings of the 2008 XXI Brazilian Symposium on Computer Graphics and Image Processing, pp. 103–110, 12–15 Oct 2008. https://doi.org/10.1109/sibgrapi.2008.31
3. Pritch, Y., Rav-Acha, A., Peleg, S.: Nonchronological video synopsis and indexing. IEEE Trans. Pattern Anal. Mach. Intell. **30**(11), 1971–1984 (2008). https://doi.org/10.1109/tpami.2008.29
4. Ji, Z., Su, Y., Qian, R., Ma, J.: Surveillance video summarization based on moving object detection and trajectory extraction. In: ICSPS, 2010 2nd International Conference on Signal Processing Systems, vol. 2, pp. 250–253, 5–7 July 2010
5. Lee, Y.J., Ghosh, J., Grauman, K.: Discovering important people and objects for egocentric video summarization. In: Proceedings of the 2012 IEEE Conference on Computer Vision and Pattern Recognition (CVPR), pp. 1346–1353, 16–21 June 2012
6. Wang, F., Ngo, C.-W.: Summarizing rushes videos by motion, object and event understanding. IEEE Trans. Multimed. **14**(1), 79–81 (2012)

7. Peng, J., Xiao-Lin, Q.: Keyframe-based video summary using visual attention clues. IEEE Trans. Multimed. **17**(2), 64–73 (2010). https://doi.org/10.1109/mmul.2009.65
8. Khosla, A., Hamid, R., Lin, C.-J., Sundaresan, N.: Large-scale video summarization using web-image priors. In: Proceedings of the 2013 IEEE Conference on Computer Vision and Pattern Recognition, pp. 2698–2705, 23–28 June 2013. https://doi.org/10.1109/cvpr.2013.348
9. Lu, Z., Grauman, K.: Story-driven summarization for egocentric video. In: Proceedings of the 2013 IEEE Conference on Computer Vision and Pattern Recognition, pp. 2714–2721, 23–28 June 2013. https://doi.org/10.1109/cvpr.2013.350
10. Kim, H., Yoon, J., Kim, T.Y., Paik, J.: Video summarization using feature dissimilarity. In: International Conference on Electronics, Information and Communications (ICEIC). IEEE (2016)
11. Jiaxin, W., Zhong, S.-H., Jiang, J., Yang, Y.: A novel clustering method for static video summarization. J. Multimed. Tools Appl. **76**(7), 9625–9641 (2017)
12. Brox, T., Bruhn, A., Papenberg, N., Weickert, J.: High accuracy optical flow estimation based on a theory for warping. In: Proceedings of European Conference on Computer Vision (2004)
13. Open Video Database. https://sites.google.com/site/vsummsite/download. Accessed 21 Feb 2017
14. Furini, M., Geraci, F., Montangero, M., Pellegrini, M.: STIMO: STIll and MOving video storyboard for the web scenario. Multimed. Tools Appl. **46**(1), 47–69 (2010). https://doi.org/10.1007/s11042-009-0307-7
15. Bay, H., Tuytelaars, T., Van Gool, L.: SURF: Speeded up robust features. In: Leonardis, A., Bischof, H., Pinz, A. (eds.) Computer Vision—ECCV 2006. Lecture Notes in Computer Science, vol 3951. Springer, Berlin, Heidelberg (2006)

Decision Support System for Black Classification of Dental Images Using GIST Descriptors

Prerna Singh and Priti Sehgal

Abstract One of the well-known pathology in the world is dental caries. Dental caries is also called as dental cavities. The prior detection of dental caries helps in decreasing the dental disease rate. Patient care has been improved due to medical image mining. In this paper, abnormal dental images have been classified into various classes based on Black's classification. Graphics and intelligence-based script technology (GIST) descriptor has been used to extract significant information from the dental images. Feature reduction is done using marginal Fisher analysis and then Wilcoxon signed-rank test is used as feature selection method. The classification techniques such as decision tree, fuzzy Sugeno, probabilistic neural network, K-nearest neighbor, AdaBoost and naïve Bayes are used for classifying the major and reduced features. According to the results, AdaBoost classifier can best diagnose infected tooth using Black's classification with the classification accuracy of 90, 92% sensitivity and 90% specificity.

Keywords Marginal Fisher analysis · Decision tree · AdaBoost · Gist descriptors

1 Introduction

Dental caries or cavity is an infection that is found in teeth. The dental cavities are of different colors, such as yellow to black. Caries can be found either (i) on a smooth surface or (ii) on pits and fissures. Caries can be classified into five classes according to Black classification [1]. The Black classification classifies dental caries based on the location of caries in the teeth of the patient. The Class I caries are located on the posterior teeth. The posterior teeth are molar and premolar. The Class I cavity is found on the occlusal surface, buccal surface and lingual surface. The Class II caries are initiated on the near and far surface of the premolar and the molar teeth. The Class

P. Singh (✉)
Department of Computer Science, University of Delhi, Delhi, India
e-mail: prerna.singh@jimsindia.org

P. Sehgal
Department of Computer Science, Keshav Mahavidyalaya, University of Delhi, Delhi, India

B. Pati et al. (eds.), *Advanced Computing and Intelligent Engineering*,
Advances in Intelligent Systems and Computing 1082,
https://doi.org/10.1007/978-981-15-1081-6_29

III caries are originated in the near and far surface on the incisor and canine teeth. The Class IV caries are established on the near and far of canine and incisor teeth along with the incisal angle. The Class V caries are located on any tooth whether anterior and posterior but on the cervical third of the tooth. The Black classification is done to find the type of affected tooth, the location of cavity or the tooth surface involved in infection. In this paper, we classify the dental images based on Black's classification using the GIST descriptors [2]. The information of the image [3] is provided with the GIST descriptor. The image is separated into grid of size 4×4. The GIST descriptors do not require any form of segmentation as they represent the low-level representation of the science. The width of the GIST descriptors ranges from 32 to 128 pixels. The GIST descriptor shows good results in image retrieval [4].

The paper is structured as follows: Sect. 2 describes related works, Sect. 3 discusses the work proposed for the identification of different classes according to Black's classification, Sect. 4 defines the various classifiers used in our work. Section 5 discusses the result of the technique that we have proposed, and Sect. 6 covers the conclusion.

2 Related Works

The detection and classification of dental caries has been a wide area of research these days. Ramzi et al. proposed a new method for finding and categorization of dental caries for X-ray images using deep neural network [5]. ALbahbah et al. introduced a tooth caries detection approach constructed on back propagation (BP) neural network for examining dental X-ray images [6]. Sehgal and Singh proposed an automatic caries discovery system based on radon transformation and discrete cosine transformation (DCT) for identifying dental caries using dental X-ray images [7]. Olsen et al. [8] has planned dental caries detection system using advance image processing technique and C4.5 decision tree classifier to perfectly recognize the caries. Berdouses et al. [9] projected a computer-aided automatic system for discovering caries lesion using five classifiers. Saravanan et al. proposed a novel method to identify caries using power spectral and histogram investigation. This was done by concentration of pixels with regards to histogram [10]. Sikiric et al. [11] proposed a novel approach for classifying traffic scenes using GIST descriptors and SVM classifiers. The results show that GIST descriptors are sufficient for the purpose of classification of different types of traffic signals. Moudni et al. [12] proposed the recognition system using SURF and GIST descriptors for extracting feature vector. The results proved that GIST descriptor was more efficient. Oujaoma et al. [13] used Walsh transform and GIST descriptor with Bayesian network as a classifier for recognition of Tifinagh characters adopted by the Moroccan Royal Institute of Amazigh Culture (IRCAM).

The results showed that GIST descriptors had amazing classification rate. Tin et al. [14] did crater detection using GIST descriptor with random forest classification, and the results gave an accurate crater recognition rate. Hence, the efficacy of the GIST descriptors has been used for feature extraction in this paper. Further, marginal Fisher analysis is used as the feature reduction method to detect the features from the image and the Wilcoxon signed-rank test is used to determine the most suitable features which are subject to the classifier to classify the dental X-ray images into different classes.

3 Proposed Work

Detection and classification of dental caries into different classes is done using Black's classification. The work we have proposed along with the block diagram is shown in Fig. 1. Dental images are divided into training and testing dataset in order to check the accurateness and significance of the classification. The dental images are subjected to GIST descriptor and hence large number of features is extracted. Later marginal Fisher analysis is used for reducing the features and Wilcoxon signed-rank test is used to select the best feature. After training, the relevant features are fed to the following classifiers such as decision tree (DT), AdaBoost, K-nearest neighbor (KNN), fuzzy Sugeno (FS), probabilistic neural network (PNN), naïve Bayes (NB), to determine which classifier produces the most accurate result.

The proposed technique is described in the following steps

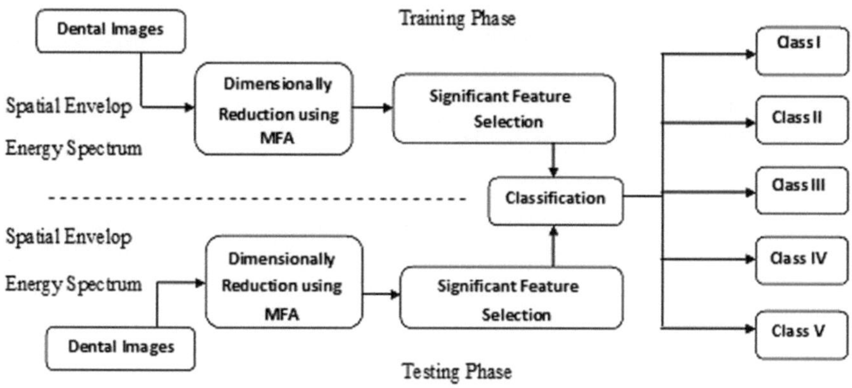

Fig. 1 Block diagram of proposed technique using GV's Black classification

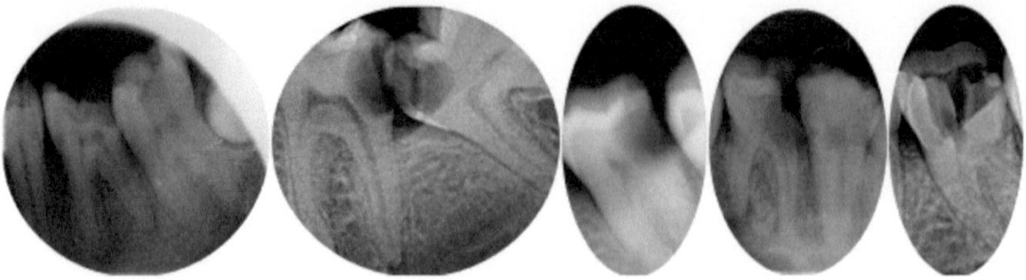

Fig. 2 G.V. Black classification of class I–V, respectively

3.1 Data Acquisition

A total of 400 images of teeth having dental caries captured by the dental X-ray machine were collected from a dental clinic. We took 80 images for each class I–V of G.V. Black classification for the proposed research work. Figure 2 shows the images of G.V. Black classification from class I to class V, respectively.

3.2 Feature Extraction Using GIST Descriptors

The information about different parts of the image is provided by GIST descriptor. The preprocessed images are subjected to GIST descriptor which represents the image in a spatial envelope energy spectrum. They concentrate on the image shape and on a correlation between the surface's outline and their properties. The structure of an image is well-defined as the three-dimensional envelop.

3.3 Dimensionality Feature Reduction

Dimensional reduction techniques [15] are used to transform data from high dimensions to lower dimensions during data analysis. Some of the methods used for feature reduction are principal component analysis (PCA), linear discriminant analysis (LDA) and marginal Fisher analysis (MFA). Marginal Fisher analysis (MFA) has been used as the reduction technique in this work. PCA is an unsupervised algorithm and it focuses on discriminant feature extraction. The supervised method, LDA is used for discriminating by reducing the within-class and the between-class ratio. LDA focuses on object reconstruction. In case of pattern classification, LDA outperforms PCA [16]. In case if there are inadequate amount of samples, then the PCA and LDA may fail. The reason we use MFA is to overcome the limitation of PCA and LDA.

3.3.1 Marginal Fisher Analysis (MFA)

Marginal Fisher analysis is constructed on graph-embedded framework. It is an organized learning procedure. It is used for dimensionality reduction process. It can model between-class separability and within-class compactness [17, 18]. The summation of distance between each sample and its neighbor is called intra-class separability, and the summation of distance between boundary points and adjacent points of dissimilar classes is called inter-class separability. Marginal Fisher analysis provides us good simplification capability and delivers the better accuracy by considering the data of class label [19]. MFA further outperforms PCA and LDA for handling marginal samples.

3.4 Wilcoxon Signed-Rank Test for Feature Selection

In order to delete the large number of irrelevant features, we use Wilcoxon signed-rank test. With only the best feature we can improve the accuracy of the classifier and provide cost-effective and robust model. It gives with perceptions into the fundamental data generator processes [20]. We have used Wilcoxon signed-rank test method in feature selection. It is a non-parametric hypothesis test that is used when two related samples are compared with each other [21]. The data that is grouped belong to the class that is same and is measured on an ordinary scale [22–24]. Wilcoxon signed-rank test aids in selecting the significant features which further improve the performance of the classifiers models and thus provide fast models.

4 Classification

With the help of classification model, we can predict the accuracy of the model. In this paper, we did our research on 400 dental images, that is, 80 images per class I to class V. Tenfold cross-validation algorithm was used to build robust model. Image file containing 400 images is divided into tenfold, that is, ten equal parts, where each fold contains 40 images. Nine parts are used for training purpose and one-tenth is used for testing purpose in the first run. This is repeated for 10 iterations and different one-tenth reserved for testing in each fold. Parameters performed in it, such as specificity, accuracy, sensitivity, are all computed for each fold. The classifier's performance of all the parameters is measured using the average value.

The following classifiers have been used for classification in the proposed work:

Decision tree
The decision tree is a classifier which is organized into the leaf and the root node [25]. Decision tree is drawn based on the certain set of the predefined rules. It follows the greedy top-down approach. The representation of the decision tree consists of a

structure which is like a tree. The rectangular boxes represent the test condition whose outcomes are represented in the form of branch and the final outcome is represented by final node at the bottom which represents the final class for which the decision tree is drawn.

Naïve Bayes
Naïve Bayes classifiers are drawn from the Bayesian network and they are also called as probabilistic learning classifiers [25]. It is based on the assumption that the impact of the value of an entity of a particular class will not be affected by the value of the other entity of the same class.

AdaBoost
AdaBoost is a first practical boosting algorithm. With the help of AdaBoost algorithm we can increase the accuracy of the certain learning method [25]. The AdaBoost work as accordingly; first, it designates the identical value to all the training examples denotes the dissemination of weight. It uses the training example to verify, thereby increasing the value of wrong classified examples. Thus, a final value distribution is generated.

K-nearest neighbor
One of the easiest of the classifiers is the K-nearest classifier (K-NN) [26]. It is based on the approach of discovering the data that is not identified with the help of the data that has been identified before. K-NN after identifying the data point, further divides the data point with the help of neighbor which are nearest and more than one. A test set and a training set are used in the K-NN model, and the K-NN identifies the elements of the training set that are identical to the test set. The main criteria for using K-NN are the following: the set of predefined objects, a metric is calculated which is based on the distance of the objects and the number of the nearest neighbor is calculated.

Probabilistic Neural Network classifier
A probabilistic neural network (PNN) has replaced the sigmoid activation function [27]. PNN is a neural network that can map any type of the pattern as the input with any type of classification. It is defined as a four-layer network.

Fuzzy Sugeno
Fuzzy Sugeno model [28] is better than the other two models, Mamdani and Tsukamoto, in terms of diagnosis of tuberculosis. Fuzzy Sugeno method was better in terms of the accuracy of the proposed approach.

5 Results

A total of 400 dental tooth X-ray images were acquired from a dental clinic. The dataset consisted of 80 images belonging to each class. The various classes of G.V Black dental images (class I–V), as shown in Fig. 2, are subjected to GIST descriptor

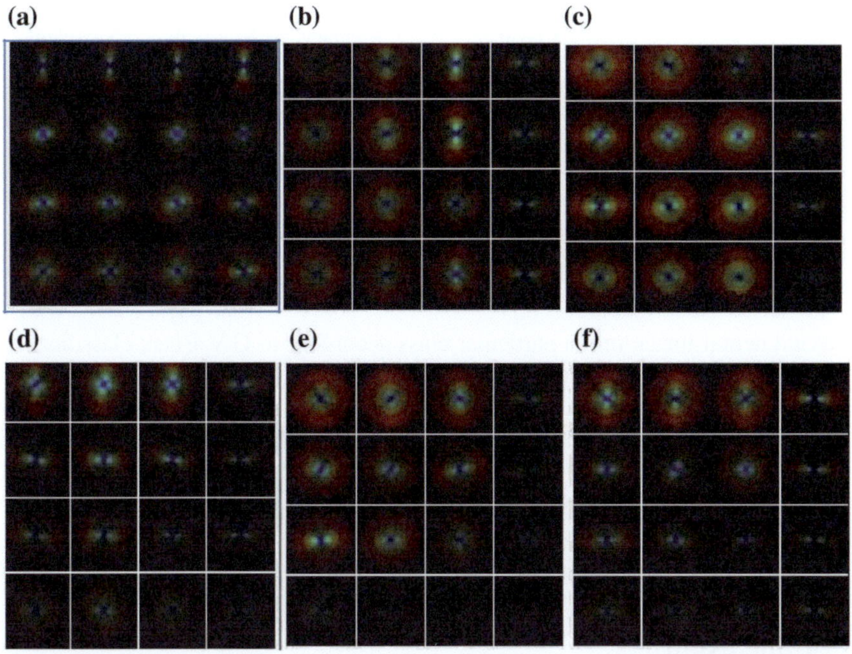

Fig. 3 GIST descriptor of dental images: **a** Normal, **b** Black Class I, **c** Class II, **d** Class III, **e** Class IV, **f** Class V

algorithm. Various features are detected from the spatial energy spectrum, as shown in Fig. 3. The total features detected for various classifications of infected dental teeth are 512. These features are further reduced using marginal Fisher analysis and best features are chosen using Wilcoxon signed-rank test.

MATLAB has been used for the implementation of GIST descriptor and marginal Fisher analysis algorithm. The performance of the different classifiers without Wilcoxon signed-ranked test is shown in Table 1. The maximum accuracy is of the AdaBoost classifier with 88% followed by PNN at 82%. Further, by applying the Wilcoxon signed-rank test the performance of the classifier is improved, as shown in Table 2. It is found that AdaBoost classifier is the best classifier to classify the

Table 1 Evaluation of classifiers using G.V. Black classification of dental images without Wilcoxon signed-rank test

Classifier	Accuracy	Sensitivity	Specificity
Decision tree	80	90	76
Fuzzy Sugeno	78	80	75
KNN	80	82	78
Naïve Bayes	81	89	81
PNN	82	87	82
AdaBoost	88	90	88

Table 2 Evaluation of classifiers using G.V. Black classification of dental images with Wilcoxon signed-rank test

Classifier	Accuracy	Sensitivity	Specificity
Decision tree	88	95	86
Fuzzy Sugeno	87	90	85
KNN	85	89	82
Naïve Bayes	88	89	90
PNN	89	92	90
AdaBoost	92	93	93

infected dental image into a particular class according to G.V Black classification as it has the maximum accurateness of 92%, sensitivity of 93% and specificity of 92%. Wilcoxon signed-rank test for feature selection is important for the classification. The number of features selected for the classification plays a significant role in increasing the performance of the classifier. After marginal Fisher analysis for feature reduction and Wilcoxon signed-rank test for feature selection, 17 features are selected. AdaBoost classifier gives best performance using the best 17 features.

6 Conclusion

The proposed technique is used to classify 400 infected dental images into various different classes according to G.V. Black classification model. Features detected using GIST descriptors are innumerable. They are further reduced using marginal Fisher analysis, and only a few significant features are selected with the help of Wilcoxon signed-rank test. Those important features are subjected to different classifiers. It is observed that AdaBoost is the best classifiers to classify the infected teeth image into different classes. The proposed technique can also be implemented in dental clinics and dental hospital and help the budding dentist to further classify the infected teeth into different classes. The technique is very accessible and computerized.

Acknowledgements The dental X-ray images used in the study is taken from Smart Smile Dental clinic and Implant Centre, New Delhi.

References

1. Powers, J.M., Sakaguchi, R.: Craig's Restorative Dental Materials, 12th edn. C.V. Mosby (2006) (VitalBook file)
2. Oliva, A., Torralba, A.: Modeling the shape of the scene: a holistic representation of the spatial envelope. Int. J. Comput. Vis. **42**(3), 145–175 (2001)
3. Koutsouri, G.D., Berdouses, E., Tripoliti, E.E.: Detection of occlusal caries based on digital image processing. IEEE (2013). ISSN 978-1-4799-3163-7

4. Douze, M., Jegou, H., Sandhawalia, H., Amsaleg, L., Schmid, C.: Evaluation of GIST descriptors for web-scale image search. In: Proceedings of the ACM International Conference on Image and Video Retrieval CIVR, Santorini, pp. 1–8. IEEE, Greece (2009)
5. Ali, R.B., Ejbali, R., Zaied, M.: Detection and classification of dental caries in x-ray images using deep neural network. In: The Eleventh International Conference on Software Engineering Advances (ICSEA), pp. 223–227 (2016). ISBN 978-1-61208-498-5
6. ALbahbah, A.A., Bakry, H.M., Abd-Elgahany, S.: Detection of caries in panoramic dental x-ray images using back propagation neural network. Int. J. Electron. Commun. Comput. Eng. **7**(5), 250–256 (2016). ISSN (Online) 2249-071X
7. Sehgal, P., Singh, P.: Automated caries detection based on radon transformation and discrete cosine transformation. In: 8th International Conference on Computing, Communication and Networking Technologies (2017)
8. Olseen, G.F., Brillant, S.S., Primeaux, D., Najarian, K.: An image processing enabled dental caries detection system. In: IEEE International Conference on Multimedia and Expo (2009)
9. Berdouses, E.D., Koutsouri, G.D., Tripoliti, E.E., Mastopoulas, G.K., Oulis, C.J., Fotiadis, D.J.: A computer-aided automated methodology for the detection and classification of occlusal caries from photographic color images. Comput. Biol. Med. **62**, 871–875 (2015) (Elsevier)
10. Saravanan, T., Raj, M.S., Gopalakrishnan, K.: Identification of early caries in human tooth using histogram and power spectral analysis. Middle-East J. Sci. Res. 871–875 (2014)
11. Sikiric, I., Brkic, K., Segvic, S.: Classifying traffic scenes using the GIST image descriptors. In: Proceeding of the Croatian Computer Vision Workshop, pp. 19–24 (2013)
12. Moudni, H., Er-rouidi, M., Oujatura, M., Bencharef, O.: Recognition of Amazigh characters using SURF and GIST descriptors. In: IJACSA Special Issue on Selected Papers from Third International Symposium on Automatic Amazigh Processing, pp. 41–44 (2013)
13. Oujaoma, M., Minaoui, B., Fakir, M.: Walsh, texture and GIST descriptors with Bayesian networks for recognition of Tifinagh characters. Int. J. Comput. Appl. **81**(12), 39–46 (2013)
14. Tin, J., Li, H., Jia, X.: Carter detection based on GIST features. IEEE J. Sel. Top. Appl. Earth Obs. Remote Sens. **8**(1), 23–29 (2015)
15. Wang, S., Yan, S., Yang, J., Zhou, C., Fu, X.: A general exponential framework for dimensionality reduction. IEEE Trans. Image Process. **23**(2), 920–930 (2014)
16. Wang, Z., Sun, X., Sun, L., Huang, Y.: Semi-supervised kernel Marginal Fisher analysis for face recognition. Sci. World J. (Article ID 981840), 13 (2013)
17. Yang, W., Liu, S., Jin, T., Xu, X.: An optimization criterion for generalized marginal fisher analysis on under sampled problems. Int. J. Autom. Comput. **8**(2), 193–200 (2011)
18. Yan, S., Xu, D., Zhang, B., Zhang, H., Yang, Q., Lin, S.: Graph embedding and extensions: a general framework for dimensionality reduction. IEEE Trans. Pattern Anal. Mach. Intell. **29**(1), 40–51 (2007)
19. Jiang, L., Xuan, J., Shi, T.: Feature extraction based on semi-supervised kernel Marginal Fisher analysis and its application in bearing fault diagnosis. Mech. Syst. Signal Process. **41**, 113–126 (2013)
20. Saeys, Y., Inza, I., Larranaga, P.: A review of feature selection techniques in bioinformatics. Bioinform. Rev. **23**(19), 2507–2517 (2007)
21. Lopes, N.: Comparing machine learning algorithms with the Wilcoxon Signed Rank Test. Information available at: http://www.uc.pt/fctuc/dei/statisticalHypothesis/noel. Accessed 04 Oct 2015
22. Hwang, T., Sun, C.H., Yun, T., Yi, G.S.: FiGS: a filter-based gene selection workbench for microarray data. BMC Bioinform. **11** (2010). Information available at: http://www.biomedcentral.com/1471-2105/11/50. Accessed on 04 Oct 2015
23. Natarajan, S., Lipsitz, S.R., Fitzmaurice, G.M., Sinha, D., Ibrahim, J.G.: An extension of the Wilcoxon rank sum test for complex sample survey data. J. R. Stat. Soc.: Ser. C (Appl. Stat.) **61**(4), 653–664 (2014)
24. Yuan, Y., van Allen, E.M., Omberg, L., Wagle, N., Amin-Mansour, A.: Assessing the clinical utility of cancer genomic and proteomic data across tumor types. Nat. Biotechnol. **32**, 644–652 (2014)

25. Suri, B., Kumar, M.: Performance evaluation of data mining techniques. In: Information and Communication Technology for Sustainable Development. Lecture Notes of Networks and Systems, vol. 9, pp. 375–383 (2017)
26. Rajendran, P., Madheswaran, M.: Novel fuzzy association rule image mining algorithm for medical decision support system. Int. J. Comput. Appl. **1**(20), 87–94 (2010)
27. Specht, D.F.: Neural Netw. **3**(1), 109–118 (1990) (Elsevier)
28. Sari, W.E., Wahyunggoro, O., Fauziate, S.: A comparative study on fuzzy Mamdani-Sugeno-Tsukamoto for childhood tuberculosis diagnosis. In: Advances of Science and Technology for Society, American Institute of Physics Conference Paper, vol. 1755 (2016)

Pattern Analysis of Brain Functional Connectivity Parameters After Removal of Artifactual Motifs from EEG During Meditation

Laxmi Shaw and Aurobinda Routray

Abstract The actual picture of brain network and their functional connectivity analysis is a challenging task due to large data dimensionality and inherent dynamics as well as nonlinear property of electroencephalogram (EEG). This nonlinearity and nonstationarity are sometimes introduced in EEG, because of various physiologic and non-physiologic artifacts. To overcome this, the noninvasive EEG data are used to find the functional brain network in missing data of EEG which are obtained after removal of the artifactual motifs. This paper deals with the problem of extracting the functional brain connectivity in missing EEG data that can be used for both data analysis and the classification problem such as brain–computer interface (BCI) implication. Three significant parameters, namely, transitivity, characteristics path length, and centrality are considered from the list of brain functional connectivity metrics. The comparison has been performed between continuous and discontinuous (artifactual motifs removed) data. The results have been shown that the missing data's brain functional connectivity metrics are also following the same pattern of continuous data. The results show the implication of the framework in different brain regions involving in specific relaxation technique of short Kriya Yoga meditation.

Keywords EEG · Meditation · Motifs · Functional connectivity · Brain networks

1 Introduction

The major challenge of characterizing the active neural substrate underlying human consciousness has captured the cutting-edge interest of many neuroscientists and neuroresearchers [1, 2]. It has covered the altered states ranging from sleep, meditation,

Supported by Govt. of India, MHRD funded project.

L. Shaw (✉) · A. Routray
Department of Electrical Engineering, Indian Institute of Technology Kharagpur, Kharagpur 721302, India
e-mail: laxmishaw1983@gmail.com
URL: http://www.iitkgp.ac.in/

B. Pati et al. (eds.), *Advanced Computing and Intelligent Engineering*, Advances in Intelligent Systems and Computing 1082,
https://doi.org/10.1007/978-981-15-1081-6_30

hypnosis, anesthesia, coma disorder of consciousness, and many other neurological diseases. Out of these, meditation has an enormous impact on reducing stress [3] which is very much essential for the fast-living lifestyle [4, 5]. In this study, we have proposed a brain connectivity framework to measure two essential aspects of connectivity study. First, to validate the functional brain connectivity parameters in a more complex scenario. Second, quantitative assessment of the performance of the connectivity metrics. The idea is that to remove the motifs-based artifacts (artifacts with similar patterns) [6] from the EEG signals obtained from the meditators while performing short Kriya Yoga meditation [7]. The motifs which are identified as very similar subsequences in long EEG time-series are removed and the functional connectivity parameters are estimated in the same artifactual motifs removed EEG signals. In such cases of EEG, the motifs are nothing but the artifacts generated due to various physiological and non-physiological events [6]. After complete removal of those artifactual motifs, the disrupted EEG is used for the connectivity study.

Most of the literature has found out the functional connectivity in the continuous EEG data [8, 9]. The analysis of disrupted and discontinuous EEG time series and their connectivity estimation is very rarely addressed. This study is the assumption that the analysis of the discontinuous EEG time series may also provide insight into the interactive brain dynamics in a different cognitive condition such as meditation.

The remaining section of the paper is organized as follows. Section 2 contains the details about the methodology employed by us. It also discusses the mathematical aspects of three brain functional connectivity parameters and their implication in both continuous and missing samples in the time domain. Section 3 presents the results and discussion about the inference obtained from the brain network parameters in continuous and missing samples of EEG time series. Section 4 concludes our observations. The complete pipeline of the proposed framework has been shown in Fig. 1.

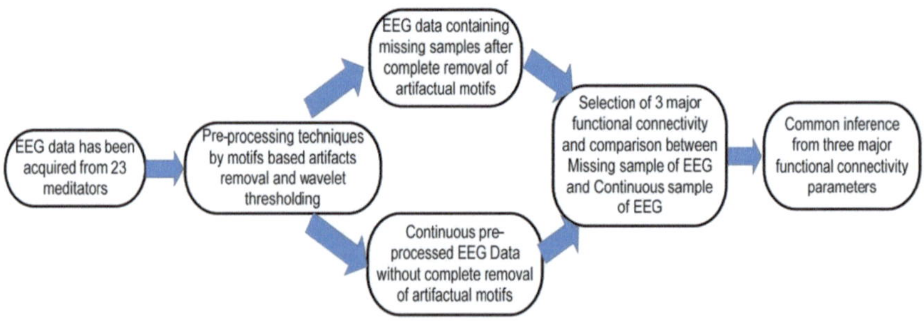

Fig. 1 The pipeline of the entire flow of the proposed framework for inferring brain connectivity

2 Materials and Methods

2.1 Participants, Experiment Design, and Data Acquisition

In this study, a total of 23 long-term Kriya Yoga practitioners were participated with an average age of 32.43 years (S.D. = 9.09 years, range = 23–56 years). EEG data were recorded when participants were performing guided Kriya Yoga meditation. Out of 23 meditators, 17 male and 6 female subjects have participated. The meditators were practicing such relaxation technique since last 15–20 years. All meditators are given written consent, and this study was approved by the ethical committee of Indian Institute of Technology Kharagpur. This study is the extension of the work. However, for the completion of the concept, we have mentioned few lines for the experiment design and acquisition procedure. The data has been acquired under proper experimental design and study protocol [6]. The total duration of the data has been collected for approximately 10 min. However, for this study, we have considered 4.17 min of data for each subject at a sampling frequency of 256 Hz. The EEG has been obtained using 64 channel RMS Victa device and with the 10–20 international standard of placement of electrodes. Out of 64, 48 electrodes are selected based on the symmetricity of the electrodes' position and considering five primary brain lobes shown in Fig. 2, for the functional connectivity study in missing sample of EEG. The 48 selected electrodes are Fp2, Fz, Fp1, AFz, AF4, AF8, AF7, AF3, F8, F4, F3, F7, FC1, FC2, FC3, FC4, C4, Cz, C3, CP3, CP4, FCz, CP1, CP2, P4, Pz,P3, CP5, CP6, CPz,P7, P8, O2, Oz, O1, POz, PO4, PO3, PO8, PO7, T7, FT7, FC6, TP8, T8, FT8, FC5, and TP7. Artifactual motifs are similar and repetitive artifacts, which occurred at irregular intervals in EEG time series. The intermittently disrupted EEG time series are obtained after complete removal of artifactual motifs from the continuous EEG data [6].

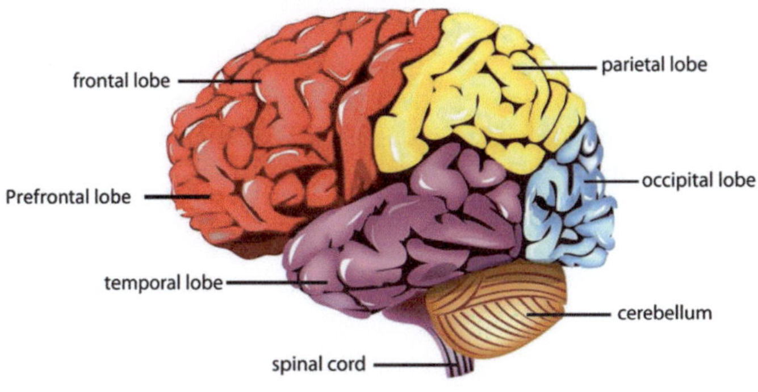

Fig. 2 Essential brain lobes and their localization based on the EEG electrodes [10]

2.2 EEG Recordings, Data Preprocessing, and Analysis

The pipeline of the entire framework has been mentioned in the schematic diagram in Fig. 1 including the signal analysis methodology. The whole data analysis comprises few most important steps which are discussed below as in the highlighted points:

- First of all, the raw EEG data has been passed through motifs-based artifacts removal [6] preprocessing techniques to remove unwanted noise and physiologic artifacts from the EEG data.
- After complete removal of the artifacts, the continuous EEG signals have become disrupted and discontinuous with some irregular blank patches as shown in Fig. 4 called EEG data with missing samples.
- The preprocessing techniques directly follow the method mentioned in the article [6]. However, we have recommended few lines for the preprocessing of the raw EEG signals for the completeness of the work presented in this study.
- The functional connectivity also has some structure [11]. This is the most apparent in the use of descriptive measures of graph analysis such as path length, node degree, and centrality [6, 12, 13]. These are the most frequently used functional connectivity measures in search of so-called small-world structure in brain networks [14].

2.3 Three Major Brain Network Parameter

All the brain network parameters are expressed with significant basic network specifications, namely, path length, centrality, and degree of the nodes. In this study, we have considered three vital parameters: (1) transitivity, (2) characteristics path length, and (3) closeness centrality and betweenness centrality with their mathematical details mentioned below. Any brain network parameters are defined with some essential elements. Among them, node degree is one.

Node degree—It is one of the most essential and elementary measures of the brain network and often indicated by k'. Node degree is the number of links connected to the node [15]. In directed networks, in-degree is the number of inward links and the out-degree is the number of outward links.

Transitivity—It is the normalized expression of the clustering coefficient [12]. It is not defined for the individual node in the brain network. Mathematical details of transitivity is given below [15]:

$$T = \frac{\sum_{i \in N} 2t_i}{\sum_{i \in N} k_i(k_i - 1)}$$

where t_i is the number of triangles around node i [12] and k_i is the degree of node i and t_i is given by

$$t_i = \frac{1}{2} \sum_{j,h \in N} a_{ij} a_{ih} a_{jh}$$

where a_{ij} is connection status between node i and node j, i.e., a_{ij} = 1, if there is a connection between node i and j else a_{ij} = 0.

Characteristic path length—It is the average shortest path length between all pairs of nodes in the network. It is the basic measure of functional integration and is generally defined by characteristics path length (L) [12]. The mathematical expression for this is as follows:

$$L = \frac{1}{n} \sum_{i \in N} L_i$$

where L_i is the average distance between node i as well as all other nodes and is given by the following equation:

$$L_i = \frac{\sum_{j \in N, j \neq i} d_{ij}}{n - 1}$$

where d_{ij} is the shortest distance between node i and j.

Centrality—Centrality can be assessed by various measures. Among them, the most commonly used measures are of two types. They are mentioned as follows:

Closeness centrality—It is the distance function that allows to determine nodes which are close to other nodes. Mathematical details of closeness centrality of node i is given by [12]

$$L_i^{-1} = \frac{n - 1}{\sum_{j \in N, j \neq i} d_{ij}}$$

where d_{ij} is the shortest distance between nodes i and j.

Betweenness centrality—The fraction of all shortest paths in the network that pass through a given node. The mathematical expression of betweenness centrality of node i is given by

$$b_i = \frac{1}{(n-1)(n-2)} \sum_{h.j \in N, h \neq j, h \neq i, j \neq i} \frac{\rho_{hj}(i)}{\rho_{hj}}$$

where ρ_{hj} is the number of shortest paths between h and j, and $\rho_{hj}(i)$ is the number of shortest paths between h and j that pass through i.

3 Results and Discussions

Many graphical representations and network parameters have been found out [12, 16] to achieve a better understanding of brain networks. In this case study, we have compared the three major brain network parameters between regular EEG time series and discontinuous EEG time series with artifactual motifs removed EEG (EEG times series have blanked space), and their corresponding figures are shown in Figs. 3 and 4, respectively. Few typical examples of artifactual motifs are shown in Fig. 5. Here,

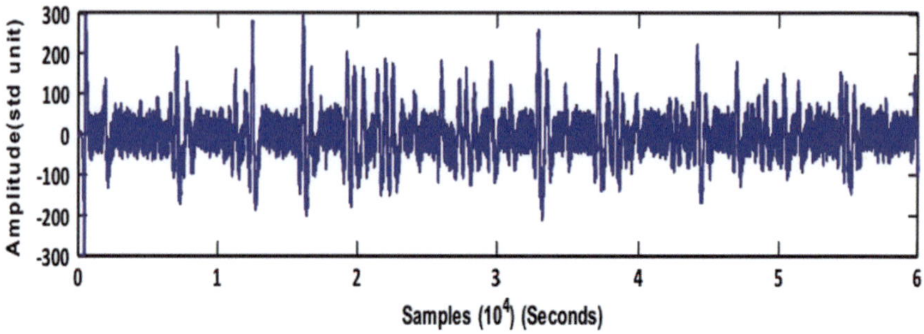

Fig. 3 A sample EEG time series without any intermittent missing samples

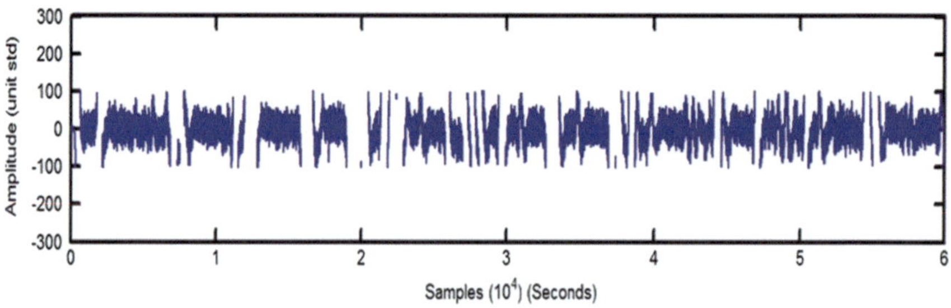

Fig. 4 The disrupted EEG signal showing intermittent missing samples within after removal of the artifactual motifs

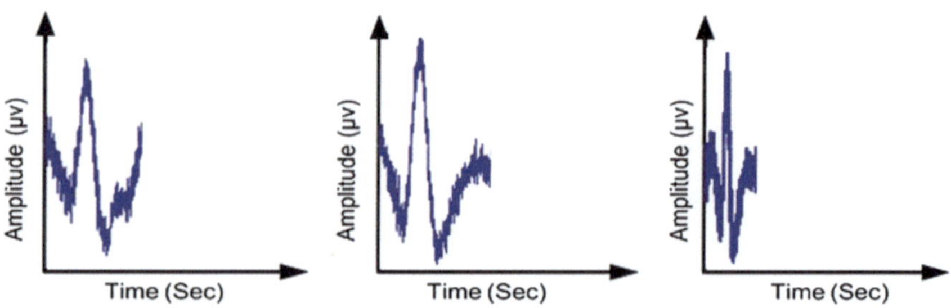

Fig. 5 Typical artifactual motifs in EEG time series

the work presented has focused on the framework of the brain network analysis. So far, all the functional connectivity and brain network analysis have been performed on the continuous preprocessed EEG data which has still some inherent noise within itself. However, the comparison of brain functional connectivity parameter between continuous and discontinuous EEG time series is hardly addressed. Figure 5 shows some typical artifactual motifs patterns found in an EEG time series. Those artifacts patterns are generated from eye movements and muscular activity. Components containing these artifacts were removed from the brain signals of all the meditator subjects. Remarkably, we found that the brain network properties, i.e., the characteristics path length and transitivity between same typical EEG time series containing missing and continuous samples follow typical trends shown in Figs. 6 and 7, respectively. From this pattern, it appears that more mindfulness a subject is the stronger the connection but also weaker the small worldliness. The effect of small worldliness is determined by the characteristic path length which is shown in Fig. 6. Several findings from the EEG study support the assumption that the meditative state seems to be unique in terms of patterns of brain activity [11, 14, 15]. On func-

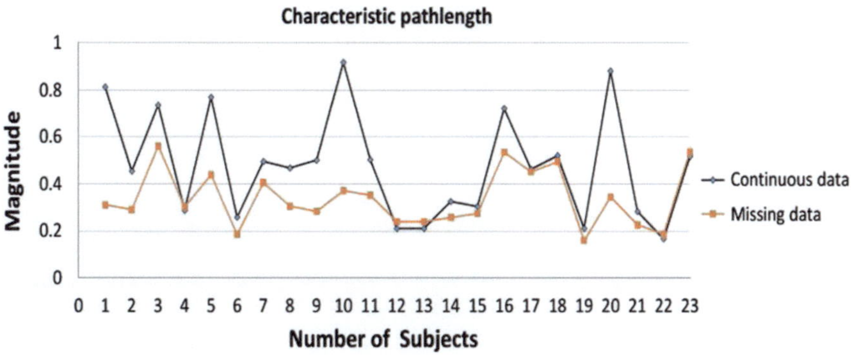

Fig. 6 Characteristics path length of 23 meditators in disrupted and continuous EEG data

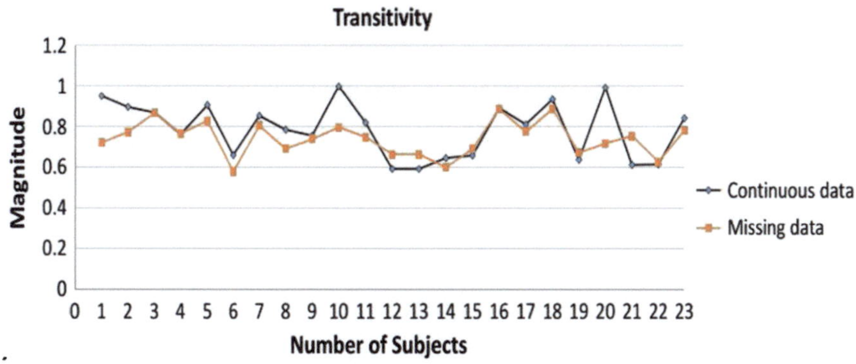

Fig. 7 Transitivity of 23 meditators in disrupted and continuous EEG data

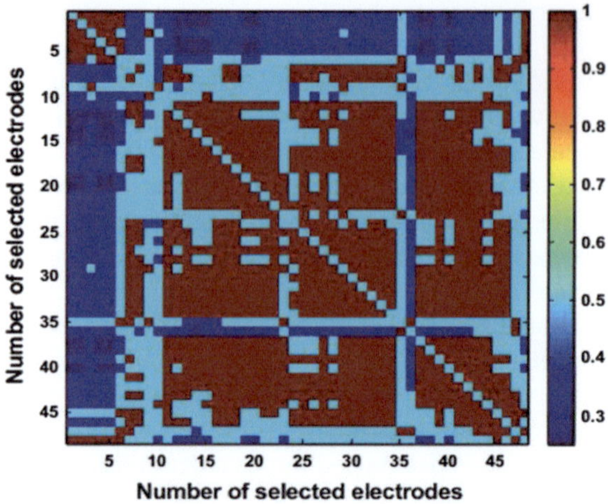

Fig. 8 Closeness centrality of one of the typical meditators in continuous EEG data

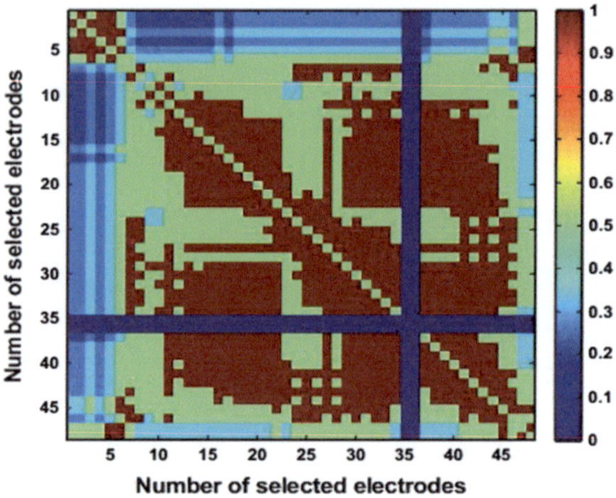

Fig. 9 Closeness centrality of one of the typical meditators in disrupted EEG data

tional connectivity estimation level, we found that significant correlation in same missing and continuous EEG time series. The closeness centrality property indicates the closeness of the shortest path in the functional brain network which is shown in Figs. 8 and 9 for the same EEG time series for the continuous and missing sample content EEG.

4 Conclusion

Due to the presence of volume conduction effect which gives rise to the spurious impact on EEG, making the functional brain connectivity is a challenging problem. Our investigation of functional brain connectivity in missing and continuous EEG data revealed an information extraction technique which may be used for the clinical conditions. The missing data has been created using the concept of entirely removed artifactual motifs. The functional connectivity has been studied using continuous preprocessed and artifact removed data to observe the effect of functional connectivity estimation in missing and continuous sample of EEG. From the results, it is pretty much clear that the missing data also follow the same trend like continuous data. So, we can conclude that to some extent of disrupted EEG, we can retrieve the functional information of the brain network. In this paper, we have tried to establish an extended understanding of underlying brain dynamics by comparing the functional connectivity between continuous and disrupted EEG data. To avoid the cross talk between the brain sites, connectivity estimation in disrupted EEG may give more insight into the connectivity in-depth stationary subspace analysis.

Acknowledgements The authors would like to thank MHRD, Government of India for funding the project "CEH-Sandhi".

References

1. Del Cul, A., Baillet, S., Dehaene, S.: Brain dynamics underlying the nonlinear threshold for access to consciousness. PLoS Biol. **5**(10), e260 (2007)
2. Nani, A., Seri, S., Cavanna, A.E.: Consciousness and neuroscience. In: Neuroimaging of Consciousness, pp. 3–21. Springer (2013)
3. Chiesa, A., Serretti, A.: Mindfulness-based stress reduction for stress management in healthy people: a review and meta-analysis. J. Altern. Complement. Med. **15**(5), 593–600 (2009)
4. Pace, T.W., Negi, L.T., Adame, D.D., Cole, S.P., Sivilli, T.I., Brown, T.D., Issa, M.J., Raison, C.L.: Effect of compassion meditation on neuroendocrine, innate immune and behavioral responses to psychosocial stress. Psychoneuroendocrinology **34**(1), 87–98 (2009)
5. Schiff, N.D., Nauvel, T., Victor, J.D.: Large-scale brain dynamics in disorders of consciousness. Curr. Opin. Neurobiol. **25**, 7–14 (2014)
6. Shaw, L., Routray, A., Sanchay, S.: A robust motifs based artifacts removal technique from EEG. Biomed. Phys. Eng. Express **3**(3), 035010 (2017)
7. Yogananda, P.: Autobiography of a Yogi. Sterling Publishers Pvt Ltd. (2003)
8. Bullmore, E., Sporns, O.: Complex brain networks: graph theoretical analysis of structural and functional systems. Nat. Rev. Neurosci. **10**(3), 186 (2009)
9. Rubinov, M., Sporns, O.: Weight-conserving characterization of complex functional brain networks. Neuroimage **56**(4), 2068–2079 (2011)
10. Thinkstocks:parts of human brain. http://www.thinkstockphotos.in/image/stock-illustration-human-brain/469537919. Accessed 30 Oct 2010
11. van den Heuvel, M.P., Sporns, O.: Network hubs in the human brain. Trends Cogn. Sci. **17**(12), 683–696 (2013)
12. Rubinov, M., Sporns, O.: Complex network measures of brain connectivity: uses and interpretations. Neuroimage **52**(3), 1059–1069 (2010)
13. Sporns, O., Chialvo, D.R., Kaiser, M., Hilgetag, C.C.: Organization, development and function of complex brain networks. Trends Cogn. Sci. **8**(9), 418–425 (2004)
14. Bassett, D.S., Bullmore, E.: Small-world brain networks. Neuroscientist **12**(6), 512–523 (2006)
15. Bullmore, E., Sporns, O.: The economy of brain network organization. Nat. Rev. Neurosci. **13**(5), 336 (2012)
16. Klonowski, W., Jernajczyk, W., Niedzielska, K., Rydz, A., Stepien, R.: Quantitative measure of complexity of eeg signal dynamics. Acta Neurobiol. Exp. **59**, 315–322 (1999)

Hyper-heuristic Image Enhancement (HHIE): A Reinforcement Learning Method for Image Contrast Enhancement

Mitra Montazeri

Abstract Conventional contrast enhancement methods such as histogram equalization (HE) have not obtained acceptable results on many different low-contrast images and they also cannot automatically handle various images. These problems are a result of specifying parameters manually for the sake of producing high-contrast images. We proposed an automatic image contrast enhancement on Hyper-heuristic. In this study, simple exploiters are proposed to improve the contrast of current image. To select these exploiters appropriately, reinforcement learning is proposed. This selection is based on the functional history of these exploiters. Having multi aim of preserving brightness, retaining the shape features of the original histogram, and controlling on the rate of over-enhancement are the achievements of the proposed method. These objectives are suitable for the application of consumer electronics. By this simulation results, it has been shown that in terms of visual assessment, absolute mean brightness error (AMBE) and peak signal-to-noise (PSNR) of the proposed method are superior to literature methods.

Keywords Image contrast enhancement · Histogram equalization · Histogram segmentation

1 Introduction

Artificial intelligence has significant effect in various fields such as data mining [1–7], pattern recognition [8–12], machine learning [13–19], and image processing [20–23]. One of the applications of image processing is image enhancement [24]. In this application, numerous image enhancement techniques have been researched such as gray-level transformation techniques and histogram processing techniques. In the

M. Montazeri (✉)
Medical Informatics Research Center, Institute for Futures Studies in Health, Kerman University of Medical Sciences, Kerman, Iran
e-mail: mitra.montazeri@gmail.com

Computer Engineering Department, Shahid Bahonar University, Kerman, Iran

B. Pati et al. (eds.), *Advanced Computing and Intelligent Engineering*,
Advances in Intelligent Systems and Computing 1082,
https://doi.org/10.1007/978-981-15-1081-6_31

first group, these methods map the gray level in the image to the new value by using transformation function such as power law transformation, logarithm transformation, etc.

In histogram processing techniques, many researches have already been studied on histogram equalization. histogram equalization (HE) is one of the method which is used in contrast enhancement, widely [25–27]. Achieving a uniform distributed histogram is the main goal of this method, which is achieved by applying the cumulative density function (CDF) on the original image [28]. Some of the problems of HE is that it may cause a washed-out appearance, intensified noise, and every undesirable artifacts. It can be verified that the mean brightness of the output image is placed at the center of the original image's gray level regardless of its mean. This property is an annoying characteristic in number of applications where brightness preservation is needed [28].

To solve the aforementioned problems different methods such as mean preserving bi-histogram equalization (BBHE) [29], equal area dualistic sub-image histogram equalization (DSIHE) [30] and recursive mean-spread histogram qualization (RMSHE) [28] have been proposed. The other method which is known as recursively separated exposure based sub image histogram equalization (RS-ESIHE) recursively applies the segregation of image histogram [31]. In this method, each updated histogram is separated more rely on their respective exposure thresholds and each sub histogram is equalized, exclusively.

Recently various approaches on genetic approach have been proposed [32–35]. In reference [32] optimizes a curve parameters by genetic algorithm (GA). In the proposed method, which is a binary GA, each chromosome represents the shape of the curve. This curve shows the relation between the gray level of input and output image to enhance the input image into good contrast. The sum of the intensities of edges is computed to determine the evaluation of the chromosome. Using GA has been proposed in [33]. In this method, chromosome structure which represents array of input gray levels is used, and each chromosome demonstrates an array of output gray levels. Each chromosome remaps the input gray levels based on specific transformation function. This process is done to evaluate how each chromosome relies on chromosome structure. In fact, the key distinction between this proposed GA and other ones is in the chromosome display. The fitness function includes the overall intensity and the number of edges.

In proposed method a hyper-heuristic image enhancement (HHIE) is based on reinforcement learning. HHIE not only converges to high-quality solutions but also searches more efficiently than their conventional counterparts [9, 36–38]. The main contribution of the proposed method is that it can handle images automatically with low and high brightness especially with high dynamic range, and its achievement includes multiple aims such as preserving brightness, retaining the shape features of the original histogram, and controlling the rate of over-enhancement. These objectives are suitable for the application of consumer electronics.

In this method, we use reinforcement learning to conduct multiple exploiters. Exploiters generate a better nomination solution at every step by mapping gray level of original images in such a way the result image has more contrast. These exploiters are designed with respect to the fact that no individual algorithm is the best one and each algorithm has its own advantages and drawbacks [39]. Besides, we need algorithms that automatically combine the strength and compensate the weakness of known exploiters. The proposed approach chooses appropriate exploiters based on the functional history of exploiters. Simulation results indicate that the proposed method performs better than the recent methods in the literature in PSNR, AMBE, and also visual assessment.

The rest of this paper is organized as follows: In Sect. 2, the proposed method will be presented. Finally, the experimental results and conclusion are discussed in Sects. 3 and 4, respectively.

2 Proposed Method

In this section, we describe the proposed hyper-heuristic image enhancement (HHIE) method that chooses exploiters appropriately by reinforcement learning. HHIE starts with the initial population of chromosome with the array of random integer numbers whose size is equal to the number of input gray levels (L). To reinforce the gray levels of dynamic range, the first and last item of the chromosome is valued to 0 and 256, respectively.

Reinforcement learning allocates subpopulation of solutions to each exploiter. The size of subpopulation depends on the functional history of exploiters. Furthermore, the exploiter with better performance has more chance to have more individuals. This performance is based on the quality of the produced solution by exploiters. Fitness function is a measurement that qualifies solutions. To evaluate the solutions, calculating the fitness function is done. Finally, evolutionary operations of GA are applied to search the solution space more effectively.

Due to the fact that each solution space region has its own specifications, proper exploiters should be elected and performed to the ongoing solution in each step. Reinforcement learning is used as a supervisor managing the choice of exploiters that should be applied at each time step. This selection is based on the existing functional history of exploiters. Exploiters cooperate at each generation in order to combine their efforts and to increase the quality of the solutions so that each exploiter would be able to search by itself (attempting on an independent basis).

2.1 *Initial Population*

Initial population includes a set of chromosomes. An example of this structure has been illustrated in Fig. 2. In this form an ordered set of random integer numbers is used. The length of this chromosome is set to L, where L shows the gray levels number in the original image. In this structure, the indices show the arrangement of gray levels in the image, for example the index 2 shows the second gray level in the

0	4	12	25	38	52	89	120	210	255

Fig. 1 An example of the chromosome structure

image and so on. In Fig. 1, the first, second, and last gray level in the image is 0, 4, and 255, respectively. In remapping, the second gray level in the original image is replaced with the value of the second element of chromosome and so on.

2.2 Evaluate the Solutions

In this study, the evaluation process has two steps: (a) remapping and (b) evaluating.

- **Remapping**: In the first step, it is needed that each chromosome maps to new image. Remapping the original gray-level image is done based on the mentioned chromosome structure. This transformation is calculated by the following expression [33]:

$$T(G(K)) = C_i(K),\ \ K = 1, 2, \ldots, L \tag{1}$$

where T is the transformation function, and G is the sorted gray level of the original image, C_i represents the ith chromosome in the population where $C_i(K)$ represents the value of kth cell. In this transformation, the kth gray level in the original image is dislocated with the kth element of ith chromosome.

- **Evaluating**: In this step, the enhanced image created by each chromosome is evaluated. This evaluation is based on the number of edges, overall intensity, and PSNR.

$$\text{fitness}(c) = \log(\log(E(I(c))))^{*} n - \text{edges}(I(c)) * P_{snr}, \tag{2}$$

In Eq. (2) fitness(c) computes the fitness value of the chromosome c and I(c) denotes the enhanced image. n − edges(I(c)) are the number of edges which are detected by the Sobel [40] edge detector. The following expression calculates the sum of intensities of edges E(I(c)) in the enhanced image:

$$E(I(c)) = \sum_i \sum_j \sqrt{\partial h_1(i, j)^2 + \partial v_1(i, j)^2}, \tag{3}$$

where

$$\begin{aligned}\partial h_1(i, j) = {} & g_1(i+1, j-1) + 2g_1(i+1, j) \\ & + g_1(i+1, j+1) - g_1(i-1, j-1)\end{aligned}$$

$$- 2g_1(i-1,j) - g_1(i-1,j+1) \tag{4}$$

$$\begin{aligned}\partial v_1(i,j) = {} & g_1(i-1,j+1) + 2g_1(i,j+1) \\ & + g_1(i+1,j+1) - g_1(i-1,j-1) \\ & - 2g_1(i,j-1) - g_1(i+1,j-1)\end{aligned} \tag{5}$$

In Eq. (2), to avoid producing unnatural images double log is used. Lastly, P_{snr} measures the peak signal-to-noise of the enhanced image. Regarding to noise expanding problem during the enhancement, PSNR quantify the quality of an enhanced image [41].

2.3 Reinforcement Learning

In the proposed method, a choice function relying on reinforcement learning is proposed to properly select and control the exploiters. In the proposed method, each exploiter has an individual weight and its value will be changed during the iterations. Based on the exploiter's weight, a proportion of population is allocated to each exploiter. In first step, all exploiters have an equal weight (Eq. (7)) to have a number of chromosomes. This weight will be changed during the generations and it can be changed based on historical functional of each exploiter. After the allocation of population to each LLHs, each one run to enhance the input image. Simple methods called exploiters were proposed. These LLHs are simple and improve the contrast of the current image. In each calling of reinforcement learning, the value of weight is updated. At first, their value is equal (Eq. (7)). The new value of weight is updated based on the number of solutions (the number of rewards) so that the LLH could be improved. The value of weight is formulated in Eqs. (6)–(8) as follows:

$$w_i(0) = 1/N_e,\ i = 1, 2, \ldots, N_e \tag{6}$$

$$w_i(t+1) = \frac{w_i(t) + R_i}{\sum_{i=1}^{N_e} w_i(t+1)}, \tag{7}$$

$$R_i = \frac{RN_i}{TN_i}, \tag{8}$$

The generated solutions by the LLHs are replaced with the old solutions respecting to the fact that these new solutions are better than before.

2.4 Low Level Heuristics

In the proposed method, two exploiters (the value of N_e is 2) are used to change the current solution. We use the domain knowledge of the image enhancement in designing the exploiters to provide a more effective search through the Hyper-heuristic approach. These exploiters are explained as follows:

- **Histogram equalization (HE)**: Histogram equalization is an approach in image processing to contrast enhancement by equalizing the image's histogram. This method enhances the global contrast of images, specifically, when the utilizable data of the image is showed by near contrast values. Through this enhancement, the intensities can be better divided on the histogram. This provides the region of lower local contrast to gain a superior contrast. Histogram equalization performs this by effectively distributing out the most repeated intensity values. The general histogram equalization formula is

$$H(v) = roand\left(\frac{cdf(v) - cdf_{min}}{(M \times N) - cdf_{min}} \times (L - 1)\right), \tag{9}$$

where cdf_{min} the minimum nonzero value of the cumulative distribution function, $M \times N$ gives the image's number of pixels (where M is width and N is height) and L is the gray levels number applied.

- **Darkness function (DF)**: In this proposed exploiter, the input gray levels images are mapped to new values in enhanced image such that it saturates 1% of data at both low and high intensities of input image. Additionally, the values are in a specific range map to values in new identified range.

Depending on the number of mapped gray level (C), the output gray levels result in one of the three situations:

- If $C > L$ then the end of this array is cut in order to reach the size of L.
- If $C < L$ then $C - L$ number of input gray levels are added to the mapped gray levels.
- If $C = L$ then the output gray levels are replaced by mapped gray levels.

As it is mentioned before, L is chromosome length.

2.5 Global Search

The global search which is used in this paper is genetic algorithm (GA). GA can have global exploration and defining new area of the solution space to recognize possible candidates, but there is no further concentration on the exploitation perspective when a possible area is recognized [42]. The GA operators which are used in this paper are as follows: (1) Selection algorithm: In the proposed method, the selection function is on a rank-based elitism roulette wheel selection [43]. If the crossover rate is called P_c and number of individuals is called N_c, $N_c * P_c$ individuals are selected by crossover operator to generate the same number of individuals from them. (2) Crossover and mutation: In the proposed method, one-point crossover is used which if two parent chromosomes are selected, the crossover operation is performed with the probability of P_c to generate two new chromosomes through the way which exchange information in a randomly cut point. Finally, to maintain our individual structure, new chromosomes are sorted in ascending order. For mutation, each individual is selected if a random number which is generated is lower than P_m (mutation rate). For each chromosome 10% of its elements are selected randomly and they mutate to a random integer number between 0 and L.

3 Experimental Results

The simulation results of the proposed method, HHIF, are presented in this section and compared to five well-known literature works, i.e., HE, BBHE, DSIHE, RMSHE and RS_ESIHE. To analyze and contrast the existing methods, we use 16 test images. Visual quality comparison of three images, i.e., 5236 and Girl are shown in Figs. 2 and 3.

To evaluate the performance of HHIE, AMBE are used [44]. AMBE is a useful for evaluating the brightness preservation degree. In addition to that the PSNR measure was applied to evaluate the appropriate output images for consumer electronic products [41]. The parameters which are used in this study are as follows: P_c is set

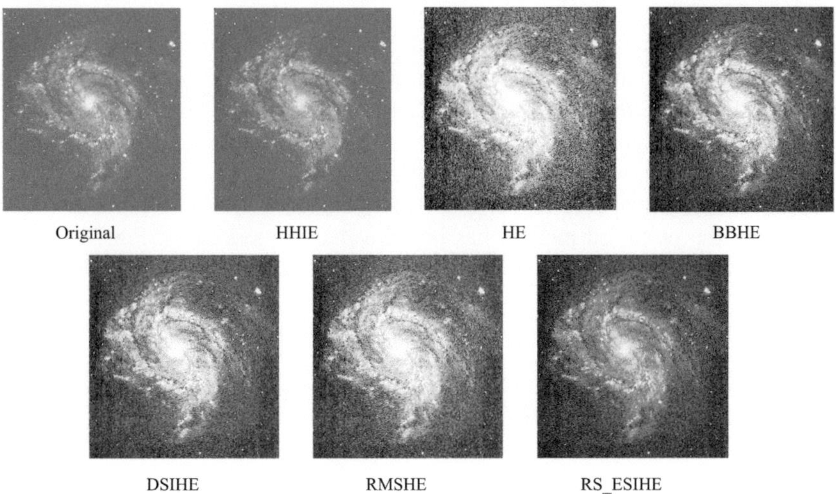

Fig. 2 Enhancement results of 5236 image

Fig. 3 Enhancement results of Cameraman image

as 0.8 and P_m is set as 0.6. Maximum number of iterations, size of population, and total number of evaluation are 50, 10, and 1000, respectively.

3.1 Performance Assessment

For comparison, accuracy measurement is necessary between the proposed method and literature work based on the AMBE and PSNR for 16 benchmark images. Table 1 shows the PSNR values of the enhancement resulted using each method on 16 test images. HHIE produces highest PSNR for all the images thus becoming the best suitable approach for producing neither noise artifacts nor over-enhancement to achieve the highest PSNR values. This would be occurring because we wish to minimize the MSE between input and enhanced image respecting maximum signal value of the image.

Applied AMBE measure for each enhanced image is listed in Table 2. It can be seen that the proposed method has higher value in most cases. Preserving the complete mean brightness, HHIE effectively achieves the lowest AMBE values in most cases.

3.2 Assessment of Visual Quality

Finally, the methods are compared based on image visual assessment. The enhanced images which are resulted after applying the HHIE and the mentioned method are

Table 1 PSNR value of different methods

Test images	HE	BBHE	DSIHE	MMSICHE	RMSHE	ESIHE	R_ESIHE	RS_ESIHE	HHIE
Woman	17.6594	17.7323	20.0595	26.7368	18.742	23.9179	23.5756	25.0405	38.01
Couple	17.1396	17.2337	17.6544	25.2949	17.2348	22.2534	20.3205	22.1283	37.95
Cameraman	19.0970	18.1902	19.2375	24.3772	18.2523	19.7901	19.5637	22.4598	40.3321
Hands	6.1279	12.464	16.697	26.8293	19.7633	21.7879	21.5227	9.1013	44.2281
Tire	9.8650	20.8037	18.9836	26.1456	20.944	22.7282	21.5101	15.6351	35.1839
Boat	16.6533	17.012	18.2456	26.4395	18.1113	21.2398	20.1272	21.3993	39.172
Pirate	17.8267	18.4634	23.8927	22.5599	23.1179	20.4761	19.8244	24.0662	24.4312
Spine	14.3123	18.9631	23.2116	24.2581	22.8882	28.6762	27.0817	24.8152	35.3504
Fractured spine	7.703	20.5208	16.5246	29.5302	20.1864	18.0367	17.9943	6.8057	45.6517
Girl	13.0797	6.5117	15.1173	25.7379	13.8731	18.2341	19.0145	18.0676	35.423
Masque	11.9487	12.8624	16.2516	21.0032	13.9668	24.6432	21.7414	20.6327	33.1256
Underwater1	15.1757	20.8658	21.1252	27.9232	20.8565	32.8844	29.0574	28.816	38.1951
Underwater2	9.6544	17.2504	14.5926	21.6197	17.1158	19.7485	19.2368	18.4372	40.4133
Underwater3	13.3998	16.7321	18.6885	20.9712	19.116	25.9784	25.8653	23.5196	28.4490
Underwater4	15.7671	26.5097	24.6876	29.9824	26.2512	25.6913	26.2144	30.3251	31.6902
5236	13.0814	15.3621	16.1403	24.4521	15.1705	15.9833	13.3981	18.1	34.7513
Average	13.6557	17.3423	18.8194	25.2413	19.0994	22.6293	21.628	20.5844	36.3973

Table 2 AMBE value of different methods

Test images	HE	BBHE	DSIHE	MMSICHE	RMSHE	ESIHE	R_ESIHE	RS_ESIHE	HHIE
Woman	14.2773	15.9377	11.8564	1.3073	18.1232	4.7679	2.4144	3.0228	**0.99**
Couple	5.5544	10.1976	3.511	3.0227	11.2417	2.8547	4.4896	3.3787	**0.41**
Cameraman	8.6955	24.0215	16.9527	4.341	24.3227	12.4296	10.2118	11.1712	**1.8860**
Hands	123.4818	16.4543	23.901	6.3288	24.8584	13.7009	14.3754	84.6875	**0.3804**
Tire	73.9839	19.2911	19.0835	6.5848	18.5675	14.8142	16.251	36.2634	**0.0417**
Boat	2.2080	21.1678	2.047	3.1063	19.7524	13.6214	15.8747	13.2618	**0.3926**
Pirate	15.8547	12.5074	11.0443	**0.8293**	12.6207	5.1933	2.3444	2.5587	4.2949
Spine	45.6574	23.3616	14.8087	6.1886	15.5702	1.4263	1.4001	11.2046	**0.9959**
Fractured spine	95.129	21.4906	19.4384	4.1157	18.7308	19.1553	18.7643	110.7071	**0.2857**
Girl	11.7799	85.0423	18.2082	3.2433	28.7108	24.7347	19.9695	21.8127	**1.0924**
Masque	43.8711	31.9291	14.0626	8.3847	36.8674	11.439	13.6961	2.9379	**0.5247**
Underwater1	41.5965	18.6814	17.9073	4.7747	18.5349	4.121	6.6603	7.4014	**0.4816**
Underwater2	72.9787	16.552	28.054	8.0968	16.5901	20.5677	19.9833	25.8791	**0.2569**
Underwater3	36.881	6.1569	8.149	7.9309	4.6908	8.2466	3.5339	2.8288	**0.5210**
Underwater4	32.2729	3.3898	10.2425	2.0598	9.4506	9.686	5.3332	**0.9209**	5.2730
5236	12.4238	17.1007	9.1633	5.4707	**0.0263**	33.5017	46.5791	21.5033	0.1362
Average	39.7904	21.4551	14.2769	4.7366	17.4162	12.5163	12.6176	22.4712	**1.1227**

demonstrated in Figs. 2 and 3. As shown in these enhanced images, HHIE has better natural looking with high-contrast images. The impacted results in contrast enhancement can be obviously perceived in Fig. 2 of 5236 image. Result of the other method enhance the noise; however, HHIE image prepares management on over-enhancement leading to good contrast enhancement consequences.

By applying HHIE, an excessive contrast of the Girl image can be seen between the background and the other human features like hair, body, and face in Fig. 3. In the other methods, the hair and clothing of the enhanced girl image are relatively dark (HE, BBHE, DSIHE, and RMSHE); the bright face is relatively bright and the background region is involved with the intensive noise in all literature methods. However, HHIE gives a relatively natural brightness enhancement on the hair and face in circled region.

4 Conclusion

In this study, we have proposed an automatic image contrast enhancement based on Hyper-heuristic. In the proposed method (HHIE), simple methods called exploiters were proposed. These exploiters improved the contrast of current image and were applied in appropriate time. We proposed a reinforcement learning to choose these exploiters in a proper time. This selection based on their functional history. HHIE is suitable for the broad variety of image with low contrast. Also, the proposed method can control different images, automatically. Having multi aim of preserving brightness, retaining the shape features of the original histogram and controlling on the rate of over-enhancement are the achievement of the proposed method. These objectives are suitable for the application of consumer electronics. In experimental results, the proposed method was applied on some standard images and it outperformed related one in three criteria: PSNR, AMBE, and visual assessment.

References

1. Madadizadeh, F., Bahrampour, A., Mousavi, S.M., Montazeri, M.: Using advanced statistical models to predict the non-communicable diseases. Iran. J. Public Health **44**(12), 1714–1715 (2015)
2. Ehtemam, H., Montazeri, M., Khajouei, R., Hosseini, R., Nemati, A., Maazed, V.: Prognosis and early diagnosis of ductal and lobular type in breast cancer patient. Iran. J. Public Health **46**(11), 1563–1571 (2017)
3. Montazeri, M., Montazeri, M., Naji, H.R., Faraahi, A.: A novel memetic feature selection algorithm, pp. 295–300
4. Montazeri, M., Naji, H.R., Montazeri, M.: Memetic feature selection algorithm based on efficient filter local search. J. Basic Appl. Sci. Res. **3**(10), 126–133 (2013)
5. Madadizadeh, F., Asar, M.E., Bahrampour, A., Montazeri, M.: Liver disease recognition: a discrete hidden markov model approach (2016)

6. Madadizadeh, F., Montazeri, M., Bahrampour, A.: Predicting the survival in breast cancer using Hidden Markov Model, pp. 228–228
7. Madadizadeh, F., Montazeri, M., Bahrampour, A.: Predicting of liver disease using Hidden Markov Model. Razi J. Med. Sci. **23**(146), 66–74 (2016)
8. Montazeri, M., Naji, H.R., Montazeri, M., Faraahi, A.: A novel memetic feature selection algorithm, pp. 295–300
9. Montazeri, M.: HHFS: Hyper-heuristic feature selection. Intell. Data Anal. **20**(4), 953–974 (2016)
10. Mitra, M., Bahrololoum, A., Nezamabadi-pour, H., Baghshah, M.S., Montazeri, M.: Cooperating of local searches based hyperheuristic approach for solving traveling salesman problem, pp. 329–332
11. Montazei, M., Baghshah, M.S., Niknafs, A.: Selecting efficient features via a hyper-heuristic approach
12. Montazeri, M., Nezamabadi-pour, H., Bahrololoum, A.: Exploring and exploiting effectively based hyper-heuristic approach for solving travelling salesman problem
13. Montazeri, M., Montazeri, M., Montazeri, M., Beigzadeh, A.: Machine learning models in breast cancer survival prediction. Technol. Health Care **24**(1), 31–42 (2016)
14. Montazeri, M., Montazeri, M.: Machine learning models for predicting the diagnosis of liver disease. Koomesh **16**(1), 53–59 (2014)
15. Abbasi, R., Montazeri, M., Zare, M.: A rule based classification model to predict colon cancer survival
16. Afzali, F., Heidari, Z., Montazeri, M., Ahmadian, L., Zahedi, M.J.: Futures studies in health: choosing the best intelligent data mining model to predict and diagnose liver Cancer in early stage. J. Health Biomed. Inform. **2**(3), 133–140 (2015)
17. Montazeri, M., Baghshah, M.S., Enhesari, A.: Hyper-heuristic algorithm for finding efficient features in diagnose of lung cancer disease (2015). arXiv preprint arXiv:1512.04652
18. Montazeri, M., Montazeri, M., Beygzadeh, A., Zahedi, M.J.: Identifying efficient clinical parameters in diagnose of liver disease. Health MED **8**(10), 1115 (2014)
19. Montazeri, M., Montazeri, M., Montazeri, M., Bahrampour, A.: Can breast cancer survival be predicted by risk factors? machine learning models, pp. 301–301
20. Montazeri, M., Nezamabadi-pour, H.: Automatic extraction of eye field from a gray intensity image using intensity filtering and hybrid projection function
21. Montazeri, M., Nezamabadi-pour, H., Montazeri, M.: Automatically eye detection with different gray intensity image conditions. Comput. Technol. Appl. **3**(8) (2012)
22. Montazeri, M., Montazeri, M., Saryazdi, S.: Eye detection in digital images: challenges and solutions (2016). arXiv preprint arXiv:1601.04871
23. Montazeri, M., Bahaadinbeigy, K., Rahnama, Z., Montazeri, M.: Comparison of the accuracy of digital image-based and patient visit-based diagnoses in an Iranian dermatology clinic. J. Basic Appl. Sci. Res. **3**(11), 28–33 (2013)
24. Montazeri, M.: Intensity adjustment and noise removal for medical image enhancement. J. Health Biomed. Inform. **3**(1), 38–47 (2016)
25. Parihar, A.S., Verma, O.P., Khanna, C.: Fuzzy-contextual contrast enhancement. IEEE Trans. Image Process. **26**(4), 1810–1819 (2017)
26. Parihar, A.S., Verma, O.P.: Contrast enhancement using entropy-based dynamic sub-histogram equalisation. IET Image Process. **10**(11), 799–808 (2016)
27. Chang, Y., Jung, C., Ke, P., Song, H., Hwang, J.: Automatic contrast-limited adaptive histogram equalization with dual gamma correction. IEEE Access **6**, 11782–11792 (2018)
28. Chen, S.-D., Ramli, A.R.: Contrast enhancement using recursive mean-separate histogram equalization for scalable brightness preservation. IEEE Trans. Consum. Electron. **49**(4), 1301–1309 (2003)
29. Kim, Y.-T.: Contrast enhancement using brightness preserving bi-histogram equalization. IEEE Trans. Consum. Electron. **43**(1), 1–8 (1997)
30. Wang, Y., Chen, Q., Zhang, B.: Image enhancement based on equal area dualistic sub-image histogram equalization method. IEEE Trans. Consum. Electron. **45**(1), 68–75 (1999)

31. Singh, K., Kapoor, R., Sinha, S.K.: Enhancement of low exposure images via recursive histogram equalization algorithms. Optik-Int. J. Light Electron Opt. **126**(20), 2619–2625 (2015)
32. Saitoh, F.: Image contrast enhancement using genetic algorithm, pp. 899–904
33. Hashemi, S., Kiani, S., Noroozi, N., Moghaddam, M.E.: An image contrast enhancement method based on genetic algorithm. Pattern Recogn. Lett. **31**(13), 1816–1824 (2010)
34. Cai, Z.-Q., Lv, L., Huang, H., Hu, H., Liang, Y.-H.: Improving sampling-based image matting with cooperative coevolution differential evolution algorithm. Soft Comput. **21**(15), 4417–4430 (2017)
35. Chen, J., Yu, W., Tian, J., Chen, L., Zhou, Z.: Image contrast enhancement using an artificial bee colony algorithm. Swarm Evol. Comput. **38**, 287–294 (2018)
36. Montazeri, M., Nezamabadi-pour, H.: Automatic extraction of eye field from a gray intensity image using intensity filtering and hybrid projection function, pp. 1–5
37. Montazeri, M., Nezamabadi-pour, H., Bahrololoum, A.: Exploring and exploiting effectively based hyper-heuristic approach for solving travelling salesman problem. In: 2011 5th Conference on the Fifth Iran Data Mining Conference (IDMC). Amirkabir University of Technology, Tehran, Iran (2011)
38. Montazei, M., Soleymani Baghshah, M., Niknafs, A.: Selecting efficient features via a hyper-heuristic approach. In: 2011 5th Conference on the Fifth Iran Data Mining Conference (IDMC). Amirkabir University of Technology, Tehran, Iran (2011)
39. Wolpert, D.H., Macready, W.G.: No free lunch theorems for optimization. IEEE Trans. Evol. Comput. **1**(1), 67–82 (1997)
40. Rosin, P.L.: Edges: saliency measures and automatic thresholding. Mach. Vis. Appl. **9**(4), 139–159 (1997)
41. Rabbani, M., Jones P.W.: Digital Image Compression Techniques. SPIE Press (1991)
42. Ang, J.H., Tan, K.C., Mamun, A.: An evolutionary memetic algorithm for rule extraction. Expert Syst. Appl. **37**(2), 1302–1315 (2010)
43. Holland, J.: Adaption in Natural and Artificial Systems. University of Michigan Press, Ann Arbor, MI (1975)
44. Kim, M., Chung, M.G.: Recursively separated and weighted histogram equalization for brightness preservation and contrast enhancement. IEEE Trans. Consum. Electron. **54**(3) (2008)

Comparative Analysis of Salt and Pepper Removal Techniques for Binary Images

Usha Rani, Amandeep Kaur and Gurpreet Josan

Abstract Binarization is the most important step in the OCR system that converts the gray level or colored images into bi-level form. In the case of degraded images, results after binarization mostly contain noises. Salt and pepper noise of different sizes is the most prevalent noise in binary images. For the better results of OCR process, it is necessary to denoise image before proceeding to the next stage. This paper conducts experiments with different existing salt and pepper noise removal methods such as median filter-based techniques and kFill algorithm-based techniques for binary document images. The statistical measures, namely, PSNR, SSIM, and EPI are used to evaluate the performance.

Keywords Salt and pepper noise · Degraded documents · Median filter · kFill filter

1 Introduction

Digitization of historical documents is the need of the day so that everyone all over the world can have access to these documents through the Internet. We know that historical documents degrade over time due to their bad storage conditions, humidity, aging, bad ink, and paper quality. Due to this, documents have wrinkles, yellow coloring of paper, ink bleed through, fading of text, smear and stains

---

U. Rani (✉)
University College Ghanaur, Punjabi University, Patiala, Punjab, India
e-mail: usha_gupta7@yahoo.co.in

A. Kaur
Department of Computer Science and Technology, Central University of Punjab, Bathinda, Punjab, India
e-mail: aman_k2007@hotmail.com

G. Josan
Department of Computer Science, Punjabi University, Patiala, Punjab, India
e-mail: josangurpreet@pbi.ac.in

B. Pati et al. (eds.), *Advanced Computing and Intelligent Engineering*,
Advances in Intelligent Systems and Computing 1082,
https://doi.org/10.1007/978-981-15-1081-6_32

type of degradations before acquisition of document. In case of data analysis and recognition, first step is the acquisition of the documents. The acquisition of documents is done either using scanner or digital cameras. Some of the degradations introduced into the document images during acquisition. Degradations such as marginal noise, clutter noise, stroke-like pattern noise, background noise, and salt and pepper noises are introduced into the scanned images. While images taken through digital cameras commonly encounter uneven illumination, color shift, and blurry text types degradations. After the acquisition, next most important step of the optical character recognition (OCR) process is binarization. Image binarization converts the gray level or a colored image into black and white image to reduce the computational overload of next stages of OCR process. The performance of next stages of OCR process that is segmentation and recognition highly depends upon the quality of binary images we get after binarization. Binarized images may suffer from salt and pepper noise a type of additive impulse noise. It looks like mixture of salt and pepper and exists in almost all binary, gray level, and colored document images. Denoising is performed as a preprocessing step before high-level image processing on binary document images. In binary images the removal of salt and pepper noise is challenging as both the noise and information-carrying pixels sharing the same values (either 0 or 1). While in case of gray scale images the noisy pixels are easier to isolate using pixel intensity difference from its neighboring pixels. There exist many salt and pepper noise removal methods. Most of the filters used to remove salt and pepper noise are order statistics nonlinear median filters like standard, weighted, center weighted, adaptive and decision-based median filters and many more [1–19]. There are also other types of filters to remove this type of noise such as kFill algorithm [20, 21] and its modifications [22–27]. The median filter-based methods can be used to denoise any kind of image binary, gray scale or colored but computational complexity of these methods is high. The kFill techniques are specifically developed low complexity algorithms to reduce impulse noise in binary images.

This paper presents the experimental study of various salt and pepper noise removal techniques for the binary document images. The paper is organized as follows: Sect. 2 elaborates on various salt and pepper removal techniques. The methodology and performance metrics are given in Sect. 3. The results are discussed and analyzed in Sect. 4. Finally, the paper is concluded in Sect. 5.

2 Salt and Pepper Noise Removal Techniques

Standard Median Filter (SMF) [1] is a nonlinear low pass filter which replaces the value of the corrupted pixel by the median of the gray levels in the neighborhood (3×3, 5×5, 7×7, etc.) including intensity value of that pixel. Although this filter is simple and removes salt and pepper noise effectively, it smoothes out the edges and distorts the corners and thin lines in the image even at low noise densities.

Center Weighted Median Filter (CWMF) [2] is the modified standard median filter which can preserve the fine details of the image. The center pixel of this filter is assigned some weight i.e. $w(0, 0) = 2P + 1$, where $P \geq 0$ and all other pixel values are set equal to one.

Adaptive Median Filter (ADMF) [3] is an enhanced form of standard median filter that accomplishes spatial processing to detect the noisy pixels in the image. It compares each pixel in the image with its neighborhood pixels and classifies it as noisy pixel if it differs from a majority of its neighbors. The size of the filter is adaptive, i.e., the neighborhood size is increased if the specific condition is not met. The intensity values of noisy pixels are then changed to the median of the pixels values in the neighborhood. Let I_{xy} be the pixel of the noisy image, P_{gr} be the greatest pixel value and P_{lo} be the lowest pixel value in the window, W be the current size of the window, $W_{\max}$ is the maximum allowed size of the window and P_{med} is the median of the pixels values in the current window. This technique works in two levels.

Level A:

(a) If $P_{\text{lo}} < P_{\text{med}} < P_{gr}$, then median value is not an impulse, so go to level B to check if the current pixel is an impulse.
(b) Else P_{med} is an impulse, the filter window size is increased and level A is repeated until the P_{med} is not an impulse so the algorithm goes to level B; or the maximum size of the window i.e. $W_{\max}$ is reached, in this case, median value is assigned to the filtered pixel.

Level B:

(a) If $P_{\text{lo}} < P_{xy} < P_{gr}$, then the current pixel is not an impulse, filtered image pixel remains unchanged, i.e., I_{xy}
(b) Else either$I_{xy} = P_{\text{lo}}$ or $I_{xy} = P_{\text{gr}}$, then pixel is noisy. This noisy pixel is assigned the median value from level A.

Adaptive Center Weighted Median Filter (ACWMF) [4] is combination of center weighted median filter [2] and adaptive median filter [3]. The window size of this filter is adaptive.

Tri-State Median Filter (TSMF) [5] uses the standard median filter and center weighted median filter to detect whether a pixel is corrupted, before applying the filtering. Let I_{xy} be the noisy image pixel value, if it is not corrupted, its value will remain unchanged, i.e., I_{xy}, otherwise the value assigned to it will be I_{xy}^{CWM} or I_{xy}^{SM} according to the following equation:

$$I_{xy}^{TSM} = \begin{cases} I_{xy}, & T_h > df_1 \\ I_{xy}^{CWM}, & df_2 \leq T_h < df_1 \\ I_{xy}^{SM}, & df_2 \leq T_h < df_1. \end{cases} \tag{1}$$

Here, I_{xy}^{CWM} is the output of center weighted median filter and I_{xy}^{SM} is the output of standard median filter. $df_1 = |I_{xy} - I_{xy}^{SM}|$ and $df_2 = |I_{xy} - I_{xy}^{CWM}|$ and T_h is the threshold value selected between 0 and 255.

kFill Algorithm [20, 21] denoises binary images by sliding a square window of size $k \times k$ pixels in a raster scan manner over the image. A square window of size $k \times k$ pixels contains $4k - 4$ periphery pixels on the boundary and the remaining $(k - 2) * (k - 2)$ pixels are core pixels. The core pixels are set to black or white based upon the values of three variables, which are calculated from window periphery pixels. The core pixels are set black or white, if following conditions are satisfied

$$(s = 1)\ \textbf{AND}\ (\text{num} > 3k - 4)\ \textbf{OR}\ [(\text{num} = 3k - 4)\ \textbf{AND}\ (t = 2)]) \quad (2)$$

Here, for fill value equal to black (white), variable num is the count of black (white) pixels in the window periphery, s is the count of connected groups of black (white) pixels in the window periphery, and t is the count of black (white) corner pixels. The value of s equal to one ensures that filling does not impair connectivity. This procedure is performed repeatedly on the image until no filling occurs.

Applied kFill Algorithm [20, 21] eliminates noise only from binary images. The kFill algorithm is unable to fill noise components, which are smaller than the core size. Applied kFill method eliminates noise components larger than one pixel using window size larger than 3×3 pixels. The core pixels are set white, when the majority of the pixels in full window are white and set black when majority of the full window pixels are black. It can eliminate noises of different sizes and shapes and does not damage the sharpness of text and graphical components.

Chinnasarn et al. [22] modified the kFill algorithm to eliminate impulse noise from binary images. But unlike kFill algorithm, this method completes the task in only one pass over the image. It can simultaneously remove both salt and pepper noise of any sizes smaller than the document objects. This algorithm works in following steps:

Step 1: Count the number of black pixels within the window core. If the number of black pixels is less than total core pixels, move to step 2. Otherwise, check whether equation (2) is satisfied or not:
(a) If Eq. (2) is satisfied, set the core pixels white, otherwise set them black. In Eq. (2), num is the number of white pixels, s is the connected components of white pixels and t is the number of white pixels at corner position in the neighborhood.

Step 2: Check whether Eq. (2) is satisfied or not:
(b) If Eq. (2) is satisfied, set the core pixels black, otherwise set them white. In Eq. (2), num is the count of black pixels, s is the connected components of black pixels and t is the count of black pixels at corner position in the neighborhood.

Premchaiswadi et al. [27] proposed method that denoises binary, gray scale and color images using the features of applied kFill filter and median filter with window sizes of 3×3 and 5×5, etc. based on the size of impulse noise. This algorithm works as follows:

- 1. Let num_blk and num_wht are the count of black and white pixels in the window core $((k - 2) * (k - 2))$.
- 2. If num_blk or num_wht in the window core are greater than half of all core pixel,s i.e., $(((k - 2) * (k - 2))/2) + 1)$, then values of all core pixels are replaced

by the median pixel value from the whole window, i.e., core and surrounding neighborhood.
- 3. If num_blk or num_wht in the window core is less than half of all core pixels, i.e., $(((k-2)*(k-2))/2)+1)$, then values of all core pixels are replaced by the median pixel value from the core.
- 4. If all core pixels are not black or white then the core is filled with the original pixel values.

3 Experimental Setup

The experiments have been conducted using MATLAB R2016a on a number of binary images. The salt and pepper noise removal performance of the filters discussed in Sect. 2 with different noise densities and different mask sizes are presented visually and also quantitatively in terms of PSNR, SSIM, and EPI. The results of kFill algorithm in this paper have been computed using single iteration.

3.1 *Performance Metrics of Denoising Algorithms*

The performance of the different denoising algorithms discussed in this paper is computed both qualitatively and quantitatively using Peak Signal Noise Ratio (PSNR), Structural Similarity Index (SSIM), and Edge Preservation Index (EPI). **PSNR** is the ratio between the maximum possible power of a signal and the power of corrupting noise. It is expressed in terms of the logarithmic decibel scale. The higher the PSNR value, the better the quality of enhanced image.

$$PSNR = 20\log_{10}\left(\frac{I_{\max}}{\sqrt{MSE}}\right), \tag{3}$$

$I_{\max} = 1$ for an binary image. **MSE** (Mean Square Error) is the square of the Euclidean distance between the original image and enhanced image.

$$MSE = \sum_{i=0}^{p-1}\sum_{j=0}^{q-1}\frac{(I(i,j) - D(i,j))^2}{p.q}, \tag{4}$$

here, I refers to original image, D is the filtered output image and p, q are the dimensions of the image. MSE value should be low for better quality of enhanced image.

SSIM (Structural Similarity Index) measures the similarity between two images using the change in structural information. Structural Information provides the idea that how the pixels in proximity inter-related with each other. Its value lies between 0 and 1. SSIM between two images u and v of common size $p \times q$ is given by the equation

$$SSIM\left(u,v\right)=\frac{\left(2m_u m_v+Z_1\right)\left(2\sigma_{uv}+Z_2\right)}{\left(m_u^2+m_v^2+Z_1\right)\left(\sigma_u^2+\sigma_v^2+Z_2\right)}. \tag{5}$$

Here, m_u average of image u, m_v average of image v, σ_u standard deviation of image u, σ_v standard deviation of image v and σ_{uv} covariance of image u and image v. $Z_1 = (Y1 * M)^2$ and $Z_2 = (Y2 * M)^2$ are variables to balance the division with infinitesimal denominator. M is the dynamic range of pixel values and $Y1 = 0.01$, $Y2 = 0.03$ by default.

EPI (Edge Preservation Index) calculates the amount of edges preserved in the image after denoising [28]. Its value lies between 0 and 1.

4 Experimental Results and Discussion

4.1 Visual Results

Experiments are conducted on a set of binary images using denoising methods discussed in Sect. 2 with different mask sizes and noise densities varying from 10 to 90%. The results of one of the images at noise density 30% are shown in Fig. 1. Visually, we are getting better results with Adaptive Median Filter (ADMF) with maximum size of window equal to 3. We noted that for up to 20% noise density, filters with mask size 3 × 3 give better results. As noise density increases, we get better results with large mask size. But, large mask size produces noise-free results at the cost of blurring the edges of image. With kFill-based filters [20–22, 27] results are better with 4 × 4 window size as compared to 3 × 3 and 5 × 5 window sizes.

4.2 Quantitative Results

The noise densities varying from 10 to 90% are added to the binary image in Fig. 1a. The PSNR, SSIM, and EPI values for different methods with different mask sizes are recorded in Tables 1, 2, and 3 respectively. The graphical representation of the results is shown in Figs. 2, 3 and 4. Figure 2 shows the PSNR results using different techniques with mask size 3 × 3. It shows that up to 70% noise density, performance of ADMF and ACWMF is almost same and the best among all the techniques. kFill, Chinnasarn et al., CWMF, and Premchaiwadi et al. shows bad performance up to 70% noise density. Chinnasarn et al. is the worst among all methods. Above 70% noise density Premchaiwadi et al. performs better than other methods. The SSIM results using different methods with mask size of 3 × 3 are shown in Fig 3. The results shows that SMF, ADMF, ACWMF, and Applied kFill give almost same performance. Chinnasarn et al. method is the worst in performance. Tables 1 and 2 shows that Chinnasarn et al. method gives better performance with mask size 4 × 4.

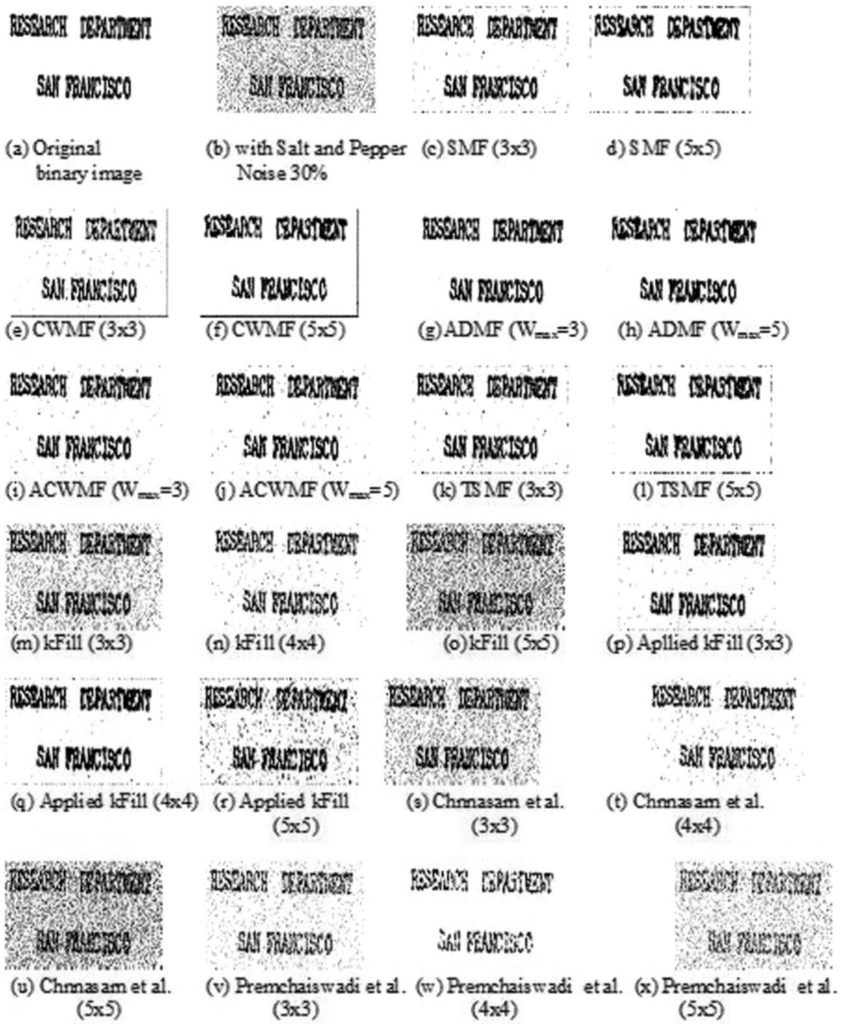

Fig. 1 The results of different salt and pepper denoising methods at noise density 30% with different window sizes

In terms of EPI, ADMF performs best (Fig. 4). Tables 1, 2 and 3 clearly shows that the performance of all methods decreases with increase in noises density.

A relationship exists between the noise density and the mask size. As noise density increases, the method with large mask size gives good results. The results show that for 10–30% noise density, mask size of 3 × 3 gives the best performance and for 40–70% noise density, mask size of 5 × 5 is the best.

Table 1 The comparison of PSNR of binary image shown in Fig. 1 with different window sizes

Method	PSNR								
	Noise densities								
	10%	20%	30%	40%	50%	60%	70%	80%	90%
SMF (3 × 3)	18.32	16.08	14.29	12.08	10.23	8.46	6.60	5.25	3.90
SMF (5 × 5)	14.54	13.75	12.64	11.79	10.71	9.67	7.98	5.96	4.4
CWMF (3 × 3)	10.89	10.75	10.1	9.1	7.78	6.4	5.1	4	3.138
CWMF (5 × 5)	8.42	8.4	8.3	8.33	8.2	7.75	6.5	5.03	3.78
ADMF (W_{max} =3)	19.30	17.00	15.01	12.70	10.56	8.62	6.90	5.50	4.00
ADMF (W_{max} =5)	14.83	14.5	13.67	12.66	12.04	10.66	8.85	6.7	4.5
ADMF (W_{max} =7)	12.08	11.8	11.6	11.2	11.01	10.59	9.25	7.4	5.2
ADMF (W_{max} =9)	10.01	10.06	9.7	9.8	9.7	9.5	8.9	7.8	5.3
ACWMF (W_{max} =3)	19.1	16.72	14.78	12.9	10.63	8.52	6.84	5.31	4.1
ACWMF (W_{max} =5)	19.27	17.08	14.8	12.76	10.54	8.54	6.97	5.35	4.1
ACWMF (W_{max} =7)	18.92	16.82	14.75	12.66	10.44	8.6	6.8	5.25	4.2
ACWMF (W_{max} =9)	19.25	17.04	14.76	12.79	10.7	8.59	6.8	5.4	4.1
TSM (3 × 3)	18.92	16.53	14.1	12.4	10.29	8.49	6.7	5.2	4.1
TSM (5 × 5)	15.12	13.94	12.5	11.64	10.62	9.46	7.97	6.03	4.6
kFill (3 × 3)	12.5	12	10.79	9.41	7.98	6.77	5.72	4.6	3.81
kFill (4 × 4)	12.73	11.9	10.49	8.7	7.1	5.9	4.9	4.23	3.1
kFill (5 × 5)	13.54	12.4	10.16	8.1	6.66	5.63	4.7	4.1	3.5
Applied kFill (3 × 3)	18.76	16.39	14.5	12.5	10.66	8.54	6.9	5.25	4
Applied kFill (4 × 4)	13.6	13.4	12.8	12.22	10.57	8.7	6.8	5.2	3.7
Applied kFill (5 × 5)	11.42	11.27	11.1	10.9	10.4	9.5	8.35	6.6	4.6
Chinnasarn et al. (3 × 3)	10.59	9.4	8.1	7.01	6.1	5.3	4.6	4.01	3.5
Chinnasarn et al. (4 × 4)	10.65	9.5	9.16	8.76	8.13	7.3	6.6	5.8	5.06
Chinnasarn et al. (5 × 5)	8.80	8.76	8.67	8.32	7.69	6.93	5.86	4.95	3.8
Premchad et al. (3 × 3)	13.10	11.50	10.50	9.45	8.40	7.64	6.80	6.13	5.50
Premchad et al. (4 × 4)	10.98	10.92	10.91	10.65	10.07	9.34	8.23	7.90	7.23
Premchad et al. (5 × 5)	8.97	9	9.1	8.84	8.54	8.14	7.1	5.6	4.3

Table 2 The comparison of SSIM of binary image shown in Fig. 1 with different window sizes

Method	SSIM								
	Noise densities								
	10%	20%	30%	40%	50%	60%	70%	80%	90%
SMF (3 × 3)	0.94	0.88	0.77	0.55	0.33	0.24	0.16	0.11	0.05
SMF (5 × 5)	0.85	0.84	0.81	0.78	0.69	0.53	0.33	0.14	0.06
CWMF (3 × 3)	0.73	0.61	0.41	0.27	0.17	0.13	0.1	0.06	0.04
CWMF (5 × 5)	0.57	0.56	0.56	0.55	0.48	0.33	0.15	0.07	0.03
ADMF (W_{max} =3)	0.96	0.89	0.76	0.58	0.38	0.23	0.16	0.09	0.05
ADMF (W_{max} =5)	0.86	0.85	0.82	0.78	0.69	0.54	0.34	0.16	0.07
ADMF (W_{max} =7)	0.75	0.74	0.73	0.71	0.69	0.66	0.50	0.26	0.10
ADMF (W_{max} =9)	0.63	0.64	0.61	0.62	0.62	0.61	0.56	0.39	0.13
ACWMF (W_{max} =3)	0.95	0.89	0.76	0.57	0.37	0.23	0.16	0.09	0.04
ACWMF (W_{max} =5)	0.95	0.89	0.75	0.58	0.38	0.23	0.16	0.11	0.05
ACWMF (W_{max} =7)	0.94	0.89	0.77	0.57	0.34	0.24	0.15	0.09	0.05
ACWMF (W_{max} =9)	0.95	0.891	0.78	0.57	0.39	0.24	0.15	0.10	0.06
TSM (3 × 3)	0.94	0.84	0.63	0.42	0.27	0.19	0.14	0.09	0.05
TSM (5 × 5)	0.87	0.84	0.8	0.75	0.66	0.51	0.32	0.13	0.06
kFill (3 × 3)	0.75	0.56	0.36	0.25	0.17	0.13	0.10	0.07	0.04
kFill (4 × 4)	0.74	0.55	0.36	0.23	0.17	0.13	0.09	0.06	0.03
kFill (5 × 5)	0.79	0.6	0.33	0.23	0.16	0.13	0.1	0.06	0.03
Applied kFill (3 × 3)	0.94	0.9	0.77	0.58	0.39	0.24	0.16	0.1	0.05
Applied kFill (4 × 4)	0.82	0.81	0.77	0.69	0.48	0.29	0.16	0.1	0.05
Applied kFill (5 × 5)	0.71	0.7	0.7	0.7	0.64	0.5	0.3	0.15	0.07
Chinnasarn et al. (3 × 3)	0.47	0.23	0.14	0.1	0.08	0.06	0.04	0.03	0.02
Chinnasarn et al. (4 × 4)	0.62	0.57	0.4	0.3	0.16	0.08	0.06	0.04	0.03
Chinnasarn et al. (5 × 5)	0.57	0.55	0.47	0.32	0.17	0.09	0.05	0.03	0.01
Premchad et al. (3 × 3)	0.74	0.54	0.36	0.24	0.15	0.11	0.07	0.06	0.03
Premchad et al. (4 × 4)	0.63	0.63	0.62	0.62	0.59	0.56	0.5	0.35	0.2
Premchad et al. (5 × 5)	0.57	0.57	0.57	0.54	0.47	0.32	0.17	0.07	0.03

Table 3 The comparison of IEP of binary image shown in Fig. 1 with different window sizes

Method	EPI								
	Noise Densities								
	10%	20%	30%	40%	50%	60%	70%	80%	90%
SMF (3 × 3)	0.85	0.77	0.67	0.58	0.51	0.38	0.27	0.23	0.20
SMF (5 × 5)	0.67	0.61	0.54	0.49	0.45	0.37	0.34	0.21	0.18
CWMF (3 × 3)	0.32	0.34	0.35	0.35	0.3	0.25	0.23	0.21	0.21
CWMF (5 × 5)	0.23	0.23	0.25	0.25	0.27	0.26	0.27	0.19	0.17
ADMF (W_{max} =3)	0.88	0.79	0.7	0.6	0.5	0.40	0.29	0.28	0.21
ADMF (W_{max} =5)	0.88	0.77	0.70	0.60	0.50	0.41	0.28	0.23	0.20
ADMF (W_{max} =7)	0.46	0.44	0.42	0.42	0.39	0.38	0.32	0.27	0.18
ADMF (W_{max} =9)	0.29	0.31	0.30	0.3	0.28	0.28	0.30	0.24	0.19
ACWMF (W_{max} =3)	0.87	0.79	0.70	0.59	0.51	0.4	0.27	0.28	0.2
ACWMF (W_{max} =5)	0.88	0.77	0.70	0.61	0.51	0.41	0.29	0.24	0.20
ACWMF (W_{max} =7)	0.87	0.79	0.69	0.60	0.50	0.41	0.27	0.23	0.20
ACWMF (W_{max} =9)	0.87	0.79	0.72	0.61	0.52	0.41	0.27	0.23	0.021
TSM (3 × 3)	0.87	0.79	0.69	0.61	0.53	0.43	0.28	0.23	0.21
TSM (5 × 5)	0.72	0.65	0.58	0.52	0.5	0.44	0.36	0.25	0.21
kFill (3 × 3)	0.445	0.46	0.46	0.45	0.35	0.313	0.29	0.25	0.22
kFill (4 × 4)	0.49	0.52	0.52	0.43	0.37	0.34	0.29	0.26	0.23
kFill (5 × 5)	0.61	0.61	0.55	0.45	0.39	0.35	0.3	0.27	0.24
Applied kFill (3 × 3)	0.86	0.78	0.7	0.6	0.51	0.4	0.28	0.22	0.20
Applied kFill (4 × 4)	0.55	0.55	0.51	0.5	0.42	0.36	0.24	0.20	0.18
Applied kFill (5 × 5)	0.32	0.32	0.34	0.35	0.34	0.33	0.31	0.26	0.19
Chinnasarn et al. (3 × 3)	0.28	0.3	0.29	0.24	0.24	0.25	0.23	0.22	0.17
Chinnasarn et al. (4 × 4)	0.29	0.29	0.31	0.32	0.31	0.28	0.23	0.21	0.19
Chinnasarn et al. (5 × 5)	0.27	0.27	0.27	0.27	0.27	0.27	0.21	0.20	0.18
Premchdi et al. (3 × 3)	0.56	0.5	0.45	0.4	0.35	0.31	0.25	0.22	0.19
Premchdi et al. (4 × 4)	0.30	0.33	0.33	0.33	0.31	0.26	0.24	0.2	0.14
Premchdi et al. (5 × 5)	0.24	0.24	0.25	0.27	0.28	0.27	0.26	0.20	0.18

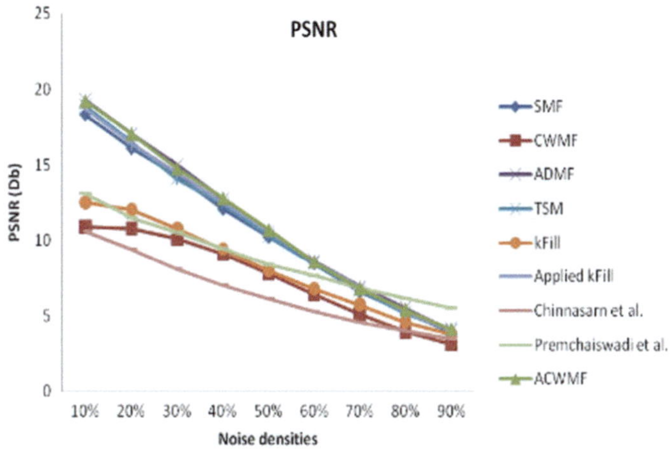

Fig. 2 Graphical presentation of PSNR values of different methods using mask size 3 × 3 with SMF, CWMF, ADMF, ACWMF, TSMF, kFill, and Applied kFill, Chinnasarn et al. and Premchaiswadi et al.

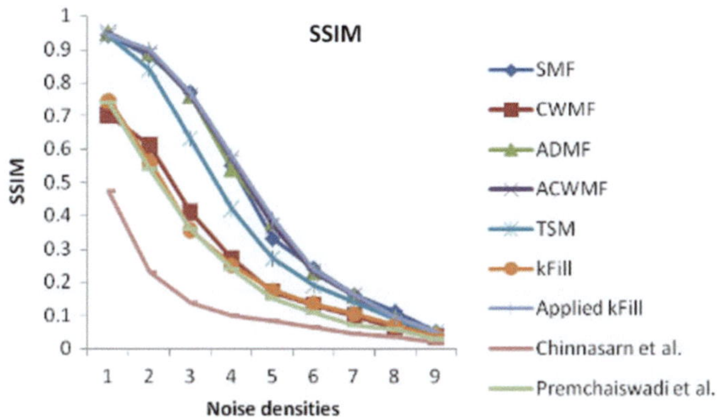

Fig. 3 Graphical presentation of SSIM values of different methods using mask size 3 × 3 with SMF, CWMF, ADMF, ACWMF, TSMF, kFill, and Applied kFill, Chinnasarn et al. and Premchaiswadi et al.

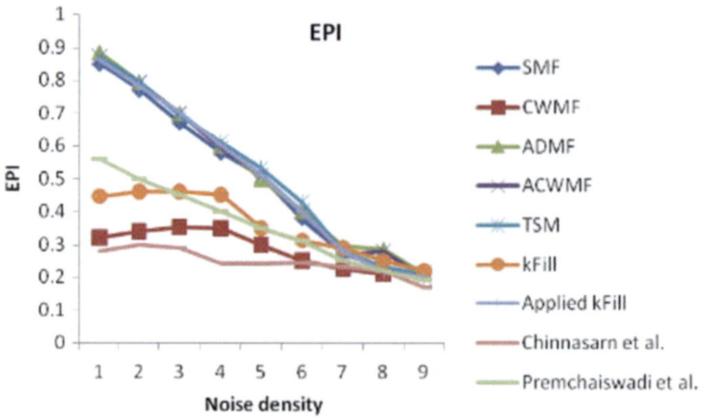

Fig. 4 Graphical presentation of EPI values of different methods using mask size 3 × 3 with SMF, CWMF, ADMF, ACWMF, TSMF, kFill, and Applied kFill, Chinnasarn et al. and Premchaiswadi et al.

5 Conclusions

This paper presents the comparative analysis of different salt and pepper removal techniques for the binary images. The performance measures, PSNR, SSIM, and EPI, are used to assess the discussed techniques with different noise densities and different mask sizes. The results are presented visually as well as graphically.

References

1. Sonka, M., Hlavac, V., Boyle, R.: Image Processing. Analysis and Machine Vision, 2nd edn. PWS (1999)
2. Sun, T., Gabbouj, M., Neuvo, Y.: Analysis of two-dimensional center weighted median filters. Multidimens. Syst. Signal Process. **6**(2), 159–172 (1995)
3. Zhao, Y., Li, D., Li, Z.: Performance enhancement and analysis of an adaptive median filter. In: International Conference on Communication and Networking, pp. 651–653 (2007)
4. Ko, S.J., Lee, Y.H.: Centre weighted median filters and their applications to image enhancement. IEEE Trans. Circuits Syst. **38**(9), 984–993 (1991)
5. Chen, T., Ma, K.K., Chen, L.H.: Tri-state median filter for image denoising. IEEE Trans. Image Process. **8**(12), 1834–1838 (1999)
6. Ansari, M.D., Singh, G., Singh, A., Kumar, A.: An efficient salt and pepper noise removal and edge reserving scheme for image restoration. Int. J. Comput. Technol. Appl. **3**(5), 1848–1854 (2012)
7. Verma, K., Singh, B.K., Thoke, A.S.: An enhancement in adaptive median filter for edge preservation. Procedia Comput. Sci. **48**, 29–36 (2015)
8. Samantaray, A.K., Mallick, P.: Decision based adaptive neighbourhood median filter. Procedia Comput. Sci. **48**, 222–227 (2015)
9. Thirilogasundari, V., Babu, S., Agatha, J.S.: Fuzzy based salt and pepper noise removal using adaptive switching median filter. Procedia Eng. **38**, 2858–2865 (2012)

10. Zeng, H., Liu, Y., Fan, Y., Tang, X.: An improved algorithm for impulse noise by median filter. AASRI Conf. Comput. Intell. Bioinform. **1**, 68–73 (2012)
11. Shresta, S.: Image denoising using new adaptive based median filter. Signal Image Process. Int. J. **5**, 4 (2014)
12. Li, Z., Liu, G., Xu, Y., Cheng, Y.: Modified directional weighted filter for removal of salt & pepper noise. Pattern Recognit. Lett. **40**, 113–120 (2014)
13. Jourabloo, A., Feghahati, A.H., Jamzad, M.: New algorithms for recovering highly corrupted images with impulse noise. Sci. Iran. **19**(6), 1738–1745 (2012)
14. Yin, L., Yang, R.: Weighted median filters: a tutorial. IEEE Trans. Circuits Syst.-II: Analog Digit. Signal Process. **43**(3), 157–192 (1996)
15. Kesharwani, A., Agrawal, S., Dhariwal, M.K.: An improved decision based asymmetric trimmed median filter for removal of high density salt and pepper noise. Int. J. Comput. Appl. **84**(8), 37–43 (2013)
16. Esakkirajan, S., Veerakumar, T., Subramanyam, A.N., PremChand, C.H.: Removal of high density salt and pepper noise through modified decision based unsymmetric trimmed median filter. IEEE Signal Process. Lett. **18**(5), 287–290 (2011)
17. Backman, V., Gurjar, R., Badizadegan, K., Itzkan, I., Dasari, R.R., Perelman, L.T., Feld, M.S.: A new fast and efficient decision-based algorithm for removal of high-density impulse noises. Signal Process. Lett. **14**(3), 189–192 (2007)
18. Leavline, E.J., Singh, D.A.A.G.: Salt and pepper noise detection and removal in gray scale images. Int. J. Signal Image Process. Pattern Recognit. **6**(5), 343–352 (2013)
19. Pushpavalli, R., Sivarajde, G.: A fuzzy switching median filter for highly corrupted images. Int. J. Sci. Res. Publ. **3**(6) (2013)
20. Story, G.A., OGorman, L., Fox, D., Schaper, L.L., Jagadish, H.V.: The right pages image-based electronic library for alerting and browsing. Computer **25**(9), 1726 (1992)
21. OGorman, L.: Image and document processing techniques for the right pages electronic library system. In: 11th IAPR International Conference on Pattern Recognition. Conference B: pattern Recognition Methodology and Systems, pp. 260263. The Hague (1992)
22. Chinnasarn, K., Rangsaneri, Y., Thitimajshima, P. Removing salt-and-pepper noise in graphics images. IEEE Computer Society Press (1998)
23. Premchaiswadi, N., Premchaiwadi, W., Pachiyankul, U., Narita, S.: Broken character identification for Thai character recognition systems. WSEAS Trans. Comput. **2**(2), 430–434 (2003)
24. Al-Khaffaf, H.S.M., Talib, A.Z., Salam, R.A.: Salt and pepper noise removal from document images. In: International Visual Informatics Conference, pp. 607–618 (2009)
25. Al-Khaffaf, H.S.M., Talib A.Z., Salam, R.A.: Removing salt and pepper noise from binary images of engineering drawings. Pattern Recogn. (2008)
26. Al-Khaffaf, H.S.M., Talib A.Z., Salam, R.A.: Enhancing salt-and-pepper noise removal in binary images of engineering drawing. IEICE Trans. E **92-D**(4), 689-704 (2009)
27. Premchaiswadi, N., Yimgnagm, S., Premchaiwadi, W.: A scheme for salt and pepper noise reduction and its application for OCR systems. WSEAS Trans. Comput. **9**(4), 351–360 (2010)
28. Sattar, F.: Image enhancement based on a nonlinear multi scale method. IEEE Trans. Image Process. **6**, 6 (1997)

An Intelligent Approach for Noise Elimination from Brain Image

Pritisman Kar and Mihir Narayan Mohanty

Abstract In general, medical images corrupt due to many causes. The brain image is essential for better and faster diagnosis by the physicians. In this work acquired brain image is considered instead of additive noise. As an intelligent technique the novel fuzzy based morphology is applied to remove the impulsive noise that occurs at the time of acquisition. After preprocessing of the image in one stage the noise is detected using Laplacian and morphological operator-based. Further the noise from the boundaries are tried to remove using modified fuzzy-based morphological operators. The operation provides clear view. The connectivity among pixels using fuzzy morphology helps to remove noise from the pixels. The proposed method performs successfully in terms of PSNR and MSE and has been presented in the result section.

Keywords Filtering · Morphology · Laplacian · Fuzzy logic · Impulsive noise

1 Introduction

Different types of noise are introduced in images during the time of its acquisition, storage, transmission, etc. As a result, visual quality of the image is degraded. Noise is created simultaneously with the image creation. Impulse Noise is generated due to faults in the sensor system during the time of acquisition as well as during transmission. The Noise reduction is an important technique and serves a preprocessing for variety of applications such as recognition of pattern, detection of edges, compression of data, segmentation of images and feature extraction. Cancelation of impulsive Noise is a tedious work as the filter designed for filtering reduces the noise

---

P. Kar · M. N. Mohanty (✉)
Department of Electronics and Communication Engineering, ITER, Siksha 'O' Anusandhan (Deemed to Be University), Bhubaneswar, India
e-mail: mihir.n.mohanty@gmail.com

P. Kar
e-mail: kar.pritisman@gmail.com

B. Pati et al. (eds.), *Advanced Computing and Intelligent Engineering*, Advances in Intelligent Systems and Computing 1082,
https://doi.org/10.1007/978-981-15-1081-6_33

component but also disturbs the useful information present in the image. The basic technique is used for removal of noise introduces blurring of images as well as the edges do not get perfectly filtered.

Several images denoising filters are proposed by researchers. These filters can broadly be divided into Spatial filter and Frequency-domain filter. Spatial filter can be divided into linear or nonlinear filter. Gaussian filter is a linear filter which is responsible to blur the image or reduce noise. It is used to reduce Gaussian noise which affects the all pixel according to a density function. As shown in [1, 2], Non-linear filter (Standard Median Filter) outstrip linear filter effectively while denoising digital images corrupted by Impulsive noise because the impuissance of the linear filter while preserving fine image details and thin edges. Standard Median Filter exploits the order statistics by substituting the current element with the middle element in a given window. Different type of filters was developed to overcome this drawback such as weighted median filter and rank condition filter.

In image denoising, where the regions are obscurely defined fuzzy, sets representation serve as a potential solution where one only has to determine specific membership to a set without allocating to a set. In Fuzzy logic, human knowledge represented as a set of fuzzy rules or definition of fuzzy set is examined. In this paper, we have created a novel iterative fuzzy adaptive filter with proper morphological parameters to dynamically remove impulsive noise while edges and fine details in the brain magnetic resonance image (brain-MRI) are preserved.

Related literature is presented in Sect. 2 and the filter design methodology is explained elaborately in Sect. 3. In Sect. 4 we have presented experimental result to buttress the effectiveness of our designed filter. Some final conclusions are drawn in the last section.

2 Related Literature

Numerous works and design have been presented by many authors. The Standard Median (SM) filter is one of the basic filtering techniques and it can outstrip the linear spatial filter in efficiency and quality of the recovered image [1, 2]. Switching-based median filtering technique to separate noise pixel from the noiseless pixel and filter only the noise pixel was introduced in [3, 4]. These methodologies could not account for the pixel corrupted partly by noise and so fuzzy technique was introduced. In [5], authors have introduced notable fuzzy logic for image denoising.

Eng. et al. proposed their Noise Adaptive Soft-Switching Median (NASM)filter [6]. In that method global and local statistic were used in the noise detection process to separate the corrupt pixel from the noise-free pixel which also includes edge pixels. In [7], a new operator for impulse noise cancelation was presented. The proposed operator was a switching median filter guided recursively by a neuro-fuzzy network which aims to detect impulses present in the digital images. The parameters of the neuro-fuzzy impulse detection method were set to optimum value using training data. In [8], to filter noisy image corrupted by impulsive noise a Progressive switching median-based filter was proposed. In [9], the author has presented a fuzzy switching median filter which is recursive as well as exclusive for cancelation of impulsive noise. S-type membership function is used which aims at estimating a noise corruption level based on the fuzzy input value. In this method the updated or filtered pixel intensity value are substituted in the noisy pixel intensity value and the noiseless pixel are used for the filtering process. For the purpose of edge detection from an image Mathematical morphology filter was introduced. First the morphology was applied to binary images where operation such as intersection and union were used. Then morphology was introduced in gray scale images where the intersection and union operation were replaced by minimum and maximum operation [10–12]. In [13], a review of the algorithm and application in grayscale image was presented. Morphology has proved to be an important tool for different image processing problems, such as elimination of noise, feature extraction and identifying the edges. In [14], basic fuzzy morphology was presented. In that method the digital images were considered as fuzzy sets. To calculate basic operation such as erosion and dilation, fuzzy concept on union and intersection operation was used. In [15], soft mathematical morphology which is an alternative to basic morphology. In that paper in place of using maxima and minima, weighted order statistics were used where the weights used for the method depends on the structuring element used. In [16, 17], morphology-based edge extraction was presented where basic operation like erosion and dilation had been used.

3 Proposed Method

In this proposed method three-fold operation is done. First of all, the raw brain-MRI image is preprocessed. Next to it the noise is detected using Laplacian operator and is explained with the window selection method. Once the noise is detected in the specific region of the image it is processed with filtering method. The design of filter uses morphological operation. Further, modified fuzzy morphology is applied to clean the noise from thin and thick boundaries of the image. The process is explained in subsequent subsection.

Image preprocessing is an important tool which is nothing but a collection of transformation applied to an original image to obtain data in a pre-specified analytic format. Preprocessing is a necessary step in order to correct different defects that affect the image during acquisition. Out of different defects, Inhomogeneity, in the MRI image occurs when the distribution function of tissue classes depends on the spatial localization of the tissue. A low pass filter is used to filter out the high frequency component present in the image. Then the image is normalized and is inhomogeneity corrected using a smoother with 3×3 kernel size. The preprocessing procedure is show in Fig. 1.

The idea behind our filter is to separate the uncorrupted pixel from the corrupted pixel using adaptive fuzzy operator-based method and to further check the corrupted pixel is actually a noise pixel or an edge pixel using fuzzy morphology accounting for

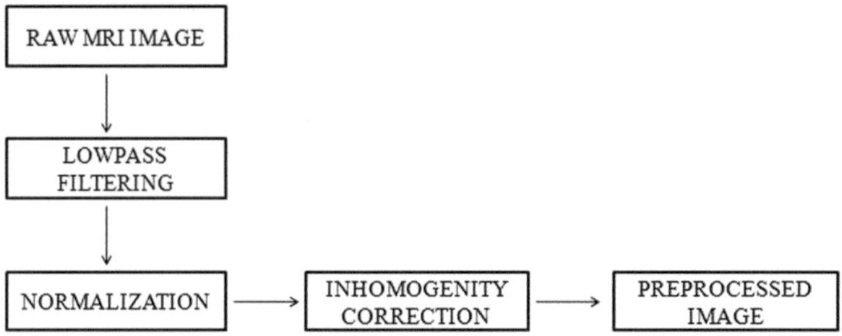

Fig. 1 Preprocessing of raw MRI image

both thin and thick edges. The edge pixels are left unfiltered and so is the uncorrupted pixel. The corrupted pixel is hence subjected to filtering based on the median value of uncorrupted pixel as well as the iteratively substituted or updated pixel values. Our proposed method is shown in Fig. 2 in a flow diagram manner.

The Noise Detection aims at detecting noise by considering the pixel value in the neighborhood using window operation. Here $v(i, j)$ which is the output of this method gives a sense of the amount of corruption in the current pixel. The Noise elimination is done by changing the current pixel value according to information from both the previous blocks.

A brain image I is defined by a matrix which has size mXn, and $I(i, j)$ is the grayscale value of the pixel present in 'i'th row and 'j'th column. This noise detection phase is again composed of two-fold method as noise detection for thin boundary and of thick boundary. It is explained as follows:

Fig. 2 Flow diagram of proposed method

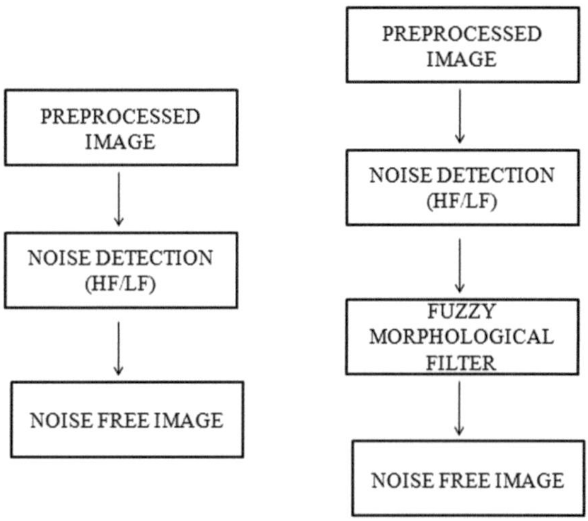

0	0	0	0	0
0	0	0	0	0
-1	-1	4	-1	-1
0	0	0	0	0
0	0	0	0	0

0	0	-1	0	0
0	0	-1	0	0
0	0	4	0	0
0	0	-1	0	0
0	0	-1	0	0

0	0	0	0	-1
0	0	0	-1	0
0	0	4	0	0
0	-1	0	0	0
-1	0	0	0	0

-1	0	0	0	0
0	-1	0	0	0
0	0	4	0	0
0	0	0	-1	0
0	0	0	0	-1

Fig. 3 Four 5 × 5 convolution kernels

1. For the Noise Detection a window of size 5 × 5 is considered for processing around the current pixel which is referred as $T(i, j)$. Four 1-dimentional Laplacian operators are taken which are volatile to different edges lying in four different orientation. These operators are shown in Fig. 3.
2. The given window is convolved with four given 1-dimentional Laplacian operators. The output matrix after convolution is trimmed to a size of 5 × 5 $\left(C_p(i, j) : p \rightarrow 1 : 4\right)$ given by Eq. (1). Then the minimum value of 4 different center element from four output trimmed matrices (from convolution) is found out as given in Eq. (2).

$$C_P(i, j) = \{|T(i, j) * K_P|\} : p \rightarrow 1 : 4 \tag{1}$$

$$v(i, j) = \min\{C_P(3, 3)\} : p \rightarrow 1 : 4\} \tag{2}$$

3. $v(i, j)$ is sensitive to impulsive noise because:

   (a) $v(i, j)$ is large when the pixel under consideration is a noise pixel as well as isolated because the absolute values of four convolutions values after trimming and considering the center pixel are themselves large.
   (b) $v(i, j)$ is small when the pixel under consideration is not a noise pixel as the absolute values of four convolution values after trimming and considering the center pixel lies close to zero.
   (c) $v(i, j)$ is small when the pixel under consideration is part of a thin edge line because the absolute value of one of the convolutions is close to zero even though all other three convolution values are large.

4. For application of fuzzy logic Bell membership function is considered and is represented as $\mu[v(i, j)] \in [0, 1]$ where $v(i, j)$ serve as the input for the fuzzy system given as per Eq. (2).
5. Now it is applied to ANFIS system for training so that it will be suitable for filtering.

$$f(x; a, b, c) = \frac{1}{1 + \left|\frac{x-c}{a}\right|^{2b}} \tag{3}$$

6. For filtering out the edges, the conventional standard median filtered output, i.e., $M_g(i, j)$ is used as input to fuzzy morphological filtering. To detect both thick and thin edges noise free image $M_g(i, j)$ is considered with structuring elements of 60–120° orientation.
7. It is processed with fuzzy dilation and erosion that is given in Eqs. (4–5). It can be used to create another operator like closing and opening of an image. The gradient operator gives the best result as shown in [16, 17] in term of edge filtering and is defined by the difference of dilation and erosion as given by Eq. (6).

$$D \oplus S = max_{[i,j]\forall S}\{d[m-j, n-k] + s[j, k]\} \tag{4}$$

$$D \Theta S = min_{[i,j]\forall S}\{d[m-j, n-k] + s[j, k]\} \tag{5}$$

$$D \Diamond S = (D \oplus S) - (D \Theta S) \tag{6}$$

Here *D*, *S* represent the digital MRI image and the structuring element (60–120°) respectively and ◇ represent the gradient operator.

In binary images erosion can be referred as the process of shrinking and dilation operation as the process of expanding. In order to introduce fuzzy morphology, the inclusion grade operator is introduced using concept of fuzzy for inclusion of set in some universe *U* which is finite, and it can be calculated as

$$R(D, S) = \inf_{x \in U}\{\min(1, \lambda(D(x)) + \lambda(S(x)))\} \tag{7}$$

In this case, $R(D, S)$ acts as an inclusion grade for $D, S \in [0, 1]^U$ where a function is defined $\lambda : [0, 1] \rightarrow [0, 1]$.

The notations are used as follows:

(a) λ is non-increasing,
(b) $\lambda(0) = 1$ & $\lambda(1) = 0$
(c) The equation $\lambda(q) = 0$ has a single solution.
(d) $\lambda(q) + \lambda(1-q) \geq 1 \quad \forall q \in [0, 1]$

The inclusion grade operator is used with erosion and dilation as given in Eqs. (8–9)

$$\zeta(D, S)(h) = R(S_H, D) \quad \forall h \in U \tag{8}$$

$$\varphi(D, S)(h) = 1 - R\left((-S)_H, D^c\right) \quad \forall h \in U \tag{9}$$

$$\nabla(D, S)(h) = \zeta(D, S)(h) - \varphi(D, S)(h) \tag{10}$$

$I(i-2,j-2)$	$I(i-2,j-1)$	$I(i-2,j)$	$I(i-2,j+1)$	$I(i-2,j+2)$
$I(i-1,j-2)$	$I(i-1,j-1)$	$I(i-1,j)$	$I(i-1,j+1)$	$I(i-1,j+2)$
$I(i,j-2)$	$I(i,j-1)$	$I(i,j)$	$I(i,j+1)$	$I(i,j+2)$
$I(i+1,j-2)$	$I(i+1,j-1)$	$I(i+1,j)$	$I(i+1,j+1)$	$I(i+1,j+2)$
$I(i+2,j-2)$	$I(i+2,j-1)$	$I(i+2,j)$	$I(i+2,j+1)$	$I(i+2,j+2)$

Fig. 4 The filter window of 5 × 5

where $\zeta(D, S)(h)$ represent fuzzy erosion and $\varphi(D, S)(h)$ represent fuzzy dilation operation and we can simply find the fuzzy gradient operator which is the filter output using Eq. (10). Further this method is extended to eliminate noise from thick regions.

8. Define $F = 1$ where the filter window size is $\left[2F + 1\ 2F + 1\right]$ and $I(i, j)$ is the current pixel of the current window of the brain image. Search for the current window's minimum and maximum intensity values referred as $W\,min$ and $W\,max$ respectively. Comparison is made for each window element with $W\,max$ and $W\,min$ value based on correlation. If any value is equal with either of them, discard that pixel. Calculate the number of the pixels not discarded in the window and assign the value *NO* to that number. If $NO = 0$ when $F = 1$, then set $F = 2$ and using the above method the median value is computed in the same way. If $NO > 0$ for the given window, then calculate the median of those *NO* gray scale intensity values and assign the median value to variable *M*. If still $NO = 0$, the average (*AVG*) of the four preceding and updated nearest neighboring pixels of the current pixel $I(i, j)$ is computed using Eq. (11). $E(i, j)$ which is a estimation parameter is defined as follows as per Eq. (12) where *M* represent the median value of *NO* elements and *AVG* represent the average value of four neighboring pixel as shown in Fig. 4.

$$AVG = 0.25 * [I(i, j-1) + I(i-1, y-1) + I(i-1, j) + I(i-1, j+1) \tag{11}$$

$$E(i, j) = \begin{cases} M & NO > 0 \\ AVG & NO = 0 \end{cases} \tag{12}$$

9. Finally the output intensity pixel $Y(i, j)$ is modified according to output of the Noise Detection method $(v(i, j))$ and fuzzy morphological operator $(\nabla(D, S)(h))$ as given in Eq. (13). The parameter 'a' and 'b' are properly tuned from observation to yield the best result.

$$Y(i, j = \begin{cases} I(i, j) + \mu[v(i, j)] * (E(i, j) - I(i, j)) & \nabla(D, S)(h)\langle a \cup \nabla(D, S)(h)\rangle b \\ I(i, j) & otherwise \end{cases} \tag{13}$$

Table 1 PSNR and MSE comparison between different contemporary filters

Method (Brain-mri)	PSNR	IMMSE
Standard median filter (3 × 3)	39.1425	7.9220
SM Filter [8]	41.8996	4.1987
NASM filter [6]	60.4257	0.0590
Proposed filter (3 × 3)	102.04	4.0566e-06

4 Results and Discussion

For our testing we have used here a Brain-MRI image which was properly pre-processed before applying our methodology. Our edge detection scheme worked excellently yielding us proper edges present in the MRI image. Table 2 displays the comparison between time needed for execution taken by median filter and our proposed filter. It is seen that the time needed to execute for median filter is the least, and that of the proposed filter is comparatively most. The experiments were run on a i5 processor @ 2.8 GHz with RAM size equal to 8 Gb.

The filtering process performed excellently by eliminating impulsive noise dynamically and iteratively which were present in the MRI image during the time of acquisition and transmission as given by Table 1. Its conspicuous upon comparing visually Fig. 5c, d that our method performs well when compared to standard median filter's output. Our proposed work is also compared with previous works as given in Tables 1 and 2. Even though the proposed method takes longer time comparatively, the filtering results are excellent.

(A) Filtering Analysis
(B) Runtime Analysis

5 Conclusion

Our proposed Fuzzy morphological (FM) filter can effectively cancel out the impulsive noise while preserving edge. The filter is sensitive to thin edges and specially to think edges where it effectively filters out the noise present in between the boundary pixel of the thick edges.

The performance of our FM filter has been compared to that of median filter. Experimental results revealed that our proposed FM filter has performed far better than the standard median filter. The FM filter performed well when compared visually with the standard median filter output.

Our proposed FM filter uses in image processing is quite generic. This methodology can be applied to variety of image which are affected by impulsive noise.

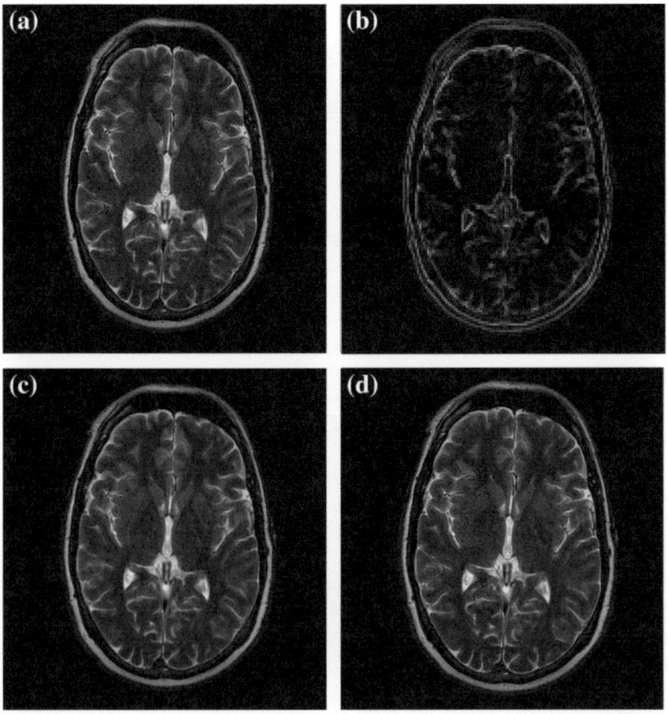

Fig. 5 **a** Brain Image (After Preprocessing). **b** Fuzzy gradient operator output. **c** Simple median filtering (PSNR-39.1425, IMMSE-7.9220). **d** Proposed filter (PSNR-102.04, IMMSE-4.0566e-06)

Table 2 Comparison of execution time

Method (Brain-mri)	Time taken to execute
Standard median filter (3 × 3)	60 ms
SM filter [8]	30.2655 s
NASM filter [6]	124.2 s
Proposed filter (3 × 3)	329 s

References

1. Pitas, I., Venetsanopoulos, A.N.: Nonlinear Digital Filters: principles and Applications, vol. 84. Springer Science + Business Media (2013)
2. Astola, J., KuosmanenP.: Fundamentals of Nonlinear Digital Filtering, vol. 8. CRC Press (1997)
3. Sun, T., Neuvo, Y.: Detail-preserving median based filters in image processing. Pattern Recognit. Lett. **15**(4), 341–347 (1994)
4. Florencio, D.A., Schafer, R.W.: Decision-based median filter using local signal statistics. Visual Commun. Image Process. **2308**, 268–276 (1994)
5. Kerre, E., Nachtegael, M., (eds.): Fuzzy Techniques in Image Processing, vol. 52. Springer-Verlag, Studies in Fuzziness and Soft Computing, New York (2000)
6. Eng, H.L., Ma, K.K.: Noise adaptive soft-switching median filter. IEEE Trans. Image Process. **10**(2), 242–251 (2001)

7. Yüksel, M.E., Bastürk, A., Besdok, E.: Detail-preserving restoration of impulse noise corrupted images by a switching median filter guided by a simple neuro-fuzzy network. EURASIP J. Appl. Signal Process. 2451–2461 (2004)
8. Wang, Z., Zhang, D.: Progressive switching median filter for the removal of impulse noise from highly corrupted images. IEEE Trans. Circuits Syst. II: Analog Digit. Signal Process. **46**(1), 78–80 (1999)
9. Qin, H., Yang, S.X.: Adaptive neuro-fuzzy inference systems-based approach to nonlinear noise cancellation for images. Fuzzy Sets Syst. **158**(10), 1036–1063 (2007)
10. Vincent, L.: Morphological grayscale reconstruction in image analysis: applications and efficient algorithms. IEEE Trans. Image Process. **2**(2), 176–201 (1993)
11. Roushdy, M.: Comparative study of edge detection algorithms applying on the grayscale noisy image using morphological filter. GVIP J. **6**(4), 17–23 (2006)
12. Davis, L.S.: A survey of edge detection techniques. Comput. Graph. Image Process. **4**(3), 248–270 (1975)
13. Shin, M.C, Goldgof, D.B., Bowyer, K.W., Nikiforou, S.: Comparison of edge detection algorithms using a structure from motion task. IEEE Trans. Syst. Man Cybern. Part B (Cybernetics) **31**(4), 589–601 (2001)
14. Sinha, D., Sinha, P., Dougherty, E.R., Batman, S.: Design and analysis of fuzzy morphological algorithms for image processing. IEEE Trans. Fuzzy Syst. **5**(4), 570–584 (1997)
15. Canny, J.: A computational approach to edge detection. In: Readings in Computer Vision, pp. 184–203 (1987)
16. Behera, S., Mohanty, M. N., Patnaik, S.: A comparative analysis on edge detection of colloid cyst: a medical imaging approach. In: Soft Computing Techniques in Vision Science, pp. 63–85. Springer, Berlin, Heidelberg (2012)
17. Jyoti, A., Mohanty, M.N., Kumar, M.P.: Morphological based segmentation of brain image for tumor detection. In: International Conference Electronics on and Communication Systems (ICECS), pp. 1–5. IEEE (2014)

Hyper-spectral Image Denoising Using Sparse Representation

Vijay Chilkewar and Vibha Vyas

Abstract Sparse representation, statistical and probabilistic approach have been used in image processing applications. Here, simple technique is used to denoise Hyper-spectral image by using sparse representation. Main focus is to transform the given image into another form which is combination of dictionary and sparse vector. So, basic statistical methods are used for the updation of dictionary and also for sparse coding. Then, probabilistic approach is used to determine new size of dictionary and this new dictionary is used to achieve denoised image and finally, peak signal to noise ratio (PSNR) is used to measure performance of denoising methods.

Keywords Trained dictionary · Sparse vector · Hyper-spectral image (HSI) · Bayesian techniques

1 Introduction

Denoising is first step in many image processing applications. Denoising plays important role before analyzing and processing of image. Hyper-spectral image analysis is very important in a wide range of applications including crop management, medical field, mineral, water quality, forest monitoring, urban area management. In recent years, there is a huge improvement in sensor technology which made available large amount of data with high spatial, spectral and temporal resolution which can be used for different applications.

In Hyper-spectral images each pixel is represented as high-dimensional vector of spectral reflectance which spans the visible and invisible bands electromagnetic spectrum. Hyper-spectral images are of very high spectral resolution [1]. The main advantage of HSI is because whole spectrum data is collected from each point, it allows all available dataset to be mined, if these plentiful information should be

V. Chilkewar · V. Vyas (✉)
College of Engineering, Pune 411005, India
e-mail: vsv.extc@coep.ac.in

V. Chilkewar
e-mail: vijayc2793@gmail.com

B. Pati et al. (eds.), *Advanced Computing and Intelligent Engineering*,
Advances in Intelligent Systems and Computing 1082,
https://doi.org/10.1007/978-981-15-1081-6_34

preserved and protected from noise. This noise can enter into image in ways like acquisition, storing, transferring process which affects the quality of image. Thus, noise removal should be the first step in any pre-processing step.

Several classes of denoising algorithms that used nonlocal means technique have given much success. When denoising an image [2], geometric features of image is extracted by using total variation (TV) method [3], while wavelet method makes use of statistical features of image and [4] the nonlocal means method makes use of the redundancy in the image texture features. However, features that extracted are all taken from noisy image itself.

But here, sparse representation plays important role in denoising of image. Recently the interest is going on in sparse modelling cause over training is avoided, easy implementation, good results, etc. Besides using in denoising, it has used in in-painting [5–7], compressive sensing [8, 9], classification [10], etc. Its been proved that most of the images represented by sparse modelling with appropriate dictionary. Most of the previous works assumed offline dictionary which is trained earlier, but in our work online dictionary is used which has several advantages in applications like classification, denoising etc.

The purpose of this paper is to denoise the given image by using sparse representation with updated dictionary. Statistical and probabilistic approach is used to train dictionary and to do sparse coding. Initially, given algorithm is applied on normal images with a patch size of 8×8 which is standard and after getting successful results, same algorithm is being used with some modifications on Hyper-spectral image (HSI) to get desired results. This method provides advantages over dictionary size, no prior threshold assumption, etc. Finally, results of this technique compared with others through PSNR and givers slightly better results by fraction of dB which also a improved as it is in dB.

2 Methodology

Initially, noise is added to given image with standard deviation varying from 5 to 25. Then, patches are extracted from given image and arranged in the form columns as shown in given Fig. 1. Initially, random dictionary is taken and then it is to be trained

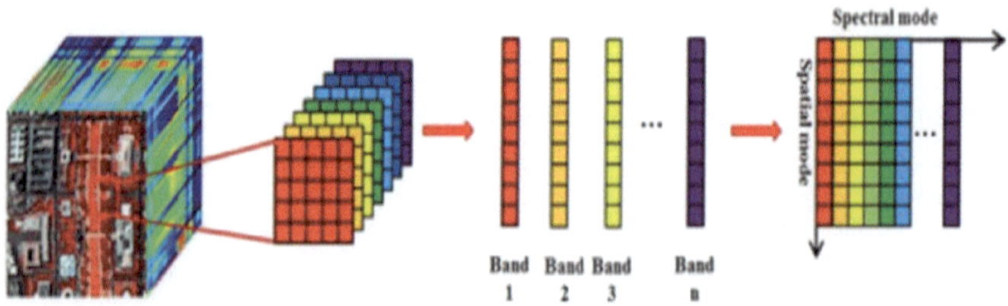

Fig. 1 Patch extraction for HSI

up to some iterations by using statistical methods and updating of sparse vector is done while updating dictionary. Finally, trained dictionary is passed through sparse coding stage to get new sparse vector and that will help us in getting final denoised image.

2.1 Dictionary Learning

Traditional techniques for sparse coding consider given signal and fixed dictionary D.

Here, main goal is represent the given signal in terms of D and α where α is sparse. All previous work is performed in following manner:

i. If D is known, α (sparse vector) is calculated by orthogonal matching pursuit (OMP) which includes basis or matching pursuits for which the stopping criteria is based on either number of iterations or sparsity level.
ii. There are no condition on size of dictionary D. So, this method is presented to overcome above limits by using trained dictionary by statistical and probabilistic approach which allows us to actually take care of sparsity level and dictionary size.

2.2 Bayesian Formulation

Here, the noise model is as below

$$X = D\alpha + N \tag{1}$$

where D is dictionary of R × K which is to be learned and also value of K need to be found which is the size of the dictionary, N is the noise to be added.

First part comes is to get trained dictionary for some iterations by reducing error and checking PSNR. To train that dictionary, statistical approach is used and for sparse coding updation too. For sparse coding, probabilistic approach means using beta-bernoulli process is used to make sparse vector as sparse as possible.

2.3 HSI Denoising

The block diagram of proposed work is shown in Fig. 2. First of all, addition of noise is done to both normal and HSI image with some standard deviation. After this, image is divided into patches of size 8 × 8 which is taken mostly while performing any image processing applications like classification, in-painting, compressive sensing,

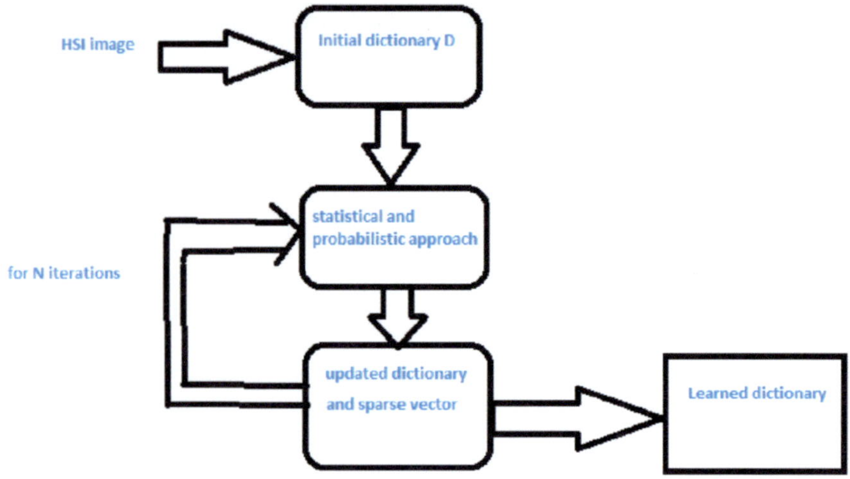

Fig. 2 Block diagram of trained dictionary and sparse calculations

etc. Suppose, if block size is q × q × n, then we are arranging patches in lexico-graphic order to obtain a matrix $(q^2) \times n$ (Fig. 2).

After getting patch of 8 × 8 block along spatial as well as spectral dimension, and then combining all such patches, matrix is formed. Now, X can be modelled as:

$$X = D\alpha \tag{2}$$

where X is a noisy matrix obtained by dividing image into patches and D is dictionary to be trained. α is sparse vector with most of the coefficients to zero value.

The given dictionary D is initially set to any random matrix which is of size depends on block size we r taking. If it is block of 8 × 8 for given HSI, then it is 64 × 256. This 256 value is taking it as a random value, but our proposed work is all about finding this value. So, initial dictionary can be random, data or can be DCT dictionary. After setting a dictionary, Bayesian, i.e., statistical and probabilistic approach is used for training of dictionary along with that one of the Gibbs sampling equation helping us in doing sparse coding. Here, for training of dictionary (D) simple statistical approach is used and same is done for sparse vector but which coefficients of sparse vector are used to represent given data by choosing columns from dictionary, this job is taken care by vector let us say Z which uses Gibbs equation which leads us to give full sparse vector with most of elements zero (Fig. 3).

Above block diagram is for denoising which is done by using sparse coding and for that Gibbs sampler is used. Let us say, vector Z as mentioned earlier is a matrix of [0, 1] which a matrix of values 0 and 1 only. This matrix actually decides which columns of D are to be used for representation of data itself. Suppose if particular

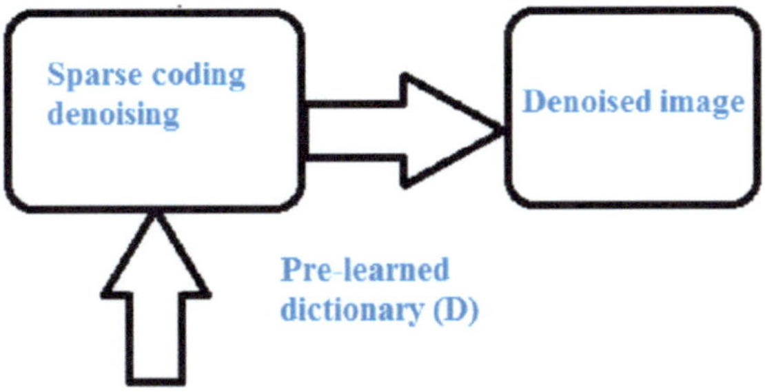

Fig. 3 Block diagram to get clean image

component of Z is 0, then corresponding column of D is not used for given signal and it is 1, corresponding column is contributing to represent given data (x). So, to get the given matrix Z, beta-bernoulli base is used which is modelled as given below:

$$Z(i) = \sum_{k=0}^{n} \text{Bernoulli } (\pi), \pi = \sum_{k=0}^{n} \beta\left(\frac{a}{K}, b(K-1)/K\right) \tag{3}$$

where a, b are constants, K is dictionary size.

Here, Z is matrix of only 0 and 1 where 1 is with probability P which is decides beta distribution and if it is 0, it's $1 - \text{p}$. These help us in recovering clean image from noisy image while simultaneously inferring D [11–13].

3 Algorithm

(1) Begin.
(2) Read HIS image of size Height * width * layer.
(3) Addition of noise standard deviation of 5–25 etc.
(4) Divide given image into patches by using block size of 2×2, 4×4, 8×8 or $N \times N$ etc.
(5) Assume initial dictionary be the random or DCT dictionary.
(6) Use Gibbs iteration and statistical methods to update or for training of dictionary and sparse vector.
(7) At the same time, sparsity is being calculated by Z matrix which will also give probability of each column of dictionary (D) by π (pi) vector. Z and π goes by Eq. (3).
(8) Perform this for several iterations up to 70 to get trained dictionary.
(9) After getting trained dictionary, finally use sparse coding to get new sparse vector.
(10) Combine both trained dictionary (D) and sparse vector (S) to get new denoised image.

(11) Finally, by looking at vector π (pi), we can tell the exact dictionary size to be used to represent this denoised image.
(12) End.

4 Experimental Setup

This experiment tested on normal images and HSI image also. Normal 2-D images like lena, house, boat, castle and then it is performed on urban HSI of size 150 × 150 with 210 spectral bands. Finally, results of this experiment compared with previous approaches and it has slightly better results than previous work.

4.1 Data Description

The Hyper-spectral image of Washington DC mall which is available online as well as Normal images are used for our experiment. First given experiment tested on normal images and then same algorithm with some modifications is applied on Hyper-spectral image (HSI). This algorithm tested on 7–8 Normal images with average PSNR varying from 25 to 30 dB and they are of size 256 × 256. Size of HSI varies from 150 × 150 with 210 spectral bands. Here, Gaussian as well as random noise is added to both type of image to test our algorithm with standard deviation varies from 5 to 30. After addition of noise, the average PSNR decreases from 30 to 20 dB.

5 Results and Figures

For this experiment, both normal as well as Hyper-spectral images with standard patch size 8 × 8 have been used and this is performed for some iterations up to 60–70 as stopping criteria. This experiment is performed on Dell laptop having windows operating system with quad core CPU with speed 1.7 GHz, 4 GB RAM (Figs. 4, 5, 6, 7, 8 and 9) (Table 1).

6 Conclusion

In this work, HSI denoising is achieved using sparse representation and dictionary learning. To denoise given image, we are following patch based method and considering HSI spectral feature, after taking patch from image they are arranged in lexico-graphic manner to retain spectral information.

Fig. 4 Clean image, noisy image (σ (standard deviation) = 25) and clean image by adaptive dictionary

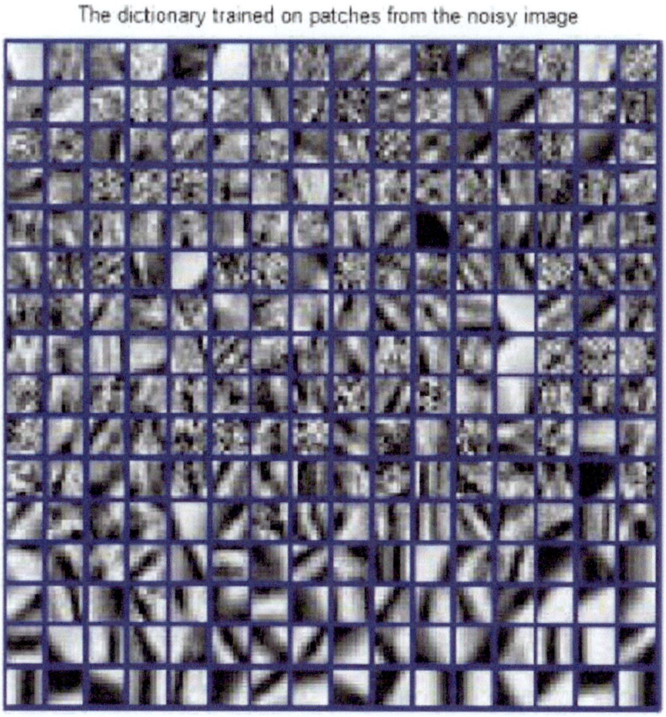

Fig. 5 Dictionary obtained from corrupted image

This algorithm applied on both normal images and Hyper-spectral images with different patch size and it is observed from table that with increasing patch size, PSNR of clean image increases and it is better than the algorithm applied by

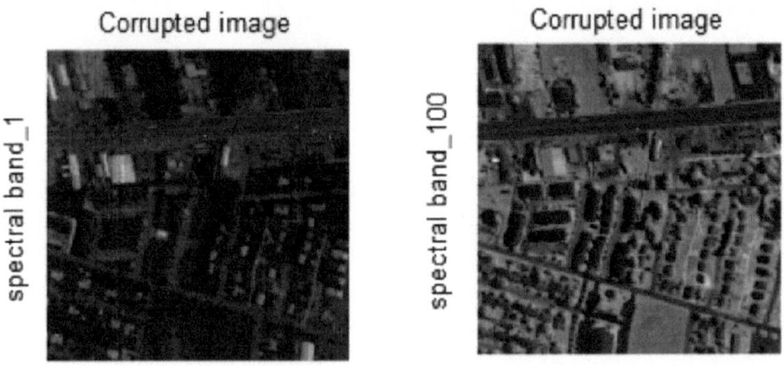

Fig. 6 HSI with addition of noise (Corrupted image with σ = 20) having PSNR = 11.05 dB for spectral band 1 and PSNR = 11.07 dB for spectral band 11

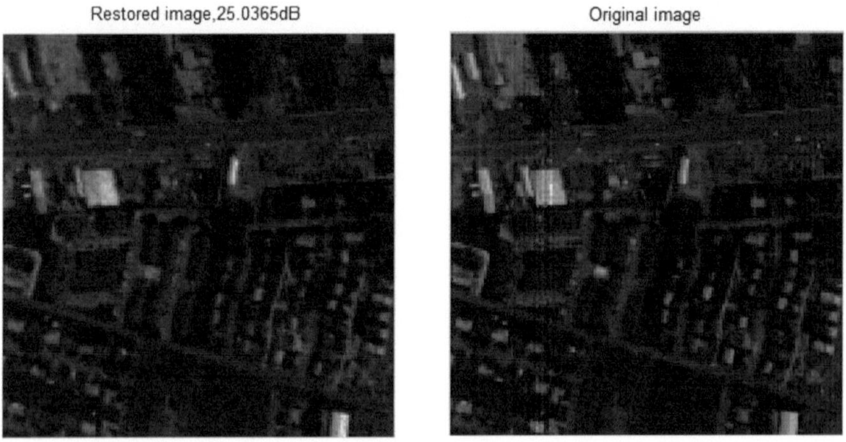

Fig. 7 Restored hyper-spectral image of spectral band 1

K-SVD. Besides PSNR, this algorithm has an advantage of getting real dictionary size to represent the clean image. It is like at the end getting clean image and inferring size of dictionary which is a better improvement considering the previous work.

Fig. 8 Restored hyper-spectral image of spectral band 100

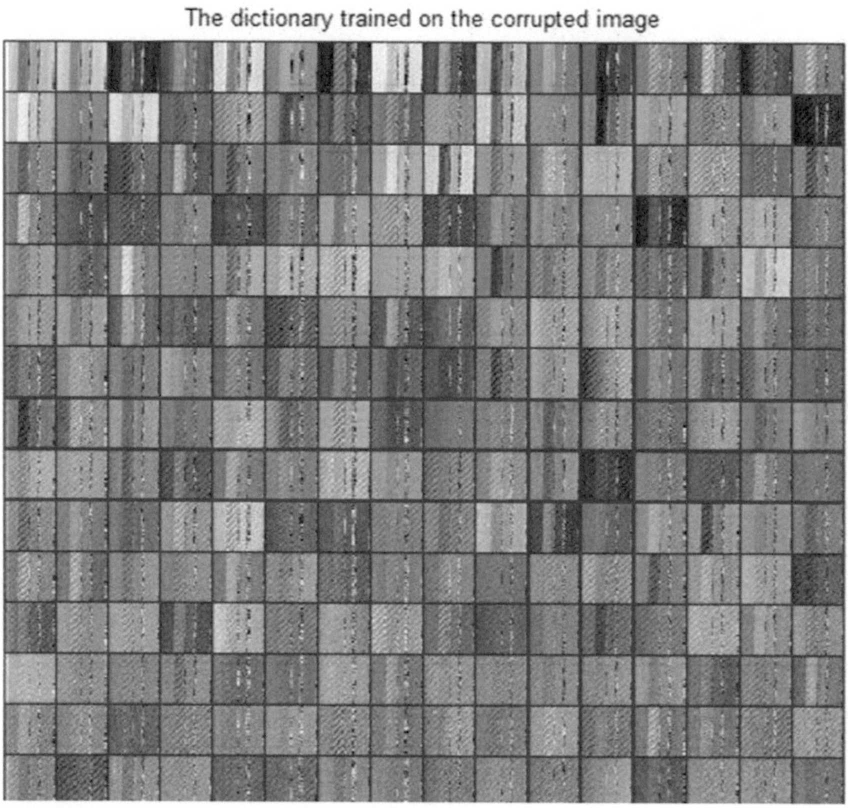

Fig. 9 Trained dictionary for given hyper-spectral image

Table 1 Table for PSNR of HSI of spectral band 100 for both noisy and clean images with respect to Patch size. Increase in patch size, increases PSNR

Index	Patch size	HSI corrupted image (PSNR in dB)	Restored image (PSNR in dB)
1	1×1	11.693	22.22
2	2×2	11.693	24.22
3	3×3	11.693	26.50
4	4×4	11.693	27.02
5	5×5	11.693	27.16
6	6×6	11.693	27.18
7	7×7	11.693	27.09
8	8×8	11.693	27.99

References

1. Yuan, Q., Zhang, L., Shen, H.: Hyperspectral image denoising employing a spectral–spatial adaptive variation model. IEEE Trans. Geosci. Remote Sens. **50**(10), 36303677 (2012)
2. Aggarwal, H.K., Majumdar, A.: Hyperspectral image denoising using spatio-spectral total variation. IEEE Geosci. Remote Sens. Lett. **13**(3) (2016)
3. Chen, G., Qian, S.E.: Denoising of hyperspectral imagery using principal component analysis and wavelet shrinkage. IEEE Trans. Geosci. Remote Sens. **49**(3), 973 979 (2011)
4. Li, A., Chen, D., Lin, K., Sun, G.: Hyperspectral image denoising with composite regularization models, Received 21 October 2015; Revised 14 April 2016; Accepted 20 April 2016
5. Aharon, M., Elad, M., Bruckstein, A.M.: K-SVD: An algorithm for designing overcomplete dictionaries for sparse representation. IEEE Trans. Signal Process. **54** (2006)
6. Elad, M., Aharon, M.: Image denoising via sparse and redundant representations over learned dictionaries. IEEE Trans. Image Process. **15** (2006)
7. Mairal, J., Elad, M., Sapiro, G.: Sparse representation for color image restoration. IEEE Trans. Image Process. **17** (2008)
8. Candes, E., Tao, T.: Near-optimal signal recovery from random projections: universal coding strategies. IEEE Trans. Inf. Theory **52** (2006)
9. Duarte-Carvajalino, J.M., Sapiro, G.: Learning to sense sparse signals: simultaneous sensing matrix and sparsifying dictionary optimization. IMA Preprint Series, p. 2211 (2008)
10. Wright, J., Yang, A.Y., Ganesh, A., Sastry, S.S., Ma, Y.: Robust face recognition via sparse representation. IEEE Trans. Pattern Anal. Mach. Intell. **31** (2009)
11. Paisley, J., Carin, L.: Nonparametric factor analysis with beta process priors. In: Proceedings of the International Conference on Machine Learning (2009)
12. Thibaux, R., Jordan, M.I.: Hierarchical beta processes and the Indian buffet process. In: Proceedings of the International Conference on Artificial Intelligence and Statistics (2007)
13. Zhou, M., Chen, H., Paisley, J., Ren, L., Sapiro, G., Carin, L.: Non-parametric bayesian dictionary learning for sparse image representations. In: Proceedings of the Neural Information Processing Systems (2009)

Automated Retinal Vessel Segmentation Based on Morphological Preprocessing and 2D-Gabor Wavelets

Kundan Kumar, Debashisa Samal and Suraj

Abstract Automated segmentation of vascular map in retinal images endeavors a potential benefit in diagnostic procedure of different ocular diseases. In this paper, we suggest a new unsupervised retinal blood vessel segmentation approach using top-hat transformation, Contrast-Limited Adaptive Histogram Equalization (CLAHE), and 2-D Gabor wavelet filters. Initially, retinal image is preprocessed using top-hat morphological transformation followed by CLAHE to enhance only the blood vessel pixels in the presence of exudates, optic disc, and fovea. Then, multiscale 2-D Gabor wavelet filters are applied to preprocessed image for better representation of thick and thin blood vessels located at different orientations. The efficacy of the presented algorithm is assessed on publicly available DRIVE database with manually labeled images. On DRIVE database, we achieve an average accuracy of 94.32% with a small standard deviation of 0.004. In comparison with major algorithms, our algorithm produces better performance concerning the accuracy, sensitivity, and kappa agreement.

Keywords Retinopathy · Blood vasculature · Retinal vessel segmentation · 2D-gabor wavelet · Top-hat transform

---

K. Kumar (✉) · D. Samal
Department of Electronics and Communication Engineering, ITER, Siksha 'O' Anusandhan (Deemed to be University), Bhubaneswar 751030, Odisha, India
e-mail: erkundanec@gmail.com
URL: https://sites.google.com/site/erkundanec/home

D. Samal
e-mail: debashishsamal@soa.ac.in

Suraj
Department of Electrical Engineering, Sardar Vallabhbhai National Institute of Technology, Surat 395007, Gujarat, India
e-mail: suraj.boom@gmail.com

B. Pati et al. (eds.), *Advanced Computing and Intelligent Engineering*, Advances in Intelligent Systems and Computing 1082,
https://doi.org/10.1007/978-981-15-1081-6_35

1 Introduction

Change in anatomical structure of retinal blood vessels (vasculature) in retina is a good indication of the presence of ophthalmic diseases, e.g., hypertension, cardiovascular diseases, diabetic retinopathy, glaucoma, etc. [1, 2]. Vascular map segmentation of fundus images has played a decisive role in assessing the change in the vasculature for severity of ocular diseases. However, periodic screening of retinal images for early recognition of the change in vascular structure can prevent major vision loss [2]. Therefore, automatic and accurate retinal vessel segmentation is the prerequisite for the initial diagnosis of retinal diseases. Extraction of the vascular map from an uneven illuminated and pigmented fundus image is a challenging problem. Besides retinal blood vessels, the presence of other structures (e.g., exudates, optic disc, fovea, red lesions) under uneven illuminated and pigmented background makes the vessel detection even more difficult. Also, the retinal vessel thickness varies in the wide range whereas thin vessels have low contrast which makes thin vessel detection more challenging [2].

Several, automatic vascular map segmentation of retinal fundus images have been proposed in literature [3–7]. Fraz et al. have done a comprehensive literature survey on retinal blood vessel segmentation in [8]. The retinal blood vessel segmentation techniques are popularly classified as: (i) supervised and (ii) unsupervised techniques. In supervised category, *k*-NN-classifier [9], Artificial Neural Network (ANN) [10], trainable COSFIER filters [11], Support Vector Machine (SVM) [12], Extreme Learning Machine (ELM) [13], Deep Neural Network (DNN) [6, 14], etc. have been explored for blood vessel segmentation as identification problem. However, supervised algorithms rely on the robust feature extraction followed by classification. In many approaches line detector [12], Gabor filters [1, 10], and Gray Level Co-occurrence Matrix (GLCM) [15] -based methods have been explored for feature extraction purpose. A feature vector is computed for each pixel using these approaches and classifier classifies the pixels as the vessel and non-vessel pixels. These approaches are time-consuming process due to training. On the contrary, in the unsupervised category, filtering methods [3, 5, 16], vasculature tracing methods [17], curvelet based [18], morphological operators [19], have been used. In these approaches, classification of the vessel or non-vessel pixels is performed without training process, i.e., training data do not contribute to finding the model parameter.

In earlier reported works under the unsupervised category, match filter has received the enormous response of the scholars due to its straightforwardness in the implementation of the technique. Match filter-based retinal vessel segmentation relies on a 2D kernel with Gaussian profile initially proposed by Chaudhuri et al. [3]. The kernel is rotated at 15° increment, and the best output of the filter for each pixel is carefully chosen to map all blood vessels orientated at different angles. After that, thresholding is applied to get binary vessel map image. Further, pruning is applied as postprocessing to improve the final identification of blood vessels. Hoover et al. [4] have used local and region-based properties for vessel segmentation where threshold probing technique is used on match filter response. However, match filter gives a strong response in terms of vessels and non-vessels edges. Zhang et al. [5] have exploited the first-order derivative of Gaussian to improve the performance of matched filter by eliminating non-vessel edges from retinal images. In literature [16, 17], multiscale match filter and its variation are suggested to identify blood vessels of different thicknesses. In [2], Zhao et al. have enhanced the retinal vessels in retinal images by utilizing 2D-Gabor wavelet filters and a Contrast-Limited Adaptive Histogram Equalization (CLAHE). After that, the processing results of region growing method and level set approach are combined to get final segmentation as a binary image. However, this approach is unable to remove non-vessel structures also takes long processing time. Roychowdhury et al. [20] have performed an unsupervised iterative process to obtain the vessels using top-hat reconstruction followed by iterative region growing method. Most of the approaches like [2, 5] fail to remove optic disc and exudates from pathological images. Also, many approaches remove these structures in postprocessing with the extra burden of computational time. Thus, an automated unsupervised blood vessel segmentation approach is needed to identify the blood vessel pixels correctly with a small false positive rate. Simultaneously, need to remove the anatomical structure other than blood vessels with high accuracy and less complexity.

We propose an entirely unsupervised approach for automatic segmentation of blood vessels to obtain the retinal vascular map. The significant contribution of this paper is to use top-hat transform followed by CLAHE for retinal image enhancement in preprocessing step. Use of top-hat transform facilitates to enhance only the blood vessels, simultaneously remove the local intensity change due to exudates, optic disc, and fovea from the background. For further image enhancement, CLAHE is applied on top-hat transformed retinal image. The preprocessed retinal image is passed through the Gabor filters bank, and the maximum outcome of the filters is chosen for each pixel. Otsu thresholding as global thresholding is applied to get a binary blood vessel structure. The proposed technique is proficient in identifying the blood vessels under uneven pigmentation and illuminance condition in the presence of exudates, optic disc, and fovea. Our proposed approach suppresses the non-vessel pixels in the preprocessing step; however, multiscale Gabor wavelet filters efficiently represent the thick and thin vessels in the retinal image. The efficacy of the presented technique is validated on the publicly available Digital Retinal Image for Vessels Extraction (DRIVE) database [9].

Rest of the paper is organized as follows. The DRIVE database and the proposed algorithm are discussed in Sects. 2 and 3, respectively. The Sect. 4 discusses the experimental results and finally, the paper is concluded in Sect. 5.

2 Materials

We validate our proposed algorithm on the DRIVE database [9] for performance evaluation. The DRIVE database is publicly available to execute a comparative study and experimental evaluation of vascular segmentation algorithms. The gold standard segmented images are provided with the database as manually labeled images. The DRIVE database contains 40 color retinal images which include 20 images in training set and 20 images in the test set. All images were captured by Canon CR5 nonmydriatic three Charge-Coupled-Device (CCD) cameras at 45° Field Of View (FOV). Each color retinal image is having a resolution of 565×584 pixels with three R, G, and B channels, and each channel is an 8-bit grayscale image. In this paper, we examined our presented algorithm on images from the test set. For test set, two subsets of manually segmented images, i.e., set A and set B, are provided. Set A as the first observer's manual segmented images are considered as gold standard which is utilized as ground truth for performance assessment. The primary objective of choosing DRIVE database is to perform a comparative analysis of the presented work with the state-of-art techniques which have been evaluated on the same database.

3 Proposed Method

3.1 Overview

In the presented work, an unsupervised blood vessel segmentation approach is proposed. Figure 1 presents the flow diagram of the proposed algorithm.

The principal idea of the proposed method is that in retinal image, vessels can be distinguished from other structures like exudates, optic disc, fovea, etc. during preprocessing. In many approaches, CLAHE is employed in preprocessing to boost the dynamic range of the retinal images besides preventing the over-amplification of noise [2]. However, CLAHE improves the contrast of images by operating on local regions rather than globally due to which vessel pixels enhance with the other structures too. Therefore, we first applied the top-hat transformation on the green channel of color retinal fundus image that intensifies only the blood vessel pixels and simultaneously suppress the other structure pixels. After that, CLAHE is applied to get the full advantage of its characteristics. In the second stage, 2D-Gabor wavelet filter is applied to the preprocessed image. Furthermore, a global Otsu thresholding method is used to get a binary segmented image.

3.2 Preprocessing

Initially, an original RGB-color retinal image is split into three channels as shown in Fig. 2. Among these three channels, the green-channel image appears having higher contrast compared to other two channel images. In green-channel image, the blood vessel pixels are visible and easily distinguishable from the background pixels. Because in our eyes, lens pigments absorb light colors differently [2, 21]. Therefore, red vessels in color retinal image are more visible in the green-channel

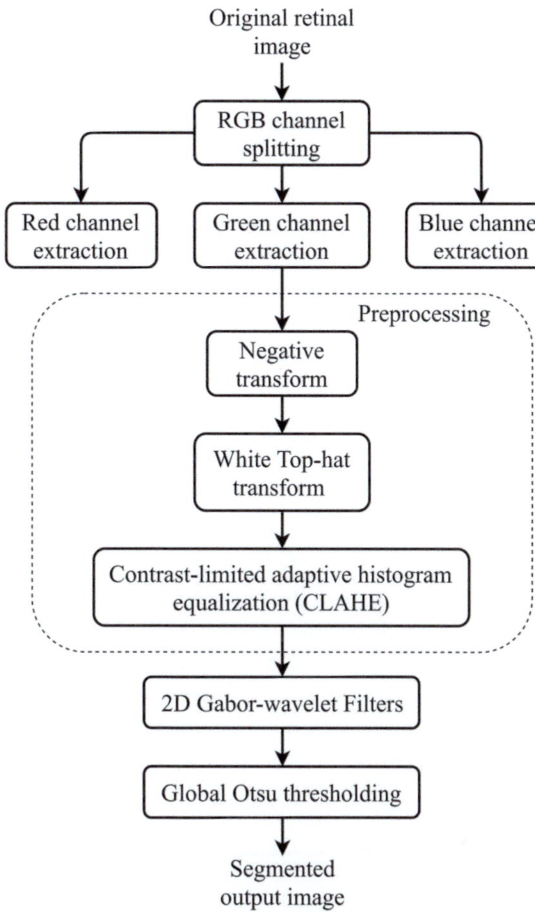

Fig. 1 Flow diagram of the proposed algorithm

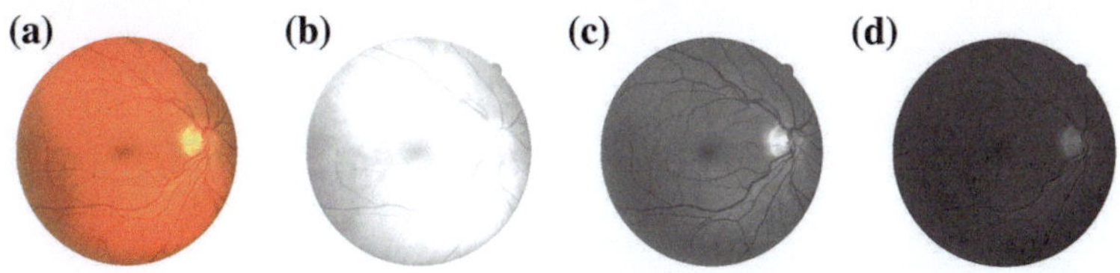

Fig. 2 Retinal image through FOV: **a** original RGB-color image, **b** red channel, **c** green channel, and **d** blue channel

image as presented in Fig. 2c; however, red-channel image is the brightest image and blue-channel image suffers from poor dynamic range as shown in Fig. 2.

In the green-channel image (I_G), the blood vessel pixels due to its intensities being close to 0 seems to be dark. Image I_G is inverted followed by superposition of fundus mask (M) to make the blood vessel pixels brighter and keep the focus on the region of interest (FOV). The inverted green-channel image through fundus mask is shown in Fig. 3b. To enhance only the vessels, white top-hat transformation is applied on inverted green-channel image (I'_G). Usually, blood vessels have small thickness compared to the other structures in retinal images. Therefore, a circular structuring element of diameter at least equal to the diameter of the thickest blood vessel is preferred for top-hat transformation. The white top-hat transform can be defined as

$$T_w(I'_G) = I'_G - I'_G \circ b. \tag{1}$$

where T_w is the transformed image of I'_G using structuring element b, and $\circ$ denotes the morphological opening operator. In image processing, the top-hat transformation is a morphological operation that highlights the object smaller than the structuring element [22]. The diameter of the structuring element is chosen 11 pixels wide as the maximum width of the blood vessel is less than 11 pixels. For the DRIVE database, the width of the blood vessel varies in the range of 1–10 pixels [23]. The diameter of the structuring element may vary for the different database having different image resolutions.

Usually, the width of widest blood vessels is less than the width of other structures, like exudates, fovea, and optic disc, which do not appear in the top-hat transformed image (T_w) as shown in Fig. 3c. Also, a homogeneous background is obtained in the transformed image. After applying the top-hat transformation, the blood vessel pixels do not achieve a good contrast compared to the background. Therefore, (CLAHE) technique is adopted for further enhancement of the processed retinal

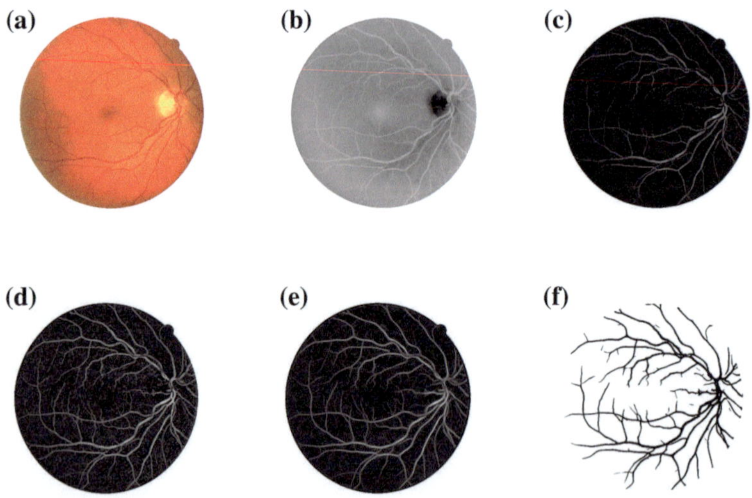

Fig. 3 Processing results at each intermediate steps of the proposed algorithm. **a** Original test image, **b** inverted green-channel image, **c** white top-hat transformed image of inverted green-channel image, **d** CLAHE processed image, **e** gabor wavelet response, **f** binary image after global thresholding

image. Figure 3d shows the CLAHE response (I_c) of the top-hat transformed image (Fig. 3c). However, CLAHE also enhances the background noise. For informative representation of the blood vessels having different thicknesses and orientation, retinal image is processed through the multiscale Gabor wavelet filters at different frequencies and orientations. Gabor wavelet also smoothes the background noise.

3.3 2D-Gabor Wavelet Filter Bank

The 2D-Gabor wavelet transformation is a tool for a complete representation of an image in terms of radial frequency and orientation [24]. To highlight the blood vessels of different widths placed at different orientations in this work, we used a bank of 2D-Gabor wavelet filters to the preprocessed retinal image. Daugman et al. [25] have proposed that ensemble of a simple cell of visual cortex can be represented as a family of 2D-Gabor wavelets. The decomposition of an image, $f = I_c$, as wavelet transform is defined as

$$(T^{wav} f)(a, \theta, x_0, y_0) = \|a\|^{-1} \iint dxdyf(x, y)\psi_\theta\left(\frac{x - x_0}{a}, \frac{y - y_0}{a}\right), \tag{2}$$

where θ is the orientation parameter of the wavelet, and a is the parameter that defines the standard deviation in x and y directions. In Eq. (2),

$$\psi_\theta(a, x, y, x_0, y_0) = \|a\|^{-1}\psi_\theta\left(\frac{x - x_0}{a}, \frac{y - y_0}{a}\right) \tag{3}$$

represents the elementary function of the 2D wavelet rotated by an angle θ. Using the Gabor elementary function, the entire family of Gabor wavelets can be generated. Lee et al. [24] have derived a specific class of 2D-Gabor wavelets which is used in this paper to obtain a set of Gabor wavelets to process the retinal images. The Gabor wavelet which satisfies the neurophysiological restraint of simple cells is defined as

$$\psi(x, y, \omega_0, \theta, K) = \frac{\omega_0}{\sqrt{2\pi}K} \exp\left(-\frac{\omega_0^2}{8K^2}\left(4(xcos\theta + y\sin\theta)^2 + (-xsin\theta + ycos\theta)^2\right)\right) \tag{4}$$
$$\cdot \left[\exp\{i\omega_0 (xcos\theta + y\sin\theta)\} - \exp\left(-\tfrac{K^2}{2}\right)\right]$$

where, θ denotes the wavelet orientation in radians and ω_0 is the radial frequency in radian per unit length. The constant K tells about the frequency bandwidth of octave where $K = \pi$ is for a frequency bandwidth of one octave and $K \approx 2.5$ for frequency bandwidth of 1.5 octaves. Each Gabor wavelet filter at radial frequency ω_0 and orientation θ is centered at $(x = 0, y = 0)$ and normalized by L^2 norm.

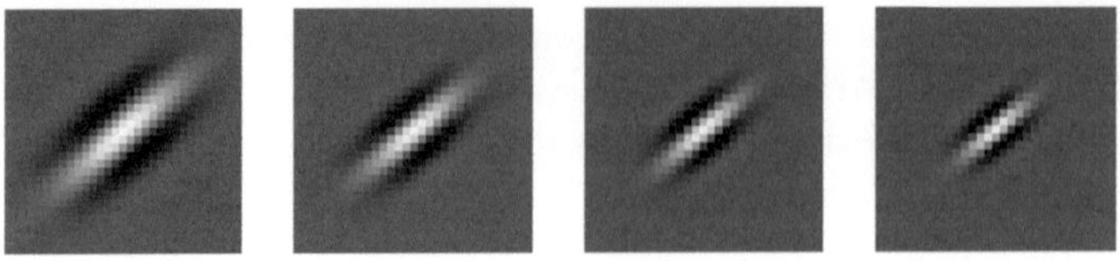

(a) $\omega_0 = 0.7$ (b) $\omega_0 = 0.9$ (c) $\omega_0 = 1.1$ (d) $\omega_0 = 1.3$

Fig. 4 Gabor wavelet filters at 50° orientation for four different radial frequencies

For each pixel in the retinal image, maximum Gabor wavelet outcome over all possible filters is stored for filtered image. If we consider $T_\psi(\omega_0, \theta, K)$ as the transformed retinal image at angular frequency ω_0 and orientation θ. Then, Gabor wavelet transformation result is obtained as

$$P(K) = \max_{\omega_0, \theta} \left| T_\psi(\omega_0, \theta, K) \right| \tag{5}$$

where θ is varied in the range of [0, 180] at an equal interval of 20°. The radial frequency is varied between [0.7, 1.5] at an interval of 0.2. Lee et al. [24] have suggested to choose K in the range of [2, 2.5]. In this paper, $K = 2.2$ is selected to accomplish better distinguishability between background and vessels. All these parameters are selected by performing few experiments on retinal images and Gabor wavelet filters. Figure 4 shows few Gabor wavelet filters from the bank of filters for different radial frequencies at orientation of 50°. The Gabor wavelet filter at $\omega_0 = 0.7$ is efficient to detect thick blood vessel. However, $\omega_0 = 1.3$ is suitable to detect thin blood vessels. The Gabor wavelet response of the preprocessed image is shown in Fig. 3e.

The hard segmented binary image as illustrated in Fig. 3f is obtained by applying Otsu thresholding on the filtered image P.

4 Results and Discussions

The presented algorithm is evaluated and compared with competitive algorithms using five different metrics: accuracy (Acc), sensitivity (Se), specificity (Sp), kappa agreement (κ), and area under the curve (A_z). All the metrics are computed using only pixels inside the FOV. Accuracy, sensitivity, and specificity are calculated using false positive (X), false negative (Y), true positive (Z), and true negative (W) values. Z denotes the number of pixels correctly identified as vessel pixels, and X denotes the number of pixels belongs to the background but wrongly identified as vessel pixel. W represents the number of pixels correctly identified as background pixels, and Y represents the number of pixels that belongs to the vessel but incorrectly assigned to background pixels. The evaluation metrics are measured using the following mathematical expressions as

$$Acc = \frac{Z + W}{Z + X + Y + W}, \qquad Se = \frac{Z}{Z + Y}, \tag{6}$$

$$Sp = \frac{W}{X + W}, \qquad \kappa = \frac{p_o - p_e}{1 - p_e}. \tag{7}$$

where p_e is the hypothetical probability of chance agreement and p_o is the relative perceived agreement. These agreement values can be calculated using the perceived data to estimate the probabilities of each observer randomly seeing all the classes. κ value varies in the range of [0, 1], where $\kappa = 0$ relates to no agreement between two rates, whereas $\kappa = 1$ relates to complete agreement between the rates. To compute A_z, Receiver Operating Characteristic (ROC) curve is attained by changing the global threshold between 1 and 0 in steps of 0.01. The global threshold divides the image into a binary image with labels 0 and 1. For each threshold, two performance measures false positive rate ($XR = 1 - Sp$) and true positive rate ($ZR = Sp$) are obtained by comparing the segmented binary image with the corresponding ground truth. Before applying the global threshold, the Gabor wavelet response values are normalized to 0–1 range.

The blood vessel segmentation outcome performances are shown in Fig. 3. All intermediate results and segmented output of the presented algorithm for a retinal image from DRIVE database are illustrated in Fig. 3. The segmentation performance of the presented algorithm on the DRIVE test data set is listed in Table 1.

The performance metrics illustrate that average segmentation accuracy, sensitivity, and specificity of the presented technique are 94.32%, 75.03%, and 97.12%, respectively. It can be noticed that a small deviation in accuracy ($\sigma = 0.0049$) is achieved. In DRIVE test data set, image 8 is a pathological image having exudates on which we have achieved 93.93 % accuracy with 0.7111 ZR. However, Zhao et al. [2] have reported 93.59 % of accuracy with 0.6524 ZR on the same image. The segmented blood vessels of image 8 are presented in Fig. 5d. It confirms that our proposed algorithm is efficient in segmenting the blood vessels in pathological images also. Average area under the curve, A_z, of 0.9524 is obtained with maximum and minimum A_z as 0.9729 and 0.9371, respectively. On DRIVE test data set, we achieved average kappa agreement 0.7374 with 0.0206 standard deviations.

A comparative analysis of the presented algorithm with competing methods is presented in Table 2 in terms of average accuracy, sensitivity, specificity, A_z, and κ agreement. The evaluation metrics values of different methods provided in the table are taken from the respective papers. Numerical results show that our proposed algorithm outperforms many unsupervised methods to DRIVE data set concerning accuracy and sensitivity. Furthermore, the proposed algorithm is better than few supervised techniques on DRIVE test data set concerning sensitivity and specificity.

Figure 5 shows the segmentation result of four retinal vessel images selected from the DRIVE test data sets. Among these four retinal images, three retinal images (Fig. 5a–c) are healthy retinal images with optic disc and fovea, whereas Fig. 5d is the pathological image having exudates, optic disc, and fovea. It can be observed that our proposed algorithm is competent in identifying thin as well as thick vessels in

Table 1 Segmented outcome of the presented work on the DRIVE test database

Image	Se	Sp	Acc	A_z	κ
1	0.8337	0.9592	0.9427	0.9687	0.7593
2	0.7781	0.9767	0.9469	0.9602	0.7837
3	0.7583	0.9639	0.9339	0.9432	0.7312
4	0.6619	0.9874	0.9440	0.9443	0.7282
5	0.7010	0.9837	0.9453	0.9468	0.7460
6	0.6975	0.9784	0.9388	0.9371	0.7279
7	0.7068	0.9737	0.9383	0.9431	0.7171
8	0.7111	0.9721	0.9393	0.9395	0.7120
9	0.7289	0.9771	0.9480	0.9525	0.7378
10	0.7021	0.9816	0.9483	0.9507	0.7354
11	0.7378	0.9660	0.9365	0.9412	0.7142
12	0.7936	0.9618	0.9408	0.9517	0.7364
13	0.6907	0.9798	0.9388	0.9481	0.7273
14	0.8094	0.9612	0.9433	0.9581	0.7386
15	0.7800	0.9672	0.9477	0.9566	0.7268
16	0.7225	0.9787	0.9451	0.9632	0.7440
17	0.7440	0.9684	0.9407	0.9451	0.7221
18	0.7920	0.9629	0.9432	0.9613	0.7300
19	0.8476	0.9713	0.9564	0.9729	0.7991
20	0.8080	0.9623	0.9459	0.9638	0.7306
Average	0.7503	0.9717	0.9432	0.9524	0.7374
Std. deviation	0.0499	0.0081	0.0049	0.0098	0.0206

Se—sensitivity, Sp—specificity, Acc—accuracy, A_z - area under the ROC curve, κ—kappa agreement

the presence of exudates, fovea, and optic disc. Besides, the proposed algorithm also preserves the connectivity of the blood vessels. The segmentation performance can be further improved by using some local adaptive thresholding technique instead of global thresholding to detect 1–2 pixel thick vessels.

5 Conclusions

Blood vessel segmentation in fundus images plays a vital role to detect different ocular diseases. To distinguish blood vessel from other structure like exudates, optic disc, fovea, etc., under uneven illuminance condition is difficult. In this paper, we have been proposed an entirely unsupervised approach for blood vessel segmentation using the retinal image and validated on DRIVE database. In preprocessing, retinal image is enhanced in the two steps. In the first step, white top-hat transform is

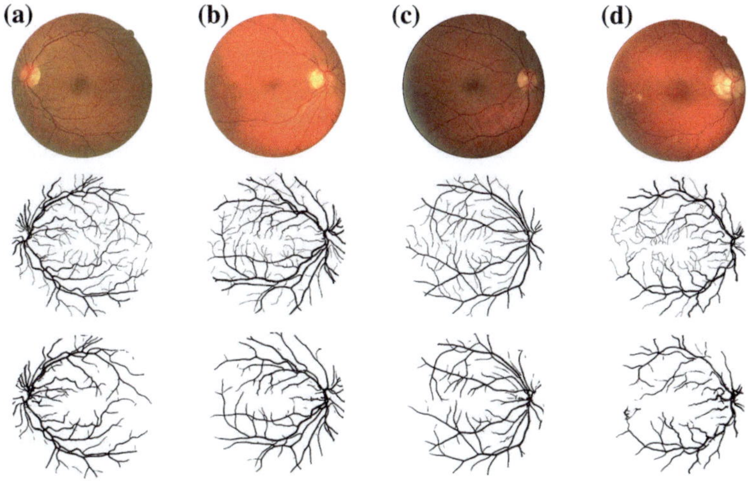

Fig. 5 Segmentation result of retinal images from the DRIVE database. First row: original image, Second row: golden standard ground truth, Third row: segmented vessel using the proposed algorithm. **a–c** Healthy retinal image with different pigmentations and illuminance, **d** pathological retinal image

Table 2 Comparative analysis of the presented work with existing approach on the DRIVE test data set with respect to golden standard ground truth image

Methods	Se	Sp	Acc (σ)	A_z	κ
Supervised methods					
Soares et al. [1]	0.7230	0.9762	0.9466 (–)	0.9614	–
Staal et al. [9]	0.7194	0.9773	0.9441 (–)	0.9520	–
Ricci et al. [12]	–	–	0.9595 (–)	0.9558	–
Lahiri et al. [26]	–	–	0.9530 (0.0030)	–	0.7090
Unsupervised methods					
2nd observer	0.7760	0.9725	0.9473 (–)	–	0.6970
Chaudhuri et al. [3][a]	–	–	0.8773 (0.0232)	0.7878	0.3357
Zana et al. [27][a]	–	–	0.9377 (0.0077)	0.8984	0.6971
Martinez-Parez et al. [28]	0.7246	0.9655	0.9344 (–)	–	–
Zhang et al. [5]	0.7120	0.9724	0.9382 (–)	–	–
Miri et.al. [18]	0.7352	0.9795	0.9458 (–)	–	–
Zhao et al. [2]	0.7354	0.9789	0.9477 (–)	–	–
Gou et al. [7]	0.7526	0.9669	0.9393 (–)	–	–
Presented method	0.7503	0.9717	0.9432 (0.0049)	0.9524	0.7374

[a]Results are taken from [29]

applied to enhance only the blood vessels and concurrently suppress all other structures which are larger than structuring element. In next step, the processed image is further enhanced using CLAHE algorithm. After preprocessing, multiscale Gabor wavelet filters are applied to emphasize the thick and thin vessels in the retinal image. A global threshold is obtained from the filtered image by using Otsu's thresholding technique. Experimental results clearly indicate that the proposed technique is efficient to recognize the blood vessels in the presence of exudates, fovea, and optic disc with the average accuracy of 94.32% with a small standard deviation of 0.0049. Moreover, the suggested algorithm in this paper is simple and easy to implement.

The outcome result of the proposed technique depends on the diameter selected as structuring element for the top-hat transform. If some other structures (like small microaneurysms) which look like vessels and have a thickness less than the width of the structuring element then our proposed method may fail to remove those structures in the segmented image. In future, the shape feature can be incorporated with Gabor wavelet response to discarding such kinds of the small structures. Furthermore, testing is to be performed on different retinal databases like STARE and CHASE_DB1 to test the robustness of the proposed method.

References

1. Soares, J.V., Leandro, J.J., Cesar, R.M., Jelinek, H.F., Cree, M.J.: Retinal vessel segmentation using the 2-D Gabor wavelet and supervised classification. IEEE Trans. Med. Imaging **25**(9), 1214–1222 (2006)
2. Zhao, Y.Q., Wang, X.H., Wang, X.F., Shih, F.Y.: Retinal vessels segmentation based on level set and region growing. Pattern Recogn. **47**(7), 2437–2446 (2014)
3. Chaudhuri, S., Chatterjee, S., Katz, N., Nelson, M., Goldbaum, M.: Detection of blood vessels in retinal images using two-dimensional matched filters. IEEE Trans. Med. Imaging **8**(3), 263–269 (1989)
4. Hoover, A., Kouznetsova, V., Goldbaum, M.: Locating blood vessels in retinal images by piecewise threshold probing of a matched filter response. IEEE Trans. Med. Imaging **19**(3), 203–210 (2000)
5. Zhang, B., Zhang, L., Zhang, L., Karray, F.: Retinal vessel extraction by matched filter with first-order derivative of Gaussian. Comput. Biol. Med. **40**(4), 438–445 (2010)
6. Liskowski, P., Krawiec, K.: Segmenting retinal blood vessels with deep neural networks. IEEE Trans. Med. Imaging **35**(11), 2369–2380 (2016)
7. Gou, D., Wei, Y., Fu, H., Yan, N.: Retinal vessel extraction using dynamic multi-scale matched filtering and dynamic threshold processing based on histogram fitting. Mach. Vis. Appl. **29**(4), 655–666 (2018)
8. Fraz, M.M., Remagnino, P., Hoppe, A., Uyyanonvara, B., Rudnicka, A.R., Owen, C.G., Barman, S.A.: Blood vessel segmentation methodologies in retinal images-a survey. Comput. Methods Programs Biomed. **108**(1), 407–433 (2012)
9. Staal, J., Abràmoff, M.D., Niemeijer, M., Viergever, M.A., Van Ginneken, B.: Ridge-based vessel segmentation in color images of the retina. IEEE Trans. Med. Imaging **23**(4), 501–509 (2004)
10. Franklin, S.W., Rajan, S.E.: Retinal vessel segmentation employing ANN technique by Gabor and moment invariants-based features. Appl. Soft Comput. **22**, 94–100 (2014)
11. Azzopardi, G., Strisciuglio, N., Vento, M., Petkov, N.: Trainable COSFIRE filters for vessel delineation with application to retinal images. Med. Image Anal. **19**(1), 46–57 (2015)

12. Ricci, E., Perfetti, R.: Retinal blood vessel segmentation using line operators and support vector classification. IEEE Trans. Med. Imaging **26**(10), 1357–1365 (2007)
13. Zhu, C., Zou, B., Zhao, R., Cui, J., Duan, X., Chen, Z., Liang, Y.: Retinal vessel segmentation in colour fundus images using extreme learning machine. Comput. Med. Imaging Graph. **55**, 68–77 (2017)
14. Sadek, I., Elawady, M., Shabayek, A.E.R.: Automatic classification of bright retinal lesions via deep network features (2017). arXiv:1707.02022
15. Rahebi, J., Hardalaç, F.: Retinal blood vessel segmentation with neural network by using gray-level co-occurrence matrix-based features. J. Med. Syst. **38**(8), 85 (2014)
16. Li, Q., You, J., Zhang, D.: Vessel segmentation and width estimation in retinal images using multiscale production of matched filter responses. Expert Syst. Appl. **39**(9), 7600–7610 (2012)
17. Sofka, M., Stewart, C.V.: Retinal vessel centerline extraction using multiscale matched filters, confidence and edge measures. IEEE Trans. Med. Imaging **25**(12), 1531–1546 (2006)
18. Miri, M.S., Mahloojifar, A.: Retinal image analysis using curvelet transform and multistructure elements morphology by reconstruction. IEEE Trans. Biomed. Eng. **58**(5), 1183–1192 (2011)
19. Hassan, G., El-Bendary, N., Hassanien, A.E., Fahmy, A., Snasel, V., et al.: Retinal blood vessel segmentation approach based on mathematical morphology. Proc. Comput. Sci. **65**, 612–622 (2015)
20. Roychowdhury, S., Koozekanani, D.D., Parhi, K.K.: Iterative vessel segmentation of fundus images. IEEE Trans. Biomed. Eng. **62**(7), 1738–1749 (2015)
21. Walter, T., Massin, P., Erginay, A., Ordonez, R., Jeulin, C., Klein, J.C.: Automatic detection of microaneurysms in color fundus images. Med. Image Anal. **11**(6), 555–566 (2007)
22. Dougherty, E.R., Lotufo, R.A.: Hands-On Morphological Image Processing, vol. 59. SPIE Press (2003)
23. Fathi, A., Naghsh-Nilchi, A.R.: Automatic wavelet-based retinal blood vessels segmentation and vessel diameter estimation. Biomed. Signal Process. Control **8**(1), 71–80 (2013)
24. Lee, T.S.: Image representation using 2D Gabor wavelets. IEEE Trans. Pattern Anal. Mach. Intell. **18**(10), 959–971 (1996)
25. Daugman, J.G.: Complete discrete 2D Gabor transforms by neural networks for image analysis and compression. IEEE Trans. Acoust. Speech Signal Process. **36**(7), 1169–1179 (1988)
26. Lahiri, A., Roy, A.G., Sheet, D., Biswas, P.K.: Deep neural ensemble for retinal vessel segmentation in fundus images towards achieving label-free angiography. In: 2016 IEEE 38th Annual International Conference of the Engineering in Medicine and Biology Society (EMBC), pp. 1340–1343. IEEE (2016)
27. Zana, F., Klein, J.C.: Segmentation of vessel-like patterns using mathematical morphology and curvature evaluation. IEEE Trans. Image Process. **10**(7), 1010–1019 (2001)
28. Martinez-Perez, M.E., Hughes, A.D., Thom, S.A., Bharath, A.A., Parker, K.H.: Segmentation of blood vessels from red-free and fluorescein retinal images. Med. Image Anal. **11**(1), 47–61 (2007)
29. Niemeijer, M., Staal, J., van Ginneken, B., Loog, M., Abramoff, M.D.: Comparative study of retinal vessel segmentation methods on a new publicly available database. In: Medical Imaging 2004: Image Processing, vol. 5370, pp. 648–657. International Society for Optics and Photonics (2004)

Spectral Smoothening Based Waveform Concatenation Technique for Speech Quality Enhancement in Text-to-Speech Systems

Soumya Priyadarsini Panda and Ajit Kumar Nayak

Abstract This work presents a spectral smoothening based concatenation technique for enhancing the quality of speech produced by Text-to-Speech systems. As, the hard waveform concatenation process may cause audible glitches at the segment boundaries affecting the overall quality of the produced speech, an optimal coupling based spectral smoothening approach is adopted to smoothen the spectral envelop of the produced speech for enhancing its quality. A number of experiments were performed to analyze the performance of the proposed technique for which different speech quality evaluation parameters are considered and the results are compared with the other concatenative techniques. The results obtained in all the experiments performed shows the effectiveness of the proposed text-to-speech conversion technique in producing high-quality results.

Keywords Speech synthesis · Waveform concatenation · Spectral smoothening · Optimal coupling · Speech quality

1 Introduction

The speech synthesis technology has shown significant progress in the last few years making it possible to provide a number of useful apps for human computer interaction [1]. The use of the Text-to-Speech (TTS) technology these days is not limited to speech synthesizers or screen readers; instead it has a large number of possible applications in designing intelligent talking computer systems, expert systems [2], language learning apps [3], pronunciation dictionaries, etc. The combination of speech processing technology with natural language processing (NLP) interfaces makes it possible to interact with mobile devices in a more convenient manner. Also,

S. P. Panda (✉)
Silicon Institute of Technology, Bhubaneswar, Odisha, India
e-mail: sppanda.cse@gmail.com

A. K. Nayak
Institute of Technical Education and Research, Bhubaneswar, Odisha, India
e-mail: ajitnayak@soauniversity.ac.in

B. Pati et al. (eds.), *Advanced Computing and Intelligent Engineering*, Advances in Intelligent Systems and Computing 1082,
https://doi.org/10.1007/978-981-15-1081-6_36

the advancement in the field of artificial intelligence enhances the technology for its better use in designing intelligent machines that can understand and produce human languages [4].

Text-to-Speech (TTS) systems take input text utterances and produce the desired spoken utterance for it. To achieve highly natural output speech majority of the available TTS systems uses the concatenative speech synthesis techniques [5, 6]. In case of a concatenative technique some pre-recorded speech samples are stored in a database and are concatenated to produce the desired output from entered text utterances [7–9]. While the concatenative techniques storing larger speech units like recorded words or syllables achieves highly natural-sounding speech, with smaller phoneme like units in database the quality of the produced speech is somehow compromised. The smaller the size of the units in the database more number of concatenations are required causing audible glitches at concatenation points [10, 11]. Therefore, smoothening the concatenation points is an important issue in concatenative speech synthesis techniques using phoneme level units in its database [12–14].

There are a number of scenarios available where the adjacent time-frames are spectrally discontinuous [15]. A number of speech enhancement algorithms are available for this purpose like- spectral restoration technique, filtering techniques, wavelet based methods and model based methods. There are a number of modern methods available these days for adjusting the spectral envelope of speech signals, which includes interpolation-based methods, optimal coupling based methods, etc. [16]. For enhancing the quality of the produced speech from concatenative technique with a limited database, smooth, adjusted, or interpolated spectral transitions are required between the concatenation points to best match the natural sound-to-sound transitions available in natural speech. The idea is to produce natural-sounding speech by concatenating recorded natural speech segments from a database. The formants and other spectral feature are then modified to match the naturally generated speech [17].

This work presents use of an optimal coupling technique to smoothen the concatenation points for producing high quality speech. The presented smoothening technique is tested for its performance under various concatenation contexts. The results of the experiments are also presented. The remainder of the paper presents the description of the WCT method in Sect. 2. Description of the adopted method in presented in Sect. 3 and the results obtained are discussed in Sect. 4. The possible future directions of the work are discussed in Sect. 5.

2 Waveform Concatenation Technique (WCT)

Majority of the available concatenative TTS systems stores language specific pronunciations in its database in any language. However, a large number of possible combinations of vowel and consonant sounds may be possible in a language. Therefore, a number of pre-recorded samples are required. To achieve uninterrupted output, all possible pronunciations in any Indian language are needed to be kept. Also, as pronunciations of a word in different Indian languages may be different, all possible

Table 1 Basic speech units considered

Vowel and consonant sound units considered						
a	o	cha	ttha	tha	pha	lla
aa	ka	chha	dda	da	ba	la
ee	kha	ja	ddha	dha	bha	sha
uu	ga	jha	nna	na	ma	ha
ae	gha	tta	ta	pa	ra	ya

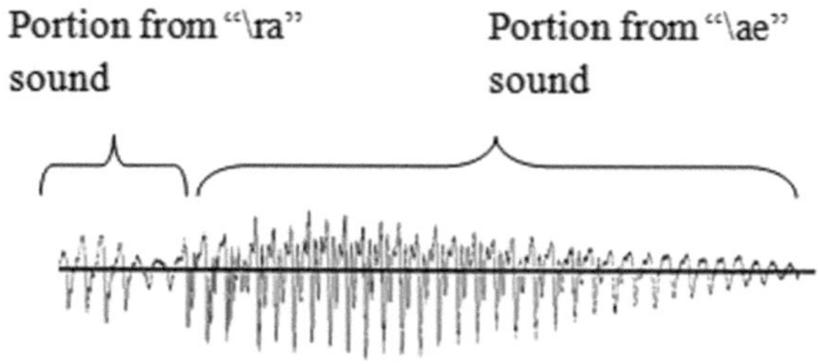

Fig. 1 Waveform concatenation process for "*/re*" (C-M) sound

pronunciations are needed to be considered. This database dependability makes a new language adaptation more difficult particularly in the Indian language context. However, WCT [16] uses a set of basic sounds. The list of considered basic vowel and consonant sound units are listed in Table 1.

A fraction-based concatenation process is followed. The fraction durations are determined dynamically from the speech data based on the vowel onset point identification technique. These fractions durations are considered for the waveform concatenation process. While the rule-based concatenative technique (RCT) uses a static fraction duration for concatenation the use of dynamic fraction durations in WCT enhances the quality of speech being produced. These dynamic fraction durations are considered for the dependent unit pairs such as Consonants attached to Matra/Fala/Halant/Consonants and the whole wavedata is used for producing the independent unit pairs like Consonants attached to Consonant/Vowel, Vowels attacched to Consonants/Vowels. Figure 1 presents the waveform concatenation method for "\re" unit using the WCT technique.

3 Spectral Smoothening Based Waveform Concatenation

This section presents the overview of the spectral smoothening based waveform concatenation technique and is shown in Fig. 2. The basic consonant and vowel sounds considered are recorded and stored in a database. The system takes input text in Indian languages. The text utterances are passes to the text pre processing phase

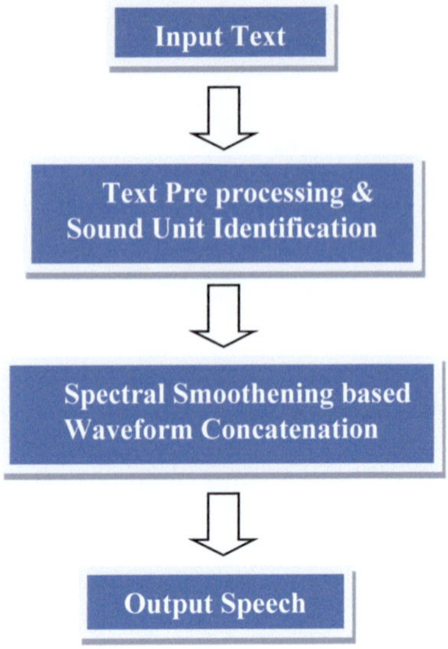

Fig. 2 The spectral smoothening based Text-to-speech conversion process

to identify the different consonant and vowel units. The speech identification phase then derives the text utterances. The written scripts are then mapped to pronounceable units. Based on the fraction percentage determined by the WCT method the desired fractions are extracted and concatenated to produce the required speech segments.

Normally in traditional concatenative techniques, the segment boundaries are fixed while, in case of the WCT a new transition frame is inserted in between the two segments being concatenated to retain the sound to sound transitions. The transition frame being created by centred average technique may not best fit in the spectral envelope of the concatenated segments. Therefore, use of the optimal coupling method makes improvement in the spectral component [17]. For the concatenation process, the segment boundary is chosen to best fit with its adjacent segments. For two segments to be concatenated, the distance measure is calculated at the boundaries for the end frame of segment-1(S1) and the beginning frame of segment-2 (S2). Figure 3 presents the example showing the changes in spectral envelop. The Algorithm for optimal coupling based smoothening is presented in Algorithm 1 [17]. After applying the spectral smoothening the final output speech is produced. This method makes formants at segment boundary to be naturally closer to each other. As the speech data is not modified in coupling, it does not introduce additional artifacts after concatenation.

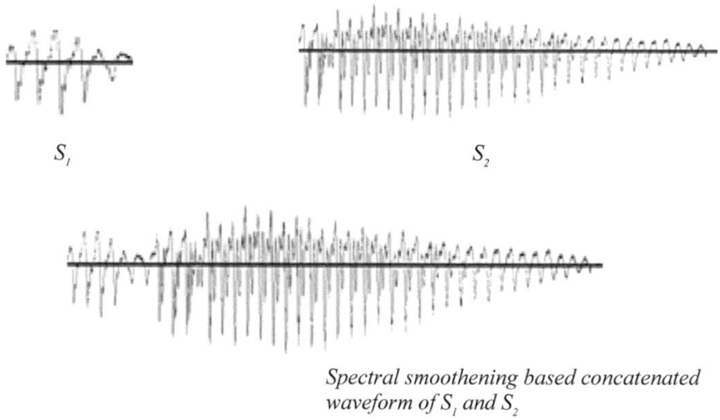

Fig. 3 Spectral smoothening at segment boundaries

Algorithm- 1:

Step-1:

For each segments S_1 to be concatenated with segment S_2, find the end frame $\{x_1, \ldots\ldots x_n\}$ of S_1 and the start frame $\{y_1, \ldots\ldots y_n\}$ of S_2

Step-2:

Compute the distance measure function by calculating the minimum i, j over any (x_i, y_j) as $Min_{i,j}\, d(x_i, y_j)$

---

4 Results Obtained

The performance of the model is analyzed with respect to fraction-based waveform concatenation and syllable-based technique covering different speech parameters. A number of experiments were performed on the output speech to evaluate the level of speech quality enhancement achieved. The speech parameters considered for the evaluation process are: Energy, auto co-relation, magnitude, and short term energy. The output speech signals are processed in a number of short frames with respect to each considered parameter. By varying the frame size, the spectral variations are analyzed with respect to the considered parameters. For this purpose the same speech units are recorded and are also produced by the two existing techniques.

The time domain representation for an example output is presented in Fig. 4. The variation in spectral component shows similarity of the produced speech compared to recorded unit for the same sound. An energy based evaluation is also carried out to evaluate the variation in energy levels. For this purpose, the energy values are calculated and plotted as the energy plot. There is a similarity also observed at energy

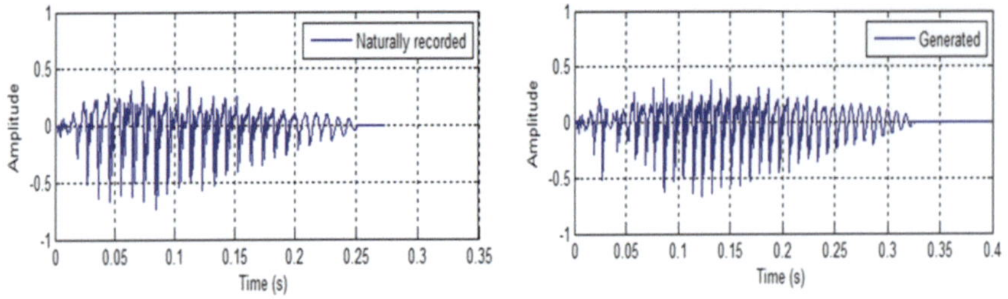

Fig. 4 Time domain analysis of "kaa" sound

component for each tests conducted. An example energy plot is presented in Fig. 5 for sample output by the proposed model and recorded unit. To better visualize the similarities the short term energy analysis is performed on different signal windows by varying the frame size from 20 to 35 ms. In each of these tests conducted, the output speech by the proposed model shows very close resemblance with the recorded units. Figure 6 presents the example short term analysis results on a 20 ms frame size.

To evaluate the spectral similarity, the time domain speech signals are converted into frequency domain and the magnitude spectrums are generated. The magnitude spectrum presents the frequency components information in speech signals. The results of the experiments performed shows similarity at frequency level. Figure 7 presents the example magnitude plot.

A subjective evaluation is also conducted by comparing the results. A five-point scale is used to evaluate the naturalness of the produced speech by a number of listeners. A number of speech samples are generated using the proposed technique.

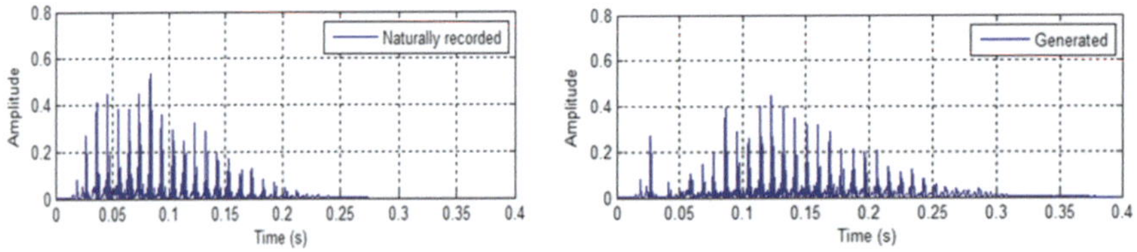

Fig. 5 Energy based analysis of "kaa" sound

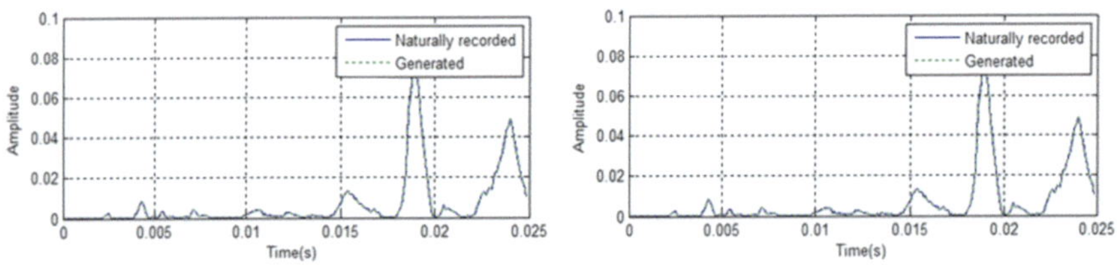

Fig. 6 Short term energy plot for "kaa" sound

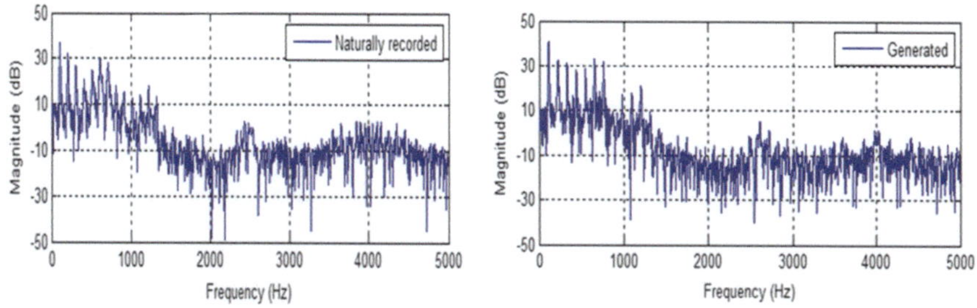

Fig. 7 Magnitude analysis for "kaa" sound

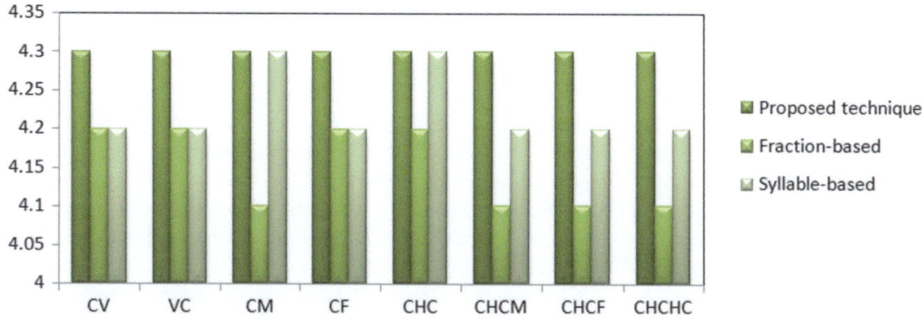

Fig. 8 MOS test results on output speech

The same sound units are also produced by the fraction-based concatenation and syllable-based techniques. The listeners are asked to give their feedback to each of the methods. The average scores recorded by listeners test for different techniques are presented in Fig. 8.

In each of the experiments performed, it has been observed that, the use of the spectral smoothening technique at concatenation points enhances the overall quality of the output speech. While the hard concatenation process being followed in fraction-based concatenation process produces audible glitches due to energy and spectral mismatch; the proposed smoothening technique, makes a smoother spectral envelope enhancing the quality of the produced speech.

5 Conclusions

This work describes a spectral smoothening based concatenation process for enhancing the output quality produced by the TTS system. The proposed model is evaluated based on a set of parameters and are compared with the existing techniques. The results show the enhancement in the quality. The model achieves remarkable results in all the experiments performed however, it may further be enhanced to achieve more naturally sounding speech units by incorporating some prosodic and intonation modeling techniques.

References

1. Feng, J., Ramabhadran, B., Hansel, J., Williams, J.D.: Trends in speech and language processing. IEEE Signal Process. Mag. (2012)
2. Panda, S.P., Nayak, A.K., Patnaik, S.: Text-to-speech synthesis with an Indian language perspective. Int. J. Grid Util. Comput. **6**(3–4), 170–178 (2015)
3. Raj, A.K., Sarkar, T., Pammi, S.C., Yuvaraj, S., Bansal, M., Prahallad, K., Black, A.W.: Text processing for text-to-speech systems in Indian languages. In: 6th ISCA Workshop on Speech Synthesis (2007)
4. http://tdil.mit.gov.in/ (2018)
5. Panda, S.P., Nayak, A.K.: Integration of fuzzy if-then rule with waveform concatenation technique for text-to-speech synthesis in Odia. In: Proceedings International Conference on Information Technology (ICIT), pp. 88–93. IEEE (2014)
6. Alías, F., Sevillano, X., Socoró, J.C., Gonzalvo, X.: Towards high-quality next-generation text-to-speech synthesis: a multidomain approach by automatic domain classification. Audio Speech Lang. Process. IEEE **16**(7) (2008)
7. OBrien, D., Monaghan, A.I.C.: Concatenative synthesis based on a harmonic model. Audio Speech Audio Process. IEEE, **9**(1), 11–20 (2001)
8. Tiomkin, S., Malah, D., Shechtman, S., kons, Z.: A hybrid text-to-speech system that combines concatenative and statistical synthesis units. Audio Speech Lang. Process. IEEE **19**(5) (2011)
9. Narendra, N.P., Rao, K.S., Ghosh, K., Vempada, R.R., Maity, S.: Development of syllable-based text to speech synthesis system in Bengali. Int. J. Speech Technol. **14**(3), 167–181 (2011)
10. http://dhvani.sourceforge.net/ (2018)
11. Panda, S.P., Nayak, A.K.: A rule-based concatenative approach to speech synthesis in indian language text-to-speech systems. In: Proceedings International Conference on Intelligent Computing, Communication and Devices, pp. 523–531. Springer, Berlin (2014)
12. Vuppala, A.K., Yadav, J., Chakrabarti, S., Rao, K.S.: Vowel onset point detection for low bit rate coded speech. Audio Speech Lang. Process. IEEE **20**(6), 1894–1903 (2012)
13. Panda, S.P., Nayak, A.K.: Automatic speech segmentation in syllable centric speech recognition system. Int. J. Speech Technol. **19**(1), 9–18 (2016)
14. Prasanna, S.R.M., Reddy, B.V.S., Krishnamoorthy, P.: Vowel onset point detection using source, spectral peaks, and modulation spectrum energies. IEEE Trans. Audio Speech Lang. Process. **17**(4), 556–565 (2009)
15. Panda, S.P., Nayak, A.K.: An efficient model for text-to-speech synthesis in Indian languages. Int. J. Speech Technol. **18**(3), 305–315 (2015)
16. Panda, S.P., Nayak, A.K.: A waveform concatenation technique for text-to-speech synthesis. Int. J. Speech Technol. **20**(4), 959–976 (2017)
17. Chappell, D.T., Hansen, J.H.L.: A comparison of spectral smoothing methods for segment concatenation based speech synthesis. Speech Commun. **36**(3–4), 343–373 (2002)

Breast Abnormality Detection Using LBP Variants and Discrete Wavelet Transform

M. K. Kaushik, Spandana Paramkusham and V. Akhilandeshwari

Abstract The second largest disease leading to the death of women is Breast Cancer. To decrease mortality rate in women breast cancer should be diagnosed early. Digital mammography screening plays vital role for early diagnosis of breast cancer. This work presents a method for two types of classification of breast tissues in mammograms. First type of classification uses texture descriptors like Local Binary Pattern (LBP) and its variants such as Local Quinary Pattern (LQP) and Local Ternary Pattern (LTP) to extract the features for the classification of mammograms into normal-abnormal. Second type of classification employs discrete wavelet transform and LBP variants to classify breast tissues into benign-malignant. SVM classifier is used for the validation of our methods. The work was carried out by considering 48 ROIs (Region of Interest) or breast tissues segmented from mammograms. We achieved 100% accuracy in classifying Region of interests into normal and abnormal using LBP and its variants. The highest accuracy of 82.19% is achieved in classifying abnormal Region of Interests into benign and malignant using discrete wavelet transform and LBP variants.

Keywords CAD · LBP · LQP · LTP · Classification · Normal · Abnormal

1 Introduction

Breast cancer is reported [1, 2] as the second leading source of cancer in women next to lung cancer. Cancer is the formation of tumor, i.e., group of cells that grow out of control because of damaged DNA. Breast cancer occurs when tumors are formed in breast tissues [3]. Mass can be categorized into benign or malignant. Benign masses are not cancerous and they will not spread to other portions of the body. Malignant

M. K. Kaushik · V. Akhilandeshwari
Department of Electronics and Communication Engineering, RGMCET, Nandyal, Kurnool (Dt), Andhra Pradesh, India

S. Paramkusham (✉)
Department of Electrical and Electronics Engineering, Bits Pilani, Hyderabad Campus, India
e-mail: spandanamadhav@gmail.com

B. Pati et al. (eds.), *Advanced Computing and Intelligent Engineering*,
Advances in Intelligent Systems and Computing 1082,
https://doi.org/10.1007/978-981-15-1081-6_37

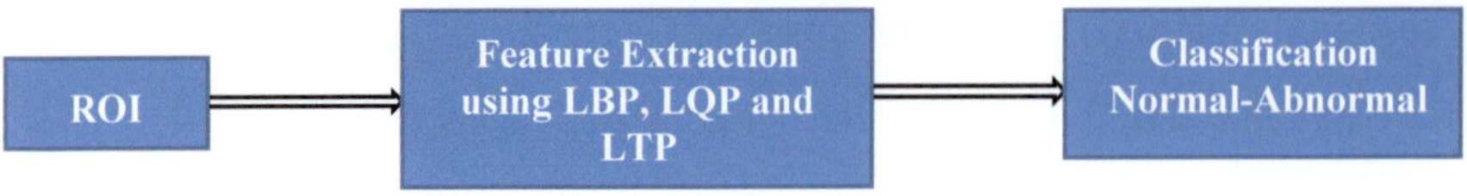

Fig. 1 Flow chart for classifying normal or abnormal from ROIs

masses are cancerous and these spread to other portions of the body. Malignant masses can grow more rapidly than benign [4]. Breast cancer usually occurs in ducts or lobules. The mortality rate can be reduced by digital mammography screening. Mammograms are produced using low energy X-rays for. It is very difficult to detect both normal and abnormal regions in mammograms if it is a dense breast. Thus, a computerized aided diagnosis (CAD) system is introduced for diagnosis of breast cancer. This CAD system gives better classification accuracy and gives radiologists a second opinion. So, with the help of this CAD system, wrong prediction of cancer can be reduced. Image processing and pattern recognition techniques are used in CAD system for detection and analysis of mammograms. The input to CAD system can be whole mammogram or any suspicious region of mammogram known as ROIs (Region of Interest). These images need to be pre-processed for noise reduction and improvement in image quality etc. Researchers have developed many feature extraction techniques for classification of mammograms. Methods for classification of mammograms into: fast finite shearlet transform [5], gray level co-occurrence and discrete wavelet transform [6], phylogenetic trees and LBP [7], LBP on curvelet coefficients [8], deep neural networks [9], fusion of discrete cosine transform and discrete wavelet transform [10], Zernike moments and SVM [11], curvelet transform [12],asymmetrical fractal analysis [13], taxnomic indexes [14] etc. In this context, LBP and its variants such as LTP and LQP are applied on breast cancer tissues and developed CAD system for categorizing the breast tissues into normal-abnormal and benign-malignant.. The flow chart in Fig. 1 shows the one stage classification of breast tissues into normal/abnormal. The flow chart for classifying abnormalities into benign/malignant using discrete wavelet transform is shown in Fig. 2. In this work, we made first attempt to use LQP and 2D-DWT for feature extraction.

1.1 Motivation and Contribution

This section includes the factors that motivated to work on the feature extraction from Mammogram ROIs. The first factor is there is need of new feature extraction methods that give high accuracy and reduce false positive cases. The second factor early detection of breast decreases mortality rate in women. Our contribution involves development of new feature extraction algorithms using LBP variants and discrete wavelet transform for classifying ROIs into benign-malignant as well as normal-abnormal.

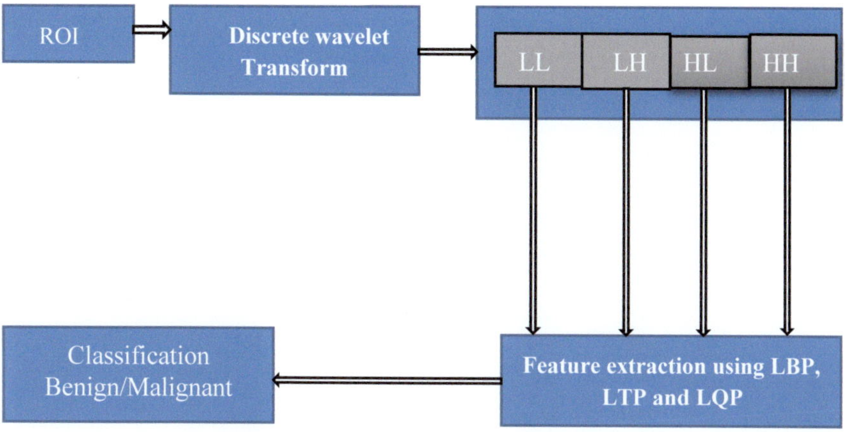

Fig. 2 Flow chart for classifying benign/malignant based on ROIs

The organization of papers is as follows: Sect. 2 concisely describes the methods of different LBP variants. Section 3 gives explanation of SVM classifier and tenfold classification. Section 4 discusses about the results and discussions followed by conclusion in Sect. 5.

2 Methods

In this section, we discuss about the LBP, LQP, and LTP feature extraction methods. For classification of mammogram we have implemented these techniques on mammogram ROIs. The ROIs have been cropped from mammograms. In the present work we have delineated ROIs into normal/abnormal and benign/malignant using the above mentioned techniques and discrete wavelet transform. Figure 3 shows normal and abnormal ROIs with mass.

LBP is a texture-based operator which labels each pixel based on the difference between the gray-level value of the center pixel and its neighbors [15, 16]. This operator effectively extracts the local and structural properties of ROIs

$$LBP_{(P,R)} = \sum_{p=0}^{p=P-1} s\left(g_p - g_c\right)2^P$$

$$s(x) = \begin{cases} 1; \text{ if x} \geq 0 \\ 0; \text{ otherwise} \end{cases} \quad (1)$$

where g_p is value of its neighbors, g_c is gray value of the center pixel, P is total number of neighbors involved and radius of the neighborhood is defined as R. Example of LBP is shown in Fig. 4.

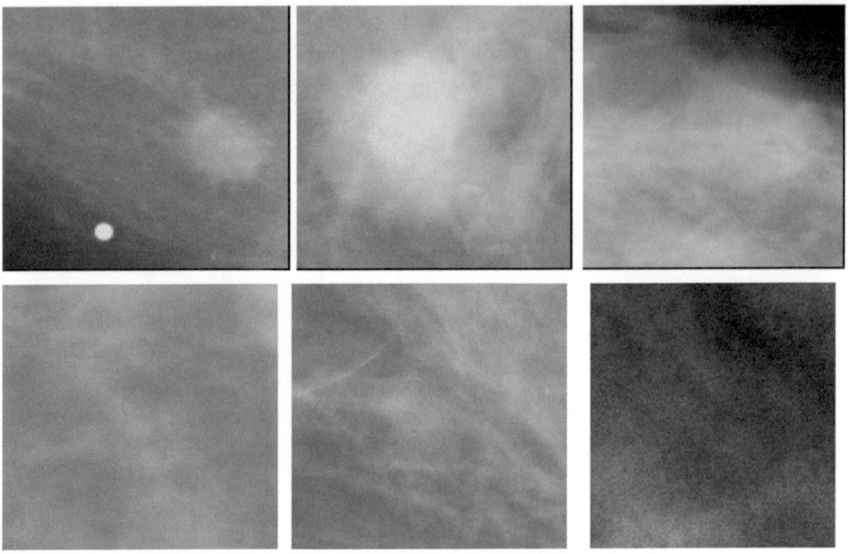

Fig. 3 IRMA Datbase containing masses,microcalcifications and normal ROIs

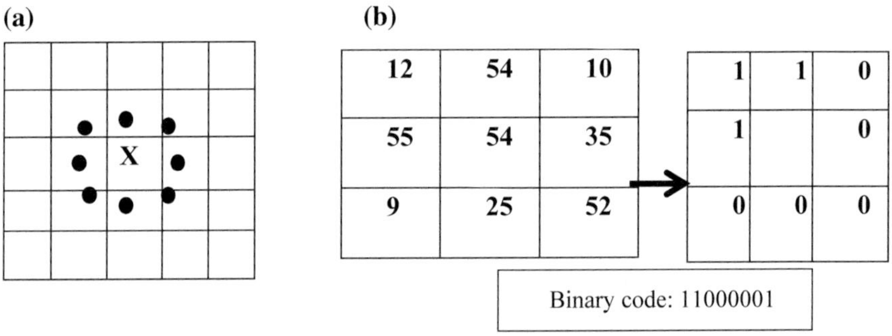

Fig. 4 **a** Circular symmetric neighbor for P = 8 and R = 1. **b** Explanation about LBP operator

2.1 *Local Ternary Pattern*

LTP labels pixel values of each image by computing the difference of center pixel and its neighbors. Based on the difference value and the user threshold, the center pixel is labeled as {1, 0,–1} [17]. The equation for the LTP coding is given below in Eq. 2

$$LTP_{(P,R)} = \sum_{p=0}^{p=P-1} s(g_p - g_c)2^p$$

where $s(x) = 1$ for $x >= \tau$

$$= 0 \text{ for } -\tau < \text{x} < \tau$$
$$= -1 \text{ otherwise} \quad (2)$$

2.2 *Local Quinary Pattern*

In LQP, the value obtained by subtracting center pixel and its neighbor pixels decides the labels for each pixel as in LBP and LTP. But in LQP, based on the difference value and with user thresholds ($\tau 1$, $\tau 2$) [18] the center pixel value is encoded into five values {–2,–1, 0, 1, 2}. LQP is given as in Eq. 3

$$\begin{aligned} \text{LQP}_{(\text{P,R})} &= \sum\nolimits_{p=0}^{p=P-1} s(g_p - g_c)2^p \\ \text{where } \text{s(x)} &= 2 \text{ for x} >= \tau \\ &= 1 \text{ for } \tau 1 < \text{x} < \tau 2 \\ &= 0 \text{ for } -\tau 1 \geq \text{x} < \tau 1 \\ &= -1 \text{ for} -\tau 2 \leq \text{x} < -\tau 1 \\ &= -2 \text{ for otherwise} \end{aligned} \quad (3)$$

In this work, different LBP variants have been implemented to extract features for distinguishing ROIs into normal/abnormal. For second type classification, first we have applied discrete wavelet transform using Daubechies wavelet to ROIs to decompose them LL, LH, HL, and HH component. Each LL, LH, HL gives information of low frequency components of image. Since, mass abnormality in ROI occupies major part of ROI and has consistent intensity value with in the mass occupied region. Hence, we acquired only LL, HL and LH components for further feature extraction. Then, we have applied LBP, LTP, and LQP variants on these wavelet components for extracting features. The extracted features are fed to SVM classifier for classification of abnormal ROIs into benign-malignant.

3 Classification

The features computed from all the above techniques are fed to SVM via tenfold cross validation to classify breast tissues into normal-abnormal and benign-malignant [19].

4 Results and Analysis

For the evaluation of proposed work, we have considered 48 ROIs taken from IRMA database [20] of which 24 are normal and 24 are abnormal ROIs. The features extracted from ROIs using LBP and its variants and 2D-DWT are fed to SVM classifier with linear kernel. LBP, LTP, and LQP have been applied on ROIs by considering multiple resolutions such as Radius (R) = 1 and neighbors (N) = 8 and Radius = 2 and neighbors = 16.

4.1 Classification of ROI into Normal/Abnormal

LBP variants achieved the classification accuracy, specificity, and sensitivity of 100% as seen in Tables 1, 2. From these tables we can observe that rotation invariant (RI) LBP achieved less accuracy compared to other variants. In overall cases, LBP average accuracy is 98% approximately. In LTP variants, we used two thresholds one is 3 and other is 5. Among LTP variants except rotational invariant (RI) and uniform (U2) LTP with threshold = 5, all the variants achieved 100% accuracy. The classification accuracy is approximately 98% (average) in overall cases. For LQP, we used two ranges of thresholds (th) one is [3, 5, 5, 7]. Among all the LQP variants, uniform rotation invariant (RIU2) LQP with R = 1, N = 8 and th = [3, 5] and R = 2, N = 16 and th = [5, 7] achieved accuracy of 100%. The evaluation parameters using LBP and its variant are tabulated in Tables 1, 2, 3, 4, 5, 6, 7, 8, 9, and 10.

Table 1 LBP-Local Binary pattern, R = 1, sampling points = 8

	(U2) (%)	(RI) (%)	(RIU2) (%)
Accuracy	100	100	100
Sensitivity	100	100	100
Specificity	100	100	100

Table 2 LBP-local Binary pattern, R = 2, sampling points = 16

	(U2) (%)	(RI) (%)	(RIU2) (%)
Accuracy	100	87.50	100
Sensitivity	100	97.5	100
Specificity	100	100	100

Table 3 LTP-local Ternary patterns, R = 1, sampling points = 8, Threshold = 3

	(U2) (%)	(RI) (%)	(RIU2) (%)
Accuracy	100	100	100
Sensitivity	100	100	100
Specificity	100	100	100

Table 4 LTP-local Ternary patterns, R = 1, sampling points = 8, Threshold = 5

	(U2) (%)	(RI) (%)	(RIU2) (%)
Accuracy	97.92	97.92	100
Sensitivity	95.83	95.83	100
Specificity	100	100	100

Table 5 LTP-local Ternary patterns, R = 2, sampling points = 16, Threshold = 3

	(U2) (%)	(RI) (%)	(RIU2) (%)
Accuracy	95.83	100	100
Sensitivity	91.67	100	100
Specificity	100	100	100

Table 6 LTP-local Ternary patterns, R = 2, sampling points = 16, Threshold = 5

	(U2) (%)	(RI) (%)	(RIU2) (%)
Accuracy	95.83	97.92	100
Sensitivity	91.67	95.83	100
Specificity	100	100	100

Table 7 LQP-local Quaternary Pattern, R = 1, sampling points = 8, Threshold = [3, 5]

	(U2) (%)	(RI) (%)	(RIU2) (%)
Accuracy	95.83	97.92	100
Sensitivity	100	100	100
Specificity	91.67	95.83	100

Table 8 LQP-local Quaternary Pattern, R = 1, sampling points = 8, Threshold = [5, 7]

	(U2) (%)	(RI) (%)	(RIU2) (%)
Accuracy	97.92	100	100
Sensitivity	95.83	100	100
Specificity	100	100	100

Table 9 LQP-local Quaternary Pattern, R = 2, sampling points = 16, Threshold = [3, 5]

	(U2) (%)	(RI) (%)	(RIU2) (%)
Accuracy	89.58	77.08	100
Sensitivity	83.33	66.67	100
Specificity	95.83	87.50	100

Table 10 LQP-local Quaternary Pattern, R = 2, sampling points = 16, Threshold = [3, 5]

	(U2) (%)	(RI) (%)	(RIU2) (%)
Accuracy	93.57	95.83	100
Sensitivity	87.50	91.67	100
Specificity	100	100	100

4.2 Classification of ROIs into Benign-Malignant

In this case, the evaluation is carried out using 24 abnormal ROIs of which 12 are benign and 12 are malignant masses. For classification of these abnormalities we have applied discrete wavelet transform using Daubhechies wavelet. It is two-step process. First step involves decomposition of ROI into LL, LH, HL and HH components. Then in second step, we have applied uniform (U2) LBP, rotation invariant (RI) LBP, uniform rotation invariant (RIU2) LBP, uniform LTP, uniform rotation invariant LTP, rotation invariant LTP, rotation invariant LQP, uniform rotation invariant LQP, and uniform LQP to wavelet components LL, LH and HL. We have applied multi-resolution LBP variants. With R = 1, N = 8 and R = 2, N = 16. For LTP, we used one threshold(th) = 3 and 5 and for LQP, we used two thresholds(th) = [3, 5]. These features extracted from LBP variants are given to SVM classifier. To validate this work 10-fold cross-validation technique using SVM classifier have been applied. In LBP, uniform LBP achieved sensitivity of 82.34%, accuracy of 82.19%, and specificity of 82.11%. In LTP, uniform rotation invariant LTP achieved best accuracy of 78.61% with th = 5, R = 2 and N = 16. In LQP, rotation invariant LQP achieved accuracy of 78.09% with [3, 5]. The evaluation parameter for classification of mass ROIs into benign/malignant are tabulated in Tables 11, 12, 13, 14, 15, 16, 17, and 18.

Table 11 LBP-Local Binary Pattern, R = 1, sampling points = 8

	(U2) (%)	(RI) (%)	(RIU2) (%)
Accuracy	82.19	71.14	79.09
Sensitivity	82.34	79.10	78.59
Specificity	82.11	63.10	79.24

Table 12 LBP-Local Binary Pattern, R = 2, sampling points = 16

	(U2) (%)	(RI) (%)	(RIU2) (%)
Accuracy	69.84	67.66	56.77
Sensitivity	35.34	48.22	29.16
Specificity	96.49	84.84	84.60

Table 13 LTP-Local Ternary patterns, R = 1, sampling points = 8, Threshold = 3

	(U2) (%)	(RI) (%)	(RIU2) (%)
Accuracy	56.33	79.83	74.46
Sensitivity	62.37	83.67	62.28
Specificity	51.63	76.56	85.80

Table 14 LTP-Local Ternary patterns, R = 1, sampling points = 8, Threshold = 5

	(U2) (%)	(RI) (%)	(RIU2) (%)
Accuracy	49.66	57.80	49.34
Sensitivity	19.27	24.99	36.65
Specificity	72.67	82.18	59.51

Table 15 LTP-Local Ternary patterns, R = 2, sampling points = 16, Threshold = 3

	(U2) (%)	(RI) (%)	(RIU2) (%)
Accuracy	51.51	51.52	66.51
Sensitivity	40.67	22.93	66.78
Specificity	58.92	71.60	65.17

Table 16 LTP-Local Ternary patterns, R = 2, sampling points = 16, Threshold = 5

	(U2) (%)	(RI) (%)	(RIU2) (%)
Accuracy	58.79	67.31	78.61
Sensitivity	44.51	56.37	81.53
Specificity	71.16	77.15	76.31

Table 17 LQP-Local Quaternary Pattern, R = 1, sampling points = 8, Threshold = [3, 5]

	(U2) (%)	(RI) (%)	(RIU2) (%)
Accuracy	56.62	77.89	72.34
Sensitivity	55.78	76.23	60.89
Specificity	50.24	72.53	75.77

Table 18 LQP-Local Quaternary Pattern, R = 1, sampling points = 8, Threshold = [5, 7]

	(U2) (%)	(RI) (%)	(RIU2) (%)
Accuracy	55.56	78.09	68.89
Sensitivity	46.61	75.46	56.65
Specificity	70.22	71.12	74.67

5 Conclusion

The proposed method uses LBP variants and 2D- DWT for classifying breast tissues. First type of classification distinguishes ROIs into normal and abnormal. Second type of classification involves delineating abnormal ROIs into benign and malignant. For first type we have applied LBP, LTP and LQP variants to extract features. For this classification, we achieved accuracy of 100% with LBP and LTP features. For second stage of classification, LBP and its variants are applied on wavelet components of abnormal ROIs. The highest accuracy of 82.19% is obtained by using uniform LBP.

In future, we mainly focus on extraction of abnormal breast tissues from whole mammograms and detecting whether that breast tissues are malignant or benign on a large database.

References

1. Li, Y., Chen, H., Cao, L., Ma, J.: A survey of computer-aided detection of breast cancer with mammography. J. Health Med. Inform. **7**(4) (2016)
2. www.Cancer.org/cancer/cancer-basics
3. Alghaib, Huda A., Scott, Melanie, Adhami, Reza R.: An overview of mammogram analysis. IEEE Potentials **35**(6), 21–28 (2016)
4. Rangayyan, Rangaraj M., Nguyen, Thanh M.: Fractal analysis of contours of breast masses in mammograms. J. Digit. Imaging **20**(3), 223–237 (2007)
5. Gedik, N.: A new feature extraction method based on multi-resolution representations of mammograms. Appl. Soft Comput. **1**(44), 128–133 (2016)
6. Beura, S., Majhi, B., Dash, R.: Mammogram classification using two dimensional discrete wavelet transform and gray-level co-occurrence matrix for detection of breast cancer. Neurocomputing **22**(154), 1–4 (2015)
7. de Sampaio, W.B., Silva, A.C., de Paiva, A.C., Gattass, M.: Detection of masses in mammograms with adaption to breast density using genetic algorithm, phylogenetic trees, LBP and SVM. Expert Syst. Appl. **42**(22), 8911–8928 (2015)
8. Bruno, D.O., do Nascimento M.Z., Ramos R.P., Batista V.R., Neves L.A., Martins A.S.: LBP operators on curvelet coefficients as an algorithm to describe texture in breast cancer tissues. Expert Syst. Appl. **5**(5), 329–340 (2016)
9. Lévy, D., Jain, A.: Breast mass classification from mammograms using deep convolutional neural networks (2016). arXiv:1612.00542
10. Uppal, M.T.: Classification of mammograms for breast cancer detection using fusion of discrete cosine transform and discrete wavelet transform features. Biomed. Res. (2016)
11. Sharma, S., Khanna, P.: Computer-aided diagnosis of malignant mammograms using Zernike moments and SVM. J. Digit. Imaging **28**(1), 77–90 (2015)
12. Gedik, N., Atasoy, A.: A computer-aided diagnosis system for breast cancer detection by using a curvelet transforms. Turkish J. Electr. Eng. Comput. Sci. **21**(4), 1002–1014 (2013)
13. Beheshti, S.M., Noubari, H.A., Fatemizadeh, E., Khalili, M.: Classification of abnormalities in mammograms by new asymmetric fractal features. Biocybern. Biomed. Eng. **36**(1), 56–65 (2016)
14. de Oliveira, F.S., de Carvalho Filho, A.O., Silva, A.C., de Paiva, A.C., Gattass, M.: Classification of breast regions as mass and non-mass based on digital mammograms using taxonomic indexes and SVM. Comput. Biol. Med. **1**(57), 42–53 (2015)
15. Ojala, Timo, Valkealahti, Kimmo, Oja, Erkki, Pietikäinen, Matti: Texture discrimination with multidimensional distributions of signed gray-level differences. Pattern Recogn. **34**(3), 727–739 (2001)
16. Ahonen, T., Hadid, A., Pietikainen, M.: Face description with local binary patterns: application to face recognition. IEEE Trans. Pattern Anal. Mach. Intell. **28**(12), 2037–2041 (2006)
17. Tan, Xiaoyang, Triggs, Bill: Enhanced local texture feature sets for face recognition under difficult lighting conditions. IEEE Trans. Image Process. **19**(6), 1635–1650 (2010)
18. Nanni, Loris, Lumini, Alessandra, Brahnam, Sheryl: Local binary patterns variants as texture descriptors for medical image analysis. Artif. Intell. Med. **49**(2), 117–125 (2010)
19. Hossin, M., Sulaiman, M.N.: A review on evaluation metrics for data classification evaluations. Int. J. Data Mining Knowl. Manag. Process **5**(2) (2015)
20. http://www.irma-project.org/index_en.php

Modified Histogram Segmentation Bi-Histogram Equalization

Mitra Montazeri

Abstract In the application of image processing, contrast enhancement is a major step. Conventional methods which are studied in contrast enhancement such as Histogram Equalization (HE) have not satisfactory results on many different low-contrast images and they also cannot automatically handle different images. These problems result of specifying parameters manually in order to produce high contrast images. In this paper, Modified Histogram Segmentation Bi-Histogram Equalization (MHSBHE) is proposed. In this study, histogram is modified before segmentation to improve the input image contrast. The proposed method accomplishes multi goals of preserving brightness, retaining the shape features of the original histogram and controlling excessive enhancement rate, suiting for applications of consumer electronics. MHSBHE avoids over-enhancement and generates images with natural improvement. Simulation results show that in terms of visual assessment, peak signal-to-noise (PSNR), average information content (entropy) and Absolute Mean Brightness Error (AMBE) the proposed method has better results compared to literature methods. The proposed method enhances the natural appearance of images especially in no static range images and the improved image is helpful in generation of the consumer electronic.

Keywords Image contrast enhancement · Histogram equalization · Histogram segmentation

---

M. Montazeri (✉)
Medical Informatics Research Center, Institute for Futures Studies in Health, Kerman University of Medical Sciences, Kerman, Iran
e-mail: mmontazeri@eng.uk.ac.ir

Computer Engineering Department, Shahid Bahonar University, Kerman, Iran

B. Pati et al. (eds.), *Advanced Computing and Intelligent Engineering*,
Advances in Intelligent Systems and Computing 1082,
https://doi.org/10.1007/978-981-15-1081-6_38

1 Introduction

Artificial intelligence has significant effect in different domains such as data mining [1–7], pattern recognition [8–12], machine learning [13–19] and image processing [20–23]. One of the applications of image processing is image enhancement [24]. In image contrast enhancement, numerous image enhancement techniques have been researched like a gray-level transformation techniques and histogram processing techniques. In the first group, these methods map the gray-level value in the image to the new one by using transformation function such as power-law transformation, logarithm transformation, etc. For example, in Ref. [25] proposed a method on the statistic image features. The proposed method which is a local, adaptive and multiscale takes the local average and local minimum/maximum in the window at the center of each pixel and then for each pixel identifies a transformation function. Another method in this group is on the 2D Taeger–Kaiser Energy Operator which is quadratic filter. This filter computes the average of the gray values at each pixel by the energy activity. A certain function transforms this value in order to enhance the pixel's contrast. After that the updated pixel is obtained by applying the reverse steps.

In histogram processing techniques, various studies have already been studied on histogram equalization. Histogram Equalization (HE) is such a method which is used in contrast Enhancement widely [26–28]. Achieving a uniform distributed histogram is the major aim of this method. This goal is done by applying the cumulative density function (CDF) of the input image [29]. One of the problems of HE is that it may cause to a washed-out looking, intensified noise and every undesirable artifacts. It can be verified that the mean brightness of the output image is placed at the center of original image gray level regardless of its mean. This property is annoying characteristic in the number of application where brightness preservation is needed [25].

To solve the aforementioned problems different methods such as mean preserving Bi-Histogram Equalization (BBHE) [30], Equal Area Dualistic Sub-Image Histogram Equalization (DSIHE) [31] and Recursive Mean-Spread Histogram Equalization (RMSHE) [29] have been proposed. Singh and Kapoor proposed [32] exposure based sub-image histogram equalization (ESIHE) method for low exposure image enhancement and to divide the image into sub-image, exposure threshold is used. Additional studies which are known as recursive exposure based sub-image histogram equalization (R-ESIHE) [33] applies, recursively ESIHE method until the exposure remnant among consecutive step is less than a certain threshold. The second method which is known as recursively separated exposure based sub-image histogram equalization (RS-ESIHE) [33] recursively applies the segregation of image histogram. In this method, each updated histogram is separated more rely on their correspondence exposure thresholds and each sub-histogram is equalized, exclusively. They have also proposed Median-Mean based sub-image clipped histogram equalization (MMSICHE) [34] algorithm, which partitions histogram based on median intensity and after that based on mean intensity, each sub-histograms are divided. Finally, each clipped sub-histograms are equalized.

Generally, methods based on the Histogram Equalization are grouped into two main groups: local and global [35]. In Global Histogram Equalization (GHE) [36], it is used the total image histogram for enhancement of input image. This method is good for equal frequency gray levels and it fails in image with very high-frequency gray levels. Because the image contrast is limited in high frequency gray levels, therefore, it leads to considerable contrast lack for gray levels with lower frequency [35]. To solve this drawback, local histogram equalization (LHE) is proposed [37–39]. In block-overlap histogram equalization [40] which is a LHE method used a windows placed on each pixel of image and HE is implemented only on sub-image that are encompassed in this windows. Then, the gray level of center pixel of window is mapped for enhancement. Shape preserving histogram modification [41] and Partially Overlapped Sub-Block Histogram Equalization (POSHE) [42] are different LHE methods. The only different between the mentioned methods and block-overlap histogram equalization is that in shape preserving histogram modification instead of rectangular window, it is used connected components and level set while in POSHE method the block size is increased in horizontal and vertical coordinate by the constant step size instead of one pixel like in block-overlap histogram equalization method. These approaches need a considerable computational cost and also it strengthens the noise of original image. Recently, combination of both LHE and GHE is proposed [43].

In this study, a Modified Histogram Segmentation Bi-Histogram Equalization (MHSBHE) is proposed. In this study, the histogram segmentation is modified based on average bins. The main contribution of MHSBHE is that it can handle images automatically with high brightness. Results of Simulation illustrate that MHSBHE outperforms recent existing methods in the literature in PSNR, entropy, AMBE and also visual assessment.

The rest of this paper is as follows: in Sect. 2 MHSBHE will be described. Finally, experimental results and conclusion are discussed in Sects. 3 and 4, respectively.

2 Proposed Method

In this section, we introduce Modified Histogram Segmentation Bi-Histogram Equalization (MHSBHE) method. MHSBHE is applied in three steps: histogram modification, histogram segmentation, sub-histogram equalization.

In first step, the histogram is modified before segmentation. In fact, this step is considerably helpful in the segmentation of histogram and is effective in brightness preservation. The past methods have not any modification in segmentation (Table 1).

In this way, the value of histogram bins is more than the average number of gray levels and they are confined to the threshold. The average value is calculated in (1) and (2):

Table 1 Quantitative analyses for six test images

Image contrast enhancement methods	Implementation steps
HE	1. HE
BBHE	1. HS based on the input image mean
	2. HE
DSIHE	1. HS based on density function
	2. HE
RMSHE	1. HS based on the input image mean, recursively
	2. HE
ESIHE	1. HC based on the average number of intensity occurrence
	2. HS based on exposure threshold
	3. HE
R-ESIHE	1. HC based on the average number of intensity occurrence, recursively until predefined threshold
	2. HS based on exposure threshold
	3. HE
MHSBHE (proposed method)	1. HS Modification
	2. HC based on exposure threshold
	3. HE

$$T_c = \frac{1}{L}\sum_{k=1}^{L} h(k) \tag{1}$$

$$h_c(k) = T_c \quad h(k) \geq T_c \tag{2}$$

where $h(k)$ and $h_c(k)$ are the input and clipped histogram, respectively.

In the second step, an exposure threshold [32] is applied to compute severity image exposure. This step splits the modified image in two sub-images, under-exposed and over-exposed sub-image. [0–1] is the normalized exposure value range. If this value is more than 0.5, it shows that the majority area of image is over-exposed and if this value is lower than 0.5 then image has majority of under-exposed area. Contrast enhancement should be done in both cases. This value is formulated as

$$exposure = \frac{1}{L}\frac{\sum_{k=1}^{L} h_c(k)k}{\sum_{k=1}^{L} h_c(k)} \tag{3}$$

where L is total gray levels number. In addition that parameter X_α (Eq. (4)) is introduced, which determines the gray level value threshold and splits the image into under and over-exposed sub-images.

This parameter obtains a value of larger (lower) than $L/2$ for exposure value lower (larger) than 0.5 for an image with a running range of 0 to L.

$$X_{\alpha} = L(1 - \text{exposure}) \tag{4}$$

Finally, in step three, HE is implemented on sub-histograms. In such process, the original image histogram is divided rely on exposure threshold value X_{α} as formulated in (4) and its results are two sub-images I_L and I_U from 0 to X_{α} gray level and $X_{\alpha} + 1$ to $L - 1$ gray level. This is known as under and over-exposed sub-images. $P_L(k)$, $P_U(k)$, $C_L(k)$, and $C_U(k)$ are related to Probability Density Function (PDF) and Cumulative Density Function (CDF) of these sub-images, respectively and they are defined in (5)–(8).

$$P_L(k) = h_c(k)/N_L, \quad k = 0 \ldots X_{\alpha}. \tag{5}$$

$$P_U(k) = h_c(k)/N_U, \quad k = X_{\alpha} + 1 \ldots L - 1. \tag{6}$$

$$C_L(k) = \sum_{k=0}^{X_{\alpha}} P_L(k), \tag{7}$$

$$C_U(k) = \sum_{k=X_{\alpha}+1}^{L-1} P_U(k), \tag{8}$$

where N_L and N_U are pixels number in sub-images I_L and I_U, respectively.

Equalization is implemented on two sub-histograms, individually. For histogram equalization, the transfer functions can be defined as

$$F_L = X_{\alpha} \times C_L \tag{9}$$

$$F_U = (X_{\alpha} + 1) + (L - X_{\alpha} + 1)C_U \tag{10}$$

F_L and F_U are the transfer functions are applied for equalizing the sub-histograms, exclusively. Finally, two sub-images combine into one full image. The enhanced image is generated by merging two transfer functions.

3 Experimental Results

The simulation results of the proposed method, MHSBHE, are presented and compared to six well-known literature works, i.e., HE, BBHE [30], DSIHE [31], RMSHE [29], ESIHE [32] and R_ESIHE [33]. To analyze these methods, 400 test images are used. Visual quality comparison of two images i.e. Road, Mass are illustrated in Figs. 1 and 2.

To measure the function of MHSBHE, Entropy is being used as image quality measure to evaluate enhanced image [33]. Higher entropy value shows greater information content is accessible in the image. Equation (11) calculates Entropy

$$\mathrm{Ent}(p) = -\sum_{I=0}^{L-1} P(I)\log P(I) \tag{11}$$

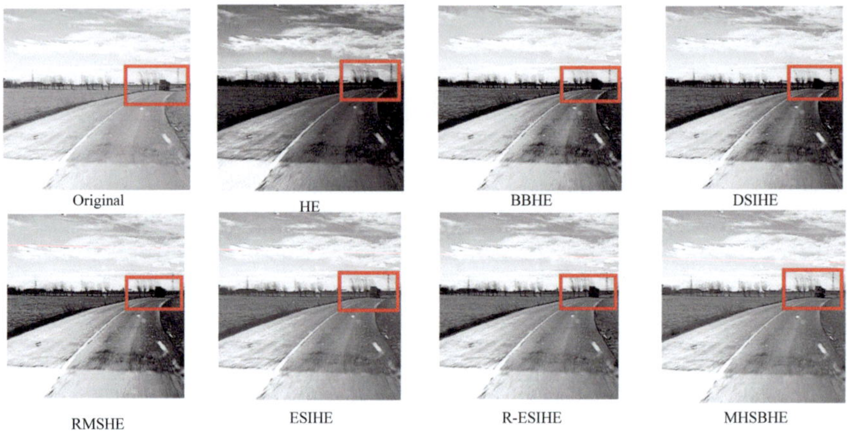

Fig. 1 Enhancement results of road image

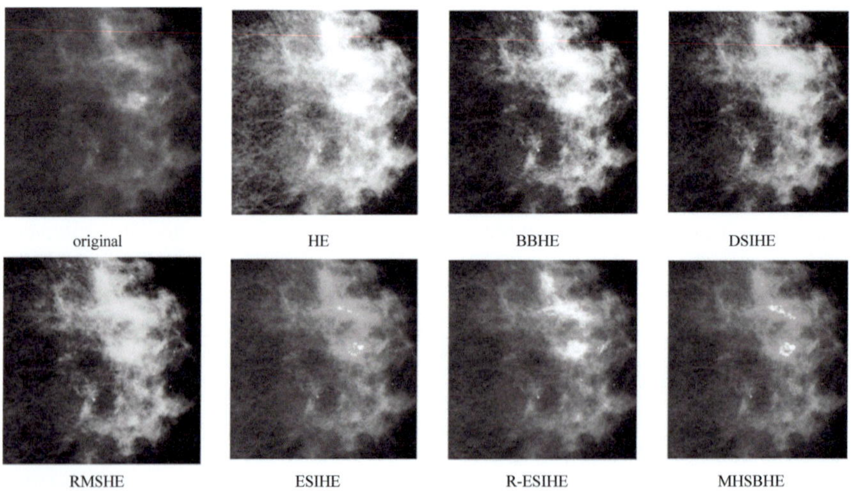

Fig. 2 Enhancement results of mass image

where $P(I)$ shows PDF of image at intensity level I.

In addition that entropy is measured in units as bits and can be as a criteria of affluence of the image details. Referred to Shannon Entropy, this entropy measures the uncertainty related to image's gray levels. An image with a superior entropy value has the affluence of details and is assumed to have superior quality.

To evaluate the performance of MHSBHE, AMBE [44] is used. AMBE is a useful in calculating the brightness preservation level. The AMBE between two input and enhanced image is computed as follows:

$$\mathrm{AMBE}(M.N) = |M\_M - N\_M| \tag{12}$$

where M and N are input and output image, respectively. Also, M_M and N_M are the mean of the two original and enhanced image, respectively. If the mean difference is less, then the input image's brightness is preserved in the enhanced image.

Lastly, P_{snr} measures the peak signal-to-noise of the enhanced image. Regarding to noise expanding problem during the enhancement, PSNR quantifies the quality of an enhanced image

$$P_{\mathrm{snr}}(I(c)) = \frac{10 \times \log_{10}(L-1)^2}{\mathrm{MSE}}, \tag{13}$$

3.1 Performance Assessment

For comparison, accuracy measurement is necessary between MHSBHE and literature work based on the PSNR, entropy, and AMBE for 400 benchmark images. Table 2 shows quantitative analyses for two test images. MHSBHE produces highest values in most cases. Beside this comparison, MHSBHE implemented on 400 images on different databases such as USC-SIPI (Misc and Sequences), USF-DM, Astronomical images, Medical images, Miscellaneous, etc. The comparison results are presented in Table 3. As it can be seen in this Table, the proposed method, MHSBHE, has better results in all measurements.

3.2 Assessment of Visual Quality

Finally, the methods are compared based on image visual assessment. The enhanced images which are resulted after applying the MHSBHE and the mentioned method are demonstration in Figs. 1 and 2. As shown in these enhanced images, MHSBHE has better natural appearance and high contrast images.

Table 2 Quantitative analyses for six test images

Test images	PSNR	Entropy	AMBE
Test image road			
HE	9.9755	5.9461	72.3376
BBHE	15.1492	0.008	27.699
DSIHE	14.693	6.8269	31.9066
RMSHE	15.6982	6.8514	26.4207
ESIHE	19.0273	6.975	22.381
R_ESIHE	18.6507	6.9619	23.0011
MHSBHE	**21.3353**	**6.9877**	**16.9791**
Test image mass			
HE	10.2326	5.8887	64.2599
BBHE	14.7022	0.0005	20.7681
DSIHE	15.651	6.5896	20.8272
RMSHE	16.2275	6.5724	17.6217
ESIHE	20.9084	6.7167	17.4432
R_ESIHE	20.054	6.6978	18.043
MHSBHE	**22.8458**	**6.7237**	**13.2481**

Table 3 Average values of quantitative analyses for 400 test images

Test images	PSNR	Entropy	AMBE
HE	14.3642	5.18160	29.45891
BBHE	16.9769	0.01453	13.94287
DSIHE	18.8017	5.78793	11.49561
RMSHE	18.6658	5.81110	12.58389
ESIHE	22.6868	5.92269	13.56786
R_ESIHE	21.8907	5.88817	11.83819
MHSBHE	**23.1671**	**5.92385**	**10.95996**

Obviously, in Fig. 1 of Road image, it is shown that the MHSBHE image improves the Truck in the image, effectively. This enhancement obviously can be seen compared to other methods. In Fig. 2, by applying MHSBHE, an extreme contrast of the results in contrast enhancement as well as natural appearance, can be obviously observed in this figure. Results of other methods enhance the noise. However, MHSBHE image manages on over-enhancement which cause to desirable contrast enhancement outputs.

Although the MHSBHE results in some images are visually comparable to literature approaches, MHSBHE gives considerably the highest PSNR, entropy and AMBE value for these images. This shows that the MHSBHE method generates enhanced images with preserving brightness, retaining the shape features of the original histogram and control over enhancement rate.

3.3 *Summary of Assessment and Discussion*

By visually inspecting the enhanced images and assessment of PSNR, entropy and AMBE measures, it can be summarized that

(i) MHSBHE technique is the best among other methods in terms of maximum signal value of the image (PSNR) and high richness of details (entropy) and the degree of brightness preservation (AMBE).
(ii) MHSBHE is robust against the noise compared to other methods which enhance noise during enhancement.
(iii) MHSBHE performs well on the images with high dynamic range with the low and high illumination.
(iv) MHSBHE generates images with high contrast enhancement and manage on over-enhancement.

In this proposed method, it produces images which are quantitatively better in quality compared to other literature methods.

4 Conclusion

In this study, the Modified Histogram Segmentation Bi-Histogram Equalization was proposed. In this study, MHSBHE was applied in three steps: histogram modification, histogram segmentation, sub-histogram equalization. The histogram segmentation was modified based on average bins. The main motivation of MHSBHE is that it can handle images automatically with high brightness.

MHSBHE is suitable for a wide variety of images with low-contrast. Also, the proposed method can control various images, automatically. This method attains multi objective of preserving brightness, maintaining the shape features of the original histogram and controlling over-enhancement rate, suiting for applications of consumer electronics.

MHSBHE eschewed over-enhancement and generated images with natural enhancement. In experimental results, the proposed method was applied on 400 standard images and it outperformed based on four criteria: PSNR, entropy, AMBE and visual assessment. In addition that the results showed that MHSBHE is applicable for consumer electronic products.

References

1. Madadizadeh, F., Bahrampour, A., Mousavi, S.M., Montazeri, M.: Using advanced statistical models to predict the non-communicačble diseases. Iranian J. Public Health **44**(12), 1714–1715 (2015)
2. Ehtemam, H., Montazeri, M., Khajouei, R., Hosseini, R., Nemati, A., Maazed, V.: Prognosis and early diagnosis of ductal and lobular type in breast cancer patient. Iranian J. Public Health **46**(11), 1563 (2017)
3. Montazeri, M., Naji, H.R., Faraahi, A.: A novel memetic feature selection algorithm, pp. 295–300
4. Montazeri, M., Naji, H.R., Montazeri, M.: Memetic feature selection algorithm based on efficient filter local search. J. Basic Appl. Sci. Res. **3**(10), 126–133 (2013)
5. Madadizadeh, F., Asar, M.E., Bahrampour, A., Montazeri, M.: Liver Disease Recognition: A Discrete Hidden Markov Model Approach (2016)
6. Madadizadeh, F., Montazeri, M., Bahrampour, A.: Predicting the survival in breast cancer using hidden Markov model, pp. 228–228
7. Madadizadeh, F., Montazeri, M., Bahrampour, A.: Predicting of liver disease using hidden Markov model. Razi J. Med. Sci. **23**(146), 66–74 (2016)
8. Mitra, M., Bahrololoum, A., Nezamabadi-pour, H., Baghshah, M.S., Montazeri, M.: Cooperating of Local Searches based Hyperheuristic Approach for Solving Traveling Salesman Problem, pp. 329–332
9. Montazeri, M.: HHFS: Hyper-heuristic feature selection. Intell. Data Anal. **20**(4), 953–974 (2016)
10. Montazeri, M., Baghshah, M.S., Niknafs, A.: Selecting efficient features via a hyper-heuristic approach (2016). arXiv:1601.05409
11. Montazeri, M., Naji, H.R., Montazeri, M., Faraahi, A.: A novel memetic feature selection algorithm, pp. 295–300
12. Montazeri, M., Nezamabadi-pour, H., Bahrololoum, A.: Exploring and exploiting effectively based hyper-heuristic approach for solving travelling salesman problem
13. Abbasi, R., Montazeri, M., Zare, M.: A Rule Based Classification Model to Predict Colon Cancer Survival
14. Montazeri, M., Baghshah, M.S., Enhesari, A.: Hyper-Heuristic algorithm for finding efficient features in diagnose of lung cancer disease (2015). arXiv:1512.04652
15. Montazeri, M., Montazeri, M.: Machine learning models for predicting the diagnosis of liver disease. Koomesh **16**(1), 53–59 (2014)
16. Montazeri, M., Montazeri, M., Beygzadeh, A., Zahedi, M.J.: Identifying efficient clinical parameters in diagnose of liver disease. Health MED **8**(10), 1115 (2014)
17. Montazeri, M., Montazeri, M., Montazeri, M.: Future studies in health care: a new approach in intelligent diagnosis of liver disease by selecting the best decision tree model
18. Montazeri, M., Montazeri, M., Montazeri, M., Bahrampour, A.: Can Breast Cancer Survival be predicted by Risk Factors? Machine Learning Models. pp. 301–301
19. Montazeri, M., Montazeri, M., Montazeri, M., Beigzadeh, A.: Machine learning models in breast cancer survival prediction. In: Technology and Health Care, no. Preprint, pp. 1–12 (20150
20. Montazeri, M., Bahaadinbeigy, K., Rahnama, Z., Montazeri, M.: Comparison of the accuracy of digital image-based and patient visit-based diagnoses in an Iranian dermatology clinic. J. Basic Appl. Sci. Res. **3**(11), 28–33 (2013)
21. Montazeri, M., Montazeri, M., Saryazdi, S.: Eye detection in digital images: challenges and solutions (2016). arXiv:1601.04871
22. Montazeri, M., Nezamabadi-pour, H.: Automatic extraction of eye field from a gray intensity image using intensity filtering and hybrid projection function. pp. 1–5
23. Montazeri, M., Nezamabadi-pour, H., Montazeri, M.: Automatically eye detection with different gray intensity image conditions. Comput. Technol. Appl. **3**(8) (2012)

24. Montazeri, M.: Intensity adjustment and noise removal for medical image enhancement. J. Health Biomed. Inform. **3**(1), 38–47 (2016)
25. Yu, Z., Bajaj, C.: A Fast and Adaptive Method for Image Contrast Enhancement. pp. 1001–1004
26. Parihar, A.S., Verma, O.P., Khanna, C.: Fuzzy-contextual contrast enhancement. IEEE Trans. Image Process. **26**(4), 1810–1819 (2017)
27. Parihar, A.S., Verma, O.P.: Contrast enhancement using entropy-based dynamic sub-histogram equalisation. IET Image Proc. **10**(11), 799–808 (2016)
28. Chang, Y., Jung, C., Ke, P., Song, H., Hwang, J.: Automatic contrast-limited adaptive histogram equalization with dual gamma correction. IEEE Access **6**, 11782–11792 (2018)
29. Chen, S.-D., Ramli, A.R.: Contrast enhancement using recursive mean-separate histogram equalization for scalable brightness preservation. IEEE Trans. Consum. Electron. **49**(4), 1301–1309 (2003)
30. Kim, Y.-T.: Contrast enhancement using brightness preserving bi-histogram equalization. IEEE Trans. Consum. Electron. **43**(1), 1–8 (1997)
31. Wang, Y., Chen, Q., Zhang, B.: Image enhancement based on equal area dualistic sub-image histogram equalization method. IEEE Trans. Consum. Electron. **45**(1), 68–75 (1999)
32. Singh, K., Kapoor, R.: Image enhancement using exposure based sub image histogram equalization. Pattern Recogn. Lett. **36**, 10–14 (2014)
33. Singh, K., Kapoor, R., Sinha, S.K.: Enhancement of low exposure images via recursive histogram equalization algorithms. Optik-Int. J. Light Electron Opt. **126**(20), 2619–2625 (2015)
34. Singh, K., Kapoor, R.: Image enhancement via median-mean based sub-image-clipped histogram equalization. Optik-Int. J. Light Electron Opt. **125**(17), 4646–4651 (2014)
35. Abdullah-Al-Wadud, M., Kabir, M.H., Dewan, M., Chae, O.: A dynamic histogram equalization for image contrast enhancement. IEEE Trans. Consum. Electron. **53**(2), 593–600 (2007)
36. Gonzalez, R.C., Woods, R.E.: Digital Image Processing, p. 954. Prentice Hall (2008)
37. Gonzalez, R.C., Woods, R.E.: Digital Image Processing, 2nd edn., pp. 85–103. Addison-Wesley,Reading, MA (1992)
38. Wan, M., Gu, G., Qian, W., Ren, K., Chen, Q., Maldague, X.: Particle swarm optimization-based local entropy weighted histogram equalization for infrared image enhancement. Infrared Phys. Technol. **91**, 164–181 (2018)
39. Zhou, X.-B., Wen, Y.-C., Jin, Y.-F., Syu, S.-S., Jou, M.-J.: P-57: A Local Histogram Framework for Contrast Enhancement. pp. 1410–1413
40. Kim, T.K., Paik, J.K., Kang, B.S.: Contrast enhancement system using spatially adaptive histogram equalization with temporal filtering. IEEE Trans. Consum. Electron. **44**(1), 82–87 (1998)
41. Caselles, V., Lisani, J.-L., Morel, J.-M., Sapiro, G.: Shape preserving local histogram modification. IEEE Trans. Image Process. **8**(2), 220–230 (1999)
42. Kim, J.-Y., Kim, L.-S., Hwang, S.-H.: An advanced contrast enhancement using partially overlapped sub-block histogram equalization. IEEE Trans. Circuits Syst. Video Technol. **11**(4), 475–484 (2001)
43. Gautam, G., Mukhopadhyay, S.: Efficient contrast enhancement based on local–global image statistics and multiscale morphological filtering. In: Advanced Computational and Communication Paradigms, pp. 229–238. Springer, Berlin (2018)
44. Kim, M., Chung, M.G.: Recursively separated and weighted histogram equalization for brightness preservation and contrast enhancement. IEEE Trans. Consum. Electron. **54**(3) (2008)

A Comparative Analysis on Filtering Techniques Used in Preprocessing of Mammogram Image

Sushreeta Tripathy and Tripti Swarnkar

Abstract Breast cancer is one of the vital causes of an increase in the rate of mortality of women in developing countries. Detection of micro-calcifications in the breast tissue by the radiologists is a significant step in the process of identification of cancer in the breast. Preprocessing in mammogram has played a vital role in identifying such micro-calcifications, masses and architectural distortion in the breast. It is important in the process of breast cancer analysis as it helps in reducing the number of false positive. The digital mammogram has emerged as the popular screening approach for early detection of masses and other abnormalities in the breast. In this manuscript, we introduced four standard filtering methods, which improve the efficiency of the model used for early detection of breast cancer. To compare the performance of the studies filter methods the mean square error and peak signal to noise ratio are widely used performance measure being considered in our work.

Keywords Mammography · Computer-Aided Diagnosis (CAD) · Breast cancer · Pre-processing · Mean Square Error (MSE) · Microcalcification · Peak Signal to Noise Ratio (PSNR)

1 Introduction

Computer technology has an incredible impact on medical imaging. Micro-calcification in the breast is a malignant tumor and gradually it spreads to distinct areas of the human body. Breast cancer is a common disease found in women but in some cases, it is also found in men. Breast screening has been used widely for

S. Tripathy (✉)
Department of Computer Science & Information Technology, Siksha 'O'Anusandhan Deemed to Be University, Bhubaneswar, Odisha, India
e-mail: sushreetatripathy@soa.ac.in

T. Swarnkar
Department of Computer Application, Siksha 'O'Anusandhan Deemed to Be University, Bhubaneswar, Odisha, India
e-mail: Triptiswarnakar@soa.ac.in

B. Pati et al. (eds.), *Advanced Computing and Intelligent Engineering*, Advances in Intelligent Systems and Computing 1082,
https://doi.org/10.1007/978-981-15-1081-6_39

detecting masses in the breast in its early stage [1]. Mammograms is the approach being widely used among the radiologists for diagnosis of breast cancer [2]. With the use of low dose x-ray in film mammography, it is painful to detect the normal and cancer tissue in the images. The digital mammography, improve the efficiency of detection of malignancy in mammograms [3]. The use of computer technology in the area of different medical image preprocessing can assists the radiologist in accurate diagnoses [4]. From the point of view of both the researcher and the radiologist, CAD systems have drawn more attention to associated challenging research and clinical application [5–8].

The digital picture normally involves image restoration. Restoration is technique of withdrawal of mortification that is induced during the picture capturing. Pictures get corrupted due to blurring as well as noise due to the electronic and photometric sources [9]. Blurring can be represented as the form of bandwidth reduction of picture, because of imperfect image formation process such as relative motion between camera and original scene or out of focus of an optical system. Noises are those unwanted signals that interfere with the original signals and reduced the visible standard of digital picture [10]. The purpose of this study was to examine the effects of divergent resolution and noise levels on job performance in digital mammography.

Removal of noise from mammogram image in medical image processing is a more challenging task today. In the diagnosis of diseases clear images are required by doctors. A lot of research work has been conducted to remove noise from mammograms for last few years. In this manuscript we have used four dissimilar filtering approaches to remove noise in mammograms and a comparative study has also been carried out to identify the foremost filter by estimate MSE and PSNR.

Rest of the manuscript contains the following sections: Sect. 2 presents the related work in the similar field. Suggested approach and preprocessing methods are shown in Sect. 3. Section 4 reflects experimental outcomes and comparison tables. Finally, conclusion and future extension of from the current study are highlighted in Sect. 5.

2 Related Literature

Various works have already been done in the past on removal of noise from mammogram image. Using different types of filters to remove noise and thus helps in the process of the image enhancement. Then in image preprocessing, the noise removal can increase the efficiency of the designed model [11]. During the transmission, process images are degraded by noise, an adequate noise reduction approach provides better perseverance by preserving vital features of a picture. Mammogram enhancement is the approach of manipulation of pixels by minimizing noise and increasing the image contrast by the use of dissimilar filtering approach [12]. Poor image contrast, noise, and special mark are found in the mammograms because of poor positioning of patients. In the process of segmentation, it is very complex to eliminate the special mark and noise exists in mammograms [13].

Filtering approaches are convenient for preprocessing which are utilized to enhance image standard, separate noise, preserving the edges, etc. According to Muller, H., et al. it concludes the adaptive median filter is suitable approach differentiate to other methods because the image standard of an adaptive median is more satisfactory than others [14].

In median filtering, morphological behavior and enhancement of contrast are applied to decrease noise and enrich the mammogram. The contrast of each region evaluation based on its discrete background [15, 16]. After preserving the edge information of suspicious zone of a l mammogram background noise is removed because it has no more importance. Most of the image processing techniques are used to delete special character and undesirable noises [17]. According to Sukhatme, N., et al. analysis was conduct between ICA and DWT, its outcome obtained are compared on the basis of PSNR. Finally, ICA technique was found to be the best for elimination of three variety of noise [18].

3 Materials and Methods

Currently, digital mammography has been used as an alternative of film mammography, to diagnosis breast cancer. It allows for the radiologist to captures and manipulates the mammograms, such that abnormal masses are easily visible [19–22].

Generally, a CAD system for breast cancer detection involves four stages which are: preprocessing, segmentation, extraction if image features and classification as depicted in Fig. 1. In this manuscript we have basically focused on improvement of image quality by using different filters which is applied at the stage of preprocessing.

Figure 2 represents the addition of different noise and filters in our experimental work on mammogram image. Resizing of image along with noise is the final input to the filters and finally produces the filtered image. Images are taken an input and

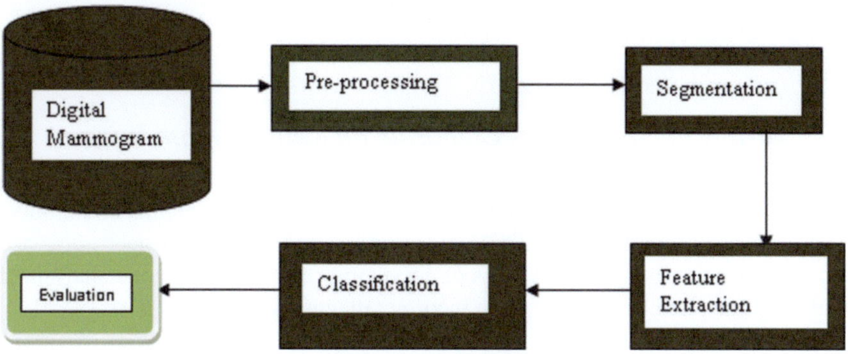

Fig. 1 CAD system used for identification of masses in the breast

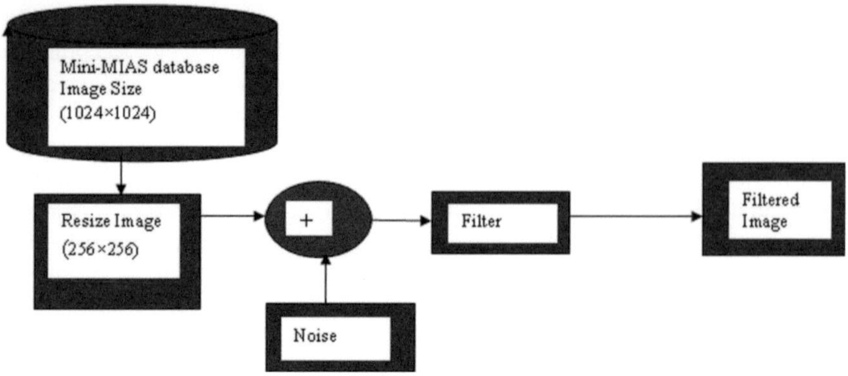

Fig. 2 Framework of the proposed techniques used in experimental analysis of mammogram in mini-MIAS database

added variety of noises one at a time and examine four filters one by one. Lastly, based on PNSR value concluded the best filter.

3.1 Data Set Used

Due to privacy issues it is too complex to acquire real patient data. Mammographic Image Analysis Society (MIAS) database is used our experiment. MIAS database of digital mammograms has generated by UK research group. Right and left breast mammograms of 161 patients are stored in this database. Its total amount comprises 322 images, which are benign, malignant and normal. It has been diminished to 200-micron pixel so entire pictures 1024 × 1024. Here 63 benign, 51 malignant, and 208 normal images are available. Further the four dissimilar types of abnormalities consider suspicious ulcer, architectural distortions, circumscribed lumps, and calcifications [23].

3.2 Variety of Noise Over- Elaborate on Mammogram Images

Noise generate while capturing the image, it is a random fluctuations of shininess or color data in images. Gaussian noise, speckle noise, salt and pepper noise and poisson noise are four different types of noises may affects the accurate analysis of mammograms, finally resulting in an incorrect diagnosis [24, 25].

3.2.1 Noise of Salt and Pepper

It is widely known as spike and impulsive noise. Here the dark pixels are replaced by bright pixel and bright pixels are replaced by unlighted pixels, which are seen as dark and color less spots in the image. Salt and pepper noises are added when the image signals are changed suddenly or sharply. It creates the basic reason to initiate tiny malpractice on the image [26].

3.2.2 Gaussian Noise

Gaussian distribution have additive natures are followed by this noise model. Normal distribution is equivalent to the probability density function, which is also known as Gaussian distribution. White Gaussian noise is the exceptional instance of Gaussian noise, where the worth is consistently statistically unbounded. White Gaussian is used in most of the applications due to its computational simplicity [27].

3.2.3 Speckle Noise

Pulse generator images quality is mainly degraded due to the speckle noises. In traditional radar, the send back indication from an entity produce variation and it guides the granular noise. Mean gray level of a confined domain will grow as a result of speckle noise creates trouble for picture interpretation in synthetic aperture radar [28].

3.2.4 Poisson Noise

While capturing the picture the poisson noise gets incorporated because of some statistical variation in assessment. It is produced mostly because of the different attributes of capturing devices like restricted number of particles namely electron in an electronic circuit which produces energy or photons on an optical appliance.

3.3 Evaluation of Parameters

Evaluation is a procedure that explores step by step a program, its motivation is to make decisions about a program. Here we have used MSE and PSNR to measure the quality of a picture.

3.3.1 Mean Square Error (MSE)

To measure any image features the mean square error is mostly used. The resultant value is always positive in nature and its best case is closer to zero. It can be defined as

$$\text{MSE} = \frac{1}{PQ}\sum_{r=1}^{P}\sum_{c=1}^{Q}(f(r,c). - f'(r,c)^2) \tag{1}$$

In the above equation f(r,c) represent original image and f'(r,c) its reconstructed image. Higher values of MSE prove the lower image quality and vies versa.

3.3.2 Peak Signal-to-Noise Ratio (PSNR)

Another picture quality performance can be measured using PSNR. It can be defined as,

$$\text{PSNR} = 20\log_{10}\left(\frac{1}{\text{MSE}}\right)\text{d}b. \tag{2}$$

From the above two Eqs. (1) and (2) we can conclude that, both the performance measures are inversely proportional to each other, i.e., PSNR value increase only on decrease of RMS and will result in the improvement of the image quality.

4 Experimental Result and Discussion

Performance study of different filters, we initiate dissimilar noises, i.e. Gaussian, salt & pepper, speckle and poisson, on mammograms. Here we are tested all the default noise value in mammograms. Then to evaluate the performance of Median Filter, Wiener Filter, Mean Filter, Laplacian Filter. We evaluate variables such as MSE and PSNR for mammograms. The demonstration supervision on MATLAB R2016b.

In this manuscript, two kinds of parameter are used for preprocessing, mostly focus MSE and PSNR. Tables 1 and 2 shows the computed value of different filters in corporate with different types of noise.

Table 1 Analysis on PSNR of different filters

Filter name	Salt and pepper noise	Poisson noise	Gaussian noise	Speckle noise
Median filter	52.9152	47.5364	44.2101	42.7884
Wiener filter	48.0416	49.1827	45.7280	43.1882
Mean filter	47.4790	47.6065	45.0130	44.6179
Laplacian filter	46.2707	46.1944	43.8091	43.6023

Table 2 Implementation analysis of filters based on MSE

Filter name	Salt and pepper noise	Poisson noise	Gaussian noise	Speckle noise
Median filter	0.3323	1.1467	2.4664	3.4217
Wiener filter	1.0208	0.7849	1.7389	3.1208
Mean filter	1.1619	1.1283	2.0501	2.2456
Laplacian filter	1.9159	1.9103	3.8996	4.1011

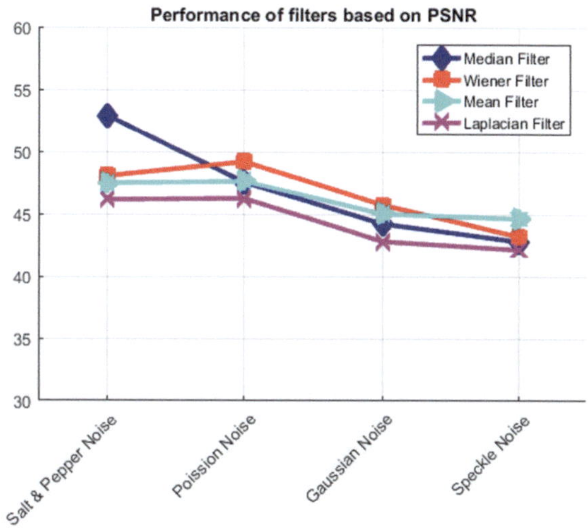

Fig. 3 Higher value of PSNR shows better performance of filters

The PSNR for the median filter is 52.9152 (mdb005) display in Table 1 and Fig. 3, which is very high while comparing with other filters. From the above survey, the median filter well performs image with salt and pepper noise while comparing with the other filter. The parameters are shown in above Table 1.

Mean square error is minute for the median filter compare to the other filters, throughput of MSE for the median filter is 0.3323 (mdb005) as shown in Table 2 and Fig. 4. For the median filter it provides better image. Less value of peak signal-to-noise ratio means that picture standard is poor.

5 Discussion

In Tables 1 and 2 if we use a median filter after adding different noise in mammogram image, then it provides better PNSR value with less MSE in case of image added with salt and pepper noise. Both the Wiener filter and mean filter provides better PNSR and less MSE value the mammogram with Poisson noise. In the case of the Laplacian filter, it will vary better PSNR and less MSE value produce, the mammogram with

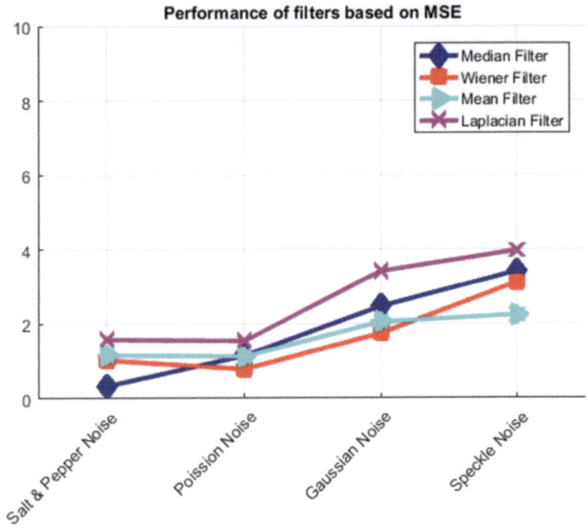

Fig. 4 Lower value of MSE shows better performance of filters

speckle noise. From the above study, we conclude that no one filter well operates on all types of noises. Out of these four types of filter, the median filter performs well on image with salt and pepper noise.

6 Conclusion

Delusion and noises influence on breast mammogram during accumulation, it affects the whole image handling and identification of disease. Elimination of noises on the preprocessing phase is another demanding job. In this manuscript we apply existing four divergent filter, i.e., MF, WF, MF, LF for removing noise in mammogram. The differentiation of these filters is analyses for 322 mammograms. From the throughput study, it concluded median filter is a suitable approach while compared with other methods, because picture standard of median filter is finer. A comparative analysis is carry out by performance of filters based on simulated output parameters PSNR and MSE. First-hand outcomes reveal that Median filter performs effectively on noise elimination of mammograms.

References

1. http://www.breastcancer.org/symptoms/understand_bc/what_is_bc
2. Amutha, S., Babu, D.R., Shankar, M.R., Kumar, N.H.: Mammographic image enhancement using modified mathematical morphology and Bi-orthogonal wavelet. In: 2011 International Symposium on IT in Medicine and Education (ITME), vol. 1, pp. 548–553. IEEE (2011)

3. Yan, Z., He, X., Liu, S., Lu, D.: An approximation-weighted detail contrast enhancement filter for lesion detection on mammograms. In: Proceedings of the 23rd Annual International Conference of the IEEE Engineering in Medicine and Biology Society, Vol. 3, pp. 2472–2475. IEEE (2001)
4. Choi, J.Y., Kim, D.H., Plataniotis, K.N., Ro, Y.M.: Classifier ensemble generation and selection with multiple feature representations for classification applications in computer-aided detection and diagnosis on mammography. Expert Syst. Appl. **46**, 106–121 (2016)
5. Rangayyan, R.M., Ayres, F.J., Desautels, J.L.: A review of computer- aided diagnosis of breast cancer: Toward the detection of subtle signs. J. Franklin Inst. **344**(3–4), 312–348 (2007)
6. Cheng, H.D., Shan, J., Ju, W., Guo, Y., Zhang, L.: Automated breast cancer detection and classification using ultrasound images: A survey. Pattern Recogn. **43**(1), 299–317 (2010)
7. Huang, Y.L., Wang, K.L., Chen, D.R.: Diagnosis of breast tumors with ultrasonic texture analysis using support vector machines. Neural Comput. Appl. **15**(2), 164–169 (2006)
8. Liu, B., Cheng, H.D., Huang, J., Tian, J., Tang, X., Liu, J.: Fully automatic and segmentation-robust classification of breast tumors based on local texture analysis of ultrasound images. Pattern Recogn. **43**(1), 280–298 (2010)
9. Saunders, R.S., Baker, J.A., Delong, D.M., Johnson, J.P., Samei, E.: Does image quality matter? Impact of resolution and noise on mammographic task performance. Med. Phys. **34**(10), 3971–3981 (2007)
10. Jalalian, A., Mashohor, S.B., Mahmud, H.R., Saripan, M.I.B., Ramli, A.R.B., Karasfi, B.: Computer-aided detection/diagnosis of breast cancer in mammography and ultrasound: a review. Clin. Imaging **37**(3), 420–426 (2013)
11. Nagaiah, K., Manjunathachari, K., Rajinikanth, T.V.: Advanced image enhancement method for mammogram analysis. In: 2016 International Conference on Recent Trends in Information Technology (ICRTIT), pp. 1–5. IEEE (2016)
12. Barash, D., Comaniciu, D.: A common framework for nonlinear diffusion, adaptive smoothing, bilateral filtering and mean shift. Image Vis. Comput. **22**(1), 73–81 (2004)
13. Müller, H., Michoux, N., Bandon, D., Geissbuhler, A.: A review of content-based image retrieval systems in medical applications—clinical benefits and future directions. Int. J. Med. Inform. **73**(1), 1–23 (2004)
14. Ramani, R., Vanitha, N.S., Valarmathy, S.: The pre-processing techniques for breast cancer detection in mammography images. Int. J. Image, Graph. Signal Process. **5**(5), 47 (2013)
15. Hussein, Z.R., Rahmat, R.W., Nurliyana, L., Saripan, M.I., Dimon, M.Z.: Preprocessing importance for extracting contours from noisy echo cardiographic images. Int. J. Comput. Sci. Netw. Secur. (IJCSNS), **9**(3), 134–137 (2009)
16. Morrow, W.M., Paranjape, R.B., Rangayyan, R.M., Desautels, J.L.: Region-based contrast enhancement of mammograms. IEEE Trans. Med. Imaging **11**(3), 392–406 (1992)
17. Lai, S.M., Li, X., Biscof, W.F.: On techniques for detecting circumscribed masses in mammograms. IEEE Trans. Med. Imaging **8**(4), 377–386 (1989)
18. Sukhatme, N., Shukla, S.: Independent component analysis based denoising of magnetic resonance images. Int. J. Comput. Appl. **54**(2) (2012)
19. Cheng, H.D., Cai, X., Chen, X., Hu, L., Lou, X.: Computer-aided detection and classification of microcalcifications in mammograms: a survey. Pattern Recogn. **36**(12), 2967–2991 (2003)
20. Tang, J., Rangayyan, R.M., Xu, J., El Naqa, I., Yang, Y.: Computer-aided detection and diagnosis of breast cancer with mammography: recent advances. IEEE Trans. Inf Technol. Biomed. **13**(2), 236–251 (2009)
21. Tosteson, A.N., Stout, N.K., Fryback, D.G., Acharyya, S., Herman, B.A., Hannah, L.G., Pisano, E.D.: Cost-Effectiveness of digital mammography breast cancer screeningcost-effectiveness of digital mammography. Ann. Int. Med. **148**(1), 1–10 (2008)
22. Khuzi, A.M., Besar, R., Zaki, W.W.: Texture features selection for masses detection in digital mammogram. In: 4th Kuala Lumpur International Conference on Biomedical Engineering 2008, pp. 629–632. Springer, Berlin (2008)
23. Mudigonda, N.R., Rangayyan, R., Desautels, J.L.: Gradient and texture analysis for the classification of mammographic masses. IEEE Trans. Med. Imaging **19**(10), 1032–1043 (2000)

24. Joseph, A.M., John, M.G., Dhas, A.S.: Mammogram image denoising filters: a comparative study. In: 2017 Conference on Emerging Devices and Smart Systems (ICEDSS), pp. 184–189. IEEE (2017)
25. Utaminingrum, F., Uchimura, K., Koutaki, G.: High density impulse noise removal by fuzzy mean linear aliasing window kernel. In: 2012 IEEE International Conference on Signal Processing, Communication and Computing (ICSPCC), pp. 711–716. IEEE (2012)
26. Bozek, J., Mustra, M., Delac, K., Grgic, M.: A survey of image processing algorithms in digital mammography. Recent Adv. Multimed. Signal Process. Commun. 631–657 (2009)
27. Kaur, R., Kaur, R.: Survey of de-noising methods using filters and fast wavelet transform. Int. J. Adv. Res. Comput. Sci. Softw. Eng. **3**(2) (2013)
28. Nettleton, D.F., Orriols-Puig, A., Fornells, A.: A study of the effect of different types of noise on the precision of supervised learning techniques. Artif. Intell. Rev. **33**(4), 275–306 (2010)

Cryptography and Information Security

Synthetic Image and Strange Attractor: Two Folded Encryption Approach for Secure Image Communication

Sridevi Arumugham, Sundararaman Rajagopalan, Sivaraman Rethinam, Siva Janakiraman, C. Lakshmi and Amirtharajan Rengarajan

Abstract Chaotic attractors are used in variety of applications especially in image encryption due to their sensitivity, confusion and diffusion properties. Such attractors are continuous form of chaos which can be described using differential equations. In this work, an image encryption technique employing Chen attractor and FPGA generated synthetic image has been discussed. Altera Cyclone II FPGA was used to generate synthetic image that consumed 438 logic elements and 34.03 mW of power consumption. The simulation was carried out on LabVIEW platform. The performance of the algorithm was evaluated through statistical (Entropy, Histogram, Correlation) and differential analyses (NPCR and UACI) for grayscale images of various sizes 128 × 128, 256 × 256 and 512 × 512 which yielded good figures.

Keywords Attractor · Chaos · Image encryption · LabVIEW · FPGA

1 Introduction

Due to the huge demand in wireless networks and tele-communication, the secure transmission of multimedia contents plays a vital role considering the various intrusions and attacks. Encryption is one such process which provides confidentiality of the information which is to be shared over public channel. From the last decade, image encryption is gaining importance for confidential sharing of images of various modalities and formats. Needless to mention that strong and secure encryption algorithms are required in image encryption domain as well. In any encryption, key occupies a center stage. Key needs to be so strong and more sensitive in order to make the encryption algorithm resistant against attacks [1]. Since the introduction of chaos, image encryption has attained fruitful growth. Standard algorithms like DES and AES can be used for image encryption. But these techniques alone are not a viable solution for digital images due to the poor correlation coefficients

S. Arumugham · S. Rajagopalan (✉) · S. Rethinam · S. Janakiraman · C. Lakshmi · A. Rengarajan
Department of ECE, School of EEE, SASTRA Deemed to be University,
Thanjavur, Tamilnadu, India
e-mail: raman@ece.sastra.edu

B. Pati et al. (eds.), *Advanced Computing and Intelligent Engineering*,
Advances in Intelligent Systems and Computing 1082,
https://doi.org/10.1007/978-981-15-1081-6_40

and speed of operation and hence additional mechanisms are required for efficient encryption [2].

Hence, many image encryption works have been addressed using various methodologies like chaotic maps [3], cellular automata [4], transforms [5], DNA sequences [6], etc. However, to accomplish good quality of encryption, it is suggested to combine different sources of confusion and diffusion. In [7], confusion of the images was carried out using hyperchaotic systems based on Lorenz and Chen and the diffusion by employing encoding and decoding rules of DNA. Li et al. suggested a RGB image encryption utilizing real and complex Lorenz hyperchaotic systems and DNA operations [8]. In another work, Complex Lorenz and Chen system-based image ciphering technique was suggested by Wang et al. wherein hamming distance was used as an important measure to increase the efficiency and performance of the encryption [9].

Considering the significance of various encryption algorithms, the proposed method utilizes the unique properties of Chen strange attractor and reconfigurable hardware generated synthetic image for encryption of grayscale images. So far, most of the image encryption works are software based and hence more possibilities for attacks and tapping. This work is a combination of hardware and software in which one of the key generation mechanisms is accomplished by FPGA. Noticeably, this encryption algorithm offers a triple layer (Confusion – Diffusion – Diffusion) protection to images. To validate this work, several analyses were performed and obtained results have been reported.

2 Preliminaries

2.1 Linear Feedback Shift Register

It is defined as a shift register in which the output depends on the input and previous state. LFSR which has feedback path from output node to input that controls the random number generation process. LFSR is constructed with primitive polynomials. Taping is the core element in LFSR to produce random bits [10, 11]. Figure 1 depicts the Generic Skeleton of LFSR.

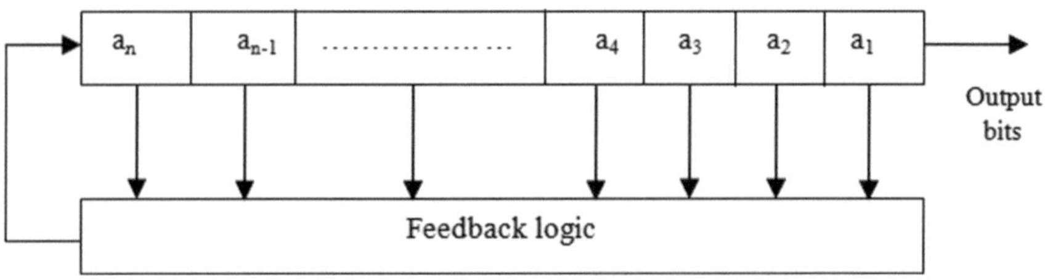

Fig. 1 Generic skeleton of LFSR

2.2 *Strange Attractors*

Strange attractors are defined and described by nonlinear differential equations. They produce butterfly pattern when subjected to number of iterations and these patterns are fractal in nature. They are defined as continuous chaos as they are more responsive to initial conditions which means even though for a small alteration in input seed can produce drastic changes in output. This property of attractors can be useful in encryption approach. Depending on the system equations, initial conditions and system patterns, different strange attractors have been proposed and employed in image encryption schemes [12, 13]. The attractor which used in this work is Chen and its system equations are presented in Table 1.

3 Proposed Methodology

Figure 2 shows the Overall Block Diagram of the proposed Image Cryptographic scheme. It performs the following operations:

- Random number generation using FPGA and formation of synthetic image
- Design of strange attractors using LabVIEW
- Image encryption with confusion and first level diffusion
- Second level diffusion with FPGA generated synthetic image.

Table 1 Strange attractors and system equations

Chen	$\frac{dr}{dt} = a(s - r)$	$\frac{ds}{dt} = (c-a)r + cs - rv$	$\frac{dv}{dt} = -bv + rs$

r, s and v are states and a, b and c are system parameters

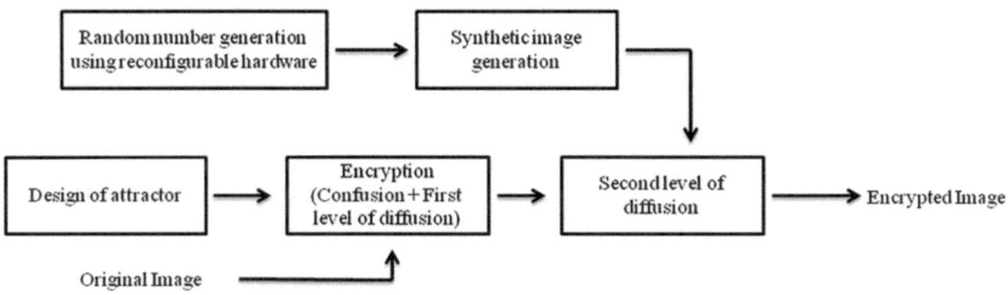

Fig. 2 Proposed work block diagram

3.1 Synthetic Image Generation

Synthetic images had been used in the past for image encryption [14]. Synthetic image can be formed using FPGA by acquiring and assembling the random bits as pixels arranged in rows and columns. In this approach, it is considered as main key to improve the encryption process. Since the software generated keys are suffering from key sensitivity and key space, FPGA generated synthetic keys have a wide span more than 2^{256} which is greater than 2^{128} so as immune to brute force attacks. Synthetic image is generated by carrying out the following tasks:

- Generation of diffused bits using LFSR
- Scrambling the diffused bits with a sampling clock generated by PLL.

LFSR architecture with a polynomial $x^{256} + x^{254} + x^{251} + x^{246} + 1$ which is shown in Fig. 3 has been developed on Altera Cyclone II EP2C35F672C6 FPGA. Every individual bits of 256 bits LFSR architecture were XORed with each other at 300 MHz and the scrambled bits were sampled using 74.25 MHz clock. Figure 4 presents the functional simulation of LFSR.

This design of LFSR requires one PLL for generating frequencies of 300 and 74.25 MHz. Among 33216 logic elements, 438 (1%) logic elements were utilized for this scheme. Noticeably, PLL consumes 34.03 mW of core dynamic power for generating 300 MHz and 74.25 MHz clocks respectively. Since the encryption has been tested on three different 256 × 256 grayscale images, synthetic image have been formed with a size of 256 × 256 which is depicted in Fig. 5a. Strength of synthetic image has been evaluated by performing entropy, correlation and histogram analyses in which the corresponding results are presented in Table 2 and Fig. 5b respectively.

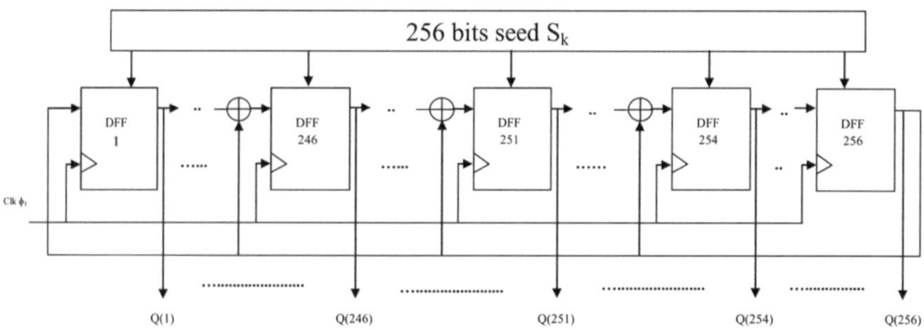

Fig. 3 Generation of random number using LFSR

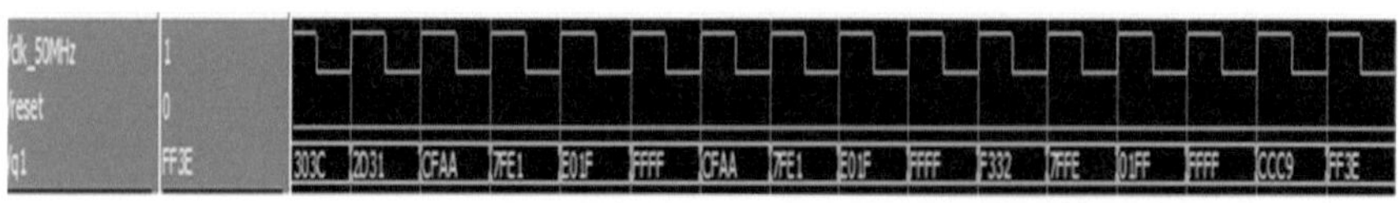

Fig. 4 Modelsim simulation of random number generation using LFSR

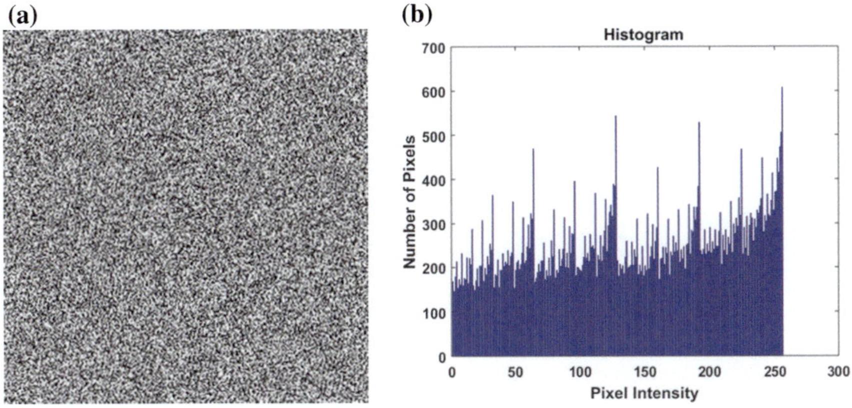

Fig. 5 **a** Synthetic image. **b** Histogram of synthetic image

Table 2 Entropy and correlation analyses

Entropy		7.9433
Correlation coefficients	Horizontal	−0.0255
	Vertical	0.0499
	Diagonal	−0.0210

3.2 Attractor-Based Encryption

Generic flow of the proposed image encryption schemes is explained as follows:

Step 1: Input: Image I ($p \times q$) which is of 8-bit depth
Step 2: Transform the image I ($p \times q$) into a row vector BV (1; p; q)
Step 3: Produce three chaotic sequences A, B and C using Chen attractor with A = (A_1, A_2, … $A_{p\times q}$); B = (B_1, B_2, … $B_{p\times q}$) and C = (C_1, C_2, … $C_{p\times q}$)
Step 4: Perform sorting for the above sequence,

$$[l_a,\ f_a] = \text{sort (A)};$$

where $[\varphi, \varphi] = \text{sort}(\varphi)$ is the indexing term; l_a,f_a denotes the new index after sorting the data A and the sorted values of A respectively.
Step 5: Perform confusion to vector BV of image A as per the following form

$$Con1(\alpha) = BV(la(\alpha)), \quad \text{where } 1 \leq \alpha \leq p \times q$$

Table 3 Attractor—system parameters and states

Attractors	System parameters		System states	
Chen	a	40	r	−0.1
	b	3	s	0.5
	c	28	v	−0.6

Step 6: Perform first stage of diffusion through XOR operation between the *Con*1 and *Y*1
$Dif1(\alpha) = Con1(\alpha) \oplus Y1(\alpha),\quad \text{where } 1 \le \alpha \le p \times q \text{ and } Y1(\alpha) = \text{mod}(Y(\alpha), 255)$

Step 7: Sort the sequence C in ascending order as follows:

$$[l_c,\ f_c] = \text{sort (C)}$$

where, l_c denotes the new index after sorting the data C and f_c provides the sorted values of C.
Step 8: Perform second stage of confusion by scrambling the *Dif*1

$$Con2(\alpha) \leftrightarrow Dif1(lc(\alpha)),\quad \text{where } 1 \le \alpha \le p \times q$$

Step 9: Produce the synthetic image *S* (*p*, *q*) of size p × q using the synthetic image generation steps explained in Sect. 3.1.
Step 10: Implement second stage of diffusion by XORing the *Con*2 with the synthetic image *S* (*p*, *q*) as follows:

$$Dif2(\alpha, \beta) = Con2(\alpha, \beta) \oplus S(\alpha, \beta),\ \text{where } 1 \le \alpha \le p;\ \ 1 \le \beta \le q$$

Since the encryption algorithm follows symmetric cryptography, same key is used for encryption as well as decryption.

Each attractor possesses a unique system state and system parameters. These are to be followed strictly to achieve proper encryption and decryption. Table 3 lists the required system parameters and system states of the Chen attractor. To decrypt the cipher image, reverse process has been carried out wherein the cipher image is input and synthetic image is key.

4 Results and Discussion

In order to verify and prove the strength of the proposed cryptographic algorithm, three test gray scale images namely Lena, Insect and Baboon were considered with three different sizes of 128 × 128, 256 × 256 and 512 × 512. The proposed algorithms were evaluated using the standard analyses such as statistical and differential analyses. Figure 6a–f shows the original and encrypted images of size 256 × 256.

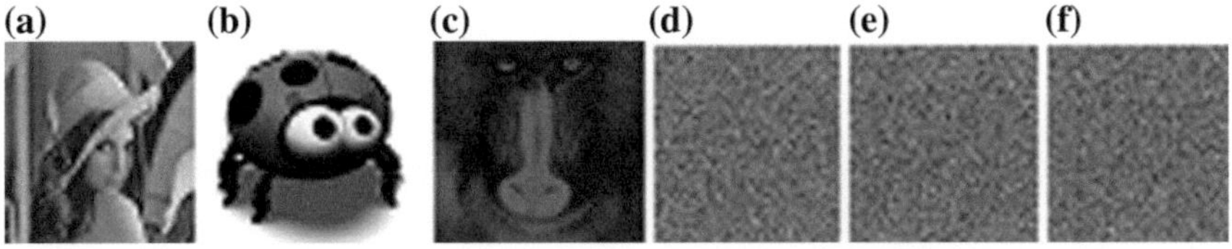

Fig. 6 Test images—original: **a** Lena; **b** Insect; **c** Baboon and test images—encrypted: **d** Lena; **e** Insect; **f** Baboon

The encryption quality was estimated with these metrics and reported for all size images as follows.

4.1 Histogram Analysis

It is a pictorial depiction of equi-distributional of pixels with regard to the total number of occurrences. It is one of the predominant properties of an algorithm to assess the capacity of proposed algorithm to withstand statistical attacks. Figure 7a–f depicts the original and cipher image histograms which resulted in an uniform distribution.

4.2 Entropy Analysis

It is a measure to verify the randomness of cipher images. In image encryption algorithms, diffusion process increases the entropy. For a grayscale image, the maximum entropy to be achieved is 8. The proposed encryption method reaches 7.9974 as maximum entropy through Chen attractor and FPGA assisted synthetic image. Entropy for all the images are given in Tables 4, 5 and 6.

4.3 Correlation Analysis

Correlation is referred as the degree of relationship between adjacent pixels of an image. To achieve better encryption quality, correlation must be as low as possible and hence statistical attacks can be prevented. Tables 4, 5 and 6 present the correlation analyses of proposed image encryption method. In addition, Fig. 8a–f showcase the correlation of original and encrypted insect image.

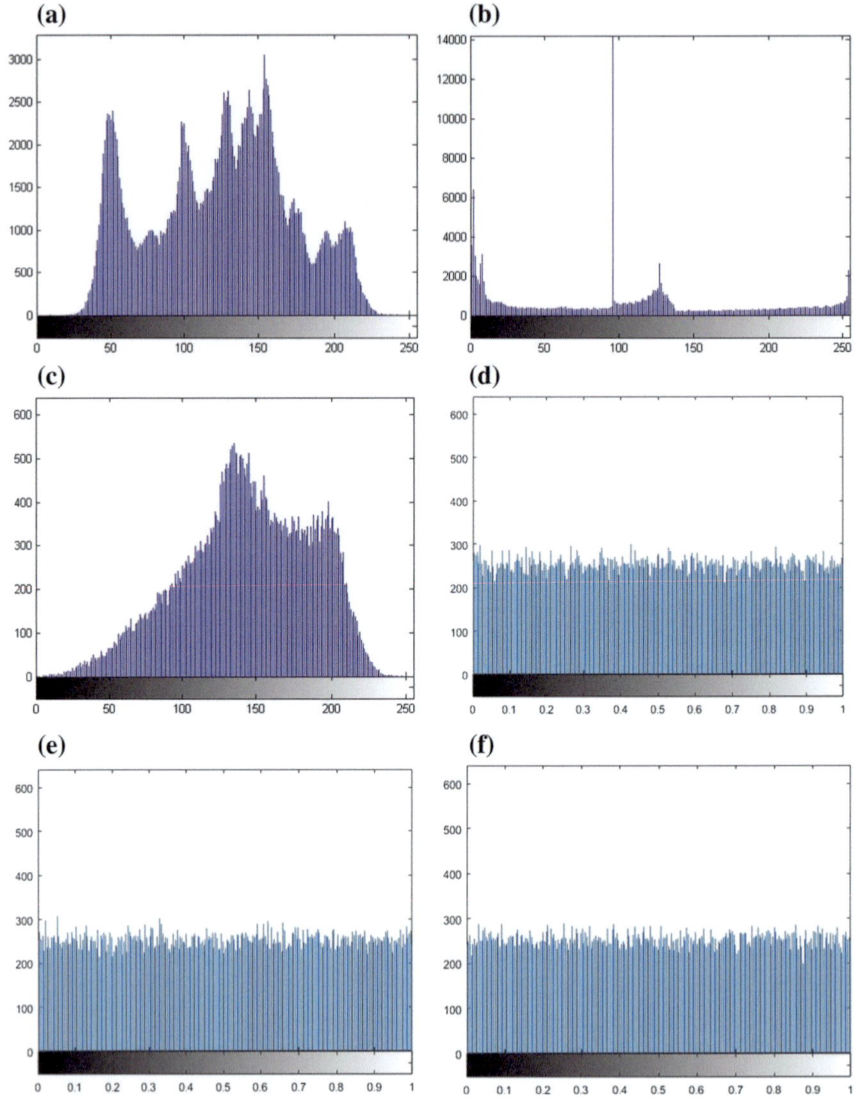

Fig. 7 Histogram—original: **a** Lena; **b** Insect; **c** Baboon and histogram—encrypted: **d** Lena; **e** Insect; **f** Baboon

4.4 Differential Analysis

NPCR and UACI are mandatory parameters to test the response of the cryptographic algorithm as well as the amount of pixel changes between any two encrypted images. Tables 7 and 8 depict the obtained differential analyses for the various cipher images.

Table 4 Correlation and entropy analysis of 128 × 128 images

Images (128 × 128)		Horizontal	Vertical	Diagonal	Entropy
Lena	Original	0.8821	0.9538	0.8475	7.4221
	Encrypted	−0.0094	0.0331	−0.0223	7.9598
Insect	Original	0.9694	0.9716	0.9470	5.6615
	Encrypted	−0.0024	0.0255	−0.0237	7.9603
Baboon	Original	0.8250	0.8319	0.7613	7.1672
	Encrypted	−0.0053	0.0269	−0.0231	7.9608

Table 5 Correlation and entropy analysis of 256 × 256 images

Images (256 × 256)		Horizontal	Vertical	Diagonal	Entropy
Lena	Original	0.9376	0.9699	0.9119	7.4436
	Encrypted	−0.0001	−0.0008	0.0029	7.9974
Insect	Original	0.9853	0.9870	0.9753	5.5102
	Encrypted	−0.0009	0.0008	−0.0038	7.9970
Baboon	Original	0.8626	0.8183	0.8090	6.6962
	Encrypted	−0.0078	0.0006	−0.0059	7.9971

Table 6 Correlation and entropy analysis of 512 × 512 images

Images (512 × 512)		Horizontal	Vertical	Diagonal	Entropy
Lena	Original	0.9783	0.9907	0.9681	7.4164
	Encrypted	0.0187	0.0491	0.0224	7.8478
Insect	Original	0.9935	0.9945	0.9893	5.4282
	Encrypted	0.0187	0.0491	0.0224	7.8478
Baboon	Original	0.8030	0.7458	0.7125	7.4128
	Encrypted	0.0125	0.0014	0.0258	7.8904

5 Conclusion

In this paper, an improved method of image encryption on grayscale images was proposed with the confluence of FPGA assisted synthetic image and strange attractors. Synthetic image was generated through Altera Cyclone II FPGA, Chen attractor has been designed using LabVIEW 2013 and image encryption was performed on software platform. Quality of the proposed methods has been validated through standard image encryption metrics. The result evidences that almost good figures can be obtained by this confluence of hardware-software co-design. The visual perception of encrypted images looks good. The implementation of color image cryptographic algorithm on FPGA will be the future work.

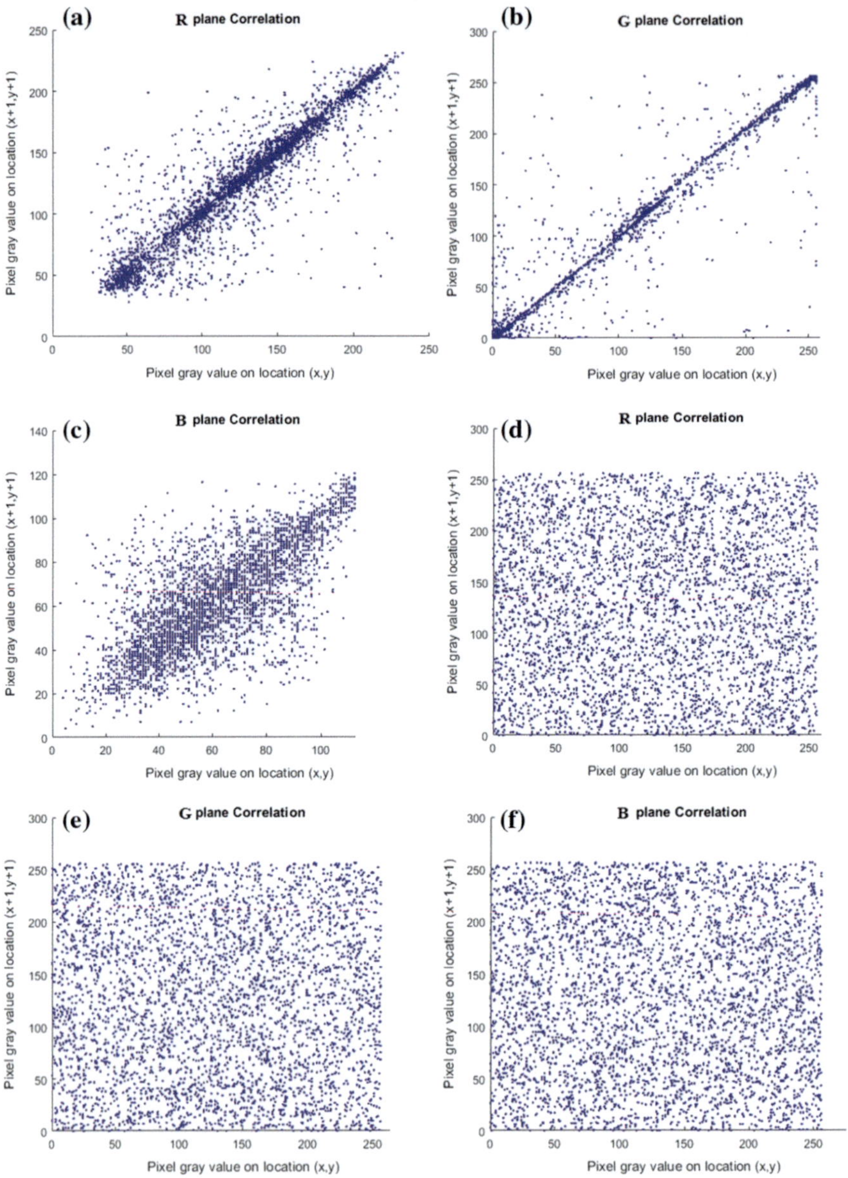

Fig. 8 Correlation coefficients original—Insect: **a** horizontal; **b** vertical; **c** diagonal and correlation coefficients encrypted—Insect: **d** horizontal; **e** vertical; **f** diagonal

Table 7 Differential analysis of 128 × 128 images

S. no.	Images (128 × 128)	NPCR	UACI
1	Lena	99.5911	34.0651
2	Insect	99.5117	39.4536
3	Baboon	99.6033	27.6859

Table 8 Differential analysis of 256 × 256 images

S. no.	Images (256 × 256)	NPCR	UACI
1	Lena	99.6201	28.7203
2	Insect	99.6048	40.0909
3	Baboon	99.5865	31.8526

Acknowledgements The authors are grateful to SASTRA Deemed University for giving support and infrastructure for performing the proposed work.

References

1. Eason, G., Noble, B., Sneddon, I.N.: On certain integrals of Lipschitz-Hankel type involving products of Bessel functions. Phil. Trans. Roy. Soc. London **A247**, 529–551 (1955)
2. Wong, K.-W., Kwok, B.S.-H., Law, W.-S.: A fast image encryption scheme based on chaotic standard map. Phys. Lett. A. **372**, 2645–2652 (2008). http://dx.doi.org/10.1016/j.physleta.2007.12.026
3. Liu, W., Sun, K., Zhu, C.: A fast image encryption algorithm based on chaotic map. Opt. Lasers Eng. **84**, 26–36 (2016)
4. Jin, J.: An image encryption based on elementary cellular automata. Opt. Lasers Eng. **50**(12), 1836–1843 (2012)
5. Liu, Z., et al.: Color image encryption by using Arnold transform and color-blend operation in discrete cosine transform domains. Opt. Lasers Eng. **2011**(284), 123–128 (2011)
6. Liu, Y., Tang, J., Xie, T.: Optics & Laser Technology Cryptanalyzing a RGB image encryption algorithm based on DNA encoding and chaos map. Int. J. Light Electron Opt. **60**, 111–115 (2014)
7. Zhang, Q., Wei, X.: A novel couple images encryption algorithm based on {DNA} subsequence operation and chaotic system. Opt. Int. J. Light Electron Opt. **124**, 6276–6281 (2013). http://dx.doi.org/10.1016/j.ijleo.2013.05.009
8. Li, X., Wang, L., Yan, Y., Liu, P.: An improvement color image encryption algorithm based on DNA operations and real and complex chaotic systems. Opt. Int. J. Light Electron Opt. **127**(5), 2558–2565 (2016)
9. Wang, Y., Wong, K.-W., Liao, X., Xiang, T., Chen, G.: A chaos-based image encryption algorithm with variable control parameters. Chaos Solitons Fractals **41**, 1773–1783 (2009). http://dx.doi.org/10.1016/j.chaos.2008.07.031
10. Zhang, H., Wang, Y., Wang, B., Wu, X.: Evolutionary random sequence generators based on LFSR. Wuhan Univ. J. Nat. Sci. **12**, 75–78 (2007). https://doi.org/10.1007/s11859-006-0196-9
11. Bhaskar, P., Gawande, P.P.D.: A survey on implementation of random number generator in FPGA. Int. J. Sci. Res. **4**, 1590–1592 (2015)

12. Wang, L., Song, H., Liu, P.: A novel hybrid color image encryption algorithm using two complex chaotic systems. Opt. Lasers Eng. **77**, 118–125 (2016)
13. Al-najjar, H.M.: Multi-chaotic image encryption algorithm based on one time pads scheme. Int. J. Comput. Theory Eng. **4**(3), 4–7 (2012)
14. Rajagopalan, S., Sivaraman, R., Upadhyay, H.N., Rayappan, J.B.B., Amirtharajan, R.: ON-chip peripherals are ON for chaos–an image fused encryption. Microprocess. Microsyst. **61**, 257–278 (2018)

Role of Soft Outlier Analysis in Database Intrusion Detection

Anitarani Brahma and Suvasini Panigrahi

Abstract With the rapid development of World Wide Web and E-commerce, concern of security is a very sensitive issue in this modern era of information and communication technology. A lot of financial and brain effort has been invested in this problem and still requires serious attention due to the increasing threats. Database centered Intrusion Detection is a prominent field in this research circumference. Concept of outlier analysis in data mining can automate this intrusion detection process with higher accuracy. In this research, we present the role of soft outlier analysis in Database-centered Intrusion Detection while comparing its performance with its counterpart hard outlier analysis which ultimately enhances its productivity by improving the accuracy and reducing the false positive costs.

Keywords Outlier detection · Clustering · Local outlier factor · Optics · Fuzzy C-mean clustering · Back Propagation Neural Network

1 Introduction

Due to huge usage of Information Technology in our society in online marketing, online banking, stock, email; dealing with the databases related to credit card transaction, health-care, and personal-financial information increases and criminal records related to these databases increases as well, which needs high level of security against a number of threats. The increasing new automated hacking tools along with prevailing vulnerability information on the web exploit the increasing number of intrusive activity toward databases. To handle such problems, the Intrusion Detection Systems (IDSs) of database of this era should be reliable, adaptive, extensible and to have a low cost of maintenance.

A. Brahma (✉) · S. Panigrahi
Department of Computer Science & Engineering, VSSUT, Burla, India
e-mail: brahmaanita00@gmail.com

S. Panigrahi
e-mail: suvasini26@gmail.com

B. Pati et al. (eds.), *Advanced Computing and Intelligent Engineering*,
Advances in Intelligent Systems and Computing 1082,
https://doi.org/10.1007/978-981-15-1081-6_41

Database Intrusion Detection Systems (DIDSs) finds fraud accesses and in genuine actions and can supplement effective method to reduce intrusion action and rejection for future malicious access. However, mainly, previous researches focus on intrusion detection on network and host or in operating system level and after having a rigorous literature survey, very few researches has been found with database intrusion detection and some of them can even deal with insider threat as this is more vulnerable than the outsider threat [1]. Normally, there are two models of intrusion detection: Anomaly Intrusion Detection and Misuse Intrusion Detection. Anomaly Intrusion detection model the normal behavior of the user and deviation from the normal behavior of the current incoming transaction are said to be suspicious. Whereas misuse detection declare the transactions to be suspicious by comparing it with a set of trained known intrusion patterns. Rather, more often we do not have either labeled or known intrusion patterns readily available [2].

Early research efforts in the area of intrusion detection limited to network- and host-based. However this early investigation toward DIDS was cast in the framework of hard computing based model or in statistical model. Christina et al. applied frequent data mining techniques for modeling normal profiles in misuse database intrusion detection [3]. Lee et al. derive regular expressions from SQL sentences and use them to represent the normal behavior of user [4]. Chung et al. uses the access pattern of users of database to build the user profile in misuse database intrusion detection system [5]. All the above methods use the dependency relationships among data items or relation of query statement; however they do not give more emphasis on the result data of the queries; hence their methods may lead to high false positive rate or they can be applied only to specific environment. However, the properties like tractability, robustness, low solution cost and reliability of soft computing make it suitable for applying it into the field of intrusion detection. The researchers of IDS community exploit the generalization capabilities of soft computing that help in detecting both misuse and anomaly detection. Evolutionary computing tool, artificial neural networks and fuzzy logic are the branches of soft computing. Cho et al. applied Self-Organizing Map and fuzzy logic for calculating optimal audit data generation and to take final decision regarding database intrusion detection [6]. Panigrahi et al. clearly stated that by embedding fuzzy logic in database intrusion detection significantly reduce the false alarm [7]. Hence, we base our method on soft clustering based outlier detection which can minutely detect the intrusive access in database and negligible false positive rate of the intrusion detection system.

Generally the malicious access toward database is accomplished by using data dependency relationship present in the database, typically, any modification of data made simultaneous updating on interrelated data, and, process of mining of data enhances the adaptability of the IDS as it discovers the patterns of intrusions and dependency among the data items from the real time dataset. Data mining equipped with components such as data transformations, normalization, data reduction, model deployment and cooperative distributed detection which is a complex engineering endeavor. The concept of outlier is derived from the Data mining which is nothing but an observation that deviates from others so much that it creates a suspect boundary on itself as it may be generated by different mechanism. Outlier detection is an efficient process which can detect the abnormal database transactions called outlier which ultimately helps in data security or in turn in DIDS. Many data mining algorithm uses clustering algorithms as a side product for finding the outlier; However the clustering algorithm groups the data based on the similarity or any distance measures that can be used for data reduction, classification, etc., in knowledge discovery in data mining process. The data points that do not lie in any of the clusters are treated to be outlier and they are the prime emphasis while considering intrusion detection based on outlier detection.

In this investigation, we exploit the clustering based outlier detection to detect intrusions in database and to increase the accuracy of the proposed model. Outlier analysis lies in the concept of cluster analysis where the data are divided into meaningful groups. Grouping of data into clusters is done in such a way that similarity of data within cluster (called intra-cluster similarity) is maximized while similarity between clusters (called inter-cluster similarity) is minimized. We assume the available clustering algorithms into two categories: hard clustering and soft clustering so as for the outlier analysis in database intrusion detection process. Hard clustering create a sharp boundary between the clusters where the data points may or may not belong to a cluster; and the data points that do not belong to any of the clusters are identified to be outlier as their behavior is very much different from others. In this research, the intrusive activity is considered to be outlier as their behavior is slightly or magnificently different from other access toward the database. Rather, in soft clustering the data points belong to all of the clusters with some membership value and the outliers are found by applying objective function.

Next section describes in detail about the concepts used in the investigation. Section 3 describes the proposed approach. Section 4 shows experimental results on biometric data. And then we conclude the paper by discussing future work.

2 Background

In this section, we formalize the concepts used in the remainder of the paper. Typically, statistics community have seriously analyzed and studied the process of detecting outlier. The data points are modeled by the user by using some statistical distribution and the outliers are determined on how they appear in relation to the postulated model. The main loop hole in these approaches is that, the users often have less knowledge regarding the data distribution. The detection of outlier mechanism is exploited in the area of research for detecting the intrusion or the malicious access in the database as the outliers are often exhibiting behavior outside the range of what is considered malicious. Zhong et al. proposed Q-clustering algorithm to determine clusters based on the similarities among queries and implement it to detect the intrusions in the database. The high time complexity and problem of importing the model in real life application are the main disadvantages of this algorithm [8]. An approximate role/user oriented profile developed by Khan et al. to detect insider attack in DBMS [9]. Kim et al. applies Local Outlier Factor (LOF), a density-based algorithm to efficiently detect the outliers of the database activities. Still, it fails to encounter the problem for real time database monitoring [10].

We consider the above-discussed outlier process as hard computing approaches as there is a strict boundary between the clusters and the determination of outliers while the applied data set are precise and certain. The Fuzzy C-Mean (FCM) clustering algorithm can handle the problem of time complexity, sharp boundary of determination of outliers in database activity and for real time database access monitoring. The usage of FCM algorithm in database intrusion detection process is exploited against a biometric dataset and experimental results is compared with one hard clustering process. The LOF, FCM algorithm are studied in the following Sects. 2.1 and 2.2 respectively.

2.1 Outlier Analysis Through Local Outlier Factor Algorithm

For many Knowledge Discovery process (KDD), extracting anomaly are more interesting than the common data. LOF algorithm is used to identify the anomaly by measuring the local deviation of a given data point with respect to its neighbors called local density. Local density is nothing but the degree of being an outlier and local state that the degree depends on how isolated the object is with respect to the surrounding neighborhood which is given by K-nearest Neighbors [10]; K-nearest neighbors is the number of objects reachable within k-distance. By comparing local density of an object to the local densities of its neighbors, one can identify regions of similar density, and points that have a substantially lower density than their neighbors are considered to be outliers [11, 12].

Procedure: LOF_Outlier Detection

1. For each data point p, k-distance (p) is to be calculated, which is the Euclidean distance to the k-th- nearest neighbor of p.
2. For each data record p, calculate the reachability distance to data record q.

   Reachability distance: The reachability distance of p from q is the true distance between p and q and at least k-distance of q which can be determined through Eq. (1)

$$\text{Reachability distance (p, q)} = \max\{\text{k} - \text{distance (q)}, \text{d(p, q)}\} \tag{1}$$

   where d(p, q) = Euclidean distance of p from q.

3. Compute Local Reachability Density (LRD) of data record of p.

   Local Reachability Density (*LRD*): It is the inverse of the average reachability distance of the object P from its neighbors as defined in Eq. (2)

$$\text{LRD (p)} = 1 \Big/ \frac{\sum \text{reachability distance (p,q)}}{|\text{N}_\text{k}(\text{p})|} \tag{2}$$

where $N_k(p)$ = k-nearest neighbors of p

4. Compute the Local Outlier Factor (LOF) of data record p.

   Local Outlier Factor (*LOF*): The LOF value of A is determined as the average local reachability *density of the neighbors* divided by the object's own local reachability as defined in (3)

$$\mathrm{LOF}\,(\mathrm{p}) = \frac{\sum \frac{\mathrm{lrd(q)}}{\mathrm{lrd(p)}}}{|\mathrm{N_k(p)}|} \tag{3}$$

   And based on the LOF value, LOF algorithm finds the anomalous data points with respect to its neighbors. If the resultant value of LOF of a given point is significantly larger than 1, then it is considered to be outlier.

2.2 *Fuzzy C-Mean Algorithm*

The FCM algorithm first execute the c-mean clustering algorithm to form clusters and then outliers are determined by identifying small clusters which holds lesser points than half of the average number of points in the k clusters. Then temporarily removing the outliers in the form of small clusters, the c-mean algorithm re-execute; and if noticeable decrease in the objective function value, then the points are said to be outliers [13].

This algorithm works by assigning membership value to each data points with respect to each cluster center. More the data point nearer to the cluster, more is its membership value and more being the chance to become the part of that cluster.

Hence, the objective of FCM algorithm is to

$$\text{Minimize } J(U, V) = \sum_{i=1}^{n}\sum_{j=1}^{c} \mu ij^{m}||xi - vj||$$

where

m is any real number greater than 1.
μij is the degree of membership of xi in cluster j.

To find out the cluster, the following steps are to be followed:

Step 1: Input $X = \{x_1, x_2, x_3, \ldots, x_n\}$
Step 2: Choose cluster centers $V = \{v1, v2, v3 \ldots, v_c\}$
Step 3: Calculate the fuzzy membership value
$\mu_{ij} = 1/\sum_{k=1}^{c}\left(\frac{||xi-cj||}{||xi-ck||}\right)^{\left(\frac{2}{m}-1\right)}$ $1 < i <= n$, Cj is the d-dimension center of cluster.
Step 4: Compute the fuzzy centers V_j

$$V_j = \sum_{i=1}^{n} \mu ij^m \cdot xi \Big/ \sum_{i=1}^{k} \mu\, ij^m \quad i < j <= c$$

Step 5: Repeat steps 3 and 4 until the minimum J value is achieved. (Where J is the objective function)

3 Proposed Approach

In this approach, we extensively studied fuzzy clustering process and applied it as outlier detection techniques in DIDS process. The characteristics of outlier resembles with characteristics of an intruder whose behavior is very much different from the normal user. Hence, in our research, outlier detection is categorized as anomaly detection to detect the suspicious behavior from the large behavioral set.

In this research, we have generalized outlier detection method in two categories: hard outlier detection and soft outlier detection process. The clustering process through which outliers are detected, if they create a hard boundary between clusters that means the data points may or may not belong to a cluster, are said to be hard clustering and detection procedure is hard outlier detection process. And if, the data points belong to any of the clusters with some membership value is called soft clustering and process of outlier detection is soft outlier detection.

In this investigation, Local Outlier Factor (LOF) is treated as hard outlier intrusion detection method and Fuzzy Clustering is as soft outlier intrusion detection procedure. And clearly from the result, soft outlier intrusion detection through Fuzzy C-mean clustering (FCM) proved to be more accurate than LOF-based outlier intrusion detection.

To demonstrate its performance, we have used a real biometric dataset of an institution to detect the intrusive activity. Biometric features of all users include <institution id, card number, in time, out time, average working hour, location>.

Whole system is monitored through database administrator and central database administrator. Now intruder can temper the database or want to modify the database with genuine or in genuine card number to facilitate his intentions. And our intention is to identify that intruder or its intrusive access toward the database. The intruder most probably is an insider who knows the total system of information flow. In a large organization, it is a common issue and need to be solved. To handle such situation, we analysed the normal access pattern to the biometric system by taking a model developed through FCM_BackPropagation learning model and identified the malicious access; The performance of the system is measured through FPR, false positive rate, accuracy, elapsed time matrices and compared against its counterpart LOF_DIDS.

All the above transactional features are tried to model the normal behavior of biometric database users and based on these features we model the DIDS and transactions or users which deviate the normal behavior are encountered to be malicious and in this research we consider them as outliers which are identified through several clustering algorithm and their performance is analyzed. Our research investigation is based on the assumption that the outliers are the abnormal database transactions that are suspected of having generated by different mechanism or are fetched for malicious intentions and are considered to be serious security threats. For example, abnormalities in geographical location, time of accessing data, frequency of access and amount of accessibility often imply high risk and are suspected to be malicious. In this regard, we exploited two types of clustering (hard and soft clustering) algorithm to determine the outliers or the malicious database transactions.

Following Sects. 3.1–3.2 describe three processes for Biometric_DIDS and finally comparisons are made based on their performance.

3.1 LOF_DIDS

Step1: Input the biometric dataset
Step2: Preprocessing the Data

- Z-score Normalize the dataset
- 10-fold Cross Validation the data
- Partition the whole dataset to training and testing the model

Step 3: Cluster the training data points and find out the outlier by determining the LOF value of each point in the dataset.
Step 4: Train the LOF_DIDS clustering model
Step 5: Test the designed model through test data

3.2 FCM_Backpropagation_DIDS

FCM algorithm assigns fuzzy membership value (μij) to each data points with respect to each cluster center and the data point whose have more membership value, i.e., the data points which are more closure to the clusters and have more chances to being part of the cluster. The membership function parameters of fuzzy logic system are tuned by Back Propagation Neural Network [14] is used. Using this learning technique, we can get optimized membership function parameters optimized and reduced error rate to get accurate results. Table 1 describes the algorithmic concept of FCM based outlier detection process that is depicted in Fig. 1.

After extracting the outliers from the biometric dataset by applying FCM algorithm, the outliers are analyzed by Type-1 fuzzy logic System which has mainly three components: Fuzzifier, Defuzzifier, and FIS. The fuzzifier converts the real data to

Table 1 Algorithm: DIDS_FCM_Backpropagation

Step 1: Input Dataset
Step 2: Process the Data before processing
    Step 2.1: Data Cleaning
    Step 2.2: Data Transformation through Normalization
Step 3: Generate a generalized train and test data set by using 10-fold cross validation
Step 4: Cluster the training dataset by using FCM.
Step 5: The outliers are determined by identifying small clusters which contains lesser points than half of the average number of points in the k clusters.
Step 6: Then temporarily removing the outliers in the form of small clusters, the c mean algorithm re-execute;
Step 7: And if noticeable decrease in the objective function value, then the points are said to be outliers.
Step 8: Analyse the outliers by using Type-1 fuzzy-logic system.
    Step 8.1: Run Back Propagation learning algorithm
    Step 8.2: Fuzzify the outlier
    Step 8.3: Input the fuzzified data to Fuzzy Inference System (FIS)
    Step 8.4: FIS takes the decision of malicious or genuine based on the generated rules.
    Step 8.5: Defuzzify the output from FIS

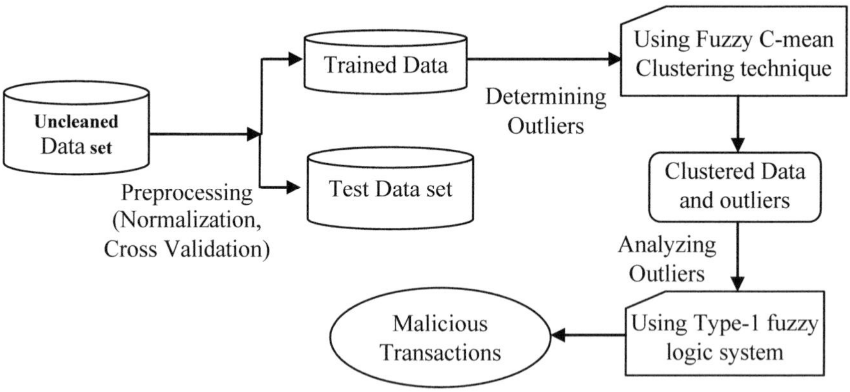

Fig. 1 FCM_Backpropagation DIDS

fuzzy data by applying them to triangular membership function; Defuzzifier is the reverse procedure of fuzzification which converts the fuzzy data to real data through Centroid method; Sugeno FIS generates several rules from the extracted outliers and take decisions regarding intrusions based on the rules. The parameters used in FIS are predicted through Back Propagation Learning algorithm [14]. The proposed DIDS model is initially trained to determine the outliers and then the outliers are analyzed to be identified as intrusions based on FCM_Backpropagation learning model.

4 Result Analysis

In this section we illustrate the results outcome and analyze the performance of the above discussed algorithms. When intrusions are not correctly detected by the intrusion detection model, then they are false negatives and false positives are normal transactions which are identified as intrusions. False positives and false negatives are treated to be dangerous for any proposed model and so they are to be minimized. Equivalently true positives and true negatives should be maximized as they are counted to be best for a model. Overall, False positive rate (FPR), True positive Rate (TPR) and accuracy are three performance matrices for any proposed model and are calculated by using the formula specified in Table 2.

From Fig. 2, we can clearly state that FCM_Backpropagation_DIDS is giving better performance with respect to all these three performance measure. By referring to the resulted Table 3, we can see FCM shows better performance in case of elapsed time than its counterpart LOF algorithm. By varying minpts and epsilon value, number of outliers and elapsed time is calculated which shows the performance of LOF-based DIDS; while comparing the elapsed time of all the three DIDS model, FCM_Backpropagation based DIDS gives better result.

Table 2 Performance measures

Performance measures	Procedure	Notations used
Accuracy	(TP + TN)/(TP + FP + FN + TN)	TP: Correctly predicted positive values TN: Correctly predicted negative values FP: Incorrectly classified positive samples FN: Incorrectly classified negative samples P: Total number of positive samples N: Total number of negative samples
TPR	TP/P	
FPR	1-(TN/N)	

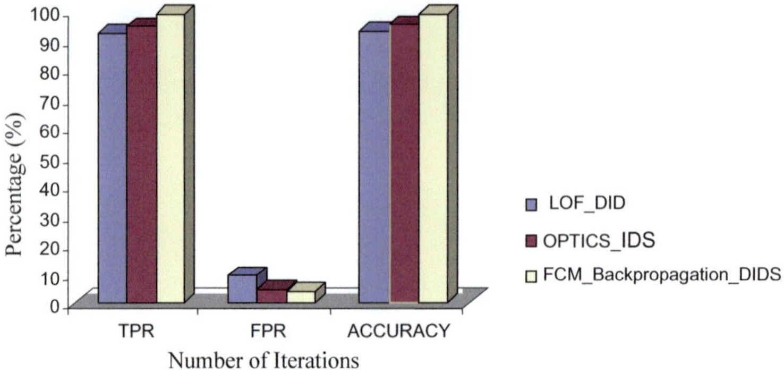

Fig. 2 Performance of three models

Table 3 Performance of the three DIDS models

	minpts	epsilon	Elapsed time (ms)	Count_outlier
LOF_DIDS	50	0.5	42.317621	5085
LOF_DIDS	100	10	41.317621	1414
LOF_DIDS	150	10	41.344283	2865
LOF_DIDS	150	50	42.20668	1219
Fcm_Backpropagation_DIDS	NA	NA	37.3	1013

5 Conclusions

In this paper, we have applied the data mining concept for identifying outliers for database intrusion detection. In addition to that, we made a comparison between hard and soft clustering based outlier detection. From the result analysis, we clearly state that outliers which are assumed to be malicious are correctly identified through fuzzy or soft outlier analysis process. In future, we will analyze the impact of the swarm intelligence to optimize the parameters used in anomaly detection and evaluate the detection rate and false alarm rate of the proposed database intrusion detection model.

References

1. Mathew, S., Petropoulos, M., Ngo, H.Q., Upadhyaya, S.: A data-centric approach to insider attack detection in database systems. LNCS **6307**, 382–401 (2010)
2. Abadeh, M.S., Habibi, J., Lucas, C.: Intrusion detection using a fuzzy genetics-based learning algorithm. J. Netw. Comput. Appl. **30**, 414–428 (2007)
3. Christina, Y.C., Michael, G., Karl, L.: DEMIDS: a misuse detection system for database systems. In: Proceedings of the Third Annual IFIP TC-11 WG 11.5 Working Conference on

Integrity and Internal Control in Information Systems, pp. 158–178. Amsterdam, Netherlands (1999)
4. Lee, S.Y., Low, W.L., Wong, P. R.: Learning fingerprints for a database intrusion detection system. In: ESORICS 2002. Lecture Notes in Computer Science, No. 2502, pp. 264–280. Springer (2002)
5. Christina, Y.C., Michael, G., Karl, L.: Demids: a misuse detection system for database systems. In: Third International IFIP TC-11 WG11.5 Working Conference on Integrity and Internal Control in Information Systems, pp. 159–178 (1999)
6. Cho, S.: Incorporating soft computing techniques into a probabilistic intrusion detection system. IEEE Trans. Syst., Man Cybern.-Part C Appl. Rev. **32**, 154–160 (2002)
7. Panigrahi, S., Sural, S.: Detection of database intrusion using two-stage fuzzy system. LNCS **5735**, 107–120 (2009)
8. Zhong, Y., Zhu, Z., Qin, X.: A clustering method based on data queries and its application in database intrusion detection. In: Proceedings of The Fourth International Conference on Machine Learning and Cybernetics, pp. 2096–2101 (2005)
9. Khan, M.I., Sullivan, B., Foley, S.N.: A semantic approach to frequency based anomaly detection of insider access in database management system. In: CRiSIS 2017, LNCS 10694, pp. 18–28 (2018)
10. Kim, S., Cho, N.W., Lee, Y.J., Kang, S., Kim, T., Hwang, H., Mun, D.: Application of density-based outlier detection to database activity monitoring. Inf. Syst. Front. **15**, 55–65 (2013)
11. Panigrahi, S., Sural, S., Majumdar, A.K.: Two-stage database intrusion detection by combining multiple evidence and belief update. Inf. Syst. Front. (2010)
12. Breunig, M.M., Kriegel, H., Ng, R.T., Sander, J.: LOF: identifying density-based local outliers. In: Proceedings of the ACM Sigmod, International Conference on Management of Data Dalles TX (2000)
13. Zhu, L., Chung, F., Wang, S.: Generalized C-means clustering algorithm with improved fuzzy partitions. IEEE Syst. Man Cybern. Soc. **39**(3), 578–591 (2009)
14. Jang J.S.R.: Neuro-fuzzy and soft computing, 1st edn. PHI/Pearson Education, New Delhi, (2004)

Grey-Level Text Encryption Using Chaotic Maps and Arnold Transform

Vasudharini Moranam Ravi, Nishi Prasad, C. Lakshmi, Sivaraman Rethinam, Sridevi Arumugham and Amirtharajan Rengarajan

Abstract Data in today's world is ubiquitous and often regarded more valuable than the paper currency itself. More often than not, sensitive data needs to be transmitted from a sender to a receiver in an encrypted format. This paper proposes a novel secret message transmission method through image encryption. The proposed system highlights four techniques; Bit Plane Splicing (BPS), Arnold Transform, a combination of logistic and tent maps. The original text is obtained after decryption, reinforcing its efficiency, security and reliability of this scheme. The security of this scheme has been rigorously analysed through image metrics like Correlation, Entropy, NPCR and UACI, histogram deviation and are tabulated for reference. The results show good security performance against threats. Chosen plaintext attack has been carried out, on two sets of different plain texts, and the results obtained have been favourable.

Keywords Image encryption · Bit Plane Splicing · Arnold transform · Logistic mapping · Tent mapping

1 Introduction

Data encryption, however advantageous though has its drawbacks, which are covered for by encryption of images. The latter offering high data storage and high redundancy. Multimedia data possess high redundancy and strong correlation between pixels but not textual data. There is a constant concern over digital media like audio, video and text being easy to append, modify, transmit, and redistribute. This begs to question the reliability of the data being transmitted. This paper employs Text

V. M. Ravi · N. Prasad · C. Lakshmi (✉) · S. Rethinam · S. Arumugham · A. Rengarajan
School of EEE, SASTRA Deemed to be University, Thanjavur, Tamil Nadu, India
e-mail: lakshmi_c@ece.sastra.edu

C. Lakshmi
Department of ECE, School of EEE, SASTRA Deemed to be University, Thanjavur, Tamil Nadu, India

B. Pati et al. (eds.), *Advanced Computing and Intelligent Engineering*, Advances in Intelligent Systems and Computing 1082,
https://doi.org/10.1007/978-981-15-1081-6_42

transmission via image encryption with two levels of diffusion, carried out through Logistic and Tent map and Arnold Transform. An image is created of the text that is to be transmitted. And on this image, Arnold Transform and chaotic maps are performed.

Image encryption is employed to produce a garbled and inarticulate image from the original image. This is done to boost the power to resist an attack to procure the original image, thereby enhancing its security. The data that is generated in large amounts is confidential and private and must be maintained as such. The information is protected by using a secret key, that converts it into some unintelligible form.

This paper utilizes makes use of Arnold Transform. It is simple, periodic and invertible. It was conceptualized by V. I. Arnold in the field of ergodic theory. It aims to decrease the intelligibility of the encrypted image in contrast to the original image. It is defined for encrypting digital two-dimensional N × N images as

$$\begin{bmatrix} x' \\ y' \end{bmatrix} = \begin{bmatrix} 1 & 1 \\ 1 & 2 \end{bmatrix} \begin{bmatrix} x \\ y \end{bmatrix} (modN)$$

Here, (x, y) and (x', y') are pixel coordinates of the original and scrambled image. Let x, y correspond to the original image, which becomes x', y' after Arnold scrambling is applied. As this is defined for square images, this system has been remodelled to suit M × N images as well. Arnold Transform is cyclic in nature. That is, the original image is obtained after several cycles, and this is dependent on image size. The period is related to the size of the image, but no direct relation has been found yet. Like for a 256 × 256 image, the period found is 192. Min et al. [1] suggested a method to carry out Arnold Transform for non-squared images (M × N images).

Jafar et al. [2] proposed a scheme based on Reversible Data Hiding (RDH) using two dual images to enhance embedding capacity and complexity. Shang et al. [3], proposed a method for digital image block location scrambling. It improved the Arnold Transform and employed Logistic Map to initiate sequences of parameters. The results were consistent and met with the security requirements for image encryption.

The drawback with using the Arnold Transform is that it can be applied to images of size N × N only. Second, its periodicity suggests that this method is unsecure because this transform is invertible. So, the proposed method proposes to substitute the scrambled values back into the original image before the next iteration is performed. Therefore, Arnold Transform itself is insufficient to carry out image encryption.

Chaotic based encryption was invented in 1989. Chaotic map properties include initial conditions, system parameters and sensitive dependence. Chaos-based encryption techniques are favoured owing to their speed, complexity, security and computational power. Chaotic maps are established based on initial conditions and closely follow cryptographic principles like confusion and diffusion. The two used are detailed below.

The logistic map is a simple non-linear function but broadly used by researchers. The one-dimensional chaos logistic equation is given as follows:

$$X_{n+1} = \mu X_n(1 - X_n) \tag{1}$$

Here, $\mu \in [0, 4]$ and $X \in [0, 1]$. Here, μ is the branch parameter and X is the initial parameter. The latter signifies that the function stays in a chaotic state and displays randomness. The disadvantage of Logistic Map is that it is sensitive to initial conditions leading to widely diverging outcomes. But it is widely used because of its efficiency in encryption schemes.

The tent map, used apart from a Logistic map that generates periodic and chaotic behaviour is defined by

$$T(x) = \begin{cases} \mu X_n, & X_n < 1/2 \\ \mu(1 - X_n), & 1/2 \geq X_n \end{cases} \tag{2}$$

This is the most important class of maps in the interval [0, 1] to itself.

Khanzadi et al. [4] brought about a novel system for image encryption integrating chaotic Logistic Map and Tent Map with the gyrator transform. Both these maps generate a random sequence of bits and then the original image is encrypted using the gyrator transform.

Som and Kotal [5] presented a system that uses Arnold Transform to scramble the plaintext image. Then chaotic sequences like the 1-D Logistic map are applied to the scrambled image thereby encrypting it and further processing them to integers. This paper was able to prove that the system provides for a very strong keyspace, robust with high sensitivity to secret keys and the desired entropy value.

Kori et al. [6] reviewed digital colour encryption with a combination of chaotic logistic and tent map. They explored and detailed concepts of chaotic maps. The encryption process involved pixel-wise shuffling with the use of a tent map and chaotic map based pixel substitution.

Safi and Maghari [7] proposed the system that obtains two keys from double logistic maps, which are XORd and the resultant value, is operated with the recombined image. This technique provides small iteration time and high security. It also highlights the disadvantage of using only the Logistic map namely blank windows, uneven distribution of sequences and production of a weak key.

Khot and Awati [8] proposed a system based on RDH in an encrypted image based on Reversible Image Transformation. Proposed for the purpose to ensure that the cloud could not add any additional data to the encrypted image, so to tamper with it.

The drawback of most chaotic maps is that it suffers from known plaintext attack. Through the system proposed, with the same set of security keys, a different encrypted image can be generated, each time the same image was used. Image encryption through chaotic maps, the image can be transmitted over the internet or a public domain and guarantees security in the communication. Based on the two chaotic maps and Arnold Transform, we can obtain an almost different encrypted image for the same set of keys and the original image (made of the text). This is highlighted in Zhou et al.'s [9] proposed system that was able to generate a new encrypted image for each of the same original images. Experimental results were found to be excellent.

In the proposed method, bit plane slicing is also employed, to split images into binary planes. Arnold Transform is applied on each bit plane. Post recombining all the slices together. A key is generated by the combined operation of Logistic and Tent map. Sometimes, data security with chaotic maps may be overestimated, plaintext attacks could be applied. So the key length must be large and sensitive. The combination of the two chaotic maps and XOR operation provide resilience against brute force attacks as a variegated keyspace can be built. The results also show sufficient randomness and correlation. This is computationally feasible, secure, and efficient.

The proposed methodology is given in Sect. 2, followed by security analysis in Sect. 3, Results and Discussion in Sect. 4, Acknowledgements in Sect. 5 and Conclusion in Sect. 6.

2 Proposed Methodology

As mentioned above, the image is encrypted with two operations. First, bit-plane slicing, and performing Arnold Transform on all the slices. Subsequently after recombination, executing a key generation through chaotic and tent mapping and finally performing a bit XOR operation with the key and recombined matrix.

Step 1: Obtain the given text in a 128 × 128 character text, $\mathbf{P_{ij}}$. And convert it to ASCII (number array), to yield a two-dimensional matrix or an image, $\mathbf{H_{ij}}$.
Step 2: Perform bit plane slicing and obtain eight two-dimensional matrices accordingly. $\mathbf{Q} = \{\mathbf{Q_1}, \mathbf{Q_2}, \ldots, \mathbf{Q_8}\}$.
Step 3: Perform Arnold Transform on each slice $\mathbf{Q_n}$, and then recombine all the slices and the resultant matrix is $\mathbf{E_{ij}}$.
Step 4: Generate keys $\mathbf{K_1}$ and $\mathbf{K_2}$ through chaotic and tent mapping respectively.
Step 5: Carry out a bitwise XOR operation with the values obtained from $\mathbf{K_1}$ and $\mathbf{K_2}$, and the resultant value is operated with $\mathbf{E_{ij}}$ to obtain encrypted matrix $\mathbf{C_{ij}}$.
Step 6: Find entropy and MSE of the final encrypted image.

The decryption process involves performing a bit XOR with the encrypted matrix, and then bit plane slicing. On these slices, inverse Arnold Transform is applied. Finally recombined and then converted to the character text. This is highlighted below (Fig. 1).

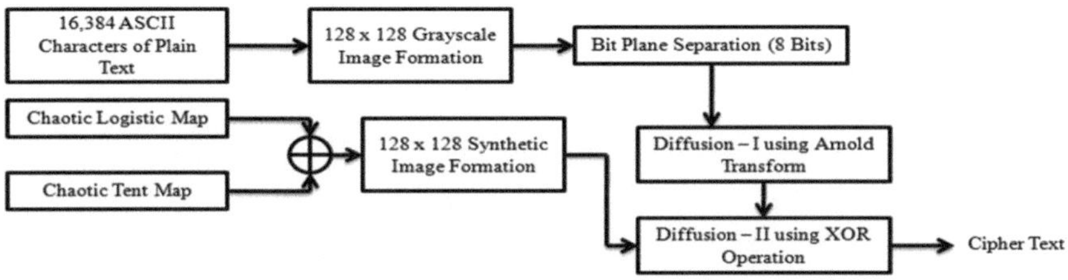

Fig. 1 The block diagram representing the process of image encryption

3 Security Analysis

As mentioned earlier, the system proposes a novel image encryption technique for text transmission using chaotic maps and Arnold Transform. Upon decryption, the original text is retrieved back. The scheme is therefore proven to be efficient and reliable. The system is both carried out in probabilistic time, making it computationally feasible owing to its short iteration time. With the use of the two chaotic logistic maps and Arnold Transform, the scheme is reversible and secure. The system parameters and the initial conditions are susceptible to change in chaotic maps. Therefore, the encrypted image can be different for the same original text and key variants. This system also is resilient to brute force attacks owing to the strong keyspace. This means the attacker would take close to 10^{34} time to crack the code. This number is greater than the 2^{256} and therefore is resistant against brute force attack. A combination of chaotic and logistic maps, make up for the disadvantage of the sole logistic map that has an uneven distribution of sequences. Chosen plain text attack was carried out between two sets of plaintexts and the corresponding cipher texts obtained, were observed and displayed in the next section.

4 Results and Discussion

To evaluate the proposed algorithm, the secret message is taken as the length of 16,384 alphanumeric characters and then made to form a 2-D matrix with the text. Each of the three represents Text1, Text2, Text3 respectively. The three texts-to-images are illustrated below (Fig. 2).

4.1 Correlation Analysis

Correlation refers to how closely the encrypted and original images are to be. The correlation coefficient is supposed to be low indicating that the original and encrypted images are uncorrelated. Correlation coefficient is given by Eq. 3.

Fig. 2 The three images represent the 2D matrices of the texts **a** text 1 **b** text 2 **c** text 3

Table 1 Correlation analyses

Images		Horizontal	Vertical	Diagonal
Text1	Original	0.0116	0.0116	0.0028
	Encrypted	0.0082	0.0082	0.0004
Text2	Original	0.0083	0.0083	0.0107
	Encrypted	0.0047	0.0047	0.0109
Text3	Original	0.0067	0.0067	0.0087
	Encrypted	0.0022	0.0022	0.0066

$$E(x) = \frac{1}{N}\sum_{i=1}^{N} x_i$$

$$D(x) = \frac{1}{N}\sum_{i=1}^{N} (x_i - E(x))^2 \tag{3}$$

$$Cov(x, y) = \frac{1}{N}\sum_{i=1}^{N} (x_i - E(x))(y_i - E(y)) \; \gamma_{xy} = \frac{Cov(x, y)}{\sqrt{D(x)}\sqrt{D(y)}}$$

where,

x, y Grey-level values of two adj. pixels,
N Total number of pixels,
γ_{xy} Cross correction of two pixels.

Correlation coefficients for the encrypted and original image are given in Table 1.

4.2 Entropy Analysis

The amount of information that must be coded for by a compression algorithm. A statistical measure of randomness used to represent the image itself. The value of an encrypted value is less than the ideal value, 7.5 in this case. Entropy is given by,

Table 2 Entropy analyses

Images		Entropy
Text1	Original	5.9491
	Encrypted	6.0169
Text2	Original	5.9501
	Encrypted	6.0180
Text3	Original	5.9506
	Encrypted	6.0191

$$E = -\sum_{i=0}^{n} p(x_i) * \log_2 x_i \tag{4}$$

where

E Entropy,
x_i i-th Grey level value of N-level image,
p Probability of each Grey level in the image

Table 2 is representative of the entropy values for the encrypted image that deviates from the ideal value 7.

4.3 *Differential Analysis*

NPCR and UACI are called as differential analyses to evaluate the pixel changes. A Number of Pixels Change Rate (NPCR) is defined as a common metric to check the result of the change of one pixel over the entire image. Indicative of the percent change of the difference pixels between two images. Higher NPCR values are preferred for ideal encryption. Unified Average Changing Intensity (UACI) represents the intensity difference averaged between the two images. This is indicative of how, when the plain text changes slightly, the ciphertext changes significantly. The values for UACI should be in the 33% bracket.

$$NPCR = \frac{\sum_{k=1}^{H \times W} B_k(pixel\ value(i, j))}{H \times W} \times 100\% \tag{5}$$

$$UACI = \frac{\sum_{k=1}^{H \times W} |C1(i, j) - C2(i, j)|}{H \times W \times 255} \times 100\% \tag{6}$$

$$B_k(pixel\ value(i, j)) = \begin{Bmatrix} 0, if\ C1\ (i, j) = C2(i, j) \\ 1, otherwise \end{Bmatrix} \tag{7}$$

Here, $C1$ and $C2$ are encrypted images. H and W are height and width of the images. Table 3 depicts the obtained NPCR and UACI values for encrypted images.

Table 3 Differential analyses

Image	NPCR	UACI
Text1	99.9939	31.8097
Text2	98.9346	32.0608
Text3	99.9146	31.7062

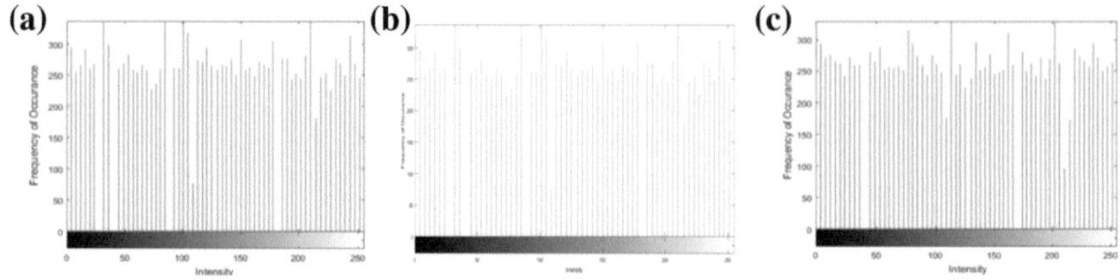

Fig. 3 Histogram analysis for encrypted **a** Text1 **b** Text2 **c** Text3

5 Statistical Analyses

An image histogram is the graphical representation of intensity levels allocated for each pixel in an image. The histograms for Text1, Text2, and Text3 are given as Fig. 3.

6 Chosen Plaintext Attack Analysis

Algorithms that implement diffusion and XOR operations must undergo this analysis to test if it can withstand chosen plaintext attack. This is performed by the equation given below

$$C1(i, j) \oplus C2(i, j) = \mathrm{P1}(i, j) \oplus \mathrm{P2(i, j)} \tag{8}$$

$C1$ and $C2$ are the ciphertext results obtained from applying the algorithm on two plain text images P1 and P2. Let the left-hand side of the equation be denoted as **R1**, and the right-hand side be denoted as **R2**. If this equation is satisfied, that means the algorithm cannot endure chosen plaintext attack (Table 4).

Table 4 Chosen plaintext attack analysis

Correlation between R1 and R2	Result of XOR with $C1$ and $C2$ (R2)
Result of XOR with P1 and P2 (R1)	0.00706

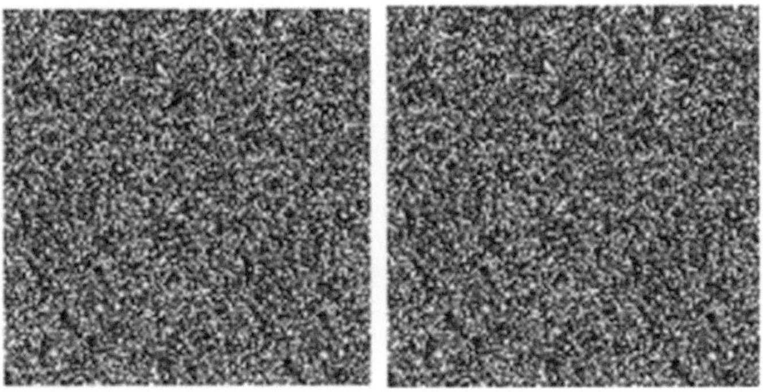

Fig. 4 XOR operation between: **a** the plain texts 1 and 2 **b** cipher Texts 1 and 2

From the calculations carried out, this equation is not satisfied and is not consistent with the values achieved. So, the algorithm can resist plaintext attack.

Figure 4 gives the pictorial representation of the results obtained after performing chosen plaintext attack.

7 Conclusion

This paper proposed a system which involved text transmission through image encryption using bit plane slicing, Arnold Transform, a combination of chaotic Logistic and Tent maps. This system is found to be reversible, as the text is obtained or decrypted and is found to be the same as the text sent or encrypted. An extensive security analysis was carried out in the form of histogram analysis and chosen plaintext attack. The results have been tabulated and discussed.

Acknowledgements The authors wish to express their gratitude to SASTRA Deemed University for providing the infrastructure and access through the Research & Modernization Fund (Ref. No: R&M/0026/SEEE-010/2012–13) to carry out the proposed research work.

References

1. Min, L., Ting, L., Yu-Jie, H.: Arnold transform based image scrambling method. In: Multimedia Technology (ICMT 2013), pp. 1309–1316 (2013)
2. Jafar, I.F., Darabkh, K.A., Al-Zubi, R.T., Saifan, R.R.: An efficient reversible data hiding algorithm using two steganographic images. Signal Process., 98–109 (2016)
3. Shang, Z., Ren, H., Zhang, J.: A block location scrambling algorithm of digital image based on Arnold transformation. In: Proceedings of the 9th International Conference for Young Computer Scientists, pp. 2942–2947 (2008)

4. Khanzadi, H., Eshghi, M. Borujeni, S.E.: Image encryption using random bit sequence based on chaotic maps. Arab. J. Sci. Eng. **39**(2), 1039–1047 (2014)
5. Som, S., Kotal, A.: Confusion and diffusion of grayscale images using multiple chaotic maps. In: National Conference on Computing and Communication Systems (NCCCS) (2012)
6. Kori, P., Dubey, R., Richhariya, V.: Survey on double phase image encryption and decryption using tent map and chaotic logistic map. Int. J. Sci., Eng. Technol. Res. (IJSETR) **4**(11) (2015)
7. Safi, H.W., Maghari, A.Y.A.: Image encryption using Double Chaotic Logistic Map. In: 2017 International Conference on Promising Electronic Technologies (ICPET) (2017)
8. Khot, A.T., Awati, C.J.: Data hiding in encrypted image by reversible image transformation. Int. J. Adv. Eng. Res. Sci. **3**(12) (2016)
9. Zhou, Y., Bao, L., Chen, C.L.P.: A new 1D chaotic system for image encryption. Sig. Process. **97**, 172–182 (2014)

Application of Deep Learning for Database Intrusion Detection

Rajesh Kumar Sahu and Suvasini Panigrahi

Abstract In this paper, we have suggested a deep learning model aimed at effective detection of malicious transactions in a database system. This method focuses on exploiting the user normal behavior, data dependencies, and data sensitivity of a transaction to predict intrusions. Currently, we have used different kinds of neural networks according to their strengths of predicting the intrusion according to the type of data such as sequential or featured data. For experimental evaluation, we have used a recurrent neural network for sequence data and feed-forward with back propagation for other attributes, together creating a hybrid deep learning model which works effectively to predict the database intrusions.

Keywords Database intrusion detection · Signature based detection · Anomaly based detection · Deep learning · Neural network

1 Introduction

In present days, there is a wide range of methods to secure vital information but frequently fail. Much efforts being taken by researchers towards increasing intrusion efficiency but they are on the network intrusion detection or on the operating system level unable to detect intrusions at the database level which can cause data corruption and can disturb the data integrity due to malicious activities. Thus, detecting malicious transactions which are done by a Database Intrusion Detection System can help at data damage assessment and fast recovery.

R. K. Sahu · S. Panigrahi (✉)
Department of Computer Science & Engineering, Veer Surendra Sai University of Technology, Burla 768018, Odisha, India
e-mail: spanigrahi_cse@vssut.ac.in

R. K. Sahu
e-mail: rksahu.1214@gmail.com

B. Pati et al. (eds.), *Advanced Computing and Intelligent Engineering*,
Advances in Intelligent Systems and Computing 1082,
https://doi.org/10.1007/978-981-15-1081-6_43

1.1 Techniques Used in Intrusion Detection

Decision Trees: Classification or Decision Tree is a technique used mainly for problems involved in classification. This technique creates a tree which has a decision parameter at each node from the class leveled training datasets. The modeled decision trees can be used to classify the testing or unseen data to its expected class.

- Entropy function:

$$E(S) = \sum_{i=1}^{c} -\left(p_i \log \frac{p_i}{2}\right). \tag{1}$$

- Information Gain:

$$\mathrm{Info}_A(D) = -\sum_{i=1}^{v} \frac{|D_j|}{|D|} \times \mathrm{Info}(D_j) \tag{2}$$

$$\mathrm{Gain}(A) = \mathrm{Info}(D) - \mathrm{Info}_A(D). \tag{3}$$

Support Vector Machines: This is a supervised learning technique used for classification and prediction. It creates a one or more hyperplanes to separate two classes. The more the distance of hyperplane to the nearest data point, lower the further classification errors. A hyperplane can be written as follows:

$$W \cdot X + b = 0 \tag{4}$$

where $W = \{w_1 \ldots w_n\}$ are weight vectors, $A = \{A_1 \ldots A_n\}$, b is a constant, and $X = \{x_1 \ldots x_n\}$ are values attribute values.

Multilayer Perceptron: It is a feed-forward artificial neural network with back propagation in Fig. 1. Back Propagation is used to calculate the gradient which is used to calculate weights. The gradient decedent optimization is commonly used in back propagation to adjust the weight of neurons by calculating the gradient of the loss function. The chain rule is expressed as shown in Eq. (5). The activation functions used are Sigmoid function and Relu functions as given in Fig. 2:

$$\frac{\partial E_{total}}{\partial w_5} = \frac{\partial net_{o1}}{\partial w_5} \times \frac{\partial out_{o1}}{\partial net_{o1}} \times \frac{\partial E_{total}}{\partial out_{o1}}. \tag{5}$$

Recurrent Neural Networks: Recurrent Neural Network is applied to sequential featured attributes. In a normal artificial neural network, we assume that all inputs

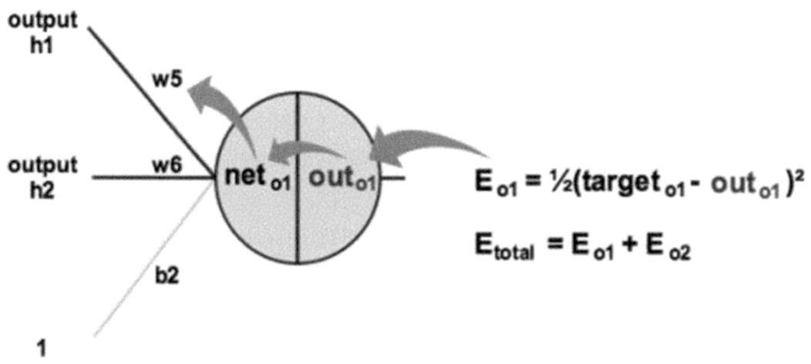

Fig. 1 Back propagation

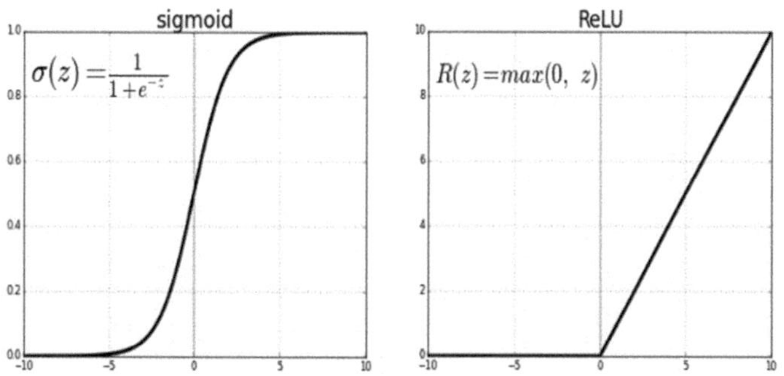

Fig. 2 Activation functions

and outputs are unrelated. In RNNs, the output is dependent on the previous output computed. In another way, RNNs have a "memory" which records previous information. Practically, RNNs can only record a few steps back (Fig. 3).

LSTM networks are a type of RNN. It has special units in with the standard units. It includes a memory cell which can maintain states in memory for long periods of duration. A set of gates is used to regulate when do information enters the memory, when its output, and when its forgotten. This architecture lets them learn longer term dependencies.

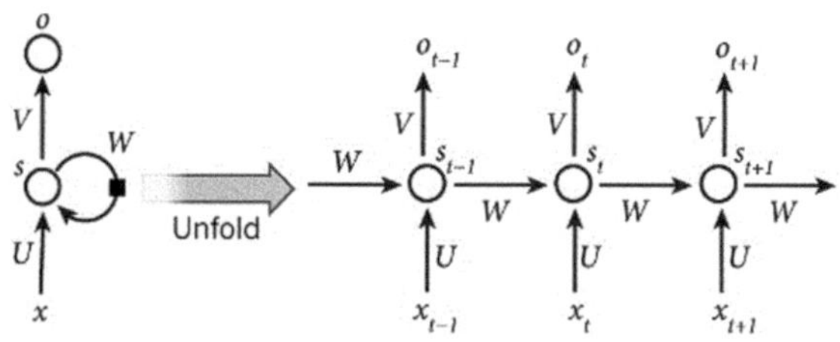

Fig. 3 Recurrent neural network node

1.2 Related Work

There are many works done in case of intrusion detection systems. Here is a brief conclusion of techniques used and proposed in different papers on intrusion detection. Detecting data dependencies among data items in a database is a feature used to find a transaction malicious or genuine [1]. Different classifiers according to the strength of detecting different attack types are merged together to create a hybrid intelligent model for NIDs [2]. Comparing SVM and C4.5, the later gives better results in case of Network Intrusion Detection [3]. The IDS should focus on anomaly detection for better future results [4]. A SQL injection attack can be detected malicious or genuine depending on the nature of the operation, the sensitivity of data fetched [6]. Recurrent Neural Network effectively does sequence classification in case of speech recognition [5] and sentimental analysis where the classification is totally based on temporal sequence data. Recurrent neural networks work better in case of intrusion detection [7].

2 Challenges

2.1 Problem Discussion

The challenges are discussed and how our proposed solution attempts to solve it has also been discussed in the next section

a. Nowadays more and more data being accumulated in the data centers thus making data corruption attempts invisible which is a costly affair to get back the integrity of the data. Apart from network intrusion detection and detection of intrusion at operating system level another additional layer of security to detect intrusions at the database level can be very useful to prevent data corruption and restore the data integrity disturbed due to intrusion attempts.
b. The main challenge or problem is to detect the unseen database intrusions attempts which are called Anomaly Based Database Intrusion Detection Systems. With an increase in computation power and a huge amount of data available we can use Deep Leaning Models to efficiently detect the Database Intrusions.
c. The Database Intrusion can be detected by analyzing user's behavior which generates a transaction behavior. The features which map behavior of a transaction are Sequence order of table ids which has been retrieved by the transaction and Sequence order of attribute ids which has been retrieved by the transaction. Thus we have used Recurrent Neural Network to create a sequence classification model to predict the sequence is malicious or genuine.
d. The Database Intrusion has a feature that the intruder either want to extract a huge amount of data (number of reads and write operations) or want to tamper sensitive

data (write operations) or want to retrieve sensitive data (read operations) which has to monitored to predict a transaction is malicious or not.

2.2 Proposed Approach

In today's scenario with a huge amount of data available and increased computation power, deep learning performs better than the other machine learning algorithms. Deep learning is getting enhanced and put into more use with the advent of GPUs. The main feature of deep learning is as plotted below (Fig. 4):

a. Recurrent Neural Network is a type of artificial neural network which helps in processing sequence data to classify or predict a sequence. It allows exhibiting temporal behavior for a time sequence. RNNs can use their memory to do sequential processing of data. RNNs gives great result in handwriting or speech recognition, these are purely ordered sequence data. The features which map behavior of a transaction are Sequence order of table ids which has been retrieved by the transaction and Sequence order of attribute ids which has been retrieved by the transaction. Deep network module for sequence classification in proposed Hybrid Model is given in Fig. 5.
b. The transaction features taken into account for detecting intrusion are Type of Operation, Least Sensitive Read, Least Sensitive Write, Medium Sensitive Read, Medium Sensitive Write, High Sensitive Read, High Sensitive Write, Transaction Gap, Attribute Sequence Ids, and Table Sequence Ids. Deep network module for feature classification in proposed Hybrid Model given in Fig. 6.

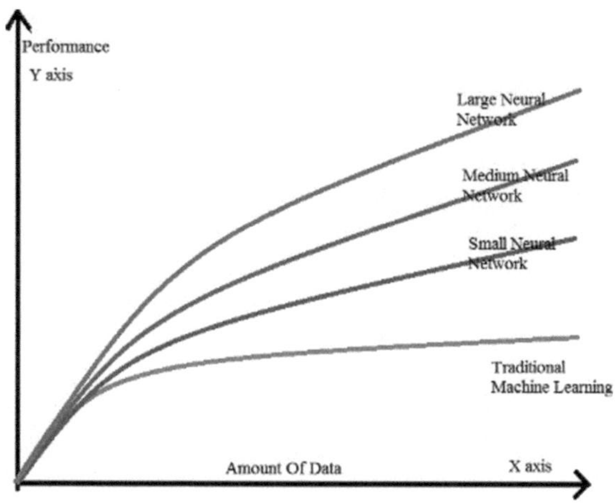

Fig. 4 Performance versus amount of data

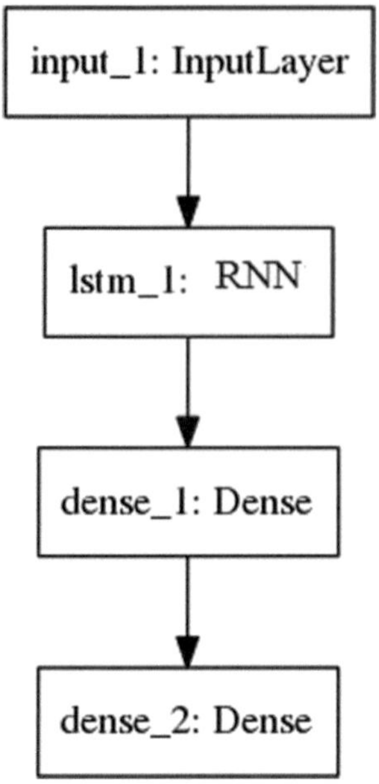

Fig. 5 Deep neural network module for sequence classification

2.3 Proposed Hybrid Model

A Hybrid Deep Neural Network Model has been proposed which is built considering all the natures of the problem discussed above. The Proposed Hybrid Model has two RNN modules used for sequence classification on table ids and attribute ids respectively and a multilayer perceptron for feature attribute classification. The output from the above three modules are feed into a multilayer perceptron and the whole model is trained with back propagation. The Proposed Model is given in Fig. 7.

3 Experimental Results

3.1 Experimental Setup

A synthetic database intrusion detection dataset [8] is used with Keras 2.0.7 as there is no availability of any realistic dataset or benchmark dataset for experimental evaluation of the proposed system. For comparison of traditional machine learning methods with multi perceptron, we have used Weka 3.8.1.

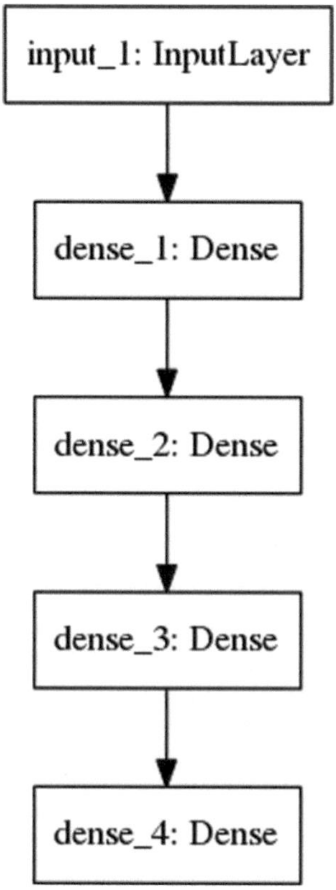

Fig. 6 Deep neural network for feature or normal attributes classification

3.2 Experimental Results

Traditional Machine Learning Versus Multilayer Perceptron: In this case, we have used only the feature attributes to classify the transactions. The comparison is done using WEKA 3.8.1. From the results achieved we get to the conclusion that Deep Learning Model in Fig. 6 works better than Traditional Machine Learning models (SVM and C4.5) only with normal feature attributes the results are compared in Fig. 9. The result of the Decision Tree we found is given in Fig. 8.

Proposed Hybrid Model Results: The accuracy of individual modules of the Hybrid Deep Neural Network proposed (Fig. 9).

I. The results of sequence classifications of Table Sequence Ids using Deep Learning Model of Fig. 5 is as given in Table 1.
II. The results of sequence classification of Attribute Sequence Ids using Deep Learning Model of Fig. 5 is as given in Table 2.
III. The results of normal feature classification of Normal **Feature Attributes** using Deep Learning Model in Fig. 6 is given in Table 3 (Fig. 10).

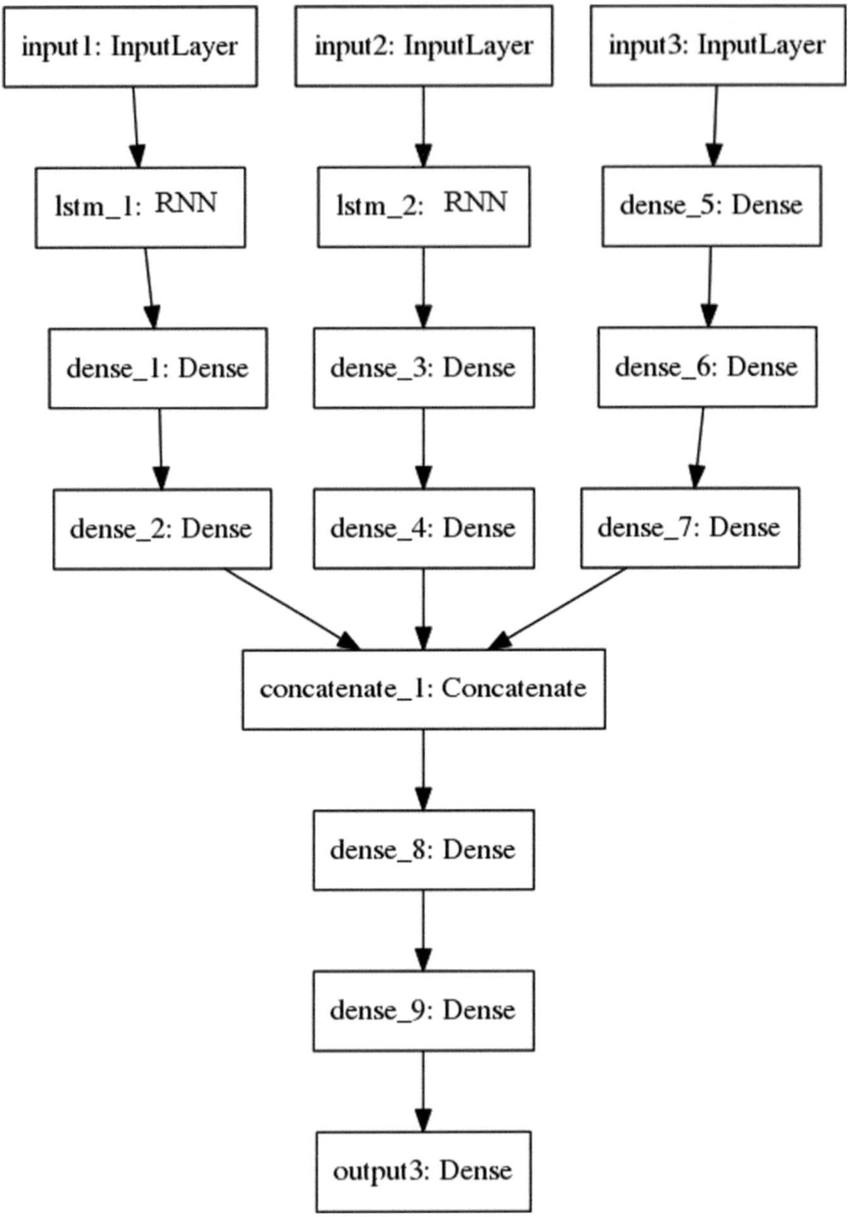

Fig. 7 Proposed hybrid deep learning model

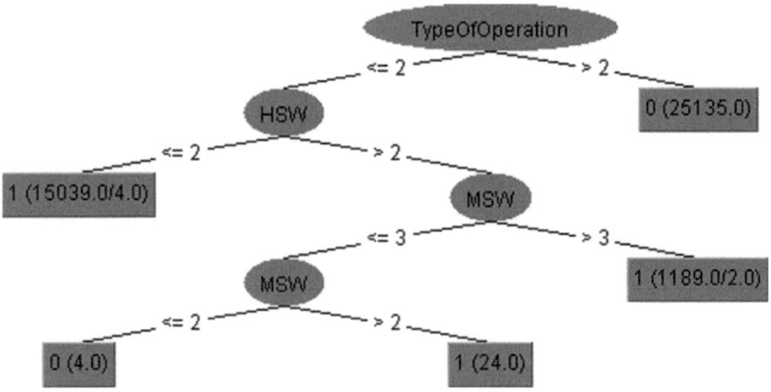

Fig. 8 Decision tree generated output

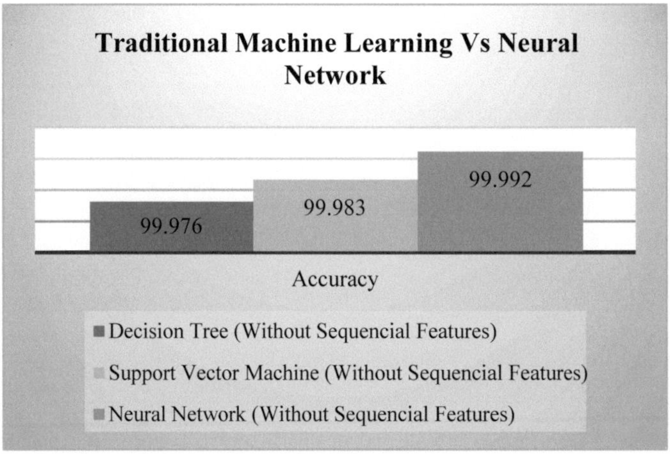

Fig. 9 Traditional machine learning versus neural network

Table 1 The result of deep learning module Fig. 5. With table sequence ids

Correctly classified instances	83.9029%
Incorrectly classified instances	16.0971%
Total number of instances	12417

Table 2 The result of deep learning module Fig. 5. With attribute sequence ids

Correctly classified instances	97.8529%
Incorrectly classified instances	2.1471%
Total number of instances	12417

Table 3 The result of deep learning module Fig. 6. With normal feature attributes

Correctly classified instances	99.9929%
Incorrectly classified instances	0.0071%
Total number of instances	12417

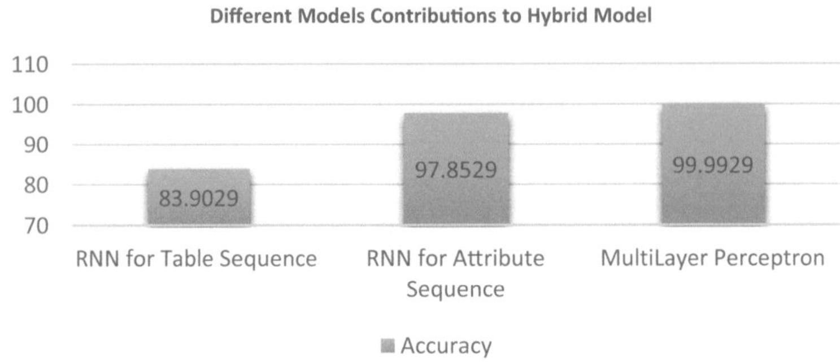

Fig. 10 Different models contributions to hybrid model

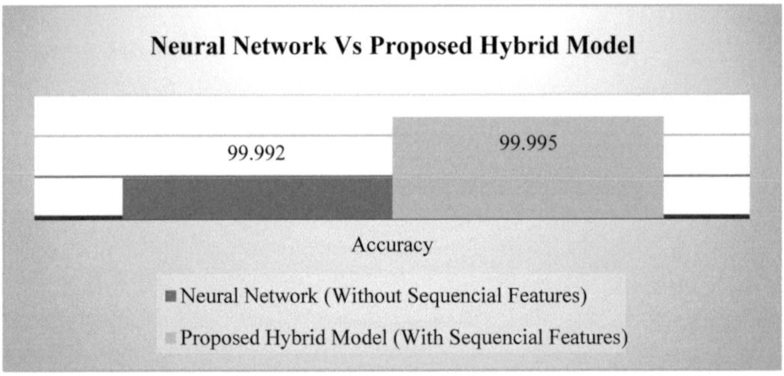

Fig. 11 Comparison of accuracy on different techniques

The Proposed Deep Learning Model in Fig. 7 which is combination of three sub deep neural nets having being trained with normal feature attributes and sequence attributes (table sequence Ids, attribute sequence Ids) and the normal neural net with only normal feature attributes are being compared with their accuracy of correctly classified instances in the Fig. 11.

4 Conclusions

A Hybrid Deep Neural Network Model for Database Intrusion Detection has been proposed in this paper. We have analyzed and found that deep learning performs better on Database Intrusion Detection Systems. We have taken different kinds of attributes to predict the behavior such as sequential attributes and normal feature attributes which can predict the nature of a database transaction. We have seen that Recurrent Neural Network does good sequential classification on sequence attributes and multilayer perceptron does a better performance on normal feature attributes. Thus, the combination of Recurrent Neural Network and Neural Network with back propagation in the proposed hybrid model effectively and more accurately detects the malicious database transactions. In future researchers may attempt to find new features which can detect intrusion and can design new deep learning models to get more sensitive and increase database intrusion detection accuracy.

References

1. Hu, Y., Panda, B.: A data mining approach for database intrusion detection. In: Proceedings of the 2004 ACM Symposium on Applied Computing, pp. 711–716. ACM (2004)
2. Peddabachigari, S., Abraham, A., Grosan, C., Thomas, J.: Modeling intrusion detection system using hybrid intelligent systems. J. Netw. Comput. Appl. **30**(1), 114–132 (2007)
3. Ektefa, M., Memar, S., Sidi, F., Affendey, L.S.: Intrusion detection using data mining techniques. In: 2010 International Conference on Information Retrieval & Knowledge Management (CAMP), pp. 200–203. IEEE (2010)
4. Denatious, D.K., John, A: Survey on data mining techniques to enhance intrusion detection. In: 2012 International Conference on Computer Communication and Informatics (ICCCI), pp. 1–5. IEEE (2012)
5. Deng, L., Chen, J.: Sequence classification using the high-level features extracted from deep neural networks. In: 2014 IEEE International Conference on Acoustics, Speech and Signal Processing (ICASSP), pp. 6844–6848. IEEE (2014)
6. Sahasrabuddhe, A., Naikade, S., Ramaswamy, A., Sadliwala, B., Futane, P.: Survey on Intrusion Detection System using Data Mining Techniques (2017)
7. Yin, C., Zhu, Y., Fei, J., He, X.: A deep learning approach for intrusion detection using recurrent neural networks. IEEE Access **5**, 21954–21961 (2017)
8. Panigrahi, S., Sural, S., Majumdar, A.K.: Two-stage database intrusion detection by combining multiple evidence and belief update. Inf. Syst. Front. **15**(1), 35–53

Design of Intrusion Detection System for Wormhole Attack Detection in Internet of Things

Snehal Deshmukh-Bhosale and S. S. Sonavane

Abstract In the upcoming future, the Internet of Things (IoT) network will take over billions of devices which are connected to each other through the Internet using IPv6 protocol. To connect the devices with each other and the Internet concept is going to be of enormous importance and highlighting aspect. Devices connected in IoT are resource constrained in terms of processing power, memory, battery life, etc. Because of these characteristics, IoT network is very prone to security invasions. We are mainly concentrating on wormhole attack, one of the major and severe attacks occurring at a network layer of IoT protocol stack. RPL along with 6LoWPAN are two major designed protocols for constrained devices in IoT. We have executed an Intrusion Detection System (IDS) which perceives the said attack and attacker using Contiki OS and Cooja simulator. We have successfully experimented the system to find true positive detection rate for detecting the wormhole attack and attackers. We have efficiently received better results for various topologies in the implemented system.

Keywords IoT · IDS · Wormhole attack · RPL · 6LoWPAN · Contiki

1 Introduction

In IoT, any attack may take place from the Internet as these devices are resource constrained and connected to the insecure Internet. These devices are connected to the Internet using IPv6 with 6LoWPAN (IPv6 Low Power Wireless Private Area Network) [1] protocol which is designed for connecting resource-constrained devices. Applications of IoT are smart city, smart healthcare, smart grid, smart home, etc.,

---

S. Deshmukh-Bhosale (✉)
Raisoni College of Engineering & Management, Wagholi, Pune, India
e-mail: sa_bhosale@yahoo.com

RMD Sinhgad School of Engineering, Warje, Pune, India

S. S. Sonavane
Dr. D. Y. Patil Technical Campus, Lohgaon, Pune, India

B. Pati et al. (eds.), *Advanced Computing and Intelligent Engineering*,
Advances in Intelligent Systems and Computing 1082,
https://doi.org/10.1007/978-981-15-1081-6_44

which have made IoT very popular and it seems that maximum of devices will be connected to the Internet in near future which makes inserting security in IoT network a very important aspect.

Protocols like DTLS [2], IPSec [3], IEEE 802.15.4 link layer security [4] have been already investigated by researchers to add the message and data security in IoT. Adding encryption or decryption or adding security algorithm is not sufficient to protect data communication within the IoT network. There must be a dedicated IDS designed to detect security attack and attackers and take corrective action in the network.

There are many IDSs available in the WSN network but they are not to be directly used for IoT network as those are developed by assuming that the network is distributed and no global ID is assigned to any devices in the network. In IoT, unlike WSN each device is assigned to a unique ID using IPv6 addressing scheme. Nodes in IoT are directly connected to the insecure Internet and these nodes are constrained in nature in many terms. Their communication is based on the latest protocols designed only for IoT, e.g., CoAP [5], RPL [6], 6LoWPAN [1], etc. These are lightweight protocols which do not include many security features.

A survey on a need of security in IoT and a survey of many security attacks taking place in various layers of the IoT protocol stack were done by authors in the papers [7, 8]. After studying many attacks in IoT, we have concluded that wormhole attack is a very severe attack in Wireless Sensor Network and so in IoT [9]. As per IoT requirement lot of work is expected to be done to detect the wormhole attack and attacker. We in this paper are focusing on wormhole attack detection using Contiki OS and Cooja simulator. The organization of the paper is as follows: Sect. 2 gives the details of IoT technologies followed by Sect. 3 which covers the details of Intrusion Detection System. Section 4 discusses the wormhole attack followed by Sect. 5 which covers system architecture and experimental setup. Section 6 concludes the paper.

2 IoT Technologies

2.1 IoT (Internet of Things)

IoT forms a heterogeneous network where constrained devices are connected to the conventional Internet with the Internet Protocol IPv6. 6LoWPAN Border Router (6LBR) plays a very important role of connecting wireless nodes to the Internet. It acts as a gateway. The structure is shown in Fig. 1.

Examples of IoT devices are a simple light bulb, smartphone, TV, washing machine, etc. With the support of IPv6, 2^{128} addresses will be assigned to IoT devices forming a huge network. These devices are heterogeneous in nature and resource constrained in terms of memory and processing capability, and hence they are extremely prone to the security attack in the network. Conventional IDS is not suitable for IoT because they have not considered the various characteristics of IoT devices like less

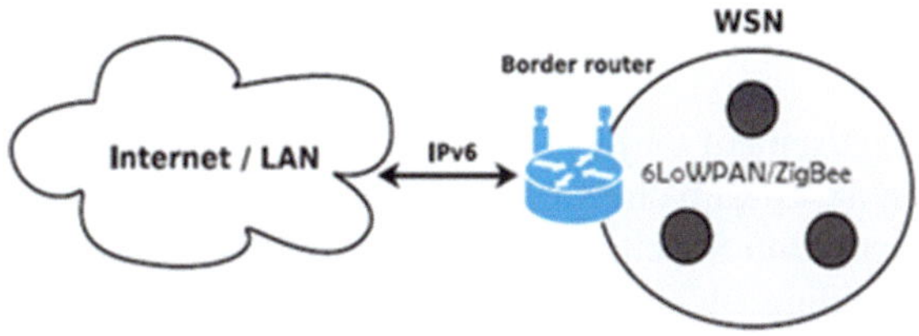

Fig. 1 Structure of IoT network

memory, less processing power, low battery life, etc. Hence the design of IDS for IoT network is a new research area in the IoT field [10, 11].

2.2 6LoWPAN (IPv6 Low Power Wireless Private Area Network)

In IoT, devices are connected to the Internet using IPv6 protocol. This is a heavy-weight protocol which is not compatible with the functioning of resource-constrained devices. 6LoWPAN is a compressed version of IPv6 used to connect the IoT nodes to the Internet. It enables routing of IPv6 packets in compressed and fragmented form. Fragmentation is required because a physical layer protocol, IEEE 802.15.4 has Maximum Transfer Unit (MTU) size that is of 127 bytes which are lesser compared to IPV6 MTU size (1280 bytes). Nodes in IoT are connected to the Internet through 6LBR as shown in Fig. 1. 6LoWPAN uses UDP, a connectionless protocol for communication. Due to resource-constrained nodes and wireless medium, it is easier to attack 6LoWPAN devices due to which many attacks take place in this network. Hence while designing IDS for IoT the network properties of 6LoWPAN must be taken into consideration [12].

2.3 Routing Protocol for Low Power and Lossy Network (RPL)

RPL protocol is developed for 6LoWPAN network of IoT. This protocol is designed for low power and lossy network using IPv6. RPL is built on directed tree graphs hence RPL has bidirectional communication. RPL creates a routing topology in the form of Destination Oriented Directed Acyclic Graphs (DODAG) which is maintained by DODAG Information Object (DIO) advertised by each node. To initiate the path between nodes and to have smooth communication between established links, RPL uses four types of the messages. They are as follows:

- **DODAG Information Object (DIO)** is used to maintain the information of routing graph.
- **DODAG Advertisement Object (DAO)** advertises the routing information from node to sink in the upward direction.
- **DODAG Information Solicitation (DIS)** is used for inclusion of new node in a network by providing DIO messages from existing node.
- **Destination Advertisement Object-Acknowledgement (DAO-ACK)** used to send acknowledgement message after DAO message is received [13].

3 Intrusion Detection System (IDS)

Intrusion Detection System is a system which detects the abnormal behavior of the network and corrects it. In WSN many IDSs are proposed which are successfully detecting the attacks taking place in the network. These IDS cannot be directly used in IoT as IoT is working with different protocols like IPv6, RPL, 6LoWPAN, etc. IDS for WSN are very heavy in terms of memory consumption which is not applicable for IoT network. IoT devices are directly connected to the insecure Internet where more chances of attacks are there. This issue is not addressed by IDS in WSN. Shahid Raza et al., have developed the first IDS for wormhole attack using Contiki OS and Cooja simulator. They have worked on Sinkhole attack and selective forwarding attack [14].

3.1 Types of IDS

There are three types of IDS which are designed for IoT. First one is **Signature-based IDS** in which a fixed pattern of attack is saved in its database. Whenever an attack takes place and its pattern matches with the saved pattern, an alarm is raised. A drawback of this IDS is that an attack with a different pattern will not be detected.

Next type is **Anomaly-based IDS** in which the system keeps on checking the behavior of the network. A threshold is set for the same. If any abnormal behavior is detected then IDS raises an alarm to detect the attack. The drawback of Signature-based IDS is eliminated in this type of IDS. But in this IDS more processing power is required to match the threshold.

One more IDS is present in IoT named as **Hybrid IDS**. This is designed to eliminate the drawbacks of both the IDS discussed previously. Practically Hybrid IDS is impractical for IoT network [15].

In our implementation, we are using Anomaly type of IDS. Using this IDS we are finding true positive detection rate. It is defined as a ratio of the number of attacks detected successfully to the total number of attacks taken place [16].

4 Wormhole Attack

In wormhole attack, a virtual link is established between two colluding nodes pretending they are directly connected to each other. They misguide other legitimate nodes to transmit packets through them. In Fig. 2, node X and node Y are physically located at long distance. But when a wormhole link as shown in the figure is established between them, they misguide the other nodes that they are having a direct link to reach each other and other nodes can send packets through them. We have elaborated it with Fig. 3.

As shown in Fig. 3, X and Y are attacker nodes who have established a link between them. Node A is a legitimate node wants to send a packet to another legitimate node B. In no attack condition, a packet will be transmitted as A-C-D-E-B as it is the shortest path. But when the attack is inserted through X and Y link, it advertises that the path through X-Y is the shortest path. Hence A chooses path as A-X-Y-B thinking

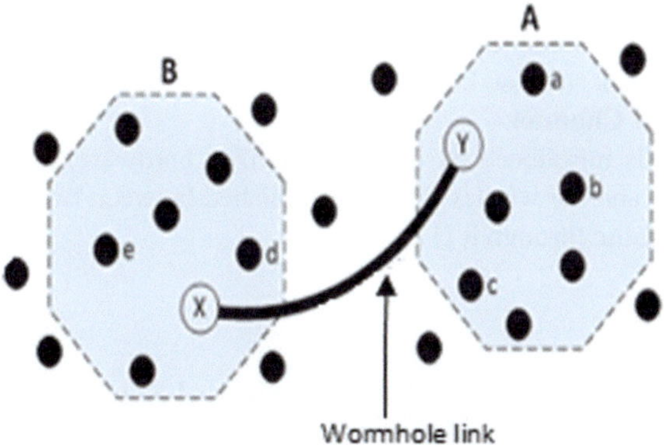

Fig. 2 Wormhole attack

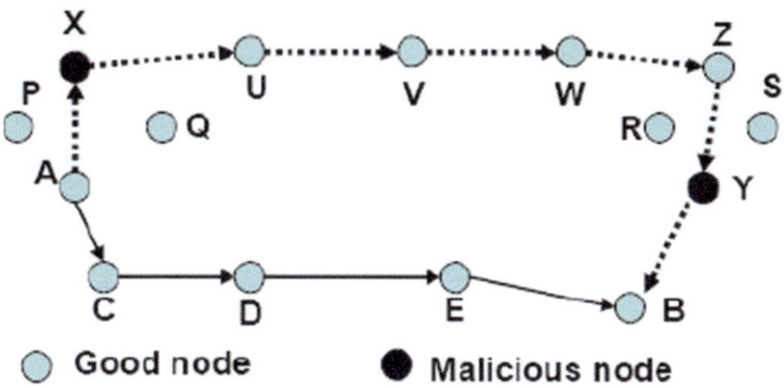

Fig. 3 Example of wormhole attack

it is the shortest path to reach to B. But the actual packet is transmitted through A-X-U-V-W-Z-Y-B which takes more time for the packet to reach from node A to node B. Thus delay in delivery of packet is observed which is an effect of wormhole attack.

Wormhole attack has four modes of operation as explained below.

i. **Packet Encapsulation**:
In this type of wormhole attack, a packet is encapsulated so that hop count between two attacker nodes will not increase though these two nodes are not directly connected to each other. Normal nodes select the Wormhole link thinking two attacker nodes are directly connected to each other. Figure 3 elaborates this mode.

ii. **Packet Relay**:
In this mode, attacker nodes relay the packet between two legitimate nodes by giving an illusion that normal nodes are neighbors. This attack can be introduced by a single attacker node also.

iii. **High Power Transmission**:
In this type, attacker node broadcasts the RREQ packet at high power level capability to attract normal nodes to overhear the RREQ packet. Like packet relay, this attack can also be launched by single attacker node.

iv. **Out of Band Channel**:
This attack is introduced by using specialized hardware. In this attack, long range wired and the wireless link is established between two attacker nodes to attract the traffic through it [17].

5 System Architecture and Experimental Setup

In the proposed system architecture, we have considered nodes as per topology explained in 5.1. 6LoWPAN Border Router (6LBR) is used to connect these nodes to the Internet through it. IDS are placed at two levels. One is at Border Router which is called as a centralized approach and other is at node level which is known as a distributed approach. The system architecture is as shown in Fig. 4.

The wormhole attack detection system is implemented using open-source operating system Contiki and its inbuilt Cooja simulator [18]. For implementation, we have taken Tmote sky nodes. In Fig. 5, Node 1 acts as a Border Router. We are evaluating true positive detection rate for detecting wormhole attack and attackers. We are using cc 2420 as radio interface and other protocols at various layers are shown in Table 1 [19, 20].

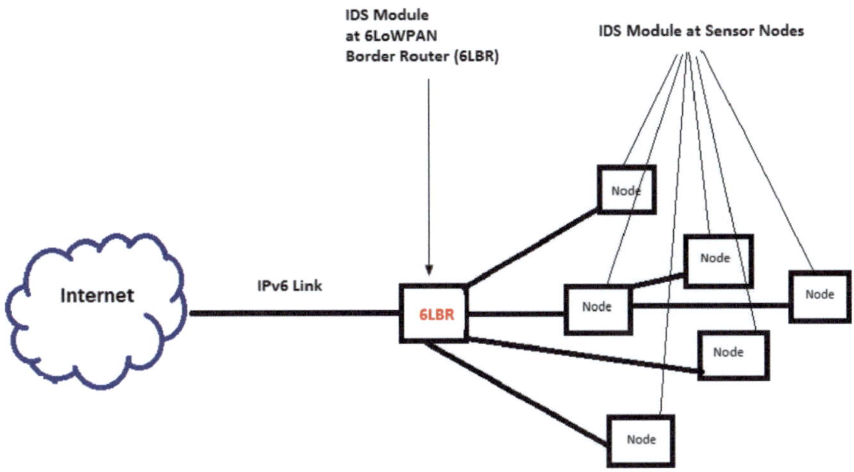

Fig. 4 System architecture

Table 1 Protocol stack in Contiki

Contiki OS layer	Protocol/Interface
Transport	UDP
Network, routing	IPv6, RPL
Adaptation	Sicslowpan
MAC	CSMA
Duty cycling	Sicslowmac
Radio	Cc2420

5.1 The Topology of Nodes for Experimentation

We are considering 8, 16, and 24 nodes (N) for our experimentation as shown in Fig. 5.

5.2 True Positive Detection Rate

True positive detection rate is a rate of successful attack detection with respect to the total attack taken place in the network. We have performed the simulation for various network topologies. Our experimental setup has given output for the attack and the attacker detected and not detected for the number of nodes (N) are 8, 16, and 24. We have used the Received Signal Strength Indicator (RSSI) characteristic to check the true positive detection rate of wormhole attack. Figure 6 gives a graphical representation of true positive detection rate for wormhole attack. From the graph, it

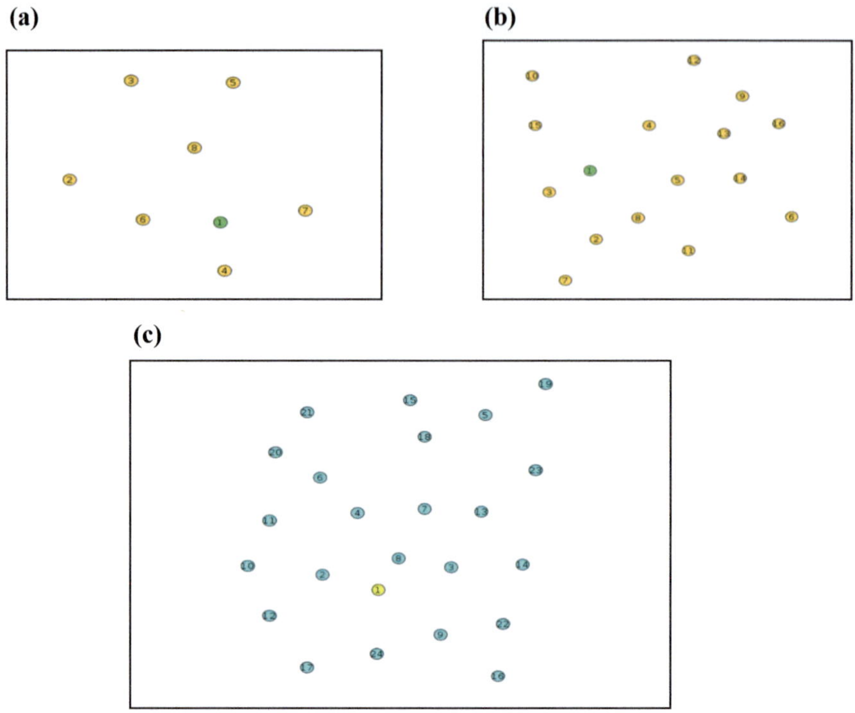

Fig. 5 **a** Network topology N = 8. **b** Network topology N = 16. **c** Network topology N = 24

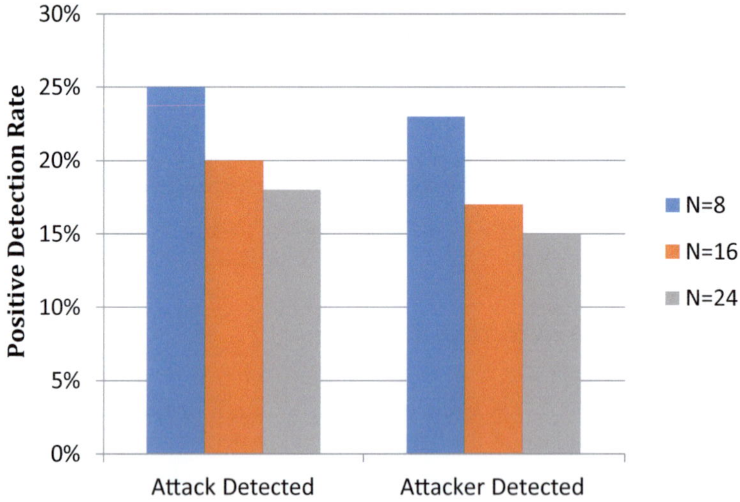

Fig. 6 True positive detection rate

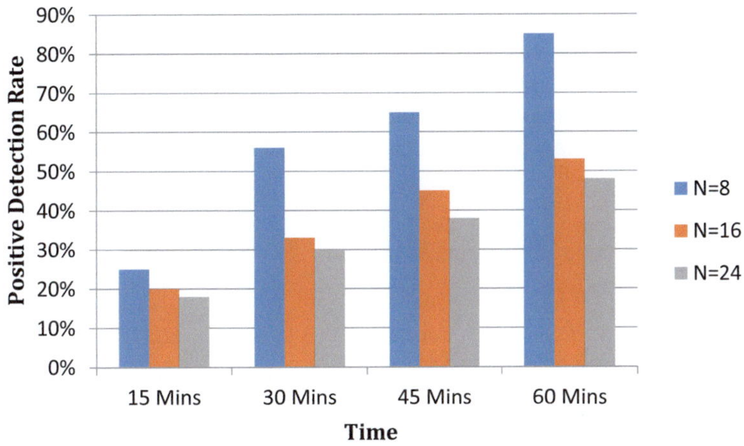

Fig. 7 Attack detection for various time slot

is observed that the positive detection rate is higher for less number of nodes in the network in the implemented architecture. In our architecture, we have placed IDS at 6BR and at the Nodes. We have considered two nodes for each topology as attacker nodes. Performance of the network is observed using RSSI values received from the nodes.

RSSI value converts the radio strength of received signal into the distance. The calculated distance using RSSI value is compared with location and range information stored with 6LBR. If the RSSI distance is coming lesser than stored information an attack is detected. One of the criteria to detect the attack is to find the transmission range of the neighbors. If the neighbors are not in transmission range then the attack is existing in the network.

We have detected the attack and the attacker. For various topologies, we have initially run the IDS for 15 min. We have got the readings as shown in Fig. 6.

We have run our system for one hour and taken readings for attack detection at intermediate levels as 15, 30, 45, and 60 min. The graph for attack detection is as shown in Fig. 7. We have observed that the detection rate is improved when we run the system for more time.

6 Conclusion

Security is a very important aspect of the Internet of Things as of today. In our work, we are addressing the same issue by considering wormhole attack. Wormhole attack is one of the severe attacks which take place at the network/routing layer of IoT protocol stack. In our work, we have designed IDS which detects the presence of wormhole attack with RSSI value as a major parameter. We have implemented said system using

Cooja simulator of open-source Contiki OS. We have calculated positive detection rate for said attack. We have considered various topologies which are running for different time slots. We have concluded from implemented experimentation that for the smaller network, attack detection rate is better than the larger network. When we run the IDS for more time better results are obtained in view of attack detection.

References

1. Hui, J., Thubert, P.: Compression Format for IPv6 Datagrams Over IEEE 802.15.4-Based Networks, RFC 6282, September (2011)
2. Kothmayr, T., Hu, W., Schmitt, C., Bruenig, M., Carle, G.: Securing the internet of things with DTLS. In: Proceedings of the 9th ACM, Conference on Embedded Networked Sensor Systems, pp. 345–346. ACM (2011)
3. Raza, S., Duquennoy, S., Chung, A., Yazar, D., Voigt, T., Roedig, U.: Securing communication in 6LoWPAN with compressed IPsec. In: 7th International Conference on Distributed Computing in Sensor Systems (DCOSS'11), Barcelona, Spain, pp. 1–8 (2011)
4. Raza, S., Duquennoy, S., Höglund, J., Roedig, U., Voigt, T.: Secure Communication for the Internet of Things—A Comparison of Link Layer Security and IPsec for 6LoWPAN, Security and Communication Networks. Wiley (2014). http://dx.doi.org/10.1002/sec.406
5. Shelby, Z., Kartke, K., Bormann, C., Frank, B.: Constrained Application Protocol (CoAP), draft-ietf-core-coap-12, October (2012)
6. Winter, T., Thubert, P., Brandt, A., Hui, J., Kelsey, R., Levis, P., Pister, K., Struik, R., Vasseur, J., Alexander, R.: RPL: IPv6 Routing Protocol for Low-Power and Lossy Networks. RFC 6550, March (2012)
7. Deshmukh, S., Sonavane, S.S.: Security Protocols for internet of Things: A Survey. ICNETS2, VIT University, Chennai (2017). 978-1-5090-5913-3/17/$31.00_c 2017 IEEE. https://doi.org/10.1109/icnets2.2017.8067900
8. Deshmukh-Bhosale, S., Sonavane, S.S.: Security Threats in 6LoWPAN and RPL Network: Mitigation Techniques and Design of IDS. In: IEEE Sponsored International Conference, August, ICCUBEA, Pune (2017)
9. Johnson, M.O., Siddiqui, A., Karami, A.: A wormhole attack detection and prevention technique in wireless sensor networks. Int. J. Comput. Appl. (0975–8887), **174**(4) (2017)
10. Jing, Q., Vasilakos, A.V., Wan, J., Lu, J., Qiu, D.: Security of the Internet of Things: perspectives and challenges. Wirel. Netw. **20**(8), 2481–2501 (2014). https://link.springer.com/article/10.1007/s11276-014-0761-7
11. Weber, R.H.: Internet of Things—new security and privacy challenges (2010). https://doi.org/10.1016/j.clsr.2009.11.008. Published by Elsevier Ltd.
12. Kushalnagar, N., Montenegro, G., Schumacher, C.: IPv6 over Low-Power Wireless Personal Area Networks (6LoWPANs): Overview, Assumptions, Problem Statement, and Goals, RFC 4919, August (2007)
13. Airehrour, D., Gutierrez, J., Ray, S.K.: Secure Routing for Internet of Things: A Survey. J. Netw. Comput. Appl. (2016). http://dx.doi.org/10.1016/j.jnca.2016.03.006
14. Raza, S., Wallgren, L., Voigt, T.: SVELTE: Real-Time Intrusion Detection in the Internet of Things. Ad hoc Netw. **11**(8), 2661–2674 (2013)
15. Ghosal, A., Halder, S.: A survey on energy efficient intrusion detection in wireless sensor networks. J. Ambient. Intell. Smart Environ. **9**, 239–261 (2017). https://doi.org/10.3233/AIS-170426
16. Le, A., Loo, J., Chai, K., Aiash, M.: A specification-based IDS for detecting attacks on RPL-based network topology. Information **7**, 25 (2016). https://doi.org/10.3390/info7020025

17. Azer, M., El-Kassas, S., El-Soudani, M.: A full of the wormhole attack. Int. J. Comput. Sci. Inf. Secur. (2009)
18. Dunkels, A., Grönvall, B., Voigt, T.: Contiki—a lightweight and flexible operating system for tiny networked sensors. In: EMNets'04, Tampa, USA, pp. 455–462 (2004)
19. Gonizzi, P., Duquennoy, S.: Hands on Contiki OS and Cooja Simulator: Exercises (Part II). https://team.inria.fr/fun/files/2014/04/slides_partI.pdf (2013). Accessed 10 Nov 2017
20. Texas Instruments: CC2420 Simple Link™ Multi standard Wireless MCUl. http://www.ti.com/lit/ds/symlink/cc2420.pdf (2016). Accessed 13 Jan 2018

Information Encoding, Gap Detection and Analysis from 2D LiDAR Data on Android Environment

Arpan Phukan, Pronam Phukan, Rohit Sinha, Shyamal Hazarika and Abhijit Boruah

Abstract Most commercial uses of LiDAR prefer high-end LiDAR systems with equally sophisticated software packages that have limited accessibility due to cost or complexity. Our purpose for developing this LiDAR point classification framework is to develop a flexible method for visualizing and processing LiDAR data that is simple and cost-effective, and yet achieves a similar degree of functionality as that of its high-end peers. To that end, we have classified data points from a LiDAR-Lite V2 (Blue Label) to represent the distance and height of obstacles, and the gaps between them. This will enable an autonomous mobile agent to determine palatable paths through the satisfactory gaps.

Keywords LiDAR-Lite V2 · Point classification · Android · 2D · 3D · Scatter plot

1 Introduction

Over the years, Light Detection And Ranging (LiDAR) has seen growing use in remote sensing and imaging, thanks to its simplicity and relatively low cost. A typical LiDAR system uses light to measure variable distances by illuminating the target with laser and processing the reflected laser pulses with a sensor. A major area of interest for LiDAR is autonomous navigation. In the past few years, it has emerged as the leading technology in safety and autonomous systems. Aside from this, LiDAR data sets have been developed for applications in urban environments (buildings, bridges, highways, etc.), for mining and geological applications, emergency management (landslides, floodplain mapping, hurricane damage assessment, etc.), land cover change and global biogeochemical cycling (biomass, ecological impacts, etc.) [6, 10, 19].

A. Phukan (✉) · P. Phukan · R. Sinha · S. Hazarika · A. Boruah
Dibrugarh University Institute of Engineering and Technology, Dibrugarh, India
e-mail: arpanphukan@gmail.com

B. Pati et al. (eds.), *Advanced Computing and Intelligent Engineering*,
Advances in Intelligent Systems and Computing 1082,
https://doi.org/10.1007/978-981-15-1081-6_45

Although there are plenty of software packages available for processing and interpreting LiDAR data, modern obstacle detection for high-end autonomous navigation packages are not accessible to most users due to cost or complexity [1, 5]. Therefore, there is a need for a software package which is simple to implement, will classify obstacles and gaps, and can incorporate additional data when they are available. The identified gaps will be classified in real time into two classes: satisfactory and unsatisfactory. The satisfactory gaps are continuously analysed and modified as required (classified as unsatisfactory) by incorporating new information about obstacles in the local environment. These satisfactory gaps will collectively form a path for the mobile agent to travel. To achieve this, we have used a fairly primitive model, LiDAR-Lite V2 (Blue Label), which costs about a third of other sophisticated LiDAR systems. However, because it lacks additional high-end features, LiDAR-Lite V2 is not considered for commercial use, which means most software packages that are available for LiDAR systems are not compatible with it. Thus a simple, toned-down package is needed to achieve similar functionality of commercial LiDAR systems.

In order to achieve these objectives, the remainder of this work is organized into the following five sections. Section 2 gives an overview of the literature relevant to this work. The hardware and circuit platform are discussed in Sect. 3. The algorithm and its real-time implementation are illustrated in Sect. 4. Section 5 presents results and conclusions, with a review and analysis of accomplished tasks with respect to the objectives, and lays the groundwork for future work.

2 Literature Review

The fundamental working principle of a LiDAR systems is laser ranging. It emits ultraviolet, visible or near infrared light from the sensor. An object within the laser footprint will generate a reflection, called a return. Differences in laser return times can then be used to calculate the distance of the target. Real-world applications of LiDAR includes autonomous driving [12], landslide investigations [9], digital elevation modelling and flood modelling [13], forest planning and management [20], oil and gas exploration [21], deployment of solar panels [17], etc to name a few. LiDARs acquire quite accurate distance information with a high range resolution and angular resolution, which is usually used in the map generating technique of the autonomous vehicles [11]. LiDAR can be used in moving vehicles for obstacle detection and collision avoidance [4]. Researchers have been studying methods to classify obstacles by using distance data to get outline of the obstacles and geometric information and subsequently use the information to classify the type of obstacles [16]. There is already a method to classify obstacles using the LIDAR intensity data [3, 8]. The method uses the probability distribution of the LIDAR intensity data and dispersion to perceive and classify obstacles. However, less complicated calculation does not generate accurate classification of obstacles based on the probability distribution of the intensity of data and dispersion. If there is a window of error in real-time response, quick planning of a path for autonomous vehicles becomes difficult as well as issues relating to safety and convenience arise. For this reason, a new obstacle classification method based on a single LiDAR is necessary, which should be simple and efficient.

3 Platform Description

An android App has been designed for this work, which takes continuous inputs from a hardware platform for 2D map generation. The hardware and Android Environment is discussed in the following subsections.

3.1 Hardware and Circuit

The hardware consists of an Arduino UNO board as the prime control circuit, with power specs of 5V, an HC-05 Bluetooth module and a Lidar-Lite V2.

Two servos driven by an Adafruit L293D motor shield v1 are used to mount the LiDAR for scanning the environment. The Servos (1 and 2 as shown in Fig. 1) are setup in such a way so that Servo 2 scans the horizontal (X-axis) plane of the environment and Servo 1 is incremented by 5° in positive direction of Y-axis to provide the height information of objects in the environment.

3.2 Android Platform

To visualize the lidar data set on a 2D scatter plot, an android app is developed which implements the android graph library [7]. The app is built in Android studio and is

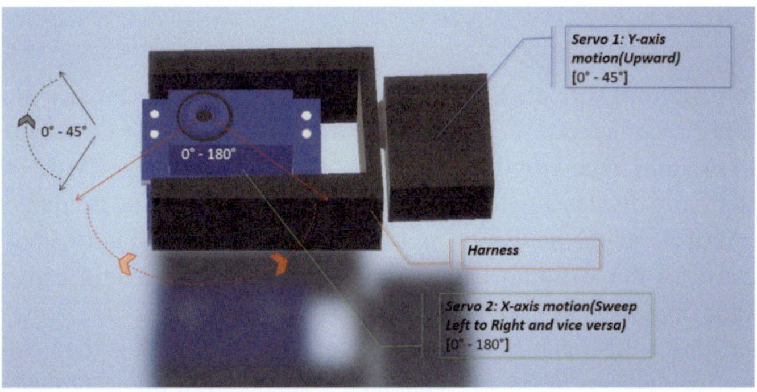

Fig. 1 Simulation of the 3D harness which supports the LiDAR

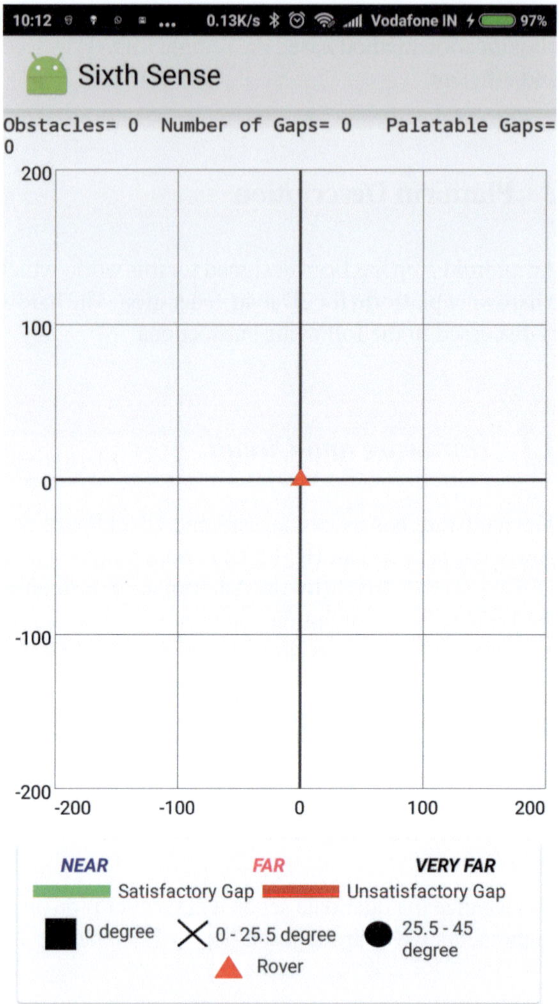

Fig. 2 User interface of the Sixth-Sense app

supported on Android version 4.4 (KitKat) or higher. This app satisfies the objective of establishing a connection between the prime control circuit (Arduino Uno) via a Bluetooth module, receiving the lidar data, analysing and classifying the relevant vertices and finally displaying the output on a 2D scatter plot. The fundamental cause for opting Android as the analysing and classifying device and not just as a display tool is because of the Arduino Uno board's space and processing capabilities are vastly lacking (Fig. 2).

The lidar senses the distance of a point of an obstacle which is then concatenated with the position and height values, obtained by the position values of the two servos, and then sent to the Android device via the HC-05 bluetooth module.

The Android app first pairs with the HC-05 Bluetooth module and then receives the string containing the distance, position and height attributes each separated by delimiters. It then separates the three values and proceeds to calculating the x, y and z values from the position and height attributes using the sin(radians), cos(radians) and tan(radians) functions. These values are then analysed and classified to be plotted in the 2D scatter plot to be viewed by the user at their discretion.

4 Algorithm and Implementation

In this algorithm, the variables x and y hold floating point values for the X and Y coordinates of the point respectively, while z holds the value of the height of the point obtained by tan(radians). The integer variables *number_of_obstacles*, *number_of_satisfactory_gaps*, *number_of_unsatifactory_gaps* count the number of obstacles, satisfactory gaps and unsatisfactory gaps in the environment of the system, respectively.

4.1 Algorithm

The algorithm used in this work is illustrated below:

Calculate x[cos(radians)], y[sine(radians)] and z[tan(radians)] coordinates of a point from lidar data
if $|x| \leq 150$ **and** $|y| \leq 150$ **then** ▶ if point lies 150 cms away from lidar, it is not considered
    if $z > 0$ **then** ▶ for points with z coordinate > 0
        if $z > 0$ **and** $z < 0.476975$ **then** ▶ elavation of point (z coordinate) from tan(0 rad) = 0 to tan(25.5 rad) = 0.476975
            for *Every satisfactory gaps* **do**
                if *point(x,y,z) lies over the line segment(satisfactory gap)* **then**
                    calculate the distance of the point(x,y) from the start and end vetices of the gap▶ We exclude the z coordinate of the point(x,y,z) to calculate the distance
                    if *either of the distance is 0 while the other is* $\geq$ *50cms* **then**
                        the point lies over one of the terminal points of the gap
                    end if
                    if *either of the distance is > 0 while the other is* $\geq$ *50cms* **then**
                        draw a RED line between the points less than 50 cms apart from each other ▶ RED line denotes unsatisfactory gaps
                        set the point(x,y) as the new termial point of the gap ▶ We exclude the z coordinate of the point(x,y,z) to be set as the new terminal point of the gap

```
            end if
            if distance of the point(x,y) from the start and end vetices of the gap are < 50cms then    ▶ Exclude the z coordinate of the point(x,y,z)
                change the gap to RED(unsatisfactory)
                number_of_unsatifactory_gaps ++    ▶ counter variable for number of unsatisfactory gaps
                number_of_satisfactory_gaps –    ▶ counter variable for number of satisfactory gaps
                set the point(x,y) as the new termial point of the gap
            end if
        else if point(x,y,z) does not lie over any gap then
            calculate the intersection
point(x_Perpendicular_Intersection, y_Perpendicular_Intersection) of the perpendicular from the point(x,y) to a gap    ▶ Exclude the z coordinate of the point(x,y,z)
            calculate the distance of the
point(x_Perpendicular_Intersection, y_Perpendicular_Intersection) from
the start and end vetices of the gap
            if either of the distance is 0 while the other is ≥ 50cms then
                the point lies over one of the terminal points of the gap
            end if
            if either of the distance is > 0 while the other is ≥ 50cms then
                draw a RED line between the points less than 50 cms apart from each other    ▶ RED line denotes unsatisfactory gaps
                set the point(x_Perpendicular_Intersection,
y_Perpendicular_Intersection) as the new terminal point of the gap
            end if
            if distance of the point(x_Perpendicular_Intersection,
y_Perpendicular_Intersection) from the start and end vetices of the gap are < 50cms then
                if the distance from the point(x,y) [ignoring the z coordinate] to the point(x_Perpendicular_Intersection,
y_Perpendicular_Intersection) < 70cms then    ▶ length of the mobile agent = 70 cms
                    change the gap to RED(unsatisfactory)
                    number_of_unsatifactory_gaps ++    ▶ counter variable for number of unsatisfactory gaps
                    number_of_satisfactory_gaps –    ▶ counter variable for number of satisfactory gaps
                end if
            end if
        end if
    end for
else if z > 0.476975 then    ▶ elavation of point (z coordinate) > tan(25.5 rad)
```

```
            do nothing
        else
            calculate distance between two consecutive points ▶ for points with z coordinate = 0
            if distance > 10 cms then ▶ if distance between two consecutive points > 10 cms
                number_of_obstacles ++ ▶ counter variable for number of obstacles
                if distance ≥ 50 then ▶ if distance between two consecutive points ≥ 50 cms
                    number_of_satisfactory_gaps ++
                    draw a GREEN line connecting the two vertices ▶ GREEN line denotes a satisfactory gap
                else
                    number_of_unsatifactory_gaps ++
                    draw a RED line connecting the two vertices
                end if
            end if
        end if
    end if
end if
if |x| ≥ 0.0 and |y| ≥ 0.0 then
    if |x| > 50.0 and |y| > 50.0 then
        if (|x| > 100.0 or |y| > 100.0) and (|x| ≤ 150 and |y| ≤ 150.0) then
            if z > 0 and z ≤ z < 0.476975 then
                plot black CROSS ▶ BLACK means the point is VERY FAR from the lidar and CROSS means the point is lower than the height of the rover
            else if z > 0.476975 then
                plot black CIRCLE ▶ BLACK means the point is VERY FAR from the lidar and CIRCLE means the point is higher than the height of the rover
            else
                plot black RECTANGLE ▶ BLACK means the point is VERY FAR from the lidar and RECTANGLE means the point is at height(z coordinate) = 0
            end if
        else if (|x| > 50.0 or |y| > 50.0) and (|x| ≤ 100 and |y| ≤ 100.0) then
            if z > 0 and z ≤ z < 0.476975 then
                plot magenta CROSS ▶ MAGENTA means the point is FAR from the lidar and CROSS means the point is lower than the height of the rover
            else if z > 0.476975 then
                plot magenta CIRCLE ▶ MAGENTA means the point is FAR from the lidar and CIRCLE means the point is higher than the height of the rover
            else
                plot magenta RECTANGLE ▶ MAGENTA means the point is FAR from the lidar and RECTANGLE means the point is at height(z coordinate) = 0
            end if
```

```
        end if
    else
        if z > 0 and z ≤ z < 0.476975 then
            plot blue CROSS    ▶ BLUE means the point is NEAR to the lidar and CROSS means the point is lower than the height of the rover
        else if z > 0.476975 then
            plot blue CIRCLE   ▶ BLUE means the point is NEAR to the lidar and CIRCLE means the point is higher than the height of the rover
        else
            plot blue RECTANGLE ▶ BLUE means the point is NEAR to the lidar and RECTANGLE means the point is at height(z coordinate) = 0
        end if
    end if
end if
Go to Step 1
```

4.2 Implementation

The objective of our project is to create a LiDAR data classifying framework to identify obstacles and gaps for the mobile agent to proceed in the environment. This is achieved through the following steps:

- Establishment of data transfer between Arduino and android via HC-05 Bluetooth module.
- Implementation of 2D interactive scatter plot in an Android device.
- Establishment of connection between LiDAR-Lite v2 and Arduino.
- Mounting of LiDAR-Lite v2 on a harness with a HS-422 servo and a HS-311 servo to generate 3D data set with Distance, Position and Height as attributes.
- Generation of Lidar-Lite v2 data set.
- Classification of received data set in the android device with Colour and Shape Encoding as Class Labels.
- Plotting of Colour Encoded Classified data set on the Scatter Plot in real time (Figs. 3, 4 and 5).

After implementation of the algorithm using the discussed platform, colour and shape information of the environment is achieved in the maps as shown in Fig. 6.

5 Results and Conclusion

The purpose of this project is to create a method for classifying LiDAR data that is simple to implement and cost-effective. To achieve this, a local 2D map was built using LiDAR-Lite V2 and Android to determine obstacles and gaps for locomotion of mobile agents in the environment.

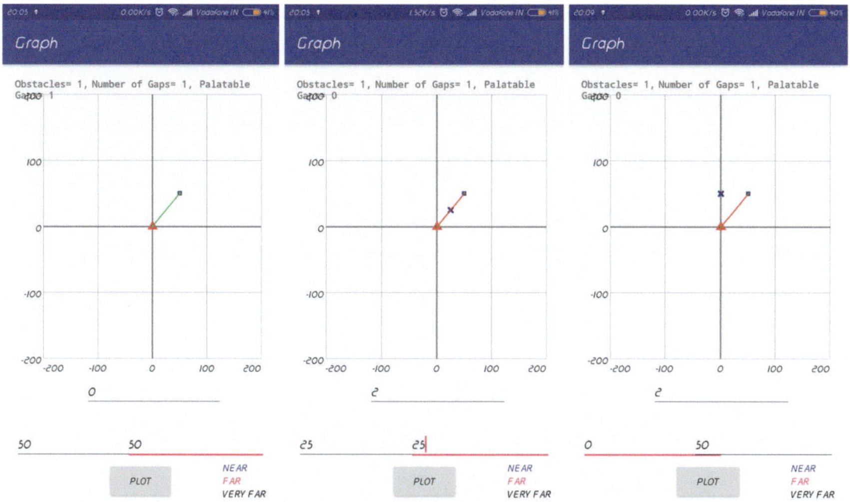

Fig. 3 (Left) Shows an obstacle at (50, 50) which results in the formation of a satisfactory gap; (Centre) shows an obstacle at height (2 units) between the earlier gap, which makes it unsatisfactory; (Right) shows an obstacle at height (2 units) and at position (0, 50) from the rover. The horizontal displacement of the obstacle from the gap is 50, which is <70. Hence, the gap is affected and it becomes non-palatable

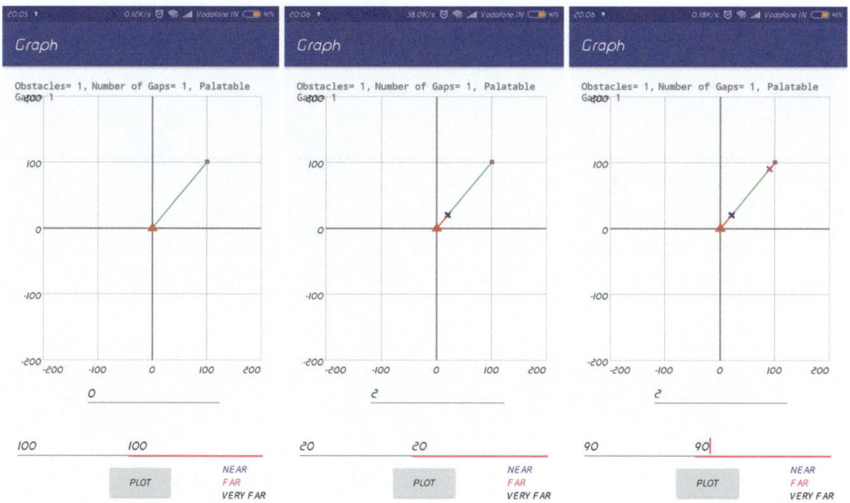

Fig. 4 (Left) Shows an obstacle at (100, 100) which results in the formation of a satisfactory gap; (Centre) shows an obstacle at height (2 units) and at position (20, 25) from the rover. This only makes part of the gap unsatisfactory as the unobstructed part of the gap is still palatable; (Right) shows another obstacle at height (2 units) and at position (90, 90) from the rover. This also makes part of the gap unsatisfactory as the unobstructed part of the gap is still palatable

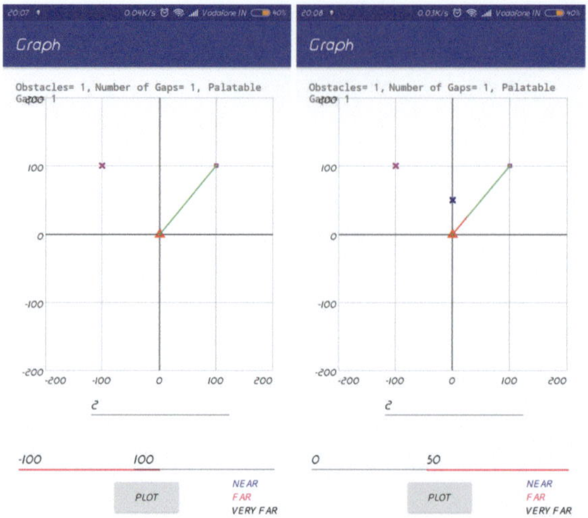

Fig. 5 (Left) Shows an obstacle at height (2 units) and at position (−100, 100) from the rover. The horizontal displacement of the obstacle from the gap is 141 (approx), which is >70. Hence, the gap remains unaffected; (Right) shows an obstacle at height (2 units) and at position (0, 50) from the rover. The horizontal displacement of the obstacle from the gap is 50, which is <70. Hence, the gap is affected and part of it becomes non-palatable

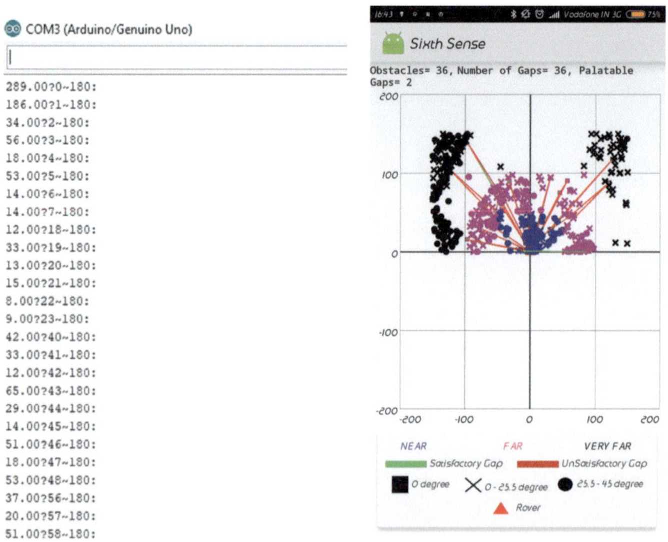

Fig. 6 (Left) Shows a part of the data sent by the Lidar to Android where '?', '~' and ':' are delimiters; (Right) shows the map plotted by the received data

Our categorizing algorithm is implemented on a data set obtained from a real-world environment, with numeric continuous attributes, generated by the LiDAR and Arduino board. This colour encoded categorized data set is finally visualized on a 2D scatter plot on our Android application with the help of Android GraphView library, developed by Jonas Gehring [7]. To minimize cost, a 2D LiDAR-Lite v2 (Blue Label) was used instead of the commercial LiDARs with an extensive set of features. As the project grows, more and more functionalities can be added to it. The benefit from the system will be that the Android app uses a fully graphical interface for the user, which displays the list of devices paired with the Android smartphone. Future scopes of this project can be concluded into the following points:

- Optimal paths searching algorithms for agent movements like Greedy [15], A* [18], Dijkstra's algorithm [2], Incremental Heuristic Search [14] can be implemented on a global map by using the local information from this implementation. The agent will then be able to decide which routes to take when moving through obstacles to reach their destination.
- Furthermore, it will make the necessary adjustments to its route if there are changes in the local environment in case the environment is dynamic.

References

1. Amadeo, R.: Google's waymo invests in lidar technology, cuts costs by 90 percent. https://arstechnica.com/cars/2017/01/googles-waymo-invests-in-lidar-technology-cuts-costs-by-90-percent/ (2017)
2. Arjun, R., Reddy, P.: Research on the optimization of dijkstra's algorithm and its applications. Int. J. Sci. Technol. Manag. **4**(1), 304–309 (2015)
3. Carballo, A., Ohya, A., et al.: People detection using range and intensity data from multi-layered laser range finders. In: 2010 IEEE/RSJ International Conference on Intelligent Robots and Systems (IROS), pp. 5849–5854. IEEE (2010)
4. Catapang, A.N., Ramos, M.: Obstacle detection using a 2d lidar system for an autonomous vehicle. In: 2016 6th IEEE International Conference on Control System, Computing and Engineering (ICCSCE), pp. 441–445. IEEE (2016)
5. Condliffe, J.: Lidar just got way better-but it's still too expensive for your car. https://www.technologyreview.com/s/609526/lidar-just-got-way-better-but-its-still-too-expensive-for-your-car/amp/ (2017)
6. Evans, J.S., Hudak, A.T., Faux, R., Smith, A.: Discrete return lidar in natural resources: recommendations for project planning, data processing, and deliverables. Remote. Sens. **1**(4), 776–794 (2009)
7. Graphview—open source graph plotting library for android. http://www.android-graphview.org/
8. Hancock, J.A.: Laser intensity-based obstacle detection and tracking. Technical report, Carnegie Mellon University (1999)
9. Jaboyedoff, M., Oppikofer, T., Abellán, A., Derron, M.H., Loye, A., Metzger, R., Pedrazzini, A.: Use of lidar in landslide investigations: a review. Nat. Hazards **61**(1), 5–28 (2012)
10. Kasperski, J., Delacourt, C., Allemand, P., Potherat, P., Jaud, M., Varrel, E.: Application of a terrestrial laser scanner (tls) to the study of the séchilienne landslide (isère, france). Remote. Sens. **2**(12), 2785–2802 (2010)

11. Kirchner, A., Heinrich, T.: Model based detection of road boundaries with a laser scanner. In: Proceedings of IEEE International Symposium on Intelligent Vehicles, pp. 93–98. Citeseer (1998)
12. Levinson, J., Askeland, J., Becker, J., Dolson, J., Held, D., Kammel, S., Kolter, J.Z., Langer, D., Pink, O., Pratt, V., et al.: Towards fully autonomous driving: systems and algorithms. In: 2011 IEEE Intelligent Vehicles Symposium (IV), pp. 163–168. IEEE (2011)
13. Li, S., MacMillan, R., Lobb, D.A., McConkey, B.G., Moulin, A., Fraser, W.R.: Lidar dem error analyses and topographic depression identification in a hummocky landscape in the prairie region of Canada. Geomorphology **129**(3–4), 263–275 (2011)
14. Liu, S.K.M.L.Y., Furcy, D.: Incremental heuristic search in artificial intelligence
15. Mannor, S., Meir, R., Zhang, T.: Greedy algorithms for classification–consistency, convergence rates, and adaptivity. J. Mach. Learn. Res. **4**, 713–742 (2003)
16. Moon, H.C., Kim, J.H., Kim, J.H.: Obstacle detecting system for unmanned ground vehicle using laser scanner and vision. In: International Conference on Control, Automation and Systems. ICCAS'07, pp. 1758–1761. IEEE (2007)
17. Redweik, P., Catita, C., Brito, M.: Solar energy potential on roofs and facades in an urban landscape. Sol. Energy **97**, 332–341 (2013)
18. Sharma, S.K., Pal, B.: Shortest path searching for road network using a* algorithm. Int. J. Comput. Sci. Mob. Comput. **4**(7), 513–522
19. Webster, T.L.: Flood risk mapping using lidar for annapolis royal, nova scotia, canada. Remote. Sens. **2**(9), 2060–2082 (2010)
20. Wulder, M.A., Bater, C.W., Coops, N.C., Hilker, T., White, J.C.: The role of lidar in sustainable forest management. For. Chron. **84**(6), 807–826 (2008)
21. Zhevlakov, A., Bespalov, V., Elizarov, V., Grishkanich, A.S., Kascheev, S., Makarov, E., Bogoslovsky, S., Il'inskiy, A.: Hydrocarbon halo-laser spectroscopy for oil exploration needs. In: Optical Sensing and Detection III, vol. 9141, p. 914125. International Society for Optics and Photonics (2014)

Attribute-Based Convergent Encryption Key Management for Secure Deduplication in Cloud

E. Silambarasan, S. Nickolas and S. Mary Saira Bhanu

Abstract Data deduplication is a methodology for eliminating duplicate copies of data by keeping the single copy of data with proper access policies, instead of maintaining duplicates of the same data. It reduces cloud storage slot and communication bandwidth. Secure deduplication is a challenge in multi-user cloud setup, where each user uses their key to secure their data, resulting in different enciphered data for the same unencrypted data. Convergent Encryption (CE) addresses this issue by fixing the single hash value of the file as the key for enciphering the content of the file. The challenge now is to store the key in the cloud by multiple users. The existing key management system that uses an encoding scheme suffers in the realistic environment. To overcome this issue, in the proposed work ABCEKM (Attribute-Based Convergent Encryption Key Management), the file key is enciphered by specifying access policies for the attributes of the file. The proposed approach is efficient and reliable for cloud storage.

Keywords Deduplication · ABE · CE · AES · Secure deduplication

1 Introduction

Cloud computing (CC) is the fastest growing technology in terms of the feature as well as in the increased number of users, which provide number of services. Its major characteristics are elasticity, on-demand basis service, pay-as-you-go service, etc. Nowadays, most of the academician and organizations are using CC, because it reduces overall capital expenditure and operational expenditure. CC provides basic services like SaaS, PaaS, and IaaS. Finally, CC provides everything as a service

E. Silambarasan (✉) · S. Nickolas · S. Mary Saira Bhanu
National Institute of Technology, Tiruchirappalli 620015, Tamil Nadu, India
e-mail: esilambarasan.me@gmail.com

S. Nickolas
e-mail: nickolas@nitt.edu

S. Mary Saira Bhanu
e-mail: msb@nitt.edu

B. Pati et al. (eds.), *Advanced Computing and Intelligent Engineering*,
Advances in Intelligent Systems and Computing 1082,
https://doi.org/10.1007/978-981-15-1081-6_46

(XaaS), where X can be data, network, storage, backup, etc. CC provides four delivery models like public, private, hybrid, and community cloud. CC is mostly preferred to store huge amounts of data. Due to the enormous increase of users in this environment, utilization of the resources is high. Especially in utilizing the storage space, multiple users, or single user stores his/her data multiple times in the cloud by which data is duplicated and the storage space is also not utilized properly. In this digital data era, IDC report says that 40 trillion GB sizes of data will reach in 2020 [5], and huge size of data stored in cloud leads to the rise in searching time as well as response time. Not only storage space, network bandwidth is also not properly utilized. Deduplication is a technique, which maintains only unique copy by eliminating duplicate copies, instead of maintaining multiple copies of the same data [11, 12], etc. It will improve storage space and communication bandwidth utilization in the cloud.

Deduplication has the following security issue. Before storing the data in the cloud storage server (CSS), service provider, or an inside attacker's views and access the data to check for the duplicates. In a cloud environment, the data breach is occurring 70% because of inside attackers and 30% by outside attackers [5]. To avoid this issue, in a single user environment, the researchers use secure deduplication which means converting original readable file to unreadable file like enciphered text. Secure deduplication improves the privacy of data. Whereas, in a multi-user environment, each user uses their own encryption key to encipher the same file, which gives the different enciphered text to the same file. When uploading this enciphered text with different key for the same file by different users, CSS considers it to be different source files from different users rather than a single common file. To overcome this issue CE scheme can be used, which eliminates the problem of different encryption key by the fingerprint (hash value) of the source file. The fingerprint is distinct for every content and reverting it to its source file is not possible. A cryptographic hash function is used to generate a hash value. The function should be a collision resistant function. Collision means two different data may generate the same hash value or H(x1||data1) is equal to H(x2||data2).

CE algorithm is mostly recommended for deduplication. Secure deduplication provides data privacy through Proof of oWnership (PoW) [3] as well as Proof of Data Possession (PDP) [3] or Proof of Data Retrievability (PDR). PDP is mainly used to check integrity of stored data in the cloud. Data owner (DO) has to be allowed to store or access PoW as well as PDP. CE-based secure deduplication eliminates duplicates of the source file, but still, duplicates are present in the form of key in cloud storage which is discussed detail in Sect. 2.

Considering the abovementioned drawbacks in secure key deduplication, this paper aims to overcome those drawbacks by proposing a ABCEKM method.

2 Problem Definition

Deduplication can be classified as file level, block level, or chunk level deduplication. Block level or file level can further be categorized into fixed size or variable size. Fixed size deduplication has boundary shifting problem where the size of the block is constant but the content of the block gets shifted between neighboring blocks if any alteration in block like insertion or deletion of data is made.

Apart from these, secure deduplication faces an important challenge, i.e., the encryption key is maintained in the user's premises may face some security problems. That is if a key stored is in a local system the privacy of data is leaked when the system is compromised or the data cannot be accessed if the system undergoes any failure. Nowadays users store their encryption key along with their enciphered file. Here the key is not stored as it is, instead the user enciphers the key by their own master key which differs from user to user. Finally, the problem is for the same copy of the enciphered file, multiple copies of enciphered key are maintained in CSS. Now, the proposed work need to remove duplicate copies of key from storage by applying the deduplication technique, which also improves efficiency and reliability of key deduplication.

Existing key management scheme generates different enciphered key for the same file which increases the duplicates if the size of the data sharing users grow. Furthermore, on continually increasing data sharing user, the security of the data, as well as key, has to be significantly enhanced [1, 2], etc.

3 Preliminaries

3.1 Convergent Encryption

Convergent encryption method [11–13], etc., is used to solve the issue in the multi-user environment by generating an encryption key from the data itself. Even if different users enciphers the same data, this method generates the same cipher text for each user. In this encryption algorithm, the key is generated from the data itself, for example, if F_s is file or a data, a cryptographic hash function $h_s = H(F_s)$ is used to generate the hash value, which will be used as an encryption key to encipher the file which can be denoted as $C_s = E_{h_s}(F_s)$ and it is stored in the CSS. But this kind of encryption algorithm is more vulnerable to brute force attack. Because the key is directly dependent on data and it can also be easily predicted.

3.2 Attribute-Based Encryption (ABE)

In multiple user environment, the proposed work use ABE [2, 7], etc., which provides secret sharing mechanism, privacy, and access control. It is a public key cryptographic technique, in which two keys are used namely, the public key and secret or private key. In ABE, attributes set are used to encipher the data and private key is used to decipher the data. The private key used is related access policy (AP): The users those who have credentials which satisfy AP can only decrypt the data. This AP not only provides access control but also provides scalability, revocation, and collision resistance. Basically, ABE is categorized into Key-Policy Attribute-Based Encryption (KP-ABE) and Cipher-Policy Attribute-Based Encryption (CP-ABE). In KP-ABE, the encipher text is generated based on attributes of data and the user's secret key is generated based on APs. In CP-ABE, the enciphered text is generated based on APs and user's secret key is generated based on attributes of the data. Most of the ABE uses bilinear pairing method for computation which is discussed in next section.

3.3 Bilinear Pairing

Bilinear pairing is a non-degenerate bilinear pairing, $e{:}G_a \times G_b \rightarrow G_T$, where G_a, G_b and G_T are finite cyclic groups of prime order p. It can be classified into three types. If $G_a = G_b$ then type 1. If $G_a \neq G_b$ then type 2. Finally, If $G_a \neq G_b$ then type 3. Difference between type 2 and 3 is type 2 is an efficiently computable homomorphic function $G_b \rightarrow G_b$ whereas type 3 is not efficiently computable homomorphic function [14].

Consider G_S be a source cyclic multiplicative group and G_T be a target cyclic multiplicative group of prime order p. Let g be the generator of the source group. A bilinear pairing map e: $G_S \times G_S \rightarrow G_T$ has the following properties:

Bilinearity
For all *i* and *j* are elements of *G*, m, and n are elements of Z_p, bilinearity says that $e(i^m, j^n) = e(i, j)^{mn}$.

Non-degeneracy
e(g, g) is not equal to *1*

Computability
There is an efficient technique to compute *e(i, j)* for all values of *i, j* that are elements of *G*.

Computational Diffie Hellman (CDH) problem.
Let *i, j* in *G* be picked at random and *g* be a generator of *G*. The difficulty of CDH is to calculate $h = g^{\log_g a \log_g b}$ given *(i, j)* as an input.

Decisional DH (DDH) problem.
Let *i, j, k, l* in *G* be picked at random. The difficulty of DDH is to agree whether $\log_i j = \log_k l$ given *(i, j, k, l)* as an input.

Deduplication faces different security issues when it is studied from different perspectives. Based on these issues many research works were proposed using ownership, retrievability/data possession and storage.

4 Related Work

Currently, many researchers are working on issues in secure cloud storage without duplicates. In this section, a discussion is made on secure deduplication, key management, and Attribute-Based Encryption (ABE) to eliminate duplicates while storing data in CSS.

4.1 Secure Deduplication

Convergent Encryption (CE), proposed by Douceur et al., is the first guide for secure deduplication and it provides both confidentiality and efficiency [10] in terms of bandwidth and storage space in cloud. The data is enciphered by hash value which is generated from the same content, which generates multiple enciphered text. The drawback of this work is that it suffers from brute force attack because the key is generated from the data itself. To overcome this drawback, a method called server side deduplication using key server was proposed by Keelveedhi et al. [11–13], etc.

4.2 Key Management Schemes

Mi et al. [4] proposed a session key-based convergent key management scheme, for dynamic convergent key updates in cross-user secure deduplication. Every time when a new user joins, the whole process is repeated once and a copy of enciphered data and key is generated newly. It does not support data owner privacy maintenance. Jin et al. proposed Dekey, in which hybrid cloud is used for secure deduplication. In this work, instead of maintaining key in a single server, the key is distributed securely among the multiple servers. It incurs small encoding overhead compared to network communication overhead in file upload and download operation [1].

4.3 ABE

Sahai and Waters [6] proposed a basic version of ABE, which was improved in Goyal et al. [7] proposed key-policy ABE (KP-ABE). Initially KP-ABE was recognized as the monotonic access structures (AS) [7]. Bethencourt et al. [8] proposed the CP-ABE, where it is secure only in the generic group model and Cheung et al. proposed that it supports only AND operation on AS [9]. Compared to KP-ABE, CP-ABE is more flexible because the AP is determined, after the user secret attributes are supplied. Cui et al. [2] proposed ABE-based secure deduplication for enciphered data, concentrated on secure deduplication of data. Here the issue is to manage CE key in CSS. Each and every user uses the individual master key to encipher his/her key and store their enciphered key and data, by the standard key management system. Here, the problem is duplicate copies of the key are stored in CSS. It was overcome by Li et al. who proposed encoding scheme based on key distribution which raises the problem of not support in realistic environment [1]. The proposed ABCEKM scheme, improves efficiency of file/block storage as well as key management in cloud environment.

5 Proposed Scheme

5.1 *Overview and Construction*

Figure 1 shows the overall system architecture. For efficient data as well as key access control, ABE is used. But in cloud environment, cloud may have multiple owners for the same file in secure deduplication, ABE cannot be directly applied. For efficient management of data as well as key in cloud and to prevent different

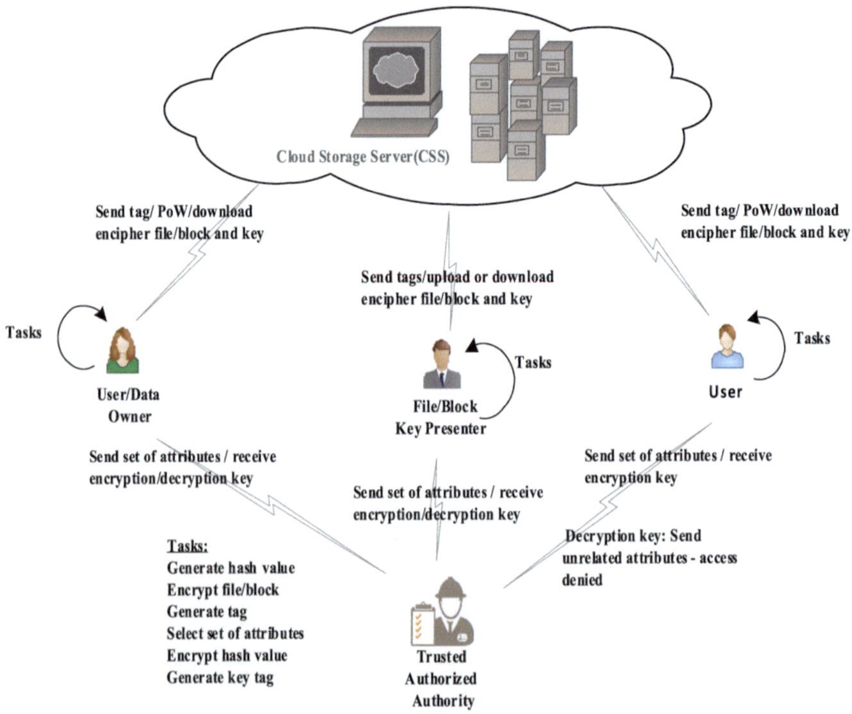

Fig. 1 Overall system architecture

copies of enciphered key, ABCEKM is proposed and it is shown in Fig. 2. Using ABCEKM, the unique copy of enciphered source file and enciphered hash value are stored in the public cloud. DO enciphers data by convergent encryption method and directly provides the rights for key access to the other users using CP-ABE.

Consider CSS are semi trusted servers. Mainly secure deduplication is to provide data privacy for multiple DOs. Privacy is needed for data as well as key in the cloud. The DO must assign the access rights for the key as well as the data. ABE-based key management is simple and also support scalability of DOs. Finally, it should be scalable, efficient, and usable to both the owners and the users.

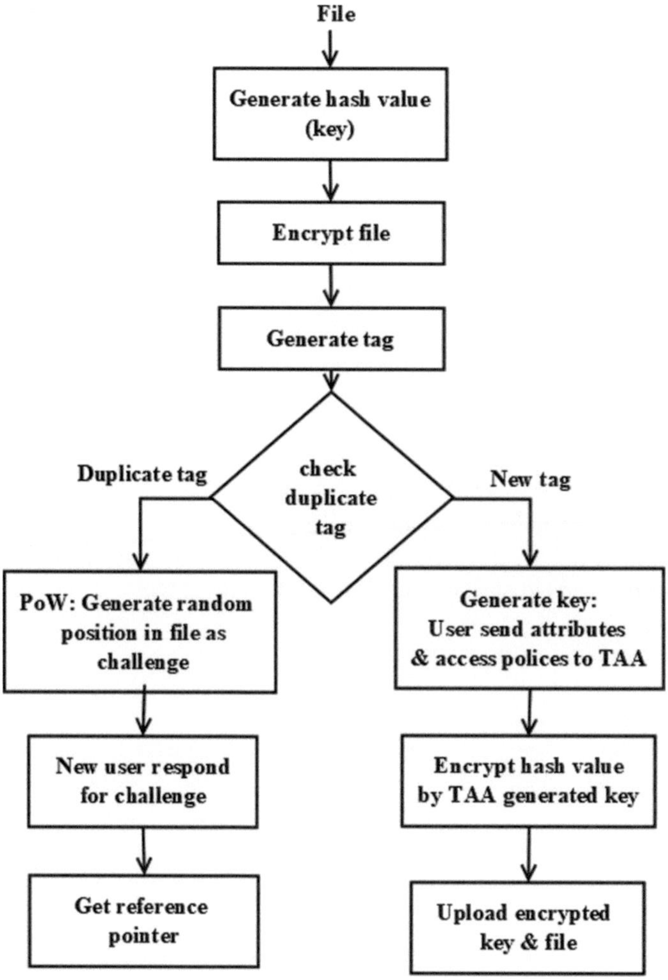

Fig. 2 Flow diagram of file level encryption

6 ABCEKM Method

ABCEKM uses both CE and CP-ABE for improving data as well as key privacy and also provides secure deduplication for both data and key. In this scheme key management risk is reduced and it is also highly scalable in terms of number of DO. AES algorithm provides high security. This proposed algorithm belongs to client-side deduplication. It is divided into two phase.

6.1 File/Block Upload

File/Block Encryption

Figure 2 shows the flow of the file encryption process and the flow of block level encryption is shown in Fig. 3. If DO wants to upload his/her file/block, first DO has to apply the cryptographic hash algorithm for example SHA256, where the hash value is generated using the entire file/block as input. Symmetric key encryption algorithm is used for encrypting the source file/block for example AES algorithm.

Algorithm: File/Block Upload

Step 1 Key generation phase
$Key = HVal = H(File)/Key_{1..n} = HVal_{1..n} = H(Block\ 1..n)$

Step 2 Encryption phase
$CT_{File} = E_h(File)/CT_{Block\ 1..n} = E_{h1..n}(Block1..n)$

Step 3 Tag generation phase
$Tag = HVal = H(CT_{File})/Tag_{1..n} = HVal_{1..n} = H(CT_{Block\ 1..n})$

Step 4 Select attributes for file/block phase
$Attr_{File} = \{A_1, A_2, ..A_k\}/Attr_{Block1..n} = \{A_2, A_2, ..A_k\}_{1..n}$

Step 5 Send generated attributes to TAA for key encryption
$ek = E_{Puk}(Key)/ek_{1..n} = E_{PuK}\{AttrBlock1..n\}(Key_{1..n})$

Step 6 Tag generation for encryption key phase
$KT = H(ek)/KT_{1..n} = H(ek_{1..n})$

Step 7 File/Block and File key/Block key upload phase.
If new tags then CSS allows the user to store his/her file/block and its key. Otherwise, CSS check his/her ownership, bychecking $H(CSS\_Random_{position}(CT_{file}))$ is equal to $H(NU\_Random_{position}(CT_{file}))/H(CSS\_Random_{position}(CT_{Blocks}))$ is equal to $H(NU\_Random_{position}(CT_{Blocks})))$ then assign reference pointer to NU else access denied

After completion of encryption, DO generates a tag for enciphered file/block and sends it to CSS which checks for duplicate tags. If the duplicate is not found in CSS database, then CSS allows the DO to upload his/her file/block. Otherwise, DO prove his/her ownership through PoW protocol. PoW protocol is challenge-response protocol, in which CSS randomly select some bytes position in file/block

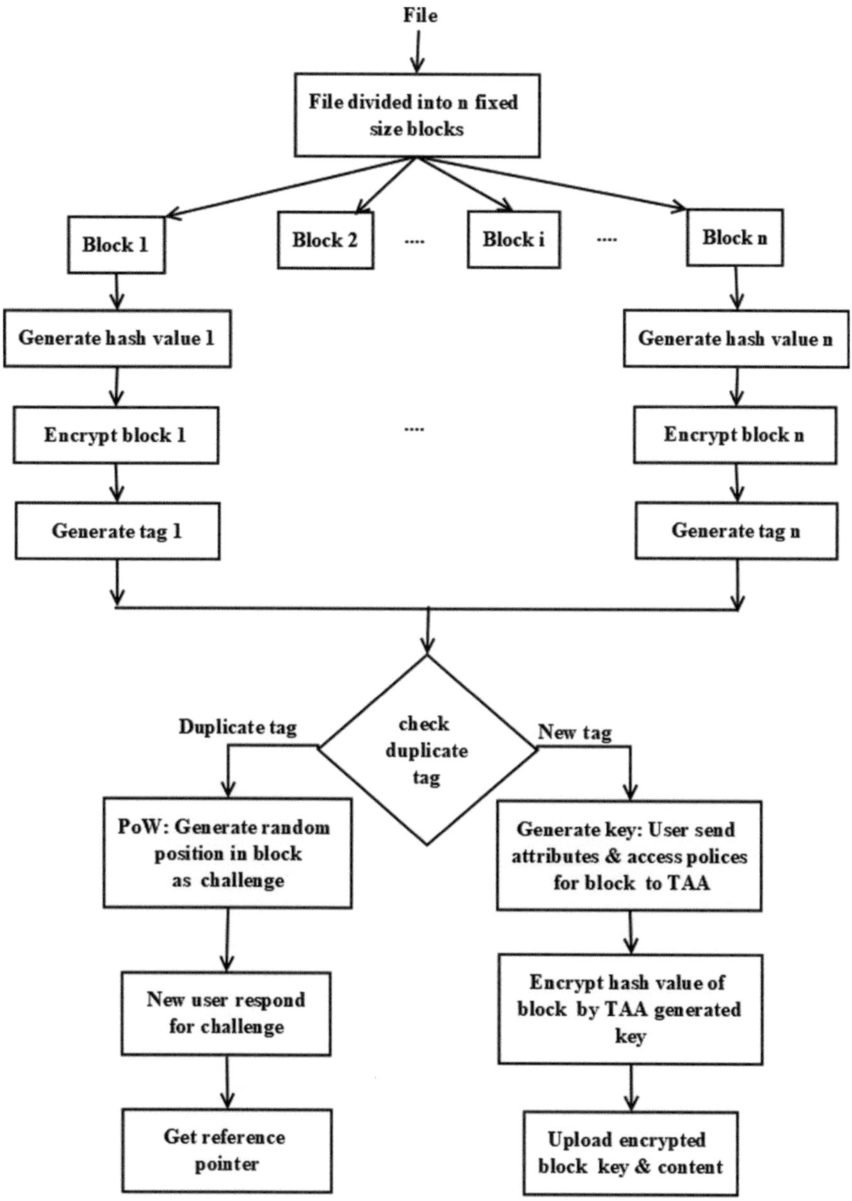

Fig. 3 Flow diagram of block level encryption

as challenges, to check data ownership of new uploader. If new uploader succeeds in his/her proof then CSS assigns reference pointer to that new uploader as DO. Otherwise, CSS will not allow accessing the file/block.

ABE-based Key management scheme
In the proposed attribute-based convergent encryption key management method for secure deduplication that is depicted in Figs. 2 and 3, the four main objects are file key presenter, trusted attribute authority (TAA), CSS, and users. In cloud, file/block key presenter store his/her key and CSS shares that key among the multiple users who are holding certain authorizations. TAA is responsible to generate a decryption key using a set of related attributes for all users. During the upload process, DO first sends the request to the CSS along with key tag *(KT)*/key tags KT_i which is related to the content of file/block key, respectively, and then enciphers the file/block key content by AS using a set of attributes. Through these process each file/block key provider generates KT/KT_i and enciphered key (ek/ek_i).

After receiving data and key storage request from new users, CSS will do equality test for key by checking the newly generated KT/KT_i with already existing KT_j in key-tag list. If newly generated KT/KT_i does not match with existing key tags then new KT/KT_i is added into key-tag list and also store the new enciphered key along with tags like $(KT/KT_i, ek/ek_i)$ in CSS. Suppose new enciphered key tag is matched with existing enciphered key (ek/ek_i) tags then CSS execute as follows.

- If the AP in $ek \subset ek'/ek_i \subset ek_i'$, new user storage request ek/ek_i is rejected: Else, the AP in $ek' \subset ek/ek_i' \subset ek_i$ CSS replaces the existing pair (KT_j', ek_j') with the new pair (KT_i, ek_i) where $KT_i = KT_j'$.
- Otherwise, CSS will not allow to store or to access the file/block.

ABE-based key management scheme consists of setup, key generation, encryption, validity-test, equality test, and decryption.

6.2 File/Block Download

File/Block download process is diagrammatically represented in Fig. 4. If DO wants to download the file/block, DO sends a request along with the tag to CSS whether to upload or download the file/block. CSS checks whether the tag exists or not. If exist then checks the option to upload or download. If download then CSS checks the requester ownership.

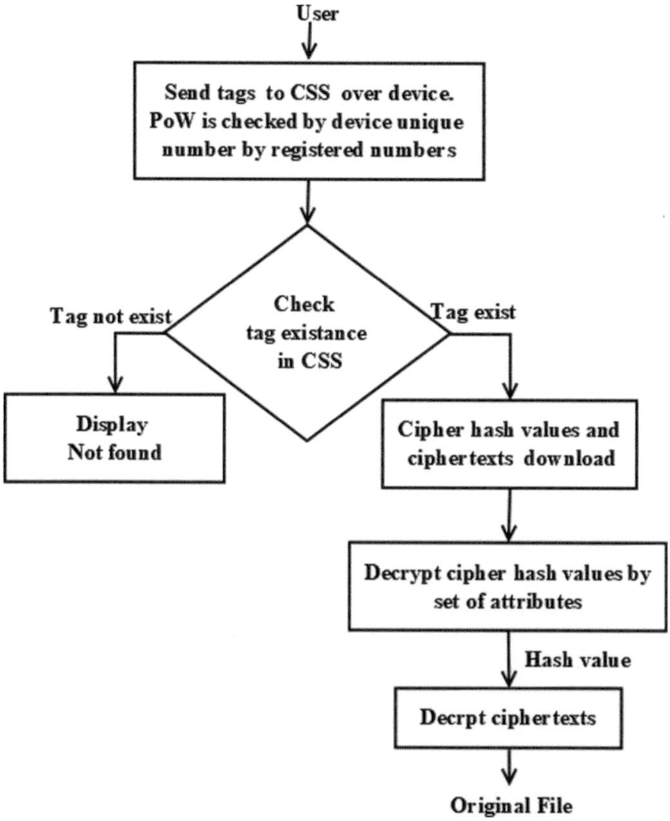

Fig. 4 Flow diagram of file/block level decryption

Algorithm: File/Block Download

Step 1 User send File/Block tags *{Tag/Tag*$_1$*..n}* and Key tags *{KT/KT*$_1$*..n}* to CSS

Step 2 If tags are present in the tag table then CSS check their ownership by his/her device number which is registered during upload phase and CSS sends both data and key in enciphered format.

Step 3 User receive a private key to decrypt enciphered key from TAA by showing valid attributes.

Step 4 Decryption of Key (Hashvalue) phase
$Key = D_{\Pr k}(ek)/Key_{1...n} = D_{\Pr k\{AttrBlock1..n\}}(ek_{1..n})$

Step 5 Decryption of File/Block Phase
$File = D_{key}(CT\,File)/Block\ 1 = D_{Key1..n}(CT\,Block\ 1..n)$

PoW works as follows: The user sends the tag to the CSS through a device which has a unique number that is registered while uploading the file/block. Through this unique device number, PoW is checked by CSS. DO proves his/her ownership then whole enciphered file/block along with enciphered key is downloaded. First, decrypt enciphered hash value (CE key) with the help of TAA, which generates attribute-based private key, if user's attribute set fulfills AS structure. Next, decrypt the enciphered file/block by using that decrypted hash value.

7 Result and Discussion

Theoretically, the computational complexity of the proposed work, compared with previous works is shown in Table 1. The proposed work has high computational complexity M*O(1) for file level deduplication but fully eliminates the duplicate copies of the key. At block level a reduced computational complexity M*O(K) is achieved when compared with existing algorithms.

8 Conclusion

The proposed work is highly scalable in a cloud environment. In this approach, attribute-based encryption is combined with a convergent encryption method in client-side secure deduplication. The proposed work has two parts: Data management and key management. In data management, the source file/block is enciphered by standard CE. CE is the best way to do secure deduplication in the multi-user cloud environment. CE key or the hash value of the source file/block is enciphered by attributes of file/block with APs. Both enciphered texts of file/block, as well as the key, are stored in CSS. The main significance of proposed work is to eliminate duplicate copies of CE key. This work also reduces storage and network bandwidth not only for content storage but also for hash value or key storage. In future, the proposed work will be analyzed experimentally and will be compared with existing approaches.

Table 1 ABE-based CE key management algorithm—computation complexity

Steps	File level deduplication		Block level deduplication	
	Master key based CEK management	Proposed attribute-based CEK management	Encoding/Decoding-based CEK management	Proposed attribute-based CEK management
Tag generation	Yes—O(1)	Yes—O(1)	Yes—O(K)	Yes—O(K)
PoW (upload)—without duplicate, if M = 0 (worst case)	No	NA	NA	NA
PoW (upload)—with duplicate, if (M = N) (average case)	No	Challenge-response model (randomly select bytes position in file)—N * O(1)	Challenge-response model (randomly select blocks(C), C < K)—N * O(C)	Challenge-response model (randomly select bytes position in block)—N * O(K)
PoW (upload)—with duplicate, if (M < N) (best case)	No	Challenge-response model (randomly select bytes position in file)—M * O(1)	Challenge-response model (randomly select blocks(C), C < K)—M * O(C)	Challenge-response model (randomly select bytes position in block)—M * O(K)
PoW (download)	No	User send tags through devices. That device unique identifier is used—O(1)	Challenge-response model (randomly select blocks -C)—O(C)	User send tags through devices. That device unique identifier is used –O(1)
Distribution of key	No	No	Yes—O(W)	No
Upload data	CK—O(1), CT—O(1)	CK—O(1),CT—O(1)	K*CT—O(K),K*W*CK—O(K*W)	K*CT—O(K), K*CK—O(K)
Download data	CK—O(1), CT—O(1)	CK—O(1),CT—O(1)	K*CT—O(K), K*X*CK—O(K*X) (X < W)	K*CT—O(K), K*CK—O(K)
Key generation to encrypt hash value	User's Master key, No third party involved	TAA generate key based on Attributes of file and access policies—O(1)	hash value is mapped into W secret shares and store it in different CSP O(K*W)	TAA generate key based on Attributes of blocks and access policies—O(K)

(continued)

Table 1 (continued)

Steps	File level deduplication		Block level deduplication	
	Master key based CEK management	Proposed attribute-based CEK management	Encoding/Decoding-based CEK management	Proposed attribute-based CEK management
Hash value generation	Yes—O(1)	Yes—O(1)	Yes—O(K)	Yes—O(K)
Encryption based on hash value	Yes—O(1)	Yes—O(1)	Yes—O(K)	Yes—O(K)
Tag generation	Yes—O(1)	Yes—O(1)	Yes—O(K)	Yes—O(K)
PoW	No	Yes—O(1)	Yes—O(K)	Yes—O(C)
Complexity (average case)	O(1)	M*O(1)	M * O(K*W)	M * O(K)

C—Number of challenges
W—Number of CSP
X—Number of key shares to form original key
K—Number of blocks
N—Number of files
M—Number of duplicate files, $M \leq N$

References

1. Li, J., Chen, X., Li, M., Li, J., Lee, P.P.C., Lou, W.: Secure deduplication with efficient and reliable convergent key management, IEEE Trans. Parallel Distrib. Syst. **25** (2014)
2. Cui, H., Deng, R.H., Li, Y., Wu, G.: Attribute-based storage supporting secure deduplication of enciphered data in cloud, IEEE Trans. Big Data (2017)
3. Chen, J., Zhang, L., He, K., Chen, M., Du, R., Wang, L.: Message-locked proof of ownership and retrievability with remote repairing in cloud. Secur. Commun. Netw. (2016). Wiley online
4. Mi Wen, Kaoru Ota, He Li, Jingsheng Lei, Chunhua Gu, Zhou Su.: Secure data deduplication with reliable key management for dynamic updates in CPSS. IEEE Trans. Comput. Soc. Syst. **2** (2015)
5. Gantz, J., Reinsel, D.: The digital universe in 2020: big data, bigger digital shadows, and biggest growth in the far east (2012)
6. Sahai, A., Waters, B.: Fuzzy identity-based encryption, in advances in cryptology—EUROCRYPT 2005, Proceedings. Lecture Notes in Computer Science, vol. 3494, pp. 457–473. Springer, Berlin (2005)
7. Goyal, V., Pandey, O., Sahai, A., Waters, B.: Attribute-based encryption for fine-grained access control of enciphered data. In: Proceedings of the 13th ACM Conference on Computer and Communications Security (2006)
8. Bethencourt, J., Sahai, A., Waters, B.: Enciphered text-policy attribute-based encryption. In: IEEE Symposium on Security and Privacy (2007)
9. Cheung, L., Newport, C.C.: Provably secure enciphered text policy ABE. In: Proceedings, ACM Conference on Computer and Communications Security (2007)
10. Douceur, J.R., Adya, A., Bolosky, W.J., Simon, D., Theimer, M.: Reclaiming space from duplicate files in a serverless distributed file system. In: ICDCS, pp. 617–624 (2002)
11. Keelveedhi, S., Bellare, M., Ristenpart, T.: Dupless: server-aided encryption for deduplicated storage. In: Proceedings of the 22th USENIX Security Symposium, USENIX Association, pp. 179–194 (2013)
12. Bellare, M., Keelveedhi, S., Ristenpar, T.: Message-locked encryption and secure deduplication. In: Annual International Conference on the Theory and Applications of Cryptographic Techniques: Advances in Cryptology—EUROCRYPT, pp. 296–312 (2013)
13. Bellare, M., Keelveedhi, S.: Interactive message-locked encryption and secure deduplication. In: Proceedings of Public-Key Cryptography, vol. 9020 (2015)
14. Galbraith, S.D., Paterson, K.G., Smart, N.P.: Pairings for cryptographers. Discret. Appl. Math. **156**, 3113–3121 (2008)

Random Invertible Key Matrix Decomposition for Classical Cryptography

Adyasha Behera, Alakananda Tripathy, Alok Ranjan Tripathy and Smita Rath

Abstract Cryptography is assumed to be an important part of the secret message composing. It is the way of securing plain text by changing it into cipher text using some algorithms. Cryptography is utilized to provide the secrecy of cipher text during transmission from an adversary. The intent of this paper is to introduce *LU* decomposition technique to generate the key matrix which makes the key more complex by enhancing the security of the message using invertible key matrix.

Keywords Cryptography · *LU* decomposition · Symmetric key · Asymmetric key · Hill cipher

1 Introduction

Cryptography is the process of hiding the original message by converting it into scrambled message. Cryptography plays an important role in science of secret writing. Cipher text is numerically analyzed which is utilized as a part of encryption and decoding process. Network security is one of the parts of cryptographic approaches. Cryptography is classified into two types based on the use of key. They are private key cryptography and public-key cryptography [1]. Secret key was used for encryption and decryption process in private key cryptography whereas two keys are used for encryption and decryption in public-key cryptography. In Secret key cryptography

---

A. Behera · A. R. Tripathy
Department of Computer Science, Ravenshaw University, Cuttack, Odisha, India
e-mail: adyashabehera01@gmail.com

A. R. Tripathy
e-mail: tripathyalok@gmail.com

A. Tripathy (✉) · S. Rath
Department of Computer Science and Engineering, Siksha 'O' Anusandhan Deemed to be University, Bhubaneswar, Odisha, India
e-mail: alakanandatripathy@soa.ac.in

S. Rath
e-mail: smitarath@soa.ac.in

B. Pati et al. (eds.), *Advanced Computing and Intelligent Engineering*, Advances in Intelligent Systems and Computing 1082,
https://doi.org/10.1007/978-981-15-1081-6_47

encryption algorithm (as in Fig. 1) use secret key on plain text and generates cipher text by performing various substitutions and transformation on plain text using the secret key. The decryption algorithm (as in Fig. 2) takes cipher text and secret key as input and generates plain text.

Security correspondence of content data is prime significance over the globe. Cryptography is one of the strategies to achieve security among all kinds of threat. Out of several block cipher Hill cipher provides an excellent use of matrices. The

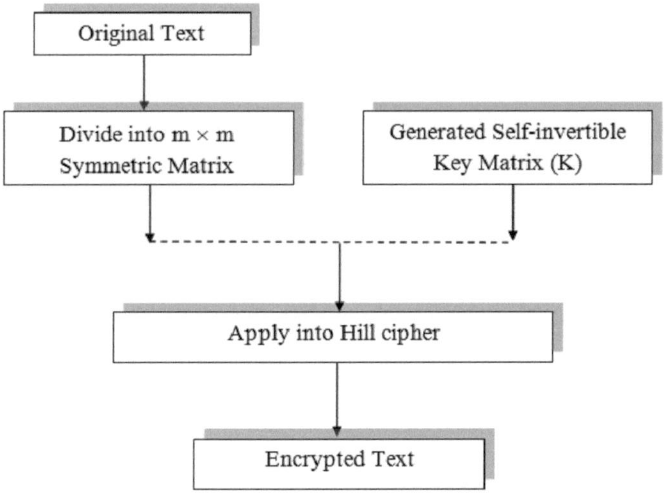

Fig. 1 Proposed encryption framework

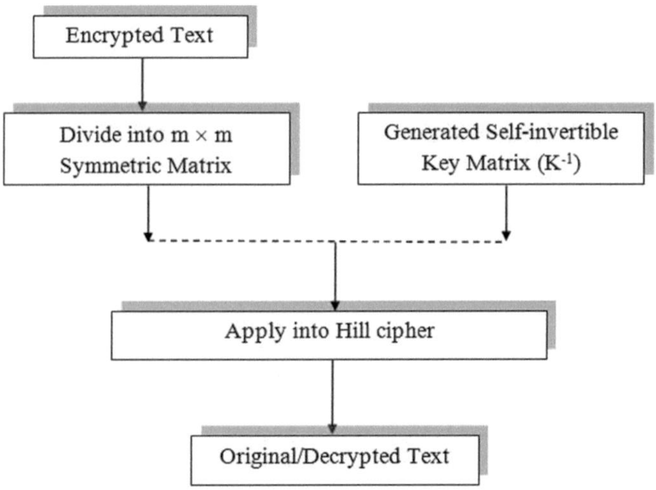

Fig. 2 Proposed decryption framework

complexity of matrix gives enough security. As, a result the complexity in key generation can be achieved. In Hill cipher algorithm, there always arises a problem to find a proper invertible matrix for encryption as well as decryption. At the same time we use *LU* matrix decomposition for getting enough number of unique keys in the key matrix. This research work illustrates to find out suitable invertible *LU* decomposed random symmetric and asymmetric matrices. The Hill cipher is a symmetric encryption against any kind of attack for the known plain text. This research work proposes random symmetric and asymmetric matrices as keys for the Hill cipher algorithm. In the proposed cryptosystem, each time keys are different for different matrices. Therefore, it is hard to find the exact keys that will use for encryption.

Great mathematician Lester Hill in 1929 developed Hill Cipher, it was the first polygraphic substitution block cipher which can process and substitute more than one character at a time [2]. Hill cipher works on the principle of linear algebra. It uses a square matrix as its secret key. Due to the complexity of the matrix, it is resilient against Ciphertext Only Attack (COA) and also immune to letter frequency analysis. In encryption algorithms like Hill cipher, it takes n successive plain text letter and key to generate cipher text [3]. The substitution is determined by n linear equation where each alphabet is assigned with numerical value starting from 0 for a to 25 for z. Each message is encrypted using the Eq. (1).

$$C = E(K, P) = PK \mod 26. \tag{1}$$

$$\begin{bmatrix} c_1 \\ c_2 \\ c_3 \end{bmatrix} = \begin{bmatrix} k_{11} & k_{12} & k_{13} \\ k_{21} & k_{22} & k_{23} \\ k_{31} & k_{32} & k_{33} \end{bmatrix} \cdot \begin{bmatrix} p_1 \\ p_2 \\ p_3 \end{bmatrix} \mod 26 \tag{2}$$

For decryption, the cipher texts use the Eq. (3) to generate plain text.

$$P = K^{-1}C \mod 26 \tag{3}$$

$$\begin{bmatrix} p_1 \\ p_2 \\ p_3 \end{bmatrix} = \begin{bmatrix} k_{11} & k_{12} & k_{13} \\ k_{21} & k_{22} & k_{23} \\ k_{31} & k_{32} & k_{33} \end{bmatrix} \cdot \begin{bmatrix} c_1 \\ c_2 \\ c_3 \end{bmatrix} \mod 26 \tag{4}$$

Although matrix gives complexity to the Hill cipher but it is a big problem to generate an invertible matrix that will be eligible for the key. In the proposed work, we have taken the *LU* decomposition of a matrix and with the help of computing facility, an invertible and secured key is produced.

Let B be a square matrix of size 3×3

$$\begin{bmatrix} b_{11} & b_{12} & b_{13} \\ b_{21} & b_{22} & b_{23} \\ b_{31} & b_{32} & b_{33} \end{bmatrix} \tag{5}$$

Then the *LU* decomposition of *B* can be written as specified by Eq. (6) where *X* represents the lower triangular matrix and *Y* represents the upper triangular matrix of *B*.

$$B = X \cdot Y. \tag{6}$$

In lower triangular matrices all the elements above diagonal are zero. In the same way, upper triangular matrices have zero entries below the diagonal as shown in Eq. (7).

$$\begin{bmatrix} b_{11} & b_{12} & b_{13} \\ b_{21} & b_{22} & b_{23} \\ b_{31} & b_{32} & b_{33} \end{bmatrix} = \begin{bmatrix} x_{11} & 0 & 0 \\ x_{21} & x_{22} & 0 \\ x_{31} & x_{32} & x_{33} \end{bmatrix} \cdot \begin{bmatrix} y_{11} & y_{12} & y_{13} \\ 0 & y_{22} & y_{23} \\ 0 & 0 & y_{33} \end{bmatrix} \tag{7}$$

But it has been observed that the *LU* decomposition of maximum matrices is not fraction free, hence it is unsuitable for the key. But this problem came with a solution that is called fraction free *LU* decomposition. We can get the fraction free *LU* decomposition of matrix *B* as: $PB = LD^{-1}U$ where *P* is a permutation matrix and *D* is a diagonal matrix.

It is a very tough task to find out fraction free *LU* decomposition manually. But with the help of advanced programming modules it can be calculated easily. As it makes the L and U matrix fraction free, the numbers in the matrix become quite big and also have mixed values like positive integer, negative integer, and zeros which gives more complexity to the key.

All matrices are not invertible. So we can not take all matrices as key matrix for Hill cipher. Invertible key matrix must be taken. If *K* is the key matrix then the gcd of $|K|$ mod 26 must be one.

In Sect. 2 we discuss some of the related work. Section 3 contains a detailed idea behind the proposed work. Section 4 explained the evaluation and comparison. Last section describes the conclusion drawn from the research work.

2 Related Work

In this section, we discuss different classical cryptographic approaches. The referred papers explained the various aspect of Hill cipher cryptography.

Dynamic keys are used in Hill cipher cryptosystem to generate a secured message for the electronic transaction system as discussed by Chowdhury et al. [4].

A three stage modified Hill cipher algorithm is proposed by Khalaf et al. [5] where each stage is considered a block cipher with a block length of 128 bits and a key length of 256 bits. The encryption process is in the three stage and the keys are generated randomly using random number generator. Hence it is more robust and provides high-level security of the data.

A different technique for encrypting text other than the conventional Hill Cipher is discussed by Thangarasu and Selva Kumar [6].

An algorithm which selects a key matrix where the elements of the matrix are random to generate a function for encrypting the image was proposed by Prerna et al. [7]. The generated self-invertible matrix is also used for decryption. This proposed method is resistance against brute force attack.

Cipher text blocks are generated using a random key matrix as introduced by Nagabhyrava [8].

Khan et al. [9], discusses generating key matrix using the orthogonal matrix for Hill cipher. This increases the security of the Hill Cipher.

Sharath Kumar, Panduranga and Naveen Kumar [10] proposes a hybrid technique for image encryption using Hill Cipher.

Mani and Mahendran [11], uses two deterministic methods, for generating the key matrix which proves to be better than traditional techniques.

Mani and Viswambari [12], proposed a deterministic method in generating the key matrix based on a magic rectangle. Higher order of key matrix is generated based on m. To generate the key matrix, first the magic rectangle of order $m \times n$is converted into magic square of order $m \times m$ from which the required key matrix is formed.

Naveenkumar and Panduranga [13], proposes Permutation and diffusion process to generation matrix using Chaos and Hill Cipher System for image encryption.

Sundarayya and Prasad [14] developed a technique to generate lower and upper triangular matrices from square matrix using decomposition. Lower triangular matrix is used as a key in encryption process and upper triangular matrix under modulation of prime number is used as key in decryption process.

3 Proposed Work

This section describes about our implemented work that will help to overcome the shortcomings of classical hill cipher. To select the key matrix, the matrix must be invertible. Although we can generate random key matrices but they do not have dynamic values in their entries. We use *LU* factorization of a matrix to solve the dynamic integer problem in the key matrix.

Out of some decomposition technique (i.e., QR, *LU*, Cholesky), *LU* decomposition is unique as it decomposed into a lower triangular matrix (X) and a triangular matrix (Y) [15]. Matrix B can be decomposed as: $B = X \cdot Y$ i.e.,

$$\begin{bmatrix} b_{11} & b_{12} & b_{13} \\ b_{21} & b_{22} & b_{23} \\ b_{31} & b_{32} & b_{33} \end{bmatrix} = \begin{bmatrix} x_{11} & 0 & 0 \\ x_{21} & x_{22} & 0 \\ x_{31} & x_{32} & x_{33} \end{bmatrix} \cdot \begin{bmatrix} y_{11} & y_{12} & y_{13} \\ 0 & y_{22} & y_{23} \\ 0 & 0 & y_{33} \end{bmatrix} \tag{8}$$

But *LU* decomposition does not generate integers in their entries. So the solution can be found by making the *LU* decomposition fraction free. A method to get fraction free LU decomposition is illustrated and hence used by us. The matrix B can be decomposed as $PB = XD^{-1}Y$ to get fraction free X and Y matrix. Where D is a diagonal matrix and P is a permutation matrix.

It is a very tough task to find out fraction free LU decomposition manually in pen and paper. But with the help of the computer it can be calculated easily. As it makes the L and U matrix fraction free, the numbers in the matrix become quite large and also have mixed values like positive integer, negative integer, and zeros which provides more complexity to the key.

All matrices are not invertible. So we cannot take all matrices as key for Hill cipher. The key matrix that taken is an invertible mod m. If K is the key matrix then the gcd of K mod q and m must be one. Only those matrices can be selected that are successfully inverted (Algorithm 1). In key matrix generation algorithm $n \times n$ random matrix is selected. Z_m is the number of alphabets that further used for modulation. The proposed algorithm is given as follow:

```
Initialize n as size of square matrix
Assign m as  Z_m
KeyGeneration(n,q)
Step 1: Set flag=0
Step 2: while flag==0 repeat
Step 3: generate random matrix M of the size of  n×n
Step 4: calculate fraction free LU decomposition of the
        matrix M
Step 5: x=det(K)
Step 6: K can be L or U
Step 7:x=x mod q
Step 8: if gcd(x,q)==1
Step 9: return K
Step 10: otherwise go to step 2
```

The flag variable is used for checking the condition (Algorithm 1). Initially, it set as 0. Inside the while loop, we take a random matrix and then its fraction free LU decomposition is calculated. Then the invertibility of the matrix is checked in steps 2–4. Then we find out the determinant of the matrix and its mod m. In step 7 it checks for its invertibility, if gcd of det (K), and mod m is 1 then, return the key K which is an invertible matrix. Otherwise the condition failed. This generated key matrix will further used for encryption propose.

3.1 Hill Cipher

As we discuss previously about the Hill cipher cryptography. In this section we describe about the algorithms as well as key generation algorithm. Given Hill Cipher Encryption Algorithm (Algorithm 2) is having the input as Plain text (P) and Key matrix (K). Cipher text (C) is calculated using PK mod 26. To generate cipher text following encryption algorithm was used:

```
Encryption(P,K)
Step 1: C=PK mod q
Step 2: return C
```

In case of Hill Cipher Decryption Algorithm (Algorithm 3) the original plain text (P) is generated by taking the $K^{-1}C\text{mod}26$ where K^{-1} is an inverse of matrix and C is the cipher text. These are previously described in the introduction. During the encryption and decryption time the key matrix needs to be invertible. To generate more number of key we use LU decomposition technique. Therefore the key can be L or U depending on the matrix multiplication. The decryption algorithm is:

```
Decryption(C,K)
Step 1: P= K^-1 C mod q
Step 2: return P
```

With the help of our proposed algorithm we can easily get the invertible key matrix. That is further decomposed and it has a higher impact in security. The next section analyzed the algorithm with an example.

4 Evaluation

To explain the previous section satisfactorily, we are explaining the whole algorithm with an example. We have taken both symmetric and asymmetric matrices. These matrices are a random 3×3 matrix. We use the key generation algorithm to find out a matrix M whose L is invertible and we take Z_{26}. M is an example of symmetric random matrix. P is the permutation matrix. D is an identity matrix. The evaluation process is explained with example in step by step manner using Eqs. (9)–(22).

$$M = \begin{bmatrix} 71 & 80 & 2 \\ 80 & 99 & 84 \\ 2 & 84 & 22 \end{bmatrix} \tag{9}$$

$$P = \begin{bmatrix} 1 & 0 & 0 \\ 0 & 1 & 0 \\ 0 & 0 & 1 \end{bmatrix} \tag{10}$$

$$L = \begin{bmatrix} 71 & 0 & 0 \\ 80 & 629 & 0 \\ 2 & 5804 & 1 \end{bmatrix} \tag{11}$$

$$D = \begin{bmatrix} 71 & 0 & 0 \\ 0 & 44,659 & 0 \\ 0 & 0 & 629 \end{bmatrix} \quad (12)$$

$$U = \begin{bmatrix} 71 & 80 & 2 \\ 0 & 629 & 5804 \\ 0 & 0 & -460,654 \end{bmatrix} \quad (13)$$

Here L is invertible so,

$$\text{Key}(K) = L = \begin{bmatrix} 71 & 0 & 0 \\ 80 & 629 & 0 \\ 2 & 5804 & 1 \end{bmatrix} \quad (14)$$

Let the message be "RAVENSHAW." Here the matrix size is 3×3. The message is divided into 3×1 unit vector.

$$P_1 = \begin{bmatrix} R \\ A \\ V \end{bmatrix} = \begin{bmatrix} 17 \\ 0 \\ 21 \end{bmatrix} \quad (15)$$

$$P_2 = \begin{bmatrix} E \\ N \\ S \end{bmatrix} = \begin{bmatrix} 4 \\ 13 \\ 18 \end{bmatrix} \quad (16)$$

$$P_3 = \begin{bmatrix} H \\ A \\ W \end{bmatrix} = \begin{bmatrix} 7 \\ 0 \\ 25 \end{bmatrix} \quad (17)$$

$$C_1 = P_1 K \mod q$$

Hence, after computing the text "RAVENSHAW" is encrypted as "LIDYVADOK."

Again we take an asymmetric matrix (N) and then compute the cipher text by using the decomposition of N. If

$$N = \begin{bmatrix} 23 & 87 & 54 \\ 56 & 15 & 67 \\ 40 & 41 & 0 \end{bmatrix} \quad (18)$$

is decomposed as

$$P = \begin{bmatrix} 1 & 0 & 0 \\ 0 & 1 & 0 \\ 0 & 0 & 1 \end{bmatrix} \quad (19)$$

$$L = \begin{bmatrix} 23 & 0 & 0 \\ 56 & -4527 & 0 \\ 40 & -2537 & 1 \end{bmatrix} \tag{20}$$

$$D = \begin{bmatrix} 23 & 0 & 0 \\ 0 & -104,121 & 0 \\ 0 & 0 & -4527 \end{bmatrix} \tag{21}$$

$$U = \begin{bmatrix} 23 & 87 & 54 \\ 0 & -4527 & -1483 \\ 0 & 0 & 261,563 \end{bmatrix} \tag{22}$$

Here L is invertible so,

$$\text{Key}(K) = L = \begin{bmatrix} 23 & 0 & 0 \\ 56 & -4527 & 0 \\ 40 & -2537 & 1 \end{bmatrix} \tag{22}$$

After calculation, the text "RAVENSHAW" is encrypted as "BCQODJFCQ." This is an example of our proposed work. Then, we try to compare symmetric matrices as well as asymmetric matrices using *LU* decomposition (Fig. 3). Where asymmetric matrix gives better results then symmetric matrix with respect to time.

By making a comparison in (Table 1) we try to prove that our proposed cryptosystem for asymmetric key is better than symmetric key. The comparison is based on some factors like size of the key, Memory requirement, Encryption/Decryption Time, block size, number of rounds, and level of confidentiality. As we can see our proposed system is hard to crack by the attacker because of the random key size that varies every moment. It also maintains the confidentiality of message.

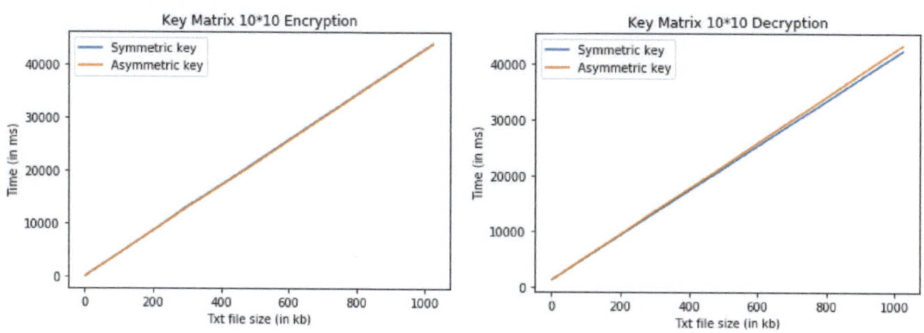

Fig. 3 *LU* decomposition for symmetric and asymmetric matrix

Table 1 Comparison between symmetric and asymmetric Hill cipher

Comparison factor	Symmetric key	Asymmetric key
Key size	Random	Random
Memory requirement	Low	Low
Encryption/decryption time	High	High
Number of rounds	1	1
Block size	Vary	Vary
Confidentiality	Low	High

5 Conclusion

The journey of our research work gives us a better clarity about the generating suitable key generation algorithm. To select a proper invertible key is always been challenging for Hill cipher algorithm. Our focus is on resolving that problem. As there is always a choice for both symmetric and asymmetric keys. Here both the keys are considered. And our proposed algorithm is much better than classical Hill cipher. The random matrix generation for given cryptographic system is more secure. The symmetric and asymmetric matrix concept gives the proposed system flexibility. This also shows that the encrypting and decrypting using symmetric key matrix is better than asymmetric key matrix.

References

1. Warjri, J., Raj, E.G.D.P.: KED-a symmetric key algorithm for secured information exchange using modulo 69. Int. J. Comput. Netw. Inf. Secur. **5**(10) (2013)
2. Panigrahy, S.K., Jena, D., Korra, S.B., Jena, S.K.: On the privacy protection of biometric traits: palmprint, face, and signature. In: International Conference on Contemporary Computing, pp. 182–193. Springer, Berlin (2009)
3. Karthikeyan, B., Chakravarthy, J., Vaithiyanathan, V.: An enhanced Hill cipher approach for image encryption in steganography. Int. J. Electron. Secur. Digit. Forensics **5**, 178–187 (2013). https://doi.org/10.1504/ijesdf.2013.058652
4. Chowdhury, S.I., Shohag, S.A.M., Sahid, H.: A secured message transaction approach by dynamic hill cipher generation and digest concatenation. Int. J. Comput. Appl. **23**(9), 25–31
5. Khalaf, A.A.M., El-karim, M.S.A., Hamed, H.F.A.: A triple hill cipher algorithm proposed to increase the security of encrypted binary data and its implementation using FPGA. In: 2016 18th International Conference on IEEE Advanced Communication Technology (ICACT), pp. 752–759 (2016)
6. Thangarasu, N., SelvaKumar, A.L.: Encryption using lester hill cipher algorithm. Int. J. Innov. Res. Adv. Eng. (IJIRAE) **2**(12), 13–17 (2015)
7. Prerna, U., Kumari, M., Shrivastava, J.N.: Image encryption and decryption using modified hill cipher technique. Int. J. Inf. Comput. Technol. **4**(17) (2014)
8. Nagabhyrava, D.H.: Efficient key generation for dynamic Blom's scheme (2014)
9. Khan, F.H., Rehan, S., Qazi, F., Agha, D.: Hill cipher key generation algorithm by using orthogonal matrix. Int. J. Innov. Sci. Mod. Eng. **3**(3) (2015)

10. Sharath Kumar, H.S., Panduranga, H.T., Naveen Kumar, S.K.: Hybrid approach for image encryption using hill cipher technique. In: International Conference on Information Processing. Springer, Berlin, Heidelberg, 200–205 (2012)
11. Mahendran, R., Mani, K.: Generation of key matrix for hill cipher encryption using classical Cipher. In: 2017 World Congress on IEEE Computing and Communication Technologies (WCCCT), pp. 51–54 (2017)
12. Mani, K., Viswambari, M.: Generation of key matrix for hill cipher using magic rectangle. Adv. Comput. Sci. Technol. 10(5), 1081–1090 (2017)
13. Naveenkumar, S.K., Panduranga, H.T.: Chaos and hill cipher based image encryption for mammography images. In: 2015 International Conference on Innovations in Information, Embedded and Communication Systems (ICIIECS), pp. 1–5 (2015)
14. Prasad, M.V., Sundarayya, P.: Symmetric key generation algorithm in linear block cipher over LU decomposition method. Int. J. Trend Sci. Res. Dev. 1(4) (2017)
15. Zhou, W., Jeffrey, D.J.: Fraction-free matrix factors: new forms for LU and QR factors. Front. Comput. Sci. China **2**(1), 67–80 (2008)

Advanced Algorithms, Software Engineering

State Space Reduction Using Bounded Search Method

Bidush Kumar Sahoo and Mitrabinda Ray

Abstract In a concurrent program, the number of state space is proportional to the amount of interleaving of threads. State space reduction is a very crucial issue during testing of the concurrent program. Many state space reduction techniques such as Delta Debugging, Hierarchical Delta Debugging, Partial Order Reduction are already proposed by researchers to enhance testing. However, the complexity issue is a big research challenge in state space. In this paper, we propose an algorithm that modifies dynamic partial order reduction technique for reduction of state space. In the modified version of the algorithm, it provides a facility of bounded function for searching and bounded value for reducing the searching time in place of backtracking as used in existing algorithms. We have evaluated our algorithm in Java Path Finder tool and observed that the reduction of states is better in our approach compared to an existing approach.

Keywords Symbolic path finder · Java path finder · Symbolic execution · Model checker · Coverage criteria

1 Introduction

In the earlier stages of developing a program, the personnel responsible usually have a low input of testing files. This happens when the application brings about new formats for input. The one testing is interested in determining whether a particular predicate on system states can be partially or fully satisfied with the countries that are reachable [1]. A determination on this, the first criteria is to calculate the number of the reachable states and evaluate their predictability. Knowing the states

B. K. Sahoo (✉) · M. Ray
Department of Computer Science and Engineering, Siksha 'O' Anusandhan University Deemed to Be University, Bhubaneswar, Odisha, India
e-mail: bidush.sahoo@gmail.com

M. Ray
e-mail: mitrabindaray@soa.ac.in

B. Pati et al. (eds.), *Advanced Computing and Intelligent Engineering*, Advances in Intelligent Systems and Computing 1082,
https://doi.org/10.1007/978-981-15-1081-6_48

that are examined or not. These states will be presented in the table. To write a test, it is not a simple task since these tests can, however, deal with all cases of the programmer [2].

Usually, programmers expend a considerable amount of time in debugging the programs. Different outcomes regularly prove that maintaining software requires more time than any other exercise of programming. There are various concurrency defects that exist in a program such as order violation, deadlock, atomicity violation, lousy composition, etc. The software testing action inherently implied for troubleshooting the framework alongside lessening the state space.

The tool for exploring the state space systems have been proved to exploring the space of the state for the systems' abstract description and set in the language of modeling [3]. When a model of an application of a system, has been thoroughly analyzed, it can be used for application implementation since it can be done using the environments for developing software. A state space tree represents the states and transitions of a program. Figure 1 illustrates a static state space tree for any Conjunctive Normal Form (CNF) formula involving 3 Boolean variables $x1, x2, x3$. Nodes of the tree correspond to variables, and edges are labeled 0 or 1 according to whether we assign FALSE or TRUE to the variable in the node. Triangular leaf nodes indicate that all three variables have been assigned truth values. The CNF is ($x1$.or. $x3$) .and. ($-x1$ .or. $x2$ .or. $x3$) .and. ($-x1$ .or. $-x2$) .and. ($x2$ .or. $-x3$).

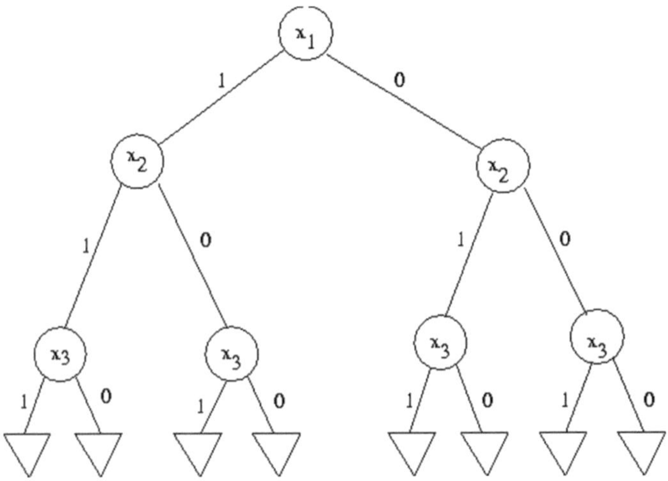

Fig. 1 A static state space tree for CNF for three variables

1.1 Motivation

In a concurrent program, the thread interleaving causes a lot more states as the thread processing is uncertain. The state space explosion is the most significant problem in the testing of the concurrent program. So, to tackle this problem, different reduction techniques are already proposed by many researchers. The reduction techniques not only reduce the state space but also check the bugs in the system. The state space reduction helps in better processing time as well as reducing the time complexity of searching of states. The limitation with backtracking method used in existing reduction techniques is that it is very slow as it is used to search the entire state space tree of the given problem [4]. The state space explosion creates havoc in dealing with states which can be unexplored. So, the attempt should be taken for reducing the time with maximum state coverage.

1.2 Objective

To the best of our knowledge, the existing techniques on state space reduction use backtracking method. Branch-and-Bound method [4] is a better searching technique in terms of minimization of search time which is used to solve the optimized problem. The objective of this paper is to reduce the state space by using the branch-and-bound technology instead of backtracking technique. As backtracking searches till the bottom, it takes more time in comparison to the branch-and-bound method.

The paper is arranged in the following format. Section 2 provides the background study of the basic concept of state space and algorithms on state space reduction. The Sect. 3 contains the literature survey on state space reduction techniques. We propose our modified algorithm in Sect. 4. Section 5 is the discussion of experimental results and discussions. Section 6 concludes the paper with future works to be done.

2 Background Study

State space reduction is very much essential in the concurrent programs. Different authors propose different algorithms for the state space reduction. There are some essential algorithms for state space reduction that are discussed later in our paper. In the below sections, we present the algorithms.

2.1 Delta Debugging (DD) Algorithm

The Delta Debugging (DD) algorithm [5] is an algorithm for recursively pruning the state space. A file that is smaller for input would make it simpler for the programmer to notice a fault in the codes. This more modest file would also occupy lesser space in the system of tracking bugs and hence it can be included in the source code for the program as a regression test. It would be easier to identify duplicate bugs with small input files. It would be better if there could be an automatic way of reducing all the file sizes. Luckily, an approach exists usually known as Delta Debugging. It is a method by Hildebrandt and Zeller which automates minimization of cases.

It operates using the below two algorithms.

Simplification: The input for inducing failure is simplified by evaluating the configurations of inputs which are small. It relies on the shape of little shortcomings to the point that it can't produce smaller arrangements that still fail [6].

Isolation: attempts to identify a configuration that is passing such that it will fail with the addition of some elements. It works a two way looking for the larger cases moving that are parts of the facts that are failing. This algorithm produces fewer initiative outputs as an aid for the programmer in debugging. The difference between the elements is not responsible individually for failing. It simply exposes any symptoms existing. Although this algorithm can be speedy than simplification, it is not convenient in big test cases since it can bring about a wrong time for running due to the time that is spent in extensive configuration testing [7].

2.2 Hierarchical Delta Debugging (HDD) Algorithm

There are several instances of data input using different definitions. In the case where input partially modified, there is a need to use this in the test case minimization. We had a few cases of data input in which HDD is errant. In cases where grammar which is context-free is important for the significance of a special language, a normal language may not be enough. The data that is set will, therefore, be nested paying the way for preference of data debugging which is standard.

In such a point where the input is going to be nested and balanced, there is a need to use it in test case minimization. Any data that is characterized by a grammar which is context-free is useful for the meaning of a certain language which is straightforward and may not render the job done. The information, therefore, can be nested giving a highway over the average Delta Debugging [8].

Certain leveled debugging of data is utilized as an abusive to organized features of the inputs of a program. In efforts to discover the parts of the inputs are prune, there is minimal motivation in discovering the parts of the input that needs to be pruned. There is reduced encouragement to arbitrarily pick areas to be divided. Instead, we work along with the limits from the top of the tree all the way to the base. By restriction to one level at any event, the smaller groups of the nodes are analyzed for minimization.

2.3 *Partial Order Reduction (POR) Algorithm*

POR is used in finding the transitions that are dependable in their operations in efforts to reduce state space. For independent transitions, they can enable or disable each other, and allow commuting of independent transitions. This characterizes the features of a likely good dependency between the changes from the same system. It is possible to give conditions that can be checked easily and ones that are enough for the development of being independent. This dependency may be the result of transitions of different activities where operations are performed on various shared objects. This behavior of systems is functioning at the same time [9].

The specific observation that is exploited using the approach may contain several paths that correspond to the different orders of execution of similar transitions. In cases where the concurrent transitions are not independent, and their executions fail to match or interfere with each other, a change in the order of their performance may fail to modify their efforts. The reduction of partial order is preferred as the most reliable technique for state space reduction process in cases of software systems that are concurrent at the level of implementation. There is general consideration of the technique of partial order reduction [10].

In POR technique the reductions are statically made, where the number of states reduced is less. To traverse the states dynamically, therefore, DPOR is used and thereby there is increased performance [9].

2.4 *Dynamic Partial Order Reduction (DPOR) Algorithm*

A partial order is a reduction that is new. It is usually an algorithm that tracks dynamically the interactions between exploits and processes of such information. It is needed for establishing the positions to backtrack, which also directs the other paths in the state space which needs to be checked. This algorithm is based on DFS technique in the state space of a given system for searching. For a transition sequence TS, the following criteria are used.

- TS_i signifies to the transition t_i;
- TS.t signifies extending TS with an extended transition t;
- Domain (TS) defines the set {1, …, n};
- Pre(TS, i), for $i \in$ Domain (TS) signifies a state s_i; and
- End (TS) determines $s_n + 1$.

DPOR algorithm minimizes the state space by use of backtracking as a method.

A transition can appear many times in a pattern of transition. It can be added with another transition in any case there exists a state in which both transitions are enabled. For instance, any two functions which utilize the same variable is co-enabled. For two transitions that are adjacent in a series of transition, may act as independent. It may be without any change of the final behavior pattern which can be swapped.

In the DPOR algorithm, the stepwise works are described below

In any case a search reaches a state, the process of exploring can be called together with stack S of which a state has been reached. Initially, the exploring method is called with a stack that is empty as an argument. End (TS) reaches the state by checking TS starting from the first state s_0. Then, for every function f, the next transition next(s, p) of each process p in state s is determined [7].

If such a transition exists, a dependency may exist in between i with next(s, p). So, there might be a need for "backtracking point" in the state Pre (TS, i), i.e., in the state prior transition execution i. 2. It is done by computing the set E of function f with a transition that is enabled in Pre (TS, i) that "happens-before" next(s, p) in the new PO ~ S. If E is empty, the implementation of all the functions in E is important to attain transition next(s, p) enabling it in the new PO ~ TS. In such a case, it is enough to include a single process of the same in E to the backtracking set related to pre (TS, i). So, if E is empty, the algorithm is not able to find a function whose execution relevance is important for next(s, p), which is to be activated from Pre (TS, i). So, the algorithm includes all allowed functions in the backtrack (Pre (TS, i)).

3 Literature Survey

State space explosion is a big problem for software testers in case of a parallel system. So, a lot many algorithms are followed by different authors discussed below. The work is by state space reduction in the concurrent program [5]. The idea behind the essential standards, partial order reductions, and its execution portrays a strategy, the POR [6]. The principal goal of this article is to give a comprehensive instinctive portray of the basic minds and presents a few techniques that can be used for execution. Moreover in our approach, we have tried to provide a reduction of searching time through MDPOR algorithm.

The process of reduction state space for the stochastic that is non-statutory can be accomplished by binary processes [11]. The first is a prior reduction in nature and can be used using stationary, the cost functions of the lower and upper limits in the deterministic system before the beginning of a trip the second one reduced the state space further on the stochastic which is non-stationary for the road network before the journey proceeds. This contribution exhibits the ability to compute the focal points of the presented techniques in the context of the real data gathered on the network of a road in the southeast of Michigan. The approach provided by us, checks all the networks of the state space tree.

The other way of dealing is by reducing memory use despite the investigation of total space. The criteria assure to focus on each reachable state at least once per a while [12]. A few times however in this situation to pick a memory performance after some time execution memory change. The method relies on the conventional idea of just abandoning additional states in our programs of known states with some probability; essentially we turn a coin to make decisions upon whether the country needs restoration or be assumed. Since storage of all states is not possible at once, it would like to occupy lesser space. In any event, it happens to store the states with the probability p; it is said to have done with P.N states in the process of inspecting N states. This is apparent and the price of resisting the same states a few times in a convincible manner. If different ways give the same state, or the framework inhibits cyclic behaviors. Moreover, MDPOR is able to provide maximum state coverage with limited time duration.

Blom provides the idea of copying the state space explosion problem in model checking [13]. The computer system needs more accurate and precise type of verification that the traditional test and simulation techniques. Hence, formal software verification approaches use instead. Model checking is an effective and efficient type of formal verification that has been used for verification of safety-critical systems in last two decades. The probability of state space explosion can be reduced by the proposed approach.

A method of program variant based was proposed to detect a group of any program bags that are similar. By testing the units for components of concurrent programs, it is possible to obtain a group of invariant programs, and they can be used as an oracle to gain poor invariants in cases where the program is online [14]. When one uses the function call graph, of the variables and uses a reduction method, to the invariants, it obtains the suspicious functioning candidates and ranks them. Due to the components' interactions, the causes are analyzed with the concurrency bugs.

In Table 1, the discussions are done based on works on state space reduction. Different authors have adopted different techniques like symbolic model checking, invariant analysis, etc., to resolve the state space explosion problem. Maximum of the approaches adopted the backtracking approach, which takes longer time. So, we have thought of replacing the backtracking by branch and bound with bounded search method in our work which is discussed in the next section.

Table 1 Work on state space reduction

Author	Feature	Pros	Cons
Larsen et al. [10]	Bound factor	The reduction is done by giving lower and upper bounds	It leaves some cases where this rule does not hold
McMillan [11]	Probability factor	The method guarantees to examine each reachable state at least once, but potentially several times	It needs to select a number of systems with various kinds of state space graphs
McMillan and Probst [12]	Symbolic model checking	It discusses on Symbolic model checking which exploits the boolean functions	The big challenge remaining is to develop such tools and algorithms with which can verify systems with larger state space
Wang et al.	Invariant technique	It analyzes the causes of concurrent bugs through invariant analysis	Some more bugs are unclear

4 Modified DPOR (MDPOR) Algorithm

In this section, we propose an algorithm, MDPOR, for state space reduction. The modified algorithm is a new version of Dynamic Partial Order Reduction (DPOR). In DPOR, backtracking method is applied for state space reduction. In backtracking, a single node is expanded multiple times from different starting nodes in a state space tree of a concurrent program. The result of a node coming across different paths in a state space tree is stored in a look-up table and can be used whenever required. Backtracking process is slow due to the expansion of nodes multiple times. Backtracking is mainly not used for the problems where final solution only matters and not the path through which it is achieved. It is not suitable for the optimization problem. So, branch-and-bound technique is preferred for optimizing state space [15].

4.1 Branch-and-Bound Algorithm

The aim of this algorithm is to find x which increases or decreases the value of a function $f(x)$, called a goal function, which includes a set K of admissible, or candidate solutions [11]. The set K is recognized for space searching or feasible region. A branch-and-bound algorithm works based on the below principles:

- It divides the search space into two parts, and then increasing $f(x)$ on the minimum areas; the process is called the branch.
- To branch, a lot would move to brute-force development of the solutions and test them all. To enhance the functioning of brute-force searching, a branch-and-bound algorithm stores record of bounds on the least that is giving, and use them to "prune" the space of the search, doing away with candidate solution to prove that it does not have the new solution [10].

To turn them into a particular algorithm for a specific optimization query needs some information representing sets of solutions for the candidate. Such a representation is called a problem case. Represent the set of solutions for the candidate of a case I by SI. The instance is in the form of three alternatives:

- Branch renders two or more examples each representing a subgroup of SI. (Typically, the subsets are disjoint to deter the algorithm from attending the same candidate more than one time, though it is not necessary. The only need for an accurate branch-and-bound algorithm is that the typical solution among SI is found in one of those subgroups.)
- Bound produces a small bound for each candidate's value in the space denoted by I, which is, linked $(I) \leq f(x)$ for all x in SI.
- The resolution shows if i render a solution of a single candidate.

By use of these, a branch-and-bound algorithm functions a recursive from top to bottom search via the instance tree created by the operation of the branch. Upon visiting an example i, it tells if the blood is lower than the normal bound for some other case that it already visited; if so, i may be discarded safely, and the process of recursion ends. This step normally used by maintaining a global variable that reads the least bound found among all evaluated instances so far [7].

The problem of reducing space is a problem of optimization. branch-and-bound criteria are the best suit for reducing space. It may affect the tree in whichever form, DFS or BFS. It searches the tree to get a solution that is optimal based on this function. Our preferred MDPOR algorithm uses branch-and-bound technique for modifying the state space.

4.2 Bounded Search

In comparison with the partial order methods, the bounded search methods are able to provide bounded coverage for the state spaces in spite of full coverage. This search measures the volume of the state space as it will not explore transitions which are exceeding the bound. The bound can be calculated by a bound evaluation function, B_v. The function $B_v(S)$ returns the bounded value for the transition sequence which is the bounded function and c is the bounded value. Bounded search requires two parameters: the bound evaluation function B_v, and the bound c, which can be any positive value. The maximum of the state space is reachable with a small context bound than is reachable with a small depth bound. This property is needed as the size of the state space grows exponentially with the bound increment.

4.3 Algorithm MDPOR

In DPOR, backtracking method is applied for state searching. In backtracking, a single node is expanded multiple times from different starting nodes in a state space tree of a concurrent program. The proposed algorithm is the revised version of Dynamic Partial Order Reduction (DPOR). The existing DPOR algorithm is providing the reduced state space, but the modified approach provides a facility of bounded function [12] for searching and bounded value for reducing the searching time. Backtracking process is slow due to the expansion of nodes multiple times. Backtracking is mainly not used for the problems where the final solution only matters and not the path through which it is achieved. It is not suitable for the optimization problem. In the modified approach, the function B_v (TS) returns the bounded value for the transition sequence TS. Bounded search requires two parameters: the bound evaluation function B_v, and the bound c, which can be any nonnegative value. The full state space is reachable with bound function B_v within bound c from initial state s_0. It recursively searches the transition that can be co-enabled with another transition. The bounded function helps in modifying the search states which is totally covered by backtracking technique [13]. The bounded value checks the number of states to be searched. In this approach, the bounded value is the upper bound which is provided for reducing the search time. The algorithm of MDPOR is provided as follows:

```
Initially: Explore (ϕ);
// given a transition sequence S//
1 Explore (TS) {
// Solution (TS) refers to S_{n+1} //
2 Suppose s = End (TS);
3 for each process p {
// TS_i refers to transition t_i //
// Domain (TS) means the set {1, . . . ,n} //
4 if ∃i = max ({i ∈ Domain (TS) | TS_i is dependent and can be co-enabled with next(s,
p) and i is not→s p}) {
// Pre (TS, i) for i ∈ Domain (TS) refers to state s_i //
5 let E = {q ∈ enabled (pre(S, i)) | q = p or ∃j ∈ Domain (TS) such that j >= i and q =
proc (TSj) and j →s p};
// TS.t denotes adding TS with an extra transition t //
        for all u ∈ enabled(final(TS)) do
        if B_v(S.next(final(S), u)) < c then //B_v is the bounded function and c is the
bounded value.
        Explore(TS.next(final(TS), u))
        end if
 end for
6 if (E ≠ϕ) then add q ∈ E to branch (Pre (TS, i));
7 else add q ∈ enabled (Pre (TS, i)) to branch (Pre(TS, i));
8 }
9 }
10 if (∃p ∈ enabled(s)) {
11 branch(s) := {p};
12 let done = ϕ;
13 while (∃ p ∈ (branch(s) \ done)) {
14 add p to done;
15 Explore (TS.next (s, p));
16 }
17 }
18 }
```

When one reaches a new state during the time of the search, the method of exploring is called by the stacks which are usually having an empty argument. Initially in line 0, the method Explore is known with the empty stack as an argument. In line 2, End (TS) represents the state reachable by executing TS from initial state s_0. Then, for every process p, the further transition next(s, p) of every process p in state s is examined in line 3. For this type of transition next(s, p), one is computed in line 4. The end transition i in TS such that the two transitions TS.i and next(s, p) are dependent, such that $i \rightarrow$ TS p. If such a transition i exist, there can be a dependent relation between i and next(s, p). Hence there is need to check a "backtracking point" in the state Pre (TS, i), i.e., in the state prior to executing the transition i. 2. In line 5, by computing the set E of processes q with an enabled transition in Pre (TS, i) is shown. The transition "happens-before" next(s, p) in the current PO ~ S. However, if E is nonempty, the execution of each process in E is necessary to reach transition next (s, p) and to make it active in the current PO ~ S. In that case, it is adequate to include all of the processes in E to the bound searching set related with Pre (TS, i). The recursive loop for bounded searching is done along with the bounded value checking condition. Conversely, if E is nonempty, the algorithm can be able to recognize a process whose execution is required for next(s, p), is to become enabled from Pre (TS, i). So, by default in line 7, the algorithm adds each of the enabled processes in backtrack (Pre (TS, i)).

Upon various computations for the new backtracking points is done in lines 3–9. The search may continue from the original state s. If the enabled processes in s exist, which is described in line 10, any one of those may be chosen to evaluate through addition to the depth bound search for set of s discussed in line 11. The existence of depth bound searching set is related to the new state s that is unexplored yet, those processes are executed accordingly the code of lines 13–15.

In the above algorithm, the depth bound searching and bound values are taken place in place of backtracking technique for providing lower bound constraint [14]. The depth bound searching is done for state space generation, and the bound phase is used for the lower bound constraint for reduction of state space. The solution part is the end part of the tree. In this approach, it will traverse throughout the tree and check for the constraints satisfaction [15]. The time complexity factor makes very little difference in favor of the modified approach. But the space complexity creates the major difference between these two approaches.

5 Experimental Results and Discussions

The proposed approach has been applied to existing following concurrent programs:

- Dining philosopher problem is one of the example of concurrent problems. It is utilized in the algorithm that is concurrent to demonstrate synchronization issues and methods for bringing a solution to them. The query is certain five philosophers sit down on a table with plates, and for each pair of a philosopher who is adjacent, forks are placed.
- The second case is race conditioning issues that happen when a pair or more threads have the accessibility of shared information and try to make changes to it simultaneously. Since the scheduling algorithm thread can swap between these threads at any given time, one cannot predict the order these threads could be accessing the information and at what time.
- The third program is meant for crossing whereby a certain group of people wishes to cross a bridge at night. Every night, a certain number of people must cross at any given time with a flashlight. There is only one flashlight that is available for one group, and hence there must be arrangements made to take back the torch for people to be able to cross.

These above discussed concurrent programs are verified on Java Path Finder (JPF) tool. JPF is a runtime tool configured combination of different components. It is set in Java environment. The input for JPF is the source code of concurrent program and output is the concurrent errors. Verification is the main feature of JPF that provides information such as time consumed for confirmation, number of threads at runtime, number of states of state space tree, and number of instructions in the source code. JPF uses POR technique by default for state space reduction. But instead of POR [7], we have applied MDPOR technique/algorithm for better state space reduction. The following steps are done for using MDPOR during verification:

Step-1: The algorithm, MDPOR, is imported in JPF through the.jpf file as shown in Fig. 2, where the concurrent input file is also linked. In this figure, the directory name and Reduction algorithm name (Bidush.MDPOR) is assigned to target, shows that JPF is taking our proposed MDPOR technique for state space reduction. The concurrent input program, Dining Philosopher problem (DiningPhil.java) is assigned to the listener.

Step-2: The verification of the.jpf file is done using JPF tool and the result is displayed in Fig. 3.

Similarly, Fig. 4 demonstrates the verification results of concurrent program, MyRacecondition, obtained by Java Path Finder tool.

The state (S) and transition (T) of the concurrent programs are provided along with the reduced state and transitions after applying the branch-and-bound algorithm in Table 2.

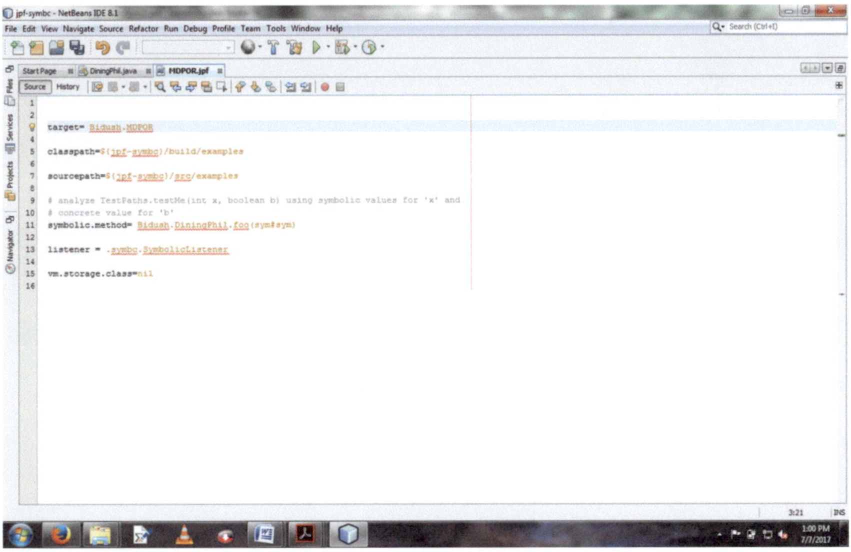

Fig. 2 Snapshot of MDPOR.jpf file

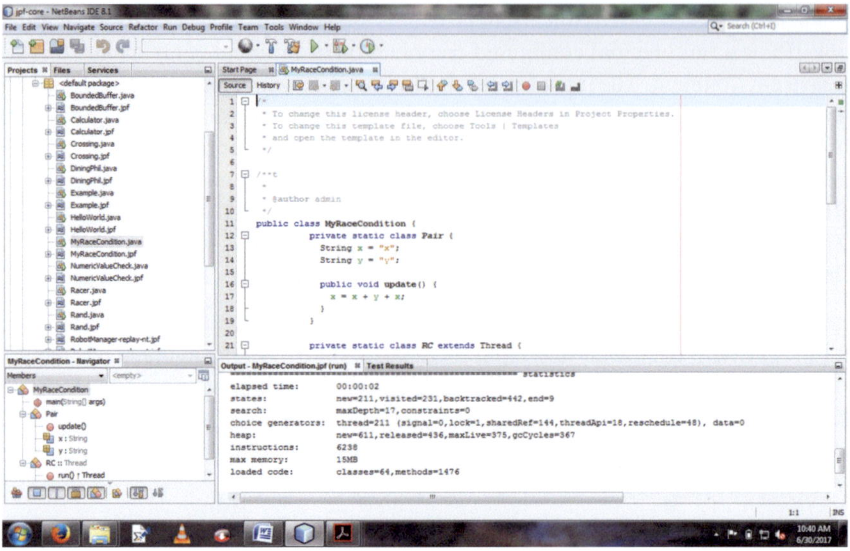

Fig. 3 Snapshot of MyRacecondition program's verification result

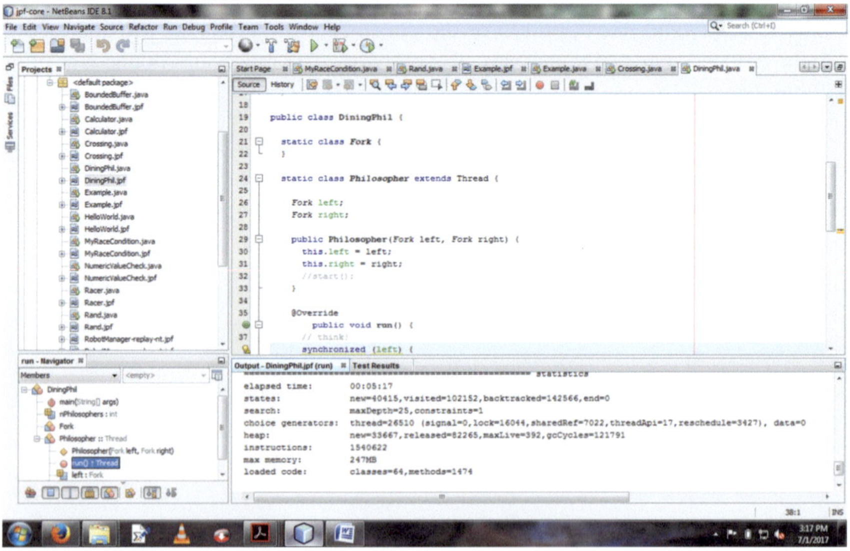

Fig. 4 Snapshot of dining philosopher program's verification result

Table 2 Comparison between the existing and modified approach

Program	Existing approach (S) (T)		Modified approach (S) (T)		No. of threads
Dining philosopher	142,566	1,540,622	130,259	1,256,322	26,510
Race condition	442	6238	366	5152	211
Crossing	1440	321,653	1341	30,1144	513
Racer	87	189	74	173	27
Producer consumer	1162	274,134	1007	24,5173	617

From these above techniques used for state space reduction, DD algorithm, and HDD algorithm are used in a stepwise manner [8]. POR and DPOR algorithms are used for reducing state space in branch reduction manner. In the proposed approach, it uses DPOR algorithm along with branch–and-bound approach to providing a restriction in expanding the number of states in the state space diagram. The invariant technique is also discussed for application in the DPOR algorithm. As in backtracking, the approach only provides repeating the process of searching more profound, but it never provides any constraints on the number of states. We found the number of states and transitions gets reduced after the modification of branch-and-bound algorithm.

6 Conclusion

Many approaches have been adopted for reducing the state space searching, to fasten the processing time and get a lesser number of test cases for testing. DD algorithm is a technique to reduce the state space to some extent, but it is not sufficient. Then POR algorithm is adopted for further state space reduction by creating equivalence classes. But in DPOR which is the advanced version of POR, provides technique through backtracking procedure. It can be further modified by our approach by giving constraints in the searching process. It reduces the number of state searching in comparison with the previous technique. The future work can be providing a lower limit constraint to the bound approach.

References

1. Blom, S., van de Pol, J.: State space reduction by proving confluence. In: CAV, vol. 2404, pp. 596–609 (2012)
2. Christofides, N., Mingozzi, A., Toth, P.: State space relaxation procedures for the computation of bounds to routing problems. Networks **11**(2), 145–164 (2011)
3. Clarke, E.M., Grumberg, O., Minea, M., Peled, D.: State space reduction using partial order techniques. Int. J. Softw. Tools Technol. Transf. **2**(3), 279–287 (2013)

4. Desrochers, M., Lenstra, J.K., Savelsbergh, M.W., Soumis, F.: Vehicle routing with time windows: optimization and approximation (No. OS-R8715). CWI. Department of Operations Research and System Theory [BS] (2017)
5. Desrochers, M., Desrosiers, J., Solomon, M.: A new optimization algorithm for the vehicle routing problem with time windows. Oper. Res. **40**(2), 342–354 (2012)
6. Duri, S., Buy, U., Devarapalli, R., Shatz, S.M.: Application and experimental evaluation of state space reduction methods for deadlock analysis in Ada. ACM Trans. Softw. Eng. Methodol. (TOSEM) **3**(4), 340–380 (2014)
7. Eagle, J.N.: The optimal search for a moving target when the search path is constrained. Oper. Res. **32**(5), 1107–1115 (2014)
8. Etessami, K., Wilke, T., Schuller, R.A.: Fair simulation relations, parity games, and state space reduction for Büchi automata. In: ICALP, vol. 2076, pp. 694–707 (2011)
9. Kim, S., Lewis, M.E., White, C.C.: State space reduction for nonstationary stochastic shortest path problems with real-time traffic information. IEEE Trans. Intell. Transp. Syst. **6**(3), 273–284 (2015)
10. Larsen, K.G., Larsson, F., Pettersson, P., Yi, W.: Efficient verification of real-time systems: compact data structure and state-space reduction. In: Real-Time Systems Symposium. Proceedings, The 18th IEEE, pp. 14–24 (2016)
11. McMillan, K.L.: Using unfoldings to avoid the state explosion problem in the verification of asynchronous circuits. In: International Conference on Computer Aided Verification, pp. 164–177. Springer, Berlin (2013)
12. McMillan, K.L., Probst, D.K.: A technique of state space search based on unfolding. Form. Methods Syst. Des. **6**(1), 45–65 (2015)
13. Pernebo, L., Silverman, L.: Model reduction via balanced state space representations. IEEE Trans. Autom. Control. **27**(2), 382–387 (2012)
14. Shih, W.: A branch and bound method for the multi constraint zero-one knapsack problem. J. Oper. Res. Soc. 369–378 (2016)
15. Tuan, H.D., Apkarian, P., Nguyen, T.Q.: Robust and reduced-order filtering: new LMI-based characterizations and methods. IEEE Trans. Signal Process. **49**, 2975–2984 (2011)

Exploring Application of Knowledge Space Theory in Accessibility Testing

Neha Gupta

Abstract Knowledge space theory has so far most predominantly found applications in e-learning and educational assessment systems. The principles of knowledge space theory can however also be used in the assessment of a system, device, or application in terms of accessibility. A precedence order among the Web Content Accessibility Guidelines (WCAG) can be discovered through careful analysis, and expert knowledge can be used to construct a knowledge structure with each knowledge state representing a set of these guidelines. This is useful not only in gaining an understanding of where a system lies on its way toward becoming universally accessible but also helps in determining the next steps for improvement of the analyzed system's accessibility from the outer fringe of the current knowledge state. This is a more feasible way of judging a system for its accessibility since it shows exactly in terms of which guidelines the assessed system is accessible. The standards for acceptable accessibility criteria specified and met by organizations or industries will thus be in terms of knowledge states instead of a numerical or graded value. This invokes comprehensiveness, clarity, and customization in practice of accessibility testing.

Keywords Accessibility · Universal design · Artificial intelligence · Knowledge-based systems · Knowledge spaces · Computer-aided analysis · Probabilistic computing · Best practices · Knowledge space theory · Accessibility testing · Inclusive technology · Probability distribution · Testing · Software testing · System testing · Adaptive systems · Probability distribution · System improvement · Precedence theory

---

N. Gupta (✉)
Easy Alliance, Brookline, MA, USA
e-mail: neha@easyalliance.org

B. Pati et al. (eds.), *Advanced Computing and Intelligent Engineering*,
Advances in Intelligent Systems and Computing 1082,
https://doi.org/10.1007/978-981-15-1081-6_49

1 Introduction

Knowledge Space Theory [1] has so far been studied and used primarily for assessment of knowledge, but its principles can also prove advantageous in assessment of systems for accessibility. The chief idea is to move from assessing the extent of accessibility of a system to finding out the exact criteria in terms of which it can be considered accessible. A precedence diagram, and subsequently a knowledge structure, can be constructed using expert knowledge. For the sake of representation, we shall consider Web Content Accessibility Guidelines 2.0 (WCAG 2.0) as an expert knowledge base and build a knowledge structure using a small excerpt of the same. The knowledge states composing this structure shall represent respective sets of WCAG guidelines. At the end of an assessment, the position of a system in the structure (its current knowledge state) shall represent the immediate guidelines satisfied by the system. Such an assessment is akin to student assessment, except that the knowledge states here represent sets of satisfiable guidelines, and the assessment is taken by a tester who/which is analyzing the system for accessibility. Depending on the type of system being assessed, these tests may be automated.

2 A Brief Overview of Knowledge Space Theory [2–4]

2.1 Precedence Theory

The idea is intuitive that some kinds of test criteria precede some others in order of their ease of fulfillment. A student may be able to solve a tough arithmetic problem only if they have mastered one or more simpler problems. A high-level accessibility guideline may be satisfied only if some corresponding lower-level prerequisite guidelines are satisfied.

Thus, it is possible to create precedence relations among testing criteria (in our case, accessibility guidelines), which can then be mapped into a precedence diagram. A precedence diagram is a directed graph in which each node represents a test criterion, and arrows represent the order of precedence among nodes. Figure 1 illustrates a simple precedence diagram. The nodes, *a*, *b*, *c*, etc., represent individual test criteria, and the directed arrows indicate the precedence orders among these criteria.

In a precedence diagram, a criterion present at an arrowhead can be satisfied only if the criterion at the corresponding arrow tail is satisfied, too.

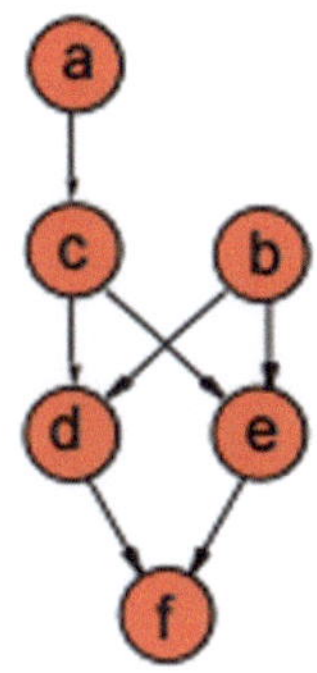

Fig. 1 A simple precedence diagram

2.2 Knowledge Structure

A knowledge structure is a directed graph constructed with reference to a precedence diagram. Unlike a precedence diagram, a knowledge structure necessarily satisfies the following principles:

1. While moving from any knowledge state to an adjacent knowledge state, there is a gain or loss of exactly one test criterion/concept.
2. Each knowledge state consists of all test criteria present in all preceding states leading to this state.
3. There is an empty state (with no test criteria met) and a final state (consisting of all test criteria) in the knowledge structure.
4. From every knowledge state present in the knowledge structure, there is a path leading to the final state (knowledge state satisfying all test criteria).
5. Each possible learning path for a knowledge structure starts from the empty state and terminates at the final state.

To build a knowledge structure, we first determine the knowledge states composing it. From the precedence diagram in Fig. 1 and using the aforementioned principle 2, we have the following set of knowledge states:

$$K = \{a, ac, b, abcd, abce, abcdef\}.$$

This knowledge structure does not satisfy the principles 1, 3, and 4 are mentioned above. Thus, we add some necessary knowledge states and derive the following set of knowledge states:

$$K = \left\{\emptyset, a, ac, b, ac, ab, abc, abcd, abce, abcde, abcdef\right\}.$$

These knowledge states are illustrated through a knowledge structure in Fig. 2.

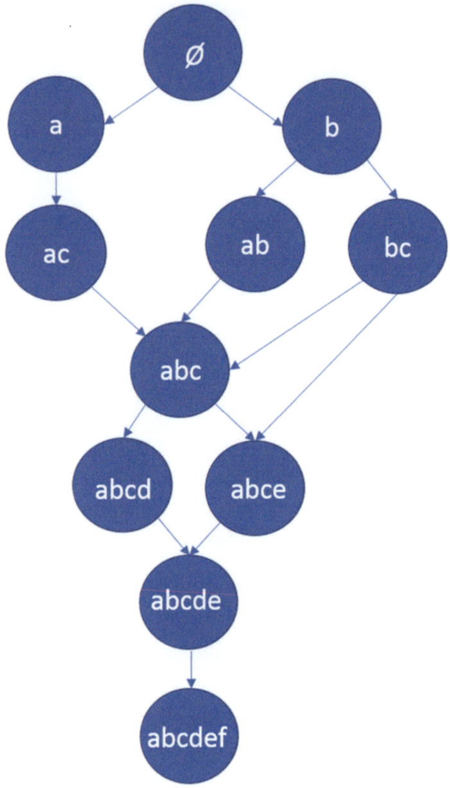

Fig. 2 A knowledge structure corresponding to the precedence diagram of Fig. 1

2.3 *Learning Paths*

As is observable from the knowledge structure in Fig. 2, the final state may be reached from the empty state by following more than one path of traversal. Three instances among the several possible paths are:

$$\text{Path } 1 = \{\emptyset, a, ac, abc, abcd, abcde, abcdef\}$$
$$\text{Path } 2 = \{\emptyset, b, ab, abc, abce, abcde, abcdef\}$$
$$\text{Path } 3 = \{\emptyset, b, bc, abc, abcd, abcde, abcdef\}$$

Note that all paths follow principle 5 of the knowledge space theory mentioned in Sect. 2.2.

2.4 Fringes

Every knowledge state in a knowledge structure has at least one immediate successor state (except for the final state), and at least one immediate predecessor (except for the empty state). By immediate, we mean that there is an increment or decrement of exactly one test criterion while moving to a successor or predecessor state. The set of test criteria which can be incremented in a knowledge state to result in one of its successors is called the outer fringe of that knowledge state. Similarly, the set of test criteria which can be decremented from a knowledge state to result in one of its predecessors is called its inner fringe.

In this way, each knowledge state can also be signified by its inner and outer fringes, given its knowledge structure. This holds great importance in accessibility testing because while the inner fringe of the current knowledge state of a system signifies the topmost accessibility guidelines satisfied by the system, its outer fringe denotes which accessibility guidelines the system can meet next to move on to a successor state.

3 Implementing Knowledge Space Theory on Accessibility Guidelines

3.1 Knowledge Base

Web Content Accessibility Guidelines (WCAG) 2.0 [5]. WCAG 2.0 is an elaborate list of guidelines compiled by World Wide Web Consortium (W3C) under its Web Accessibility Initiative. This list has been divided into four accessibility principles, viz. Perceivable, Operable, Understandable, and Robust. Each of these principles is further branched into relevant guidelines of web accessibility. For example, Principle 1—Perceivable, consists of four guidelines: 1.1. Text Alternatives, 1.2. Time-based Media, 1.3. Adaptable, and 1.4. Distinguishable. WCAG 2.0 specifies a set of success criteria for achieving each such aspect of web content accessibility.

Reason for Using WCAG 2.0 as an Example. Aside from the fact that Web Content Accessibility Guidelines have been devised after profound research and study, and have evolved over several years, the structure, classification, and level-identification in WCAG serve useful, if not ideal, for application of Knowledge Space Theory. For ease of representation in this paper, we use a closely interconnected subset of the success criteria of Guideline 1.4 (for Distinguishable web content). Table 1 briefly describes our success criteria of interest.

Table 1 A subset of WCAG 2.0 guideline 1.4—distinguishable [7]

Sr. No.	Guideline	Label for reference in this article
1.4.1	Use of color Color is not used as the only visual means of conveying information, indicating an action, prompting a response, or distinguishing a visual element	a
1.4.3	Contrast (minimum) The visual presentation of text and images of text has a contrast ratio of at least 4.5:1	b
1.4.4	Resize text Except for captions and images of text, text can be resized without assistive technology up to 200% without loss of content or functionality	c
1.4.5	Images of text If the technologies being used can achieve the visual presentation, text is used to convey information rather than images of text	d
1.4.6	Contrast (enhanced) The visual presentation of text and images of text has a contrast ratio of at least 7:1	e

3.2 Constructing a Precedence Diagram

Practically, a precedence diagram would be constructed after extensive survey with accessibility experts and devising a precedence order among success criteria for each guideline. Such a survey, for example, may present the answering individual with a set of questions, each question composed of the following elements:

1. A randomly generated pair of accessibility success criteria—A and B
2. Radio button choices: a. A precedes B; b. B precedes A; and c. The two are not related.

A precedence diagram would then be constructed after analysis of responses. Figure 3a shows a possible example of a precedence diagram constructed thus. This precedence diagram, however, is merely an illustration, and has not been constructed from analysis of expert knowledge. The labels *a*, *b*, *c*, etc., correspond to the labels for guidelines specified in Table 1.

3.3 Constructing a Knowledge Structure

Next, from the precedence diagram of Fig. 3a, we construct a knowledge structure that satisfies the Knowledge Space Theory principles specified in Sect. 2.B. The most

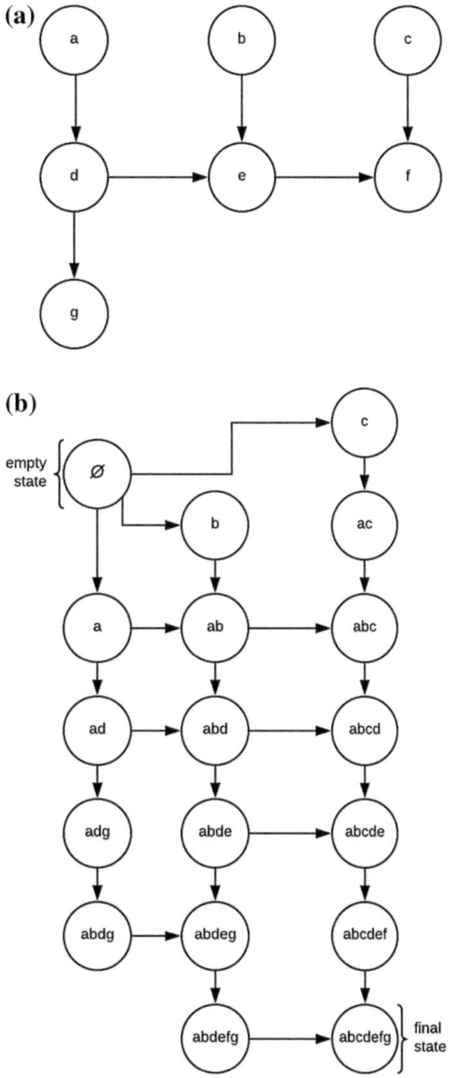

Fig. 3 **a** An illustrative precedence diagram constructed using the success criteria of Table 1. **b** Example of a corresponding knowledge structure constructed for the precedence diagram of Fig. 3a

obvious subset of knowledge states, derived by using Principle 2, is

$$K' = \{a, b, c, ad, adg, abde, abcdef\}.$$

This subset, however, does not satisfy Principles 1, 3, and 4. We, therefore, devise the following final set of knowledge states.

$$K = \{\emptyset, a, b, c, ab, ac, ad, abc, abd, adg, abcd, abdg, abde, abcde, abdeg, abcdef, abdefg, abcdefg\}.$$

Here, Ø signifies the empty state, and *abcdefg* is the final state. Numerous possible knowledge structures can be constructed for the same precedence diagram, based on expert knowledge, or feedback from assessment. The difference can be marked in the learning paths presented in different knowledge structures. An illustrative knowledge structure for the precedence diagram of Fig. 3a is shown in Fig. 3b.

Importance of Knowledge Structures in Accessibility. From the knowledge structure presented in Fig. 3b, we can see the various possible learning paths possibly taken by assessed systems to reach the final state. A webpage may reach the state ac, i.e., it may satisfy guidelines 1.4.1 and 1.4.4, by learning either of the guidelines first, followed by the other. A webpage in the empty state does not satisfy any of the success criteria specified in Table 1, and a webpage in the final state satisfies all of these success criteria.

4 Accessibility Testing Using Knowledge Structures

Once we have a knowledge structure to test a system against our set of success criteria, we can easily assess the system for accessibility by traversing this knowledge structure. In our case, a webpage can now be tested against the success criteria defined in Table 1 with the help of the knowledge structure in Fig. 3b.

4.1 Assessment Initialization

Before the assessment starts, the following parameters of the knowledge structure need to be initialized.

Initial probability of all states. Each knowledge state holds a parameter depicting its probability. This is the probability that the assessed system (in our case, a webpage) lies in that knowledge state, i.e., the probability that this knowledge state is the resultant current state of the webpage.

This step is important because the greater the accuracy of the initial values of probability assigned to knowledge states, the smaller is the number of assessment steps required to devise the system's current state.

Several factors can play an important role in deciding the initial probability for all knowledge states:

The expected probability of each WCAG success criterion taken into account. Knowledge states possessing a success criterion with a greater probability of being satisfied have a greater probability. WCAG 2.0 assigns a Level to each success criterion. The possible Level values are A, AA, and AAA, in increasing order of difficulty. Since more-difficult criteria may have lesser probability of being met by the system, a possible method is assuming a lower probability for AAA and vice versa.

The number of immediately preceding states. It can also be assumed initially that the greater the number of preceding states adjacent to a knowledge state, the greater is the chance that a system may reach this state.

Note that these assumptions are practically harmless because their purpose is to merely speed up the assessment process, and the initial probabilities shall not affect the final result. As the assessment proceeds, the values of probability for each knowledge state are revised until the probability of one state crosses a threshold maximum, resulting in that state being suggested as the system's current state.

Initial State of the Assessed System. A pointer is initialized to point at a knowledge state initially assumed as the system's current state. As the assessment proceeds, the pointer moves along the knowledge structure based on assessment principles of knowledge space theory, until there is no further (or backward) movement possible, at which point it is decided that the system belongs in the knowledge state pointed to at that instant.

It is common practice to assume the initially most probable state as the initial state of the system.

4.2 *Assessment Process*

The assessment proceeds by following the process described below:

Testing criterion at each step. At each step of assessment, the system is tested against a success criterion. There is a strictly defined set of testing criteria that can be tested while the pointer is at a particular knowledge state. This set comprises the state's outer fringe and inner fringe of success criteria.

Movement of pointer at each step. At each step, the tested criterion may either belong to the outer fringe or the inner fringe of the pointer. If it belongs to the outer fringe, a successful test (the case when the system meets the test criterion) leads the pointer to its immediate successor reached by adding that criterion, and an unsuccessful test leads to no movement. Similarly, if the tested criterion belongs to the inner fringe, an unsuccessful test leads the pointer to its immediate predecessor reached by subtracting that criterion, and a successful test leads to no movement.

Revision of state probabilities at each step. Every time a success criterion is found to be met by the system during the assessment, the probability of all knowledge states comprising that criterion is increased by a defined fraction, and the probability of all states not comprising that criterion is decreased by the same fraction. The reverse happens every time the system is unsuccessful in meeting a success criterion during the assessment.

4.3 *Assessment Termination*

The assessment terminates if one of the following conditions is met.

1. All possible criteria in the inner and outer fringe are tested, and no further movement of the pointer is possible.
2. The probability of one knowledge state exceeds a defined threshold maximum.

4.4 Assessment Result

The result of assessment is intended to specify the current state of the assessed webpage, as well as the success criteria which the webpage should aim to meet next. This is achieved by specifying the following information in the assessment result.

The Inner Fringe of the System's Current State (Guidelines Met by the System). The success criteria immediately met by the system to reach the current state speak a lot about its extent of accessibility. Since there exists a precedence order among the success criteria, the inner fringe is a concise signification that all preceding success criteria are also met by the system.

This makes a great difference in accessibility assessments because now the assessor knows exactly which accessibility guidelines are met by the assessed webpage, not just a graded or numerical extent of accessibility conformity.

The Outer Fringe (Next Steps). The immediately exceeding success criteria which can lead the system to a succeeding state signify which accessibility guidelines the webpage should aim to follow next.

4.5 A Dry Run

To help understand the process, let us take a dry run of accessibility testing using the knowledge structure of Fig. 3b. Assuming that the initially pointed knowledge state is *ad*, the possible success criteria to be tested are a union of its outer and inner fringes, i.e.,

$$S(ad) = \{g, b, d\}.$$

We test the system for g, and find out that this criterion is not satisfied. We next test the system for b, and find that this criterion is met by the webpage. The pointer, therefore, moves to *abd*. The possible success criteria to be tested in this state are

$$S(abd) = \{e, c, a, d\}.$$

Now, we may test the system for *e* and find that this criterion is not met. We may test the system for c next, and supposedly find that this criterion is not met either. On the other hand, when subsequently tested for a and d, the system may successfully meet the latter two criteria. Since there is no further or backward movement of pointer possible, we decide that the knowledge state *abd* is the system's current knowledge state.

As a result of the assessment, we say that the guidelines *d* and *b* are the furthermost guidelines satisfied by the system, inferring that all preceding guidelines in the knowledge structure have been met, too. If required by the reviewer, the result can be expanded to reveal all success criteria met by the system. Hence, we know that the system meets guidelines 1.4.1, 1.4.3, and 1.4.5 specified in WCAG 2.0. The assessment result also claims that the assessed webpage should next aim to meet criteria *c* and *e* (outer fringe of *abd*), i.e., guidelines 1.4.4 and 1.4.6 of WCAG 2.0, in order to proceed in the knowledge structure. Such an assessment gives the organization maintaining the website some feedback as to what they can do next to improve the accessibility of the assessed webpages.

5 Conclusion

5.1 *Merits of Using Knowledge Space Theory for Accessibility Testing*

There are numerous benefits of using knowledge space theory for accessibility testing:

The results are comprehensive and descriptive. At the end of assessment, a person analyzing the results does not find themselves guessing which accessibility guidelines are satisfied by the system.

The results are predictive. The assessment report predicts the best suited accessibility guidelines that the assessed system should strive to meet next so as to move forward in the knowledge structure.

The assessment system improves with time. Such an assessment system learns from its errors by improving the knowledge structure during test runs. Some missing knowledge states may be identified and added during test runs to improve accuracy, and some knowledge states which are never reached by systems may be removed for improving efficiency. Similarly, the learning paths present in the knowledge structure may be improved, too.

The assessment process is adaptive, and hence faster [6]. The success criteria tested at each step of assessment is dependent on the current pointer state, which, in turn, is dependent on the results of testing at the previous steps. This proves that a knowledge-space-theory-based assessment system is adaptive, and instead of testing the system in a predefined linear order of criteria, it speeds up the assessment process by adapting to the results at each step.

The results are accurate. No matter how much accuracy numbers and grades may hold, they act as black boxes because we cannot tell which criteria a system met to obtain such a score. Definition of a result in terms of successfully met criteria ensures accuracy and can serve as an effective standard in accessibility testing.

5.2 Future Scope

Software such as web crawlers can serve to be useful in completely automating the testing process for websites, eliminating the need for manually testing the system for different criteria.

An assessment system based on knowledge space theory keeps improving with time, by improving the knowledge structure during test runs. Although the example presented in this paper points majorly at the assessment of web content for accessibility, the only hurdle in applying the same methods to all kinds of systems (appliances, software, buildings, assistive technology, etc.,) is the definition of success criteria for testing their accessibility and a precedence order among these criteria. Once this is achieved, the same principles can be applied to any system for obtaining comprehensive, accurate, and predictive results from an adaptive and self-improving assessment system.

The purpose of this elementary paper is to point toward the unfathomable potential of knowledge space theory in building robust and intelligent accessibility testing software, encouraging further research, and extensive application in the future.

References

1. Doignon, J.-P., Falmagne, J.-Cl.: Knowledge Spaces. Springer, Berlin (1999)
2. Falmagne, et al.: The assessment of knowledge, in theory and in practice. Springer, Berlin (2006)
3. Falmagne JC., Cosyn E., Doignon JP., Thiéry N.: The assessment of knowledge, in theory and in practice. In: Missaoui, R., Schmidt, J. (eds.) Formal Concept Analysis. Lecture Notes in Computer Science, vol 3874. Springer, Berlin (2006)
4. Albert, D., Lukas, J. (eds.): Knowledge Spaces: Theories, Empirical Research, Applications. Lawrence Erlbaum Associates, Mahwah (1999)
5. World Wide Web Consortium: Web Content Accessibility Guidelines (WCAG) 2.0 (2008). https://www.w3.org/TR/WCAG20/
6. Gouli, E., Gogoulou, A., Papanikolaou, K., Grigoriadou, M.: Compass: an adaptive web-based concept map assessment tool. In: Proceedings of the 1st International Conference on Concept Mapping, Pamplona, 14–17 September 2004, pp. 295–302 (2004)
7. World Wide Web Consortium: How to meet WCAG 2.0. https://www.w3.org/WAI/WCAG20/quickref/

Unsupervised Detection of Dispersion and Merging Activities for Crowded Scenes

Manasi Pathade and Madhuri Khambete

Abstract Continuous monitoring and automatic detection of specific crowd activities such as dispersion and congestion; is extremely helpful for management at public places to avoid any possible disaster. Analysis of crowded scenes is a critical issue as it typically involves the poor resolution of objects, occlusions, and complex dynamics. In this paper, we propose a systematic, novel, and unsupervised method based on global motion analysis of people, to detect dispersion and merging events in crowded scenes. We avoid tracking of individual person as well as the use of any trained classifier while detecting the event. Our approach is tested on standard datasets as well as our own dataset. The results show the efficacy of our approach.

Keywords Motion analysis · Dispersion · Merging · Unsupervised

1 Introduction

Closed circuit television systems are popularly used for continuous monitoring in public places. Still, human intervention is necessary for understanding abnormal activities (such as destructive actions of people, sudden dispersion, and congestion) and for subsequent decision-making to avoid disasters and ensure safe environment [1].

However, when the places are overcrowded, it becomes very difficult for the human operator to monitor every individual person for suspicious behavior. Moreover, fatigue caused due to continuous monitoring of videos, less resolution of the objects, clutter, occlusions, etc., makes the task more challenging [2]. Hence, it has become very important to develop automated systems that are capable of recognizing specific events and triggering alarms for abnormal situations. Due to the substantial

M. Pathade (✉) · M. Khambete
MKSSS Cummins College of Engineering for Women, Karvenagar, Pune, India
e-mail: manasi.pathade@cumminscollege.in

M. Khambete
e-mail: madhuri.khambete@cumminscollege.in

B. Pati et al. (eds.), *Advanced Computing and Intelligent Engineering*,
Advances in Intelligent Systems and Computing 1082,
https://doi.org/10.1007/978-981-15-1081-6_50

progress over the last few decades, computer vision techniques can be efficiently used to automatically extract and analyze important attributes of the crowd for event recognition.

This paper intends to propose an automated system to recognize dispersion and merging events in a crowded scene. *Dispersion* is a condition when people who are moving coherently in a group start diverging from each other. This is an important event to be recognized as it can be an indication of crowd panic situation in response to some threat. *Merging* is also an important event to detect because it may be an initial step of a congestion situation. Congestion is said to occur when people start gathering in a particular region as a result of which the number of people in that region starts increasing and after a certain time their motion completely stops. Under such situations, people may become uneasy and violent. Thus, it is very important to detect possibility of congestion in a scene in order to avoid future disasters.

For analyzing a crowded scene, moving object detection, motion estimation, flow feature extraction, and event detection (classification) are the main tasks to be performed. To accomplish this, researchers have followed either *macroscopic* (*holistic*) or *microscopic* (*object-based*) approach [2]. Holistic methods treat the crowd as a single entity and analyze global motion characteristics. In contrast, object-based methods analyze movement of each individual separately. In high-density crowd, detection, segmentation, and tracking of each individual may become very difficult. Also, it may be difficult to discriminate people from each other due to the possible interactions between them.

Our work is based on holistic methods. It is a general observation that people in a crowd have a tendency of moving in groups while gathering or dispersing [3]. Thus, it is better to analyze the motion of a group (or to analyze the motion collectively) rather than the motion of an individual. We have used global optical flow for motion estimation. Velocity, direction, and spatial location of the moving pedestrians are used as features for further analysis.

The outline of the paper is as follows: the work done by other researchers is summarized in Sect. 2. The detailed explanation of the proposed method and the analysis of results on different datasets are given in Sects. 3 and 4, respectively. In Sect. 5, conclusion and future work are discussed.

2 Literature Survey

Various approaches are proposed in the literature in the last few years for crowded scene analysis using motion features. A detailed review of different techniques for crowded scene analysis can be found in [1, 2]. Detection of abnormal situations in crowded scenes is carried out by several methods such as using motion intensity and dynamic threshold [4], wavelet analysis of energy curve [5], and acceleration [6].

The analysis of motion features obtained by optical flow vectors is also proposed in [7] to detect congestion in the scene and avoid possibility of dangerous situation. The authors tested their method on real video footage of Loveparade, Germany (2010).

The use of spatiotemporal features is also suggested in some papers [8] for detecting anomaly in the scene. In these papers, although, abnormal situation is detected, the explicit event classification is not carried out.

Actual event classification (like walking, running, splitting, or merging, etc.,) is carried out by researchers like [9–13]. The use of FAST features and HOG descriptors for tracking the objects and recognizing different activities is proposed in [9]. Their method is tested on PETS dataset but for classification of events into different categories, they have used different thresholds. The setting of these threshold values is critical and also results in classification errors. The method proposed in [10] is based on extraction of corner features, computation of motion vectors, and histogram of orientations. The authors used these features separately for classifying different events; however, they suggest using the abovementioned features along with the spatial position for real situations. The method proposed in [11] is based on histogram of optical flow orientations computed on a grid. However, they have tested this method for only walking or running events. Different mid-level spatiotemporal features are proposed in [12] for event recognition. A Bayesian model for crowd escape detection is proposed in [13]. The authors tested their method on UMN as well as PETS datasets.

Some researchers solved the problem of event recognition by monitoring changes in flow patterns in the scene. To accomplish this, the regions possessing similar motion characteristics are segmented. For example, authors of [14] proposed a method for behavioral analysis in sparse and dense crowd. In this method, foreground is first segmented and then the regions are merged based on spatial closeness using k-means clustering. The clusters are further tracked using Kalman filtering. The dense-crowd measure is proposed which is used for recognizing active or calm crowd.

The method proposed by [3] is based on tracking each pedestrian using mean shift tracking. The trajectories obtained are then clustered based on the similarity of motion features. Then the mean speed of clusters and the number of clusters information is used for detecting abnormal situations. But, the abnormal situation is not detected during run-time; it is done after processing all the frames. A similar approach is adopted in [15] with KLT tracking, AMC clustering, and SVM classifier for detecting merging or splitting.

In [16], detection and segmentation of groups of people using dynamic active contours are proposed. These groups are further tracked for finding direction of flow and testing formation or splitting. All the methods mentioned above are suitable for low or medium density crowds. In high-density crowded scenes, tracking an individual is not possible due to severe occlusion. So, for high-density crowds, [17] presented a method for motion pattern analysis using clustering of tracklets obtained by dense point tracking. But, they have not classified the flow pattern in any specific category.

3 Proposed Method

The work proposed in this paper is based on the holistic approach. Our approach is motivated by [3]. However, the features used in our method are different than [3]. Moreover, it will be explained further that, tracking of features in subsequent frames is avoided in our approach. Due to this, the complexity of our algorithm is reduced as well as our method does not remain specific to the density of the crowd. The main steps of proposed method are shown in Fig. 1 and each step is discussed in detail below:

- *Feature extraction*:

Corner features are extracted using Harris Algorithm on every frame.

- *Motion estimation*:

This is accomplished using Optical Flow. The optical flow is computed at every corner location using Horn–Schunck algorithm [18, 19] by assuming that optical flow is smooth over the entire image. The flow vectors are obtained by minimizing the following equation:

$$E = \begin{array}{l} \iint \left(I_x V_x + I_y V_y + I_t\right)^2 \mathrm{d}x\mathrm{d}y \\ +\alpha \iint \left\{\left(\frac{\partial V_x}{\partial x}\right)^2 + \left(\frac{\partial V_x}{\partial y}\right)^2 + \left(\frac{\partial V_y}{\partial x}\right)^2 + \left(\frac{\partial V_y}{\partial y}\right)^2\right\} \mathrm{d}x\mathrm{d}y \end{array} \tag{1}$$

I_x and I_y are derivatives of the image in x and y directions. I_t is the temporal derivative. V_x and V_y are the horizontal and vertical components of flow vectors. α is the smoothness parameter. Larger the value of α, more is the smoothness of flow. The velocity estimate $[V_x\ V_y]$ for the pixel (x, y), can be obtained by:

$$\begin{aligned} V_x^{k+1} &= \bar{V}_x^k - \frac{I_x\left[I_x \bar{V}_x^k + I_y \bar{V}_x^k + I_t\right]}{\alpha^2 + I_x^2 + I_y^2} \\ V_y^{k+1} &= \bar{V}_y^k - \frac{I_y\left[I_x \bar{V}_y^k + I_y \bar{V}_y^k + I_t\right]}{\alpha^2 + I_x^2 + I_y^2} \end{aligned} \tag{2}$$

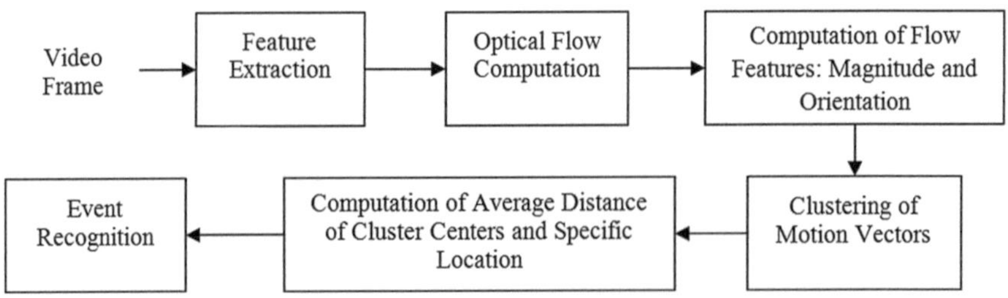

Fig. 1 Block diagram of proposed method

- *Computation of flow-features*:

For each flow vector, the velocity magnitude V_i is calculated using:

$$V_i = \sqrt{V_{xi}^2 + V_{yi}^2} \tag{3}$$

The flow vectors having non-zero velocity are only considered for further processing. This is done in order to remove stationary flow vectors.

The direction of motion is also a very important attribute because if the person starts to move in a different direction, he would more likely leave the group. Therefore, orientation (θ_i) of the significant flow vectors is also computed using [20]:

$$\theta_i = \tan^{-1}\left(\frac{V_{yi}}{V_{xi}}\right) \tag{4}$$

- *Clustering of flow vectors*:

It is already mentioned in Section I that, people in a crowd have a tendency of moving in groups. Obviously, people moving in the same group will be located close to each other as well as they will show similar motion characteristics like same speed and direction. Taking this fact into consideration, in every frame, the flow vectors are clustered based on the similarity of the following three features: spatial location, speed, and direction. The flow vectors grouped in the same cluster will have similar characteristics but they will be very dissimilar to that of the other flow vectors falling in different clusters.

Every significant flow vector is represented by its features as follows:

$$f_i = (x_i, y_i, V_i, \theta_i) \tag{5}$$

where, x_i and y_i are the spatial coordinates, V_i is the magnitude and θ_i is the orientation of ith flow vector.

The clustering of the flow vector can be accomplished using any clustering algorithm such as hierarchical clustering or k-means algorithm. However, in k-means clustering, it is required to predefine the number of clusters. To avoid this, we have used a hierarchical clustering method. Euclidean Distance measure is used to find similarity between the cluster members.

So, the flow vectors f_i and $\mathrm{f_j}$ can be grouped into one cluster if:

$$D_{f_f, f_j} = \sqrt{\left(f_{x_x} - f_{x_x}\right)^2 + \left(f_{y_1} - f_{y_1}\right)^2 + \left(f_{y_1} - f_{y_y}\right)^2 + \left(f_{\theta_1} - f_{\theta_1}\right)^2 \\ < \text{threshold}} \tag{6}$$

Each cluster is represented by its centroid; which is computed as follows:

$$C_k = \left(\frac{\sum_{m=1}^{n} x_m}{n}, \frac{\sum_{m=1}^{n} y_m}{n}, \frac{\sum_{m=1}^{n} V_m}{n}, \frac{\sum_{m=1}^{n} \theta_m}{n} \right) \quad (7)$$

where, C_k is the centroid of kth cluster having total "n" members. This means that the cluster is formed by "n" flow vectors. x_m and y_m are the spatial coordinates of mth member. V_m and θ_m are the velocity and orientation of mth member.

For every frame, after clustering the flow vectors, the average distance of cluster centers and specific location is computed. This specific location is considered to be the mean spatial position of all the clusters. The average distance of clusters from the specific location is computed as follows:

$$D_{\text{average}}(t) = \frac{\sum_{k=1}^{\text{no\_of\_clusters}} \sqrt{\left(x_{C_k} - x\right)^2 + \left(y_{C_k} - y\right)^2}}{\text{no\_of\_clusters}} \quad (8)$$

where, (x_{Ck}, y_{Ck}) is the spatial position of cluster "k" and (x, y) is the mean spatial position of all the clusters.

We have first observed the change in average distance with time. The histogram of average distance is shown in next section in Figs. 2 and 3.

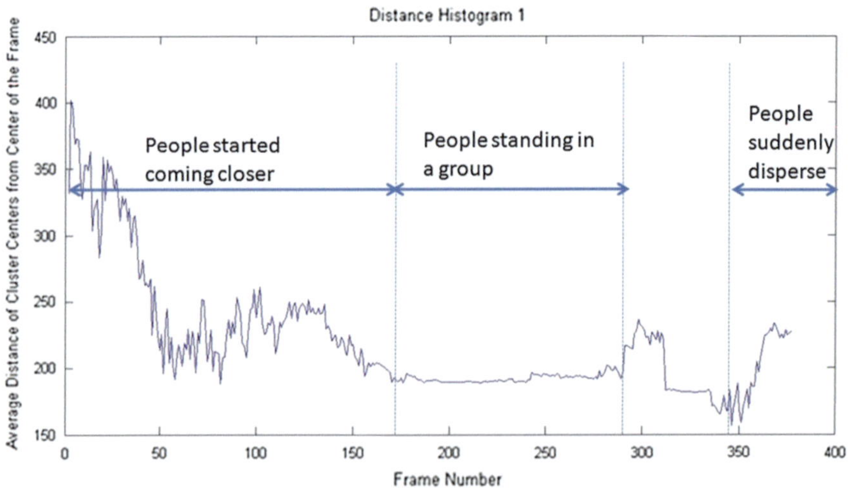

Fig. 2 Change in average distance of cluster centers from specific location (PETS dataset)

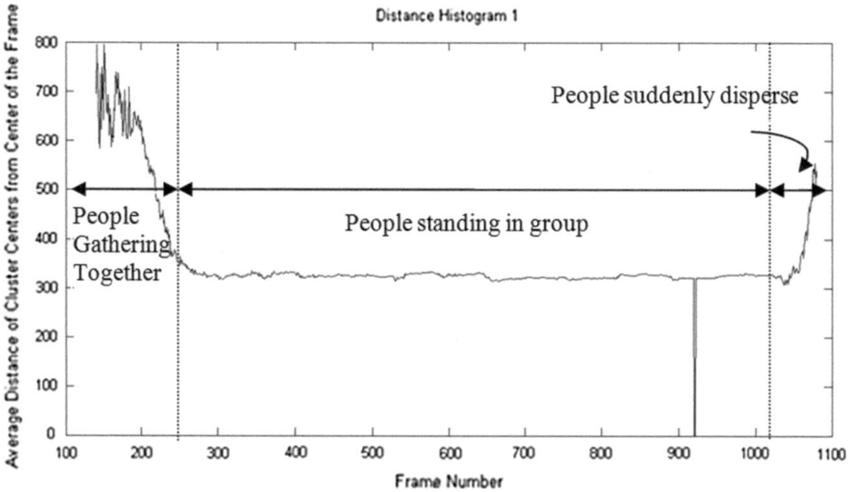

Fig. 3 Change in average distance of cluster centers from specific location (our dataset)

- *Event Recognition*:

It was found in the literature survey that, researchers have accomplished activity classification with the help of either trained or supervised classifiers such as SVM [12, 15], Baye's Classifier [13] or by setting different threshold values for different events [9]. So, we may say that these methods are not generalized methods; rather they are trained to produce good results on the specific dataset(s) only. Hence, we tried to design a generalized classifier that can be applied to any video dataset without the need of any prior training. The proposed strategy of event classification/recognition is explained in detail in this section.

Our aim is to recognize crowd dispersion or merging events in every frame. We define these events as follows:

Dispersion: It is a situation when people who are moving coherently in a group start diverging (or splitting) from each other. In some situations, people normally roaming in some areas may also disperse suddenly in different directions. This is an important event to be recognized as it can be an indication of crowd panic situation in response to some threat.

Merging: It is a situation when people start gathering in different directions. It is also an important event to detect because it may be an initial step of congestion situation.

Detection of these events is accomplished by monitoring the change in the spatial distance of each cluster centroid from the specific location for subsequent frames (computed using (8)).

It is a general observation that merging of people does not take place suddenly. Also, in many cases, dispersion or splitting of people may be quite a slow process. Sometimes, this process may not only take a few seconds but also minutes. So, it is not necessary to monitor the change in the distance for each frame (which is actually a fraction of a second). We have decided to monitor this change after few seconds. The average distance at time t (computed using (8)) is compared with the average distance at time $(t + \delta t)$.

Considering the above fact, the decision rule for event recognition is set as follows:

$$\begin{aligned} D_{\text{average}}(t) &\geq D_{\text{average}}(t + \delta t) \Rightarrow Merging\ Event \\ D_{\text{average}}(t) &< D_{\text{average}}(t + \delta t) \Rightarrow Dispersion\ Event \end{aligned} \tag{9}$$

From the above discussion, it is clear that we are neither using any trained classifier nor any model for event recognition. This is another advantage of our method due to which, our method becomes more simple and generalized. Also, the event recognition can be carried out during run-time. The results of the proposed event classification method are given in the next section.

4 Experimental Results

The proposed method is validated on PETS (S3) dataset, UMN dataset as well as our own dataset. PETS'09 dataset is widely used for Performance Evaluation of Tracking and Surveillance [21]. It consists of outdoor scenes of various activities (like walking, running, merging, and dispersing) recorded from different viewpoints (View_001, View_002, View_003 and View_004). We have tested our approach on video clips of View_001. We have manually labeled every frame, as the ground truth is not available for PETS dataset. UMN dataset (University of Minnesota) consists of 11 video clips of different indoor and outdoor scenes. We have tested our algorithm on outdoor video clips. Our dataset is captured at Cummins College of Engineering's Campus at a rate of 23fps. Our dataset also consists of various outdoor video clips of crowd walking, running, gathering, and dispersing.

As explained in the previous section, the change in the average distance of cluster centers and the specific location is monitored over time. The change in the average distance is shown in Fig. 2 (PETS dataset) and Fig. 3 (our dataset). The different activities taking place in the scene are also marked on the figures.

In both the figures, it is observed that the average distance gradually reduces when people are coming closer. When they are just standing in a group, the average distance does not change much; whereas, the distance starts increasing rapidly when people start dispersing from each other. This proves that the criteria set for event recognition is valid.

After performing the event classification, we need to represent the results in quantitative parameters. A confusion matrix is prepared after event recognition and accuracy of every event is calculated (Refer to Table 1). Accuracy is calculated using following formula:

$$\text{Accuracy} = \frac{\text{TP} + \text{TN}}{\text{TP} + \text{TN} + \text{FP} + \text{FN}} \tag{10}$$

Table 1 (%) Accuracy of the proposed method

Dataset	% Accuracy
PETS	87.22
UMN	87.23
Our dataset	85.63

5 Conclusion and Future Work

In this paper, we have presented a systematic approach for recognizing the dispersion and merging of people in a crowd. We have adopted macroscopic approach, i.e., to analyze the motion of group of people instead of motion of an individual. There are three main steps of our approach: computing optical flow, clustering of flow vectors based on spatial location, magnitude and orientation of flow vectors, and finally, event recognition by monitoring the distance of cluster centroids from a specific location. We have completely avoided tracking of corner-like features in this process.

Also, we have not used any trained classifier for event recognition. Thus, the proposed method is not only computationally simple but also it does not remain specific to the density of the crowd, although the accuracy of the method is less than the other researchers [12].

However, the severity of the dispersion may be detected by further classifying it as sudden dispersion (evacuation) or local dispersion. This would be more helpful for raising alarms at correct time instants. Also, we plan to explore more features in the future for improving accuracy of the method.

References

1. Jacques, J.C.S., Musse, S.R., Jung, C.R.: Crowd analysis using computer vision techniques. IEEE Sig. Proc. Mag. **27**(5), 66–77 (2010)
2. Teng, L., Huan, C., Meng, W., Bingbing, N., Hong, R., Shuicheng, Y.: Crowded scene analysis: a survey. In: IEEE Transactions on Circuits and Systems for Video Technology, vol. 25, no. 3 (2015)
3. Ma, J., Song, W.: Automatic clustering method of abnormal crowd flow pattern detection. Elsevier, Proc. Eng. **62**, 509–518 (2013)
4. Liu, Y., Li, X., Jia, L.: Abnormal crowd behavior detection based on optical flow and dynamic threshold. In: Proceedings 11th World Congress on Intelligent Control and Automation, pp. 2902–2906. China (2014)
5. Zhong, Z., Ye, W., Wang, S., Yang, M., Xu, Y.: Crowd energy and feature analysis. In: Proceedings on IEEE International Conference on Integration Technology, pp. 144–150. Shenzhen (2007)
6. Chen, C., Shao, Y., Bi, X.: Detection of anomalous crowd behavior based on the acceleration feature. IEEE Sens. J. **15**, 7252–7261 (2015)
7. Krausz, B., Bauckhage, C.: Automatic detection of dangerous motion behavior in human crowds. In: Proceedings 8th IEEE International Conference on Advanced Video and Signal-Based Surveillance, pp. 224–229. Klagenfurt (2011)

8. Cong, Y., Yuan, J., Tang Y.: Video anomaly search in crowded scenes via spatio-temporal motion context. In IEEE Transactions on Information Forensics and Security, vol. 8, pp. 1590–1599 (2013)
9. Garate, C., Bilinski, P., Bremond, F.: Crowd event recognition using HOG tracker. In: Proceedings 12th IEEE International Workshop on Performance Evaluation of Tracking and Surveillance, UT, pp. 1–6 (2009)
10. Albiol, A., Silla, M.J., Albiol, A., Mossi, J.: Video analysis using corner motion statistics. In: Proceedings IEEE Workshop Performance Evaluation of Tracking and Surveillance, pp. 31–37 (2009)
11. Achmad, S., Agus, H., Agfianto, P.: Grid-based histogram of oriented optical flow for analyzing movements on video data. In: Proceedings International Conference on Data and Software Engineering, pp. 114–119 (2015)
12. Silos, E., Diaz, I., Maria, E.: Mid-level feature set for specific event and anomaly detection in crowded scenes. In: Proceedings 20th IEEE International Conference on Image Processing, pp. 4001–4005, Melbourne (2013)
13. Wu, S., Wong, H., Yu, Z.: A bayesian model for crowd escape behavior detection. In: IEEE Transactions on Circuits and Systems for Video Technology, vol. 24, pp. 85–98 (2014)
14. Andersson, M., Rydell, J., St-Laurent, L., Prevost, D., Gustafsson, F.: Crowd analysis with target tracking, K-means clustering and hidden markov models. In: 15th International Conference on Information Fusion, pp. 1903–1910 (2012)
15. Rajanayaki, C., Srinivasagan, K.,Kalaiselvi, S.: An efficient method for crowd event recognition based on motion patterns. In: Proceedings IEEE International Conference on Recent Trends in Information Technology, pp. 1–6 (2014)
16. Nakhmani, A., Surana, A., Tannenbaum, A.: Macroscopic analysis of crowd motion in video sequences. In: Proceedings 53rd IEEE Conference on Decision and Control, pp. 1822–1827. CA (2014)
17. Chongjing, W., Xu, Z., Yi, Z., Yuncai, L.: Analyzing motion patterns in crowded scenes via automatic tracklets clustering. China Commun. (2013)
18. Horn, B., Schunck, B.: Determining optical flow. Artif. Intell. **17**, 185–203 (1981)
19. Barron, J., Fleet, D., Beauchemin, S., Burkitt, T.: Performance of optical flow techniques. In: Proceedings 1992 IEEE Computer Society Conf. on Computer Vision and Pattern Recognition, USA, pp. 1063–6919 (1992)
20. Rao, A., Gubbi, J., Marusic, S., Maher, A., Palaniswami, M.: Determining object directions using optical flow for crowd monitoring. In: 9th International Symposium on Visual Computing, Advances in Visual Computing (Lecture Notes in Computer Science series), vol. 8034, pp. 613–622. Springer, Berlin, Heidelberg (2013)
21. PETS Benchmark Data—Crowd Activity Dataset: www.cvg.rdg.ac.uk/PETS2009/a.html

Mapping Clinical Narrative Texts of Patient Discharge Summaries to UMLS Concepts

Swarupananda Bissoyi and Manas Ranjan Patra

Abstract Patient discharge summaries are typical unstructured text which is not amenable for processing by an automated system. For an efficient electronic healthcare system, free-form medical texts generated at various subsystems need to be mapped into standard codes like ICD-10, SNOMED-CT, etc. The Unified Medical Language System (UMLS) unifies these standard codes into a set of concepts identified by a Concept Unique Identifier (CUI). In this paper, we have used NLP techniques to map clinical narrative texts found in discharge summaries to CUIs. We have developed a Matcher algorithm to match the clinical text strings to that of UMLS and thereby extract the concepts. We achieve 70% similarity between the set of concepts generated using our Matcher algorithm to that of a gold standard tool.

Keywords Healthcare · UMLS · Discharge summaries · Natural language processing · Concept mapping · Information retrieval

1 Introduction

In a patient care system, discharge summary is an important document that describes the conditions, diagnoses, lab tests, and treatments provided to the patient. A discharge summary is a timeline-based document which tells about diagnosis during admission, patient condition during stay at the hospital, diagnosis during discharge, medications to be followed post discharge and follow-up recommendations. Discharge summary is a narrative free-text containing medical terms, phrases, and lab test parameters with their units. There is no standard way of writing a discharge summary thereby making it a typical unstructured document that is not amenable for an automated system to process.

S. Bissoyi (✉)
Department of Computer Application, North Orissa University, Baripada, India
e-mail: swarupananda.bissoyi@nou.nic.in

M. R. Patra
Department of Computer Science, Berhampur University, Berhampur, India
e-mail: mrpatra.cs@buodisha.edu.in

B. Pati et al. (eds.), *Advanced Computing and Intelligent Engineering*,
Advances in Intelligent Systems and Computing 1082,
https://doi.org/10.1007/978-981-15-1081-6_51

An automated system like a clinical decision support system would need a structured form of data with entities conforming to established medical coding systems, nomenclatures, and standardized clinical units for efficient query retrieval, processing, and information dissemination. Mapping the unstructured patient discharge summary to such structured form involves NLP-based text analysis which is addressed in this paper. Automatic assignment of standard codes to free-form clinical text is one crucial task involved here. The standard codes could belong to any of the established coding systems like ICD-9, ICD-10, SNOMED, LOINC, etc. The Unified Medical Language System (UMLS) [1, 2] tries to unify such disparate coding systems. For standard medical texts, each of these disparate coding systems will have different codes. The UMLS unifies them and maps such standard medical text strings of one single code. Those strings include even clinical texts in non-English language. Strings extracted from free-form discharge summary can, therefore, be mapped to the string of UMLS, to extract a set of UMLS unified codes, which can form a structured database for further information processing. In this paper, we devise a novel matching algorithm that tries to match the strings in UMLS to the strings extracted out of discharge summaries using simple NLP techniques. Depending on the matching score, appropriate unified code is extracted for a given clinical text.

This paper is organized as follows: First, we study the related works in this area in terms of processing discharge summaries as well as medical records and free-text clinical records such as radiology reports. Then we discuss the UMLS as a knowledge base for our work with an emphasis on how to use UMLS as a coding system. Then we present a novel algorithm that we have devised which maps clinical free-texts to concepts in UMLS. We discuss the technologies involved and the approach followed. We then evaluate our algorithm by analyzing the similarity between the UMLS concepts generated by our algorithm with that of a Gold standard tool.

2 Related Work

Many research have been carried out in processing discharge summaries to make it knowledge representable. It involves a range of activities from labeling clinical texts to performing feature extractions for classification and pattern analysis. Labeling or popularly called coding assigns standard code to narrative text found in clinical documents. When it comes to automatic code assignment to narrative medical texts, according to Stanfill [3] who has done a systemic review of automated medical coding and classification systems, discharge summaries are the most studied clinical documents. In a very early work, Larkey and Croft performed automatic assignment of ICD9 codes to the discharge summaries using kNN, Bayesian independence, and relevance feedback method [4]. Batool et al. used NLP techniques to extract concepts from discharge summaries and mapped them to SNOMED-CT [5]. Boytcheva matched Diagnoses in discharge summaries to ICD-10 Codes using SVM [6]. Zhu et al. leveraged contextual information in the patient's discharge summaries to improve information retrieval using NLP techniques [7]. Tufts-Conrad et al. automated the process of feature extraction from discharge summaries for classifying the most responsible diagnosis. They used Self Organizing Maps (SOM) an unsupervised neural network to classify against noisy input patterns [8].

Further many researchers have used other narrative clinical texts like radiology reports and other textual reports instead of discharge summaries. Crammer et al. developed a system to assign ICD-9-CM-based codes to free-text radiology report [9]. In a similar work, Farkas and Szarvas automated the construction of a rule-based ICD9 coding system [10]. Lita et al. performed a large scale ICD9 classification of patient records [11]. Suominen et al. using Machine Learning to automatically assign ICD9 codes to free-text radiology reports [12]. Goldstein et al. devised a Rule-based ICD-9-CM Coder for medical records that used BoosTexter which implements ML algorithms to boost performance of supervised learning [13]. Kiritchenko and Cherry used Lexically triggered HMMs to codify clinical narrative text to ICD-10 codes [14]. Coffman and Wharton applied NLP techniques to automatically assign ICD9 codes to radiology reports [15]. Yang and Chute developed a learning method called Expert Network for clinical classification of patient records and MEDLINE documents [16]. Pakholov et al. further developed it to develop an auto coder which used example based as well naive Bayes classification [17].

Researchers have shifted to UMLS as de facto biomedical coding system for concept mapping. Azam et al. developed Q-Map, which performed concept mining out of clinical documents using UMLS Metathesaurus [18]. They introduced a rule-driven phrase sense detection algorithm for eliminating contextual negations in clinical texts. Chen et al. applied topic modeling on online health community posts for knowledge discovery using UMLS [19]. Lee et al. normalized UMLS concepts generated from social media texts using convolutional neural networks and recurrent neural networks [20]. Among the open sources available for mapping clinical texts to UMLS concepts: Pal used NLP techniques for mapping clinical text to UMLS concepts [21]. Soldaini and Goharian used unsupervised, approximated dictionary matching algorithm for concept extraction [22]. Burckhardt created a utility tool for extraction of UMLS concept via the Consumer Health Vocabulary (CHV) [23].

3 Using UMLS as a Knowledge Base

The UMLS developed and maintained by National Library of Medicine (NLM), USA is a system that unifies more than 150 controlled biomedical vocabularies and coding systems. Started in 1986, this long term R&D project of NLM has gained significant popularity, thanks to its adaptation by industry and academia. UMLS can be viewed as a facilitator system toward development of such computer systems that is capable of understanding the semantics of the language of biomedicine and health. UMLS Knowledge Sources (database) and associated software help building information systems for creating, processing, retrieving, integrating, and aggregating biomedical and healthcare data. UMLS database and tools are produced, distributed, and maintained by NLM with an update every quarter. UMLS is available to the public free of charge.

Biomedical vocabularies or terminology systems are structured list of terms and concepts covering disorders, symptoms, findings, diagnoses, treatments, operations, drugs, allergies, hospital billing, and other healthcare administrative items. These multilingual vocabularies are developed independently by different agencies worldwide. Also each of these terminology systems has its own way of expressing the same concept. Because a large number of terms and concepts are found in these vocabularies, there was a need to link the terms together which culminated in UMLS.

The UMLS does this mapping and also enables translation of terms from one terminology system to another and vice versa. It also maintains the semantic relationship between such concepts while ensuring consistency concepts taken from various vocabularies, i.e., the original vocabulary information is never lost. UMLS is thus a compendium of several controlled vocabularies that can be used as a comprehensive thesaurus. On the other hand, it can also be viewed as a rich ontology of biomedical concepts. This makes UMLS a perfect candidate for a knowledge base accommodating almost all the concepts in the medical domain.

4 Using UMLS as a Coding System

Each of the concepts is identified by a unique identifier called Concept Unique Identifiers or CUIs found in Metathesaurus [24], a major component of UMLS. One CUI can represent a range of concepts, e.g., CUI C0024671 represents concept strings like "Mammographies", "Mammography", "Mammogram, NOS", "Mammography, NOS", and even "Radiographic examination of breast, NOS." Because of this simplification, most of the clinical text can be mapped to one of the concepts. Therefore, we intend to get a set of concept ids or CUIs for a given clinical narrative text, in our case this being the discharge summary. Thus, we devise a matching algorithm that maps the sentences to the concepts of UMLS.

5 Matching Algorithm

We intend to match the sentences of discharge summaries with the strings stored in UMLS. Note that among many languages supported by UMLS, we restrict to English sentences. However, the same strategy could be adapted for different languages. Once we find a potential match, we get the corresponding CUI for the string. To achieve it, we develop two algorithms detailed below:

5.1 Match-n-Fallback Algorithm

```
Algorithm 1: Match-n-Fallback
BuildIndex(UMLS String, UMLS CUI)
for each of the phrase in Discharge summary
{
          //perform exact match
          Search the phrase in Index for match
          if match found:
                Get the respective CUI from index
                                and assign score 1.0
          end if
          // Normalized match
          Convert phrase to lowercase and replace
            multiple spaces with one space
          Search in Index for match
          if match found:
                Get the respective CUI from index
                and assign score 0.8
          end if
          // Punctuation Normalized match
          Replace all punctuation characters
          in the phrase with spaces
          Search in Index for match
          if match found:
                Get the respective CUI from index
                and assign score 0.6
          end if
          // Sorted match
          Sort the phrase alphabetically word-wise
          Search in Index for match
          if match found
                Get the respective CUI from index
                and assign score 0.4
          end if
          // Stemmed match
          Apply porter stemming to the phrase
          Search in Index for match
          if match found:
                Get the respective CUI from index
                and assign score 0.2
          end if
          // Final fallback
          Run Transformed n-Gram Matcher algorithm
            for the phrase
}
```

The Matching strategy is depicted in Fig. 1 and the algorithm is listed in Algorithm 1. The strategy is to maximize the matching of the sentences of the discharge summary against the concepts of UMLS. First, the match is performed toward an exact match. This is likely to get hit because most of the texts written in the discharge

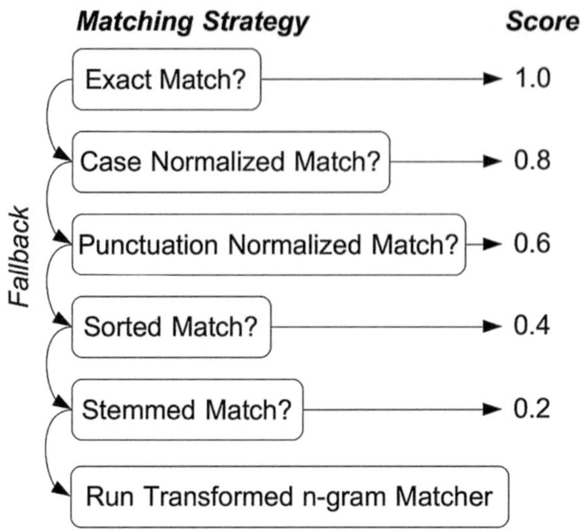

Fig. 1 Matching strategy

summaries by the physicians are routine type, thereby making it more likely to be found in UMLS. We, therefore, assign a score of 1.0 to it. In case of a miss, the sentences are case normalized, i.e., converted to lower case and matched. The potential matches are assigned score of 0.8. Further in case of fallback, we go for punctuation normalization, i.e., replace all the punctuation marks by space. For resulting matches, we assign a score of 0.6. In case of matches not found, the algorithm falls back to a strategy where we go for a sorted match, i.e., the sentence is sorted alphabetically word wise. For such matches, we assign a score of 0.4. Again, in case of fallback, we now stem the words using Porter Stemmer [25] after due removal of basic stop words. A score of 0.2 is assigned for matches resulting in this step. Finally, in case of fallback, we run our Transformed n-Gram Matcher algorithm.

5.2 *Transformed n-Gram Matcher Algorithm*

The input sentences which failed to find a match in Match-n-Fallback Algorithm are input to this Transformed n-Gram Matcher algorithm. An n-Gram is a contiguous sequence of n words from a given text sequence. n-Gram models are used to estimate the probability of the last word of an n-Gram, given its previous words. It is a predictive model, which can also be used to assign probabilities to an entire text sequence [26]. In our work, we are interested only to construct n-Grams out of the phrases for a possible matching in our index.

For a given text sequence $(w_1, w_2, \ldots, w_n)$

1-Gram (unigrams): $(w_1), (w_2), \ldots, (w_n)$
2-Gram (bigrams): $(w_1, w_2), (w_2, w_3), \ldots, (w_{n-1}, w_n)$
3-Gram (trigrams): $(w_1, w_2, w_3), (w_2, w_3, w_4), \ldots, (w_{n-2}, w_{n-1}, w_n)$ and so on.

We construct k-grams of our phrases, where k varies from $n - 1$ to 1, i.e., from higher order to lower order. For example, in the phrase "*bilateral perihilar, peribronchial thickening*": if we ignore the punctuation, the number of words is 4, i.e., $n = 4$; so we construct up to 3-grams. The 1-Grams or unigrams are: "*bilateral*", "*perihilar*", "peribronchial", and "*thickening*". The 2-Grams or the bigrams are: "*bilateral perihilar*", "*perihilar peribronchial*", and "*peribronchial thickening*". Similarly, the 3-Grams or trigrams are: "*bilateral perihilar peribronchial*", and "*perihilar peribronchial thickening*", and so on.

For each of the phrases, n-grams are constructed and matched against the index using our Match-n-Fallback algorithm. Words in n-grams that are already matched are ignored in subsequent matches. We then apply a correction to the n-gram score by a smoothing factor α which is computed as follows:

$$\alpha = \frac{|\text{words in n-Gram}|}{|\text{words in the phrase}|}$$

With this, we achieve a relatively higher score for higher order n-grams for our phrases. The algorithm is described below

```
Algorithm 2: Transformed n-Gram Matcher
    n = number of words in phrase
    for k=n-1 downto 1:
        for each k-gram constructed from the phrase
            if k-gram not matched earlier
                Run Match-n-Fallback algorithm
                with the k-gram to get the score
                If match found:
                    alpha = k / n ;
                    newScore = score * alpha ;
                    Assign newScore to the CUI fetched
                    from Index
                end if
            end if
        end for
    end for
```

6 Methodology

We ran our Matching algorithm to obtain the CUIs out of the free-text of discharge summaries. We use Metamap [27], a popular tool used for UMLS Concept Identification as our Gold standard to see how are matching algorithm is performing. MetaMap uses sophisticated NLP and computational linguistic techniques to recognize words and phrases in clinical free-text and map them to UMLS concepts. We then run a Similarity Analysis among the CUIs obtained by our Matching Algorithm and those obtained by Metamap.

6.1 Similarity Analysis

The results that we obtained are sets of CUIs. Next, we need to figure out how similar is the set obtained by our Matching algorithm to that generated by the Metamap for the same dataset. The similarity is often computed as distance metrics like Euclidean distance [28] or Cosine Distance [29]. However, these distance metrics cannot compute distance between vectors of different lengths. In our case, we need to find similarity between two sets of CUIs of arbitrary length. One possibility is to use a binary vector of all the unique CUIs of UMLS and if a CUI is present it is marked as 1 otherwise it is marked as 0 in the vector. With this, we can use distance metrics like Euclidean distance and Cosine distance. However, the number of unique CUIs runs into millions, thereby making the distance calculation computationally intensive. Therefore, for our similarity computation, we choose the Jaccard Index [30] which is a simple and popular similarity measure for sets of arbitrary length. Also called the Jaccard Similarity Coefficient, for any finite sets, it is the ratio between the number of elements in both sets to the number of elements in either of the sets. Mathematically:

$$J(A, B) = \frac{|A \cap B|}{|A \cup B|} = \frac{|A \cap B|}{|A| + |B| - |A \cap B|}$$

$$If\ A = \phi\ and\ B = \phi,\ then\ J(A, B) = 1$$

$$0 \leqslant J(A, B) \leqslant 1$$

Jaccard Index concerns the presence or absence of data. Our matcher algorithm will result in a set of CUIs for a given clinical text. With Jaccard Index, the idea is to check, whether the same clinical text is resulting in same or almost the same CUI elements in two sets generated using two different algorithms.

6.2 Dataset Used

For our experiment, we have chosen the Computational Medicine Center's (CMC) 2007 Medical NLP Challenge dataset [31] which contains radiology reports. The dataset comprises the radiology reports labeled with ICD9CM codes with 978 records for training and 976 records for testing. For our experiment, we only use the clinical texts of the report under the attribute IMPRESSION in each record.

7 Experiment

Figure 2 shows our experimental setup. We pulled the UMLS version 2017AA from the NLM website and stored it in a MySQL database. We then queried for CUIs and the corresponding Concept Name in English from the MRCONSO table, i.e., the Concept Names and Sources table of UMLS. A total 7,465,024 records were fetched and indexed using Lucene [32], a popular open-source library for efficient text retrieval. For each of the concept, we maintain the original string, punctuation removed string, further word-wise alpha sorted string and finally the stemmed string in the index for enabling efficient matching and retrieval. For each of the 976 records of the dataset, we ran Metamap to obtain 976 sets of CUIs. We implemented our

Fig. 2 Experimental setup

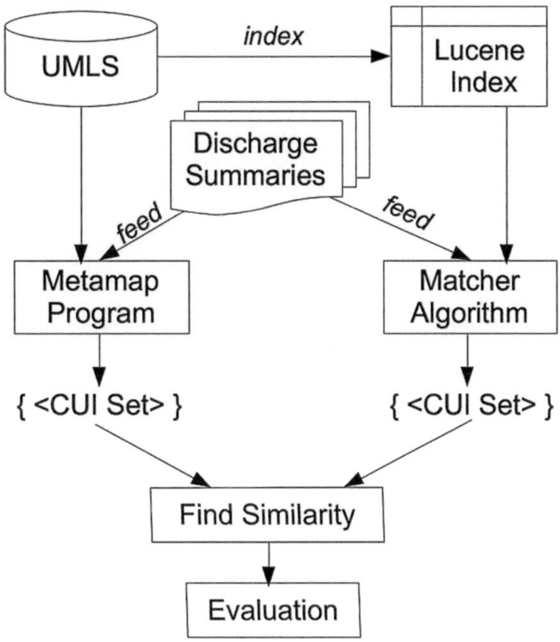

Matching algorithm using Lucene libraries and ran it against each of the records in the dataset to retrieve 976 sets of CUIs. We then used the " sets" library of R [33] to compute the similarity among the 976 pairs of sets of CUIs.

8 Results

We computed the Jaccard Similarity Coefficient between the two results, e.g., for the string from our dataset "Wheezing.Mild hyperinflation without focal pneumonia," the set of CUIs obtained using our algorithm is {C1290339, C0043144, C2945599, C0332288, C0020449} and for the same string the set of CUIs obtained using Metamap is {C0020449, C2945599, C0043144, C1290339}. The corresponding strings (STR field in MRCONSO) for the CUIs we fetched are as follows:

C1290339: Focal pneumonia
C0043144: [D]Wheezing (context-dependent category)
C2945599: Mild
C0332288: Without
C0020449: Hyperinflation.

Note that the results shown here are from SNOMEDCT Source of UMLS. Our results are almost similar to Metamap. We have retrieved a CUI C0332288 for the word "Without" whereas Metamap seems to have ignored that due to handling of negation [34].

Figure 3 shows the plotting of the 976 similarity values. We achieve a mean similarity of 0.708011 which indicates the similarity matching is almost 70% with

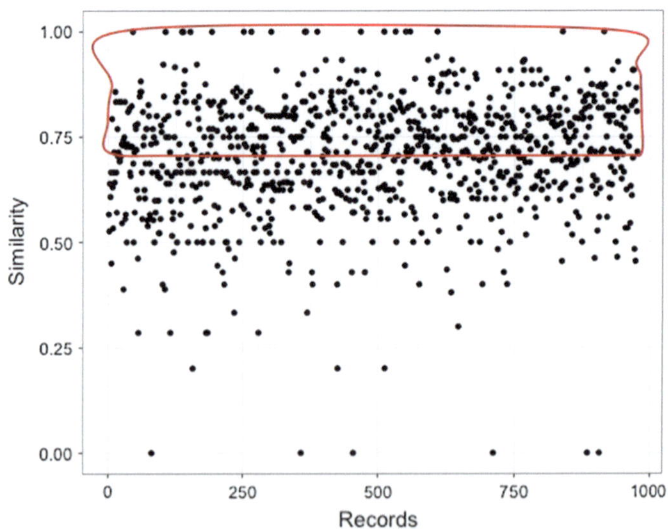

Fig. 3 Similarity index plot

respect to the gold standard MetaMap. The area enclosed in red in the plotting shows the similarity which is greater than mean similarity. It shows that our Matching algorithm with a simple match and fallback strategy could result in as much as 70% retrieval of CUIs that of Metamap.

9 Conclusion and Future Work

In this research, we have devised a novel algorithm to extract clinical narrative texts from discharge summaries and map them to UMLS concepts. We devised a simple NLP-based matching algorithm to achieve nearby efficiency of a gold standard system. This can help to process unstructured data like discharge summaries to become amenable to an automated system. Also, this algorithm can be improvised to find hidden concepts in clinical narrative texts, negated sentences, etc. We propose to extend our work by applying different machine learning techniques on the actual clinical records of patients containing lab test results, patient demographics, etc., with a view to identify similar patients and recommend appropriate medical procedures.

References

1. Unified Medical Language System. https://www.nlm.nih.gov/research/umls/
2. Bodenreider, O.: The unified medical language system (umls): integrating biomedical terminology. **32**, D267–D270 (2004)
3. Stanfill, M.H.: A systematic review of automated medical coding and classification systems (2009)
4. Larkey, L.S., Croft, W.B.: Automatic assignment of icd9 codes to discharge summaries. Technical report, University of Massachusetts Amherst, Amherst, MA (1995)
5. Batool, R., Khattak, A.M., Kim, T.S., Lee, S.: Automatic extraction and mapping of discharge summarys concepts into snomed ct. In: Conference Proceedings of the IEEE Engineering Medicine Biology Society, pp. 4195–4198 (2013)
6. Boytcheva, S.: Automatic matching of icd-10 codes to diagnoses in discharge letters. In: Proceedings of the Second Workshop on Biomedical Natural Language Processing, pp. 11–18 (2011)
7. Zhu, D.,Wu, S.T., Masanz, J.J., Carterette, B., Liu, H.: Using discharge summaries to improve information retrieval in clinical domain. In: CLEF (Working Notes) (2013)
8. Tufts-Conrad, D.J., Zincir-Heywood, A.N., Zitner, D.: Som: feature extraction from patient discharge summaries. In: Proceedings of the 2003 ACM symposium on Applied computing, pp. 263–267. ACM (2003)
9. Crammer, K., Dredze, M., Ganchev, K., Talukdar, P.P., Carroll, S.: Automatic code assignment to medical text. In: Proceedings of the Workshop on BioNLP 2007: Biological, Translational, and Clinical Language Processing, pp. 129–136. Association for Computational Linguistics (2007)
10. Farkas, R., Szarvas, G.: Automatic construction of rule-based icd-9-cm coding systems. BMC Bioinf. **9**, S10 (2008)
11. Lita, L.V., Yu, S., Niculescu, S., Bi, J.: Large scale diagnostic code classification for medical patient records. In: Proceedings of the Third International Joint Conference on Natural Language Processing: Volume-II (2008)

12. Suominen, H., Ginter, F., Pyysalo, S., Airola, A., Pahikkala, T., Salanter, S., Salakoski, T.: Machine learning to automate the assignment of diagnosis codes to free-text radiology reports: a method description. In: Proceedings of the ICML/UAI/COLT Workshop on Machine Learning for Health-Care Applications (2008)
13. Goldstein, I., Arzumtsyan, A., Uzuner, Ö.: Three approaches to automatic assignment of icd-9-cm codes to radiology reports. In: AMIA Annual Symposium Proceedings, vol. 2007, p. 279. American Medical Informatics Association (2007)
14. Kiritchenko, S., Cherry, C.: Lexically-triggered hidden markov models for clinical document coding. In: Proceedings of the 49th Annual Meeting of the Association for Computational Linguistics: Human Language Technologies-Volume 1, pp. 742–751. Association for Computational Linguistics (2011)
15. Coffman, A., Wharton, N.: Clinical natural language processing: auto-assigning icd-9 codes. Overview of the Computational Medicine Centers (2007)
16. Yang, Y., Chute, C.G.: An application of expert network to clinical classification and medline indexing. In: Proceedings of the Annual Symposium on Computer Application in Medical Care, p. 157. American Medical Informatics Association (1994)
17. Pakhomov, S., Buntrock, J., Duffy, P.: High throughput modularized nlp system for clinical text. In: Proceedings of the ACL 2005 on Interactive poster and demonstration sessions, pp. 25–28. Association for Computational Linguistics (2005)
18. Azam, S.S., Raju, M., Pagidimarri, V., Kasivajjala, V.: Q-map: clinical concept mining with phrase sense disambiguation. arXiv preprint arXiv:1804.11149 (2018)
19. Chen, D., Zhang, R., Liu, K., Hou, L.: Knowledge discovery from posts in online health communities using unified medical language system. Int. J. Environ. Res. Public Health **15**(6), 1291 (2018)
20. Lee, K., Hasan, S.A., Farri, O., Choudhary, A., Agrawal, A.: Medical concept normalization for online user-generated texts. In: 2017 IEEE International Conference on Healthcare Informatics (ICHI), pp. 462–469. IEEE (2017)
21. Salmon Run: Fuzzy String Matching Against UMLS Data. http://sujitpal.blogspot.com/2014/02/fuzzy-string-matching-against-umls-data.html
22. Soldaini, L., Goharian, N.: Quickumls: a fast, unsupervised approach for medical concept extraction. In: MedIR workshop, SIGIR (2016)
23. Planeshifter/node-chvocab: Mapping Texts to UMLS via the Consumer Health Vocabulary (CHV). https://github.com/Planeshifter/node-chvocab
24. Tuttle, M.S., Blois, M.S., Erlbaum, M.S., Nelson, S.J., Sherertz, D.D.: Toward a bio-medical thesaurus: building the foundation of the umls. In: Proceedings of the Annual Symposium on Computer Application in Medical Care, p. 191. American Medical Informatics Association (1988)
25. Porter, M.F.: An algorithm for suffix stripping. Program **14**(3), 130–137 (1980)
26. Jurafsky, D., Martin, J.H.: Speech and language processing: an introduction to natural language processing, computational linguistics, and speech recognition (2009)
27. Aronson, A.R.: Effective mapping of biomedical text to the umls metathesaurus: the metamap program. In: Proceedings of the AMIA Symposium, p. 17. American Medical Informatics Association (2001)
28. Gower, J.C.: Euclidean distance geometry. Math. Sci. **7**(1), 1–14 (1982)
29. Salton, G., Buckley, C.: Term-weighting approaches in automatic text retrieval. Inf. Process. Manage. **24**(5), 513–523 (1988)
30. Jaccard, P.: The distribution of the flora in the alpine zone. New Phytol. **11**(2), 37–50 (1912)
31. Pestian, J.P., Itert, L., Anderson, C., Duch, W.: Preparing clinical text for use in biomedical research. J. Database Manage. (JDM) **17**(2), 1–11 (2006)
32. Apache Lucene. https://lucene.apache.org/
33. Meyer, D., Hornik, K.: Generalized and customizable sets in R. J. Stat. Softw. **31**(2), 1–27 (2009)
34. Chapman, W.W., Bridewell, W., Hanbury, P., Cooper, G.F., Buchanan, B.G.: A simple algorithm for identifying negated findings and diseases in discharge summaries. J. Biomed. Inform. **34**(5), 301–310 (2001)

Enhanced Decision Tree Algorithm for Discovery of Exceptions

Sunil Kumar, Saroj Ratnoo and Renu Bala

Abstract Decision trees are the most admired and extensively used classification algorithms in data mining. These are considered accurate, easy to use, and comprehensible classifiers. Like many other classification models, decision tree classifiers ignore exceptions as noise. Exceptions are precious pieces of knowledge that are required to modify our decisions in extraordinarily rare circumstances. Since exceptions pertain to a tiny number of instances, it is a very challenging task to capture exceptions since a learning algorithm's focus is on discovering knowledge with high generalization power. This paper proposes an enhanced decision tree learning algorithm that accommodates exceptions. The working of the suggested algorithm is demonstrated on a toy example dataset. It is further applied to three datasets obtained from machine learning repository. The results reveal that the Enhanced Decision Tree Algorithm (EDTA) discovers several exceptions across the experimental datasets.

Keywords Classification · Exception · Censored production rules · Enhanced decision tree

1 Introduction

In data mining, there is a significant amount of research that focuses on designing efficient classification techniques for knowledge discovery from databases [11]. Decision tree methods are widely used and powerful tools for classifying data. Decision trees are chronological models, in which a series of simple tests on attributes is joined logically. Each test evaluates a quantitative attribute against a given threshold

S. Kumar (✉) · S. Ratnoo · R. Bala
Department of Computer Science and Engineering, Guru Jambheshwar University of Science and Technology, Hisar, Haryana, India
e-mail: skvermacse@gmail.com

S. Ratnoo
e-mail: ratnoo.saroj@gmail.com

R. Bala
e-mail: renu0805@gmail.com

B. Pati et al. (eds.), *Advanced Computing and Intelligent Engineering*, Advances in Intelligent Systems and Computing 1082,
https://doi.org/10.1007/978-981-15-1081-6_52

value or a categorical attribute against a group of its labels. The decision tree classifiers are more advantageous than other "black-box" models like neural networks and support vector machines regarding interpretability. The *If–Then* form of rules derived from decision trees are far easier to understand for decision-makers instead of the weights of a neural network model or the tedious mathematics involved for finding maximal marginal hyper-plane for support vector machines.

A decision tree is constructed from training data instances. A decision tree is formed through a top-down greedy recursive process in which tree formation starts with a root node containing the entire training data. The attribute which best partitions the training data (on criteria like Information gain or Gini-index) is opted as the splitting attribute at the root node. The training dataset is partitioned into different data subsets, which satisfy the splitting criteria of the splitting attribute. This procedure is reiterated for each subset recursively until all examples in a subset fall in a single class. Various versions of decision tree classifiers have been implemented CART, ID3, ID4, ID5, C4.5 and C5.0 [12]. All these algorithms depend on incremental development process. However, none of these algorithms can capture the exceptional instances inherent in data.

Exceptions are those rare circumstances (attribute–value conditions) in the presence of which a generalized rule may misclassify some test data tuples. For example, from a database related to flying objects, a classifier can extract a rule that if something is a bird, then it would fly. However, there are unique and extraordinary non-flying birds such as kiwi, ostrich, and penguin. Such kinds of exceptional instances are ignored by conventional decision tree algorithms in the modeling process itself or pruned at post-processing step as a machine noise.

This paper proposes an Extended Decision Tree Algorithm (EDTA) which identifies the exceptional tuples and classifies them in a suitable category. In the proposed algorithm, the discovery of exceptions is accommodated while building the decision tree from training data. We have used a sample dataset for illustration and three real-world datasets to show how the proposed algorithm discovers more accurate, complete, and interesting classification rules.

The rest of the paper is structured as follows: Sect. 2 elaborates on the related research work in this area. Section 3 illustrates the current approach through a sample dataset specially designed for this purpose. Section 4 presents the experimental results and validation of the current approach on some real-world datasets. Finally, Sect. 5 concludes the paper.

2 Related Work

Decision-makers are likely to be more comfortable with models that can be easily understood. Decision trees are comprehensible than other black box representations, like neural nets and support vector machines. Hence, several decision tree algorithms are popular and have been in use since their inception, e.g. CART, C4.5, SPRINT, SLIQ [12, 14, 16]. Decision tree algorithms have been compared to other learning algorithms in [13]. These studies show that the later versions of decision trees like C4.5 and C5.0 have less error rate and more speed. Further, many methods like fast tree-growing algorithm, [17] data partitioning, [10] and parallelization [22] have been suggested for scaling-up decision tree building process. The rainforest methods have been proposed by Gehrke et al. [8] to develop fast and scalable algorithms to build decision trees [8]. Many strategies for improving decision trees have been proposed [6, 18] and their focus has been on small modifications on the available models. In the series, a lot of ensemble-based classifiers have also been recommended [7] which brings slight improvement particularly in accuracy. There is inadequate research for upgrading the learning process of decision trees to discover knowledge which is not only accurate and comprehensible but interesting also.

Numerous techniques have been envisaged in data mining to discover interesting rules from a huge number of discovered rules [9]. The major approaches are based either on some interestingness measures to filter out uninteresting rules or depend upon the user's domain knowledge to identify unexpected rules [20]. Compton and Jansen [5] have proposed ripple-down rules for knowledge discovery. The ripple-down rules capture exceptions at many levels. Suzuki and colleagues in [19, 20] have classified exceptions in different categories. They have mined exceptions in the form of rule pairs and rule triplets for dependence modeling—a data mining job that can take several attributes as the class labels. They have discovered many exceptions that seem inappropriate for human insight and analysis. Another kind of exception (Censored Production Rules (CPRs)) has been discovered using an evolutionary approach in [4, 15]. Vashishtha et al. [21] and Bala et al. [3] have devised genetic algorithms to discover classification rules and fuzzy classification rules with intra-class and inter-class exceptions for datasets containing nominal and continuous attributes [3, 21]. A genetic algorithm approach for discovering fuzzy censored classification rules has been proposed in [2]. None of the contributions, on the discovery of exceptions mentioned so far, use decision trees as a learning method. A different type of exception handling has been carried out in [1, 18] where the rule induction algorithms have been improved to resolve the ties that emerge in particular instances of data during the rule generation procedure. The ties take place in decision tree induction algorithm when majority vote cannot determine the class label at a leaf node. To deal with this problem, an influence factor is calculated for every attribute and an update technique has been devised to provide subsequent rectification steps. Our work is different from such kind of exception handling. Our focus is on appending exceptional conditions in the presence of which the general rules cease to work, and we must decide along some alternate path.

3 The Proposed System

As described earlier, the conventional decision tree construction process does not capture exceptions and ignores these as noise. This section illustrates the proposed algorithm that captures exceptions as an essential part of the classification rules resulting from the decision tree model. To make things more comprehensible, a toy sample dataset (a small subset of the "Mushroom" dataset) is used. The toy dataset has 7 attributes and 31 instances with two class labels as "edible" and "poisonous." Assume that the attribute "odor" has the highest information gain. Hence, the splitting at the root node takes place on this attribute. Figure 1 shows the sub-tables (A to F) of the sample dataset. Each sub-table shows the portion of the sample data covered by the attribute–value pairs marked along the different branches of the decision tree with "odor" values being "a", "c", "f", "l", "n", and "p". The number given in brackets below each sub-table denotes the number of instances in the respective sub-table. All the sub-tables from "A" to "F" are pure partitions of data space either belonging to "edible" or "poisonous" class except sub-table "D".

The sub-table D has 10 number of instances covered by the attribute–value pair "odor = n" out of which 8 instances belong to "edible" class, and the rest 2 belongs to "poisonous" class. The last two instances may be noise, or these may be valuable and valid exceptions. The usual tree construction process stops at this node and concludes a rule "If the odor = n, Then Mushroom = edible" because further splitting may produce a decision tree that over fits the training data and has low generalization power. Obviously, such a decision tree model is bound to misclassify some of the instances with "odor = n" attribute–value pairs from the test data.

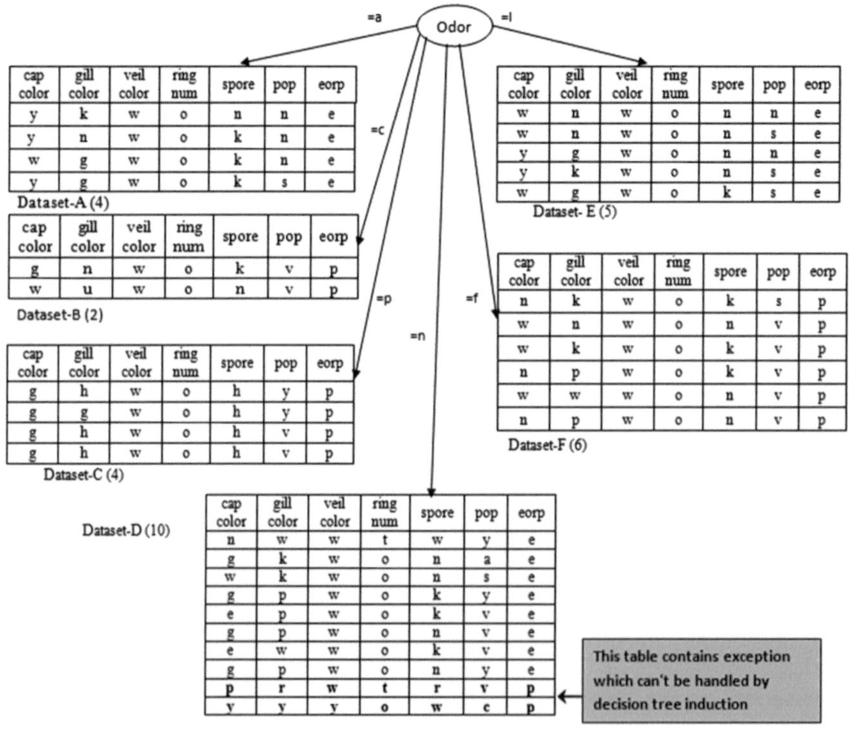

cap color	gill color	veil color	ring num	spore	pop	eorp
y	k	w	o	n	n	e
y	n	w	o	k	n	e
w	g	w	o	k	n	e
y	g	w	o	k	s	e

cap color	gill color	veil color	ring num	spore	pop	eorp
g	n	w	o	k	v	p
w	u	w	o	n	v	p

cap color	gill color	veil color	ring num	spore	pop	eorp
g	h	w	o	h	y	p
g	g	w	o	h	y	p
g	h	w	o	h	v	p
g	h	w	o	h	v	p

cap color	gill color	veil color	ring num	spore	pop	eorp
n	w	w	t	w	y	e
g	k	w	o	n	a	e
w	k	w	o	n	s	e
g	p	w	o	k	y	e
e	p	w	o	k	v	e
g	p	w	o	n	v	e
e	w	w	o	k	v	e
g	p	w	o	n	y	e
p	r	w	t	r	v	p
y	y	y	o	w	c	p

cap color	gill color	veil color	ring num	spore	pop	eorp
w	n	w	o	n	n	e
w	n	w	o	n	s	e
y	g	w	o	n	n	e
y	k	w	o	n	s	e
w	g	w	o	k	s	e

cap color	gill color	veil color	ring num	spore	pop	eorp
n	k	w	o	k	s	p
w	n	w	o	n	v	p
w	k	w	o	k	v	p
n	p	w	o	k	v	p
w	w	w	o	n	v	p
n	p	w	o	n	v	p

Fig. 1 Decision tree for the toy dataset

The decision tree model represented either as a set of rules or in tree form though accounts for such misclassification, however, takes no action to prevent such misclassifications. Hence, there is a need for a novel procedure to incorporate the discovery of valid exceptions in decision tree building process. The modified decision tree algorithm that discovers exceptions is presented in Fig. 2. The modified algorithm partitions the dataset instances into different class regions and finds the exceptional conditions as well.

```
Algorithm: Generate Decision Tree with exceptions.
Input: Training Dataset TD; Attribute List L; Splitting
       Criteria
Output: A decision tree T with Exceptions
Begin
  (1) Generate root node N
  (2) If each instance in TD falls in class Ck Then
         Return the leaf node N with class label Ck
      End-if
  (3) If L = ∅ Then// attribute list empty Treat N as leaf-node and
      label it with majority class label Cm and return
      End-if
  (4) Find the best splitting attribute (SA)based on the given
      criterion for attribute relevance
  (5) Label this node with SA
  (6) If SA is categorical and multi-way splits are allowed Then
         L ← L - SA; // delete splitting attribute from L
      End-if
  (7) For every outcome-value i of splitting attribute (SA)
      perform data partition along the branches at the node and
      extend sub-tree for each partition
       (7.1) Compute partition TDi
       (7.2) If TDi= ∅ Then // Empty partition
               (a) Produce a leaf node ni and label it with majority
                   class Cm
               (b) Produce a leaf node ni' and label it with
                   another class Cm' ≠ Cm
             Else
              Create the node produced by decision tree (TDi, L) to
              node N
       (7.3) Let node ni denotes rule Ri and node ni' denotes rule
              Ri'
       (7.4) Generate a set TP (true positive examples) covered by
              the rule Ri
       (7.5) Create a set FP (false positive examples) covered by
              the rule Ri'

       (7.6) Compute γ1, γ2 and γ3

       (7.7) If(γ1<1) Then

          If(γ1>>γ2 )&&(γ2< tE)&&(γ3 == 1) Then
             Add Exceptions to node NE
          Else
             Add a normal node N
       End if
    End for
  (8) Return N
End
```

Fig. 2 Algorithm for EDTA for exception discovery

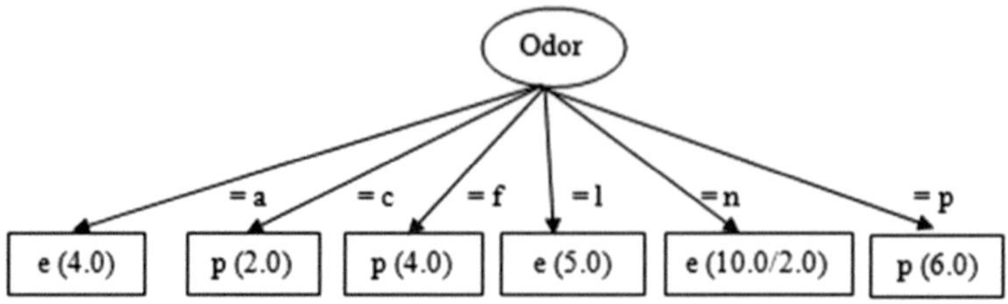

Fig. 3 A decision tree of the toy dataset

Table 1 Confusion matrix of the decision tree model

	Predicted edible	Predicted poisonous
Actual edible	(TP): $P \rightarrow D$ 17	(FN): $\neg P \rightarrow D$ 0
Actual poisonous	(FP): $P \rightarrow \neg D$ 2	(TN): $\neg P \rightarrow \neg D$ 12

The decision tree is constructed from the sample data, and exceptions are appended at each leaf node, which is not a pure partition where all the instances belong only to one of the classes. For demonstration purposes consider the decision tree (Fig. 3) constructed in WEKA tool by using the toy dataset as the training data.

The rules deduced from the tree are given below.

$$R_1 : \textit{If}\,(\text{Odor} = \text{a})\textit{Then}\,\text{edible}\ (4.0/0) \quad R_4 : \textit{If}\,(\text{Odor} = \text{l})\textit{Then}\,\text{edible}\ (5.0/0)$$
$$R_2 : \textit{If}\,(\text{Odor} = \text{c})\textit{Then}\,\text{poisonous}\ (2.0/0) R_5 : \textit{If}\,(\text{Odor} = \text{n})\textit{Then}\,\text{edible}\ (10.0/2.0)$$
$$R_3 : \textit{If}\,(\text{Odor} = \text{f})\textit{Then}\,\text{poisonous}\ (4.0/0) R_6 : \textit{If}\,(\text{Odor} = \text{p})\textit{Then}\,\text{poisonous}\ (6.0/0)$$

The numbers shown in the bracket represent the number of total instances covered by the premise of the rule and the instances that belong to minority class in the respective node. The confusion matrix resulted from these rules is given in Table 1. The symbols P and D denote premise and decision parts of the rules.

Here, follows the procedure based on which some misclassified instances by the usual decision tree model may qualify as valid exceptions to be accommodated in the decision tree. For the above decision tree model, 17 instances are True positive (TP) cases, 12 instances are True Negative (TN) cases, 2 instances are False Positive (FP), and none of the instances fall in the False Negative (FN) cases. The exceptions may be hidden either in FP or FN cases. For the example we have taken, these are hidden only in FP cases. To find out whether an instance or some instances in the FP cases qualify as an exception, we need to define three additional parameters γ_1, γ_2, and γ_3 (Eqs. 1–3) and some constraints (Eqs. 4–6).

$$\gamma_1 = \frac{TP}{TP + FP} = \frac{|P \wedge D|}{|P|} \tag{1}$$

$$\gamma_2 = \frac{FP}{TP + FP} = 1 - \gamma_1 = \frac{|P \wedge \neg D|}{|P|} \tag{2}$$

$$\gamma_3 = \frac{TN_E}{(TP + FP) \wedge E} = \frac{|P \wedge E \wedge \neg D|}{|P \wedge E|} = 1 \tag{3}$$

where

$$\gamma_1 < 1; \tag{4}$$

$$\gamma_1 \gg \gamma_2 \tag{5}$$

$$\gamma_3 = 1 \tag{6}$$

The parameters γ_1 and γ_2 signify the TP and FP rates. The symbol E is used to specify the attribute–value pair that is being tested to be qualified as an exception. The discovery of exception turns some of the FP cases into TN cases. TN_E represent those training data instances that were falling in FP before the discovery of exception(s) but now become TN cases. The decision of a rule is revised in the presence of an exception and the example which was earlier falsely classified belonging to positive class is now correctly classified belonging to negative class. The third parameter γ_3 ensures that attribute–value pair qualify as an exception only and only if it occurs in FP cases and nowhere in the TP cases of the training data. In other words, the precision of a rule augmented with exception will always be one.

For sake of an example, rule R_5 is a contender that might have exceptions. This rule has 8 instances in TP set and 2 instances in FP set out of total 31 instances. The value of γ_1 is 0.8 which less than 1 and value of γ_2 is 0.2 which is far less than γ_1. Now, the attribute–value pairs that do not already occur in the premise part are tested to see if any of these attribute–value pairs do qualify for exceptions. Let us modify the R_5 to R'_5 by appending some exceptions as follows. The modified decision tree is shown in Fig. 4.

R'_5: *If* (odor = n) *Then* (decision = e) ***Unless*** (CapColor = p) ∨ (CapColor = y) ∨ (GillColor = r) ∨ (GillColor = y) (2.0): Poisonous.

Now let us verify the values of the parameters γ_1, γ_2 and γ_3.

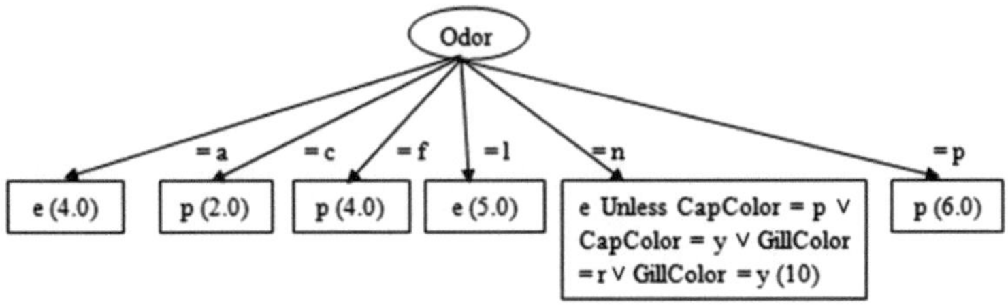

Fig. 4 Enhanced decision tree with exceptions

$$\gamma_1(R_5') = \frac{TP}{TP + FP} = \frac{8}{10} = 0.8$$

$$\gamma_2(R_5') = \frac{FP}{TP + FP} = \frac{2}{10} = 0.2$$

$$\gamma_3(R_5'('\text{CapColor} = p') = \frac{TN_E}{(TP + FP) \wedge E} = \frac{1}{1} = 1.0$$

$$\gamma_3(R_5'('\text{GillColor} = r') = \frac{TN_E}{(TP + FP) \wedge E} = \frac{1}{1} = 1.0$$

Some exceptions may cover common examples while others may cover different examples of the FP cases. The rule $R_{5'}$ with exception conditions alters the class labels of two instances from "*edible*" to "*poisonous*". This is an interesting rule which asserts that in most cases when the value of "*odor*" attribute is none (signified by the letter "*n*") *then* a mushroom is classified as "*edible*". However, If in addition to *odor* being none ('*n*'), other exceptional conditions like "*CapColor* = *p*," or "*CapColor* = *y*" or "*GillColor* = *r*" or "*GillColor* = *y*" etc. do exist in the rule as given in Fig. 4 then the class of such a mushroom becomes poisonous. The rule is interesting in the sense that it gives an opportunity to revise the decision in rare exceptional circumstances. With such a rule the decision tree predicts two additional instances of the sample dataset correctly and hence the accuracy of the classifier is also improved.

4 Experimental Setup

To test the effectiveness of the proposed algorithm we have used three datasets Zoo, Vote, and Mushroom. The Zoo dataset has 101 instances with 17 attributes and 7 values for class attributes. In vote dataset there are 232 instances and 16 attributes with two values of class labels. The Mushroom dataset has 5644 instances and 24 attributes with two values of class attribute. Mushroom and vote datasets have been preprocessed and instances with missing values have been removed from these datasets.

Table 2 displays the results for the three datasets. The first column gives the name of the datasets, the second columns lists the rules augmented with exceptions discovered (If any) by applying the proposed algorithm. The rest of the columns presents the values of parameters and accuracy of the decision tree models without and with exceptions. All the experiments have been conducted on Window platform using JAVA programming language.

In Table 2, rules that have a value of γ_1 less than 1 are candidate rules that may have exceptions. Exceptions are rare pieces of valuable knowledge and present in tiny parts of a dataset and hence have low support.

The enhancement in accuracy may only be considerable if the number of exceptions exists in several small data disjuncts. The rules with exceptions are valuable nuggets of knowledge that are interesting and prevent a classifier to make wrong

Table 2 Results of the three datasets

Dataset	Rules	$\gamma 1$	$\gamma 2$	Accuracy without exception (%)	Accuracy with exception (%)
Zoo	*If* (feathers = false ∧ milk = true) *Then* (*Decision* = mammal)	1.000			
	If (feathers = false ∧ milk = false ∧ backbone = true ∧ fins = false ∧ tail = false) *Then* (*Decision* = amphibian)	1.000			
	If (feathers = false ∧ milk = false ∧ backbone = true ∧ fins = false ∧ tail = true) *Then* (*Decision* = reptile) ***Unless*** (aquatic = true)(*Decision* = amphibian)	0.833	0.167		
	If (feathers = false ∧ milk = false ∧ backbone = true ∧ fins = true) *Then* (*Decision* = fish)	1.000			
	If (feathers = false ∧ milk = false ∧ backbone = false ∧ airborne = false ∧ predator = true) *Then* (*Decision* = invertebrate)	1.000		92.08	100
	If (feathers = false ∧ milk = false ∧ backbone = false ∧ airborne = false ∧ predator = false ∧ legs <= 2) *Then* (*Decision* = invertebrate) *If* (feathers = false ∧ milk = false ∧ backbone = false ∧ airborne = false ∧ predator = false ∧ legs > 2) *Then* (*Decision* = insect)	1.000			
	If (feathers = false ∧ milk = false ∧ backbone = true ∧ airborne = true) *Then* (*Decision* = insect)	1.000			
	If (feathers = true) *Then* (*Decision* = bird)	1.000			
Vote	*If* (physician.fee.freeze = n) *Then* (*Decision* = democrat) ***Unless*** (adoption.of.the.budget.resolution = n ∧ religious.groups.in.schools = n ∧ duty.free.exports = n) (*Decision* = republican)	0.992	0.008	96.5	99.5
	If (physician.fee.freeze = y) *Then* (*Decision* = republican) ***Unless*** (synfuels.corporation.cutback = y ∧ crime = n ∧ duty.free.exports = y) (*Decision* = democrat)	0.955	0.045		
Mushroom	*If* (odor = a) *Then* (*Decision* = e)	1.000			

(continued)

Table 2 (continued)

Dataset	Rules	$\gamma 1$	$\gamma 2$	Accuracy without exception (%)	Accuracy with exception (%)
	If (odor = c) *Then* (*Decision* = p)	1.000			
	If (odor = f) *Then* (*Decision* = p)	1.000			
	If (odor = l) *Then* (*Decision* = e)	1.000			
	If (odor = m) *Then* (*Decision* = p)	1.000			
	If (odor = n ∧ spore-print-color = b) *Then* (*Decision* = e)	1.000			
	If (odor = n ∧ spore-print-color = h) *Then* (*Decision* = e)	1.000			
	If (odor = n ∧ spore-print-color = k) *Then* (*Decision* = e)	1.000			
	If (odor = n ∧ spore-print-color = n) *Then* (*Decision* = e)	1.000			
	If (odor = n ∧ spore-print-color = o) *Then* (*Decision* = e)	1.000			
	f (odor = n) *Then* (*Decision* = e) ***Unless*** (spore-print-color = r)(*Decision* = p)	0.980	0.020	98.52	100
	If (odor = n ∧ spore-print-color = w ∧ gill-size = b) *Then* (*Decision* = e)	1.000			
	If (odor = n) *Then* (*Decision* = e) ***Unless*** (spore-print-color = w ∨ gill-spacing = c) (*Decision* = p)	0.922	0.078		
	If (odor = n) *Then* (*Decision* = e) ***Unless*** (spore-print-color = w ∧ gill-spacing = w ∧ population = c) (*Decision* = p)	0.974	0.026		
	If (odor = n ∧ spore-print-color = w ∧ gill-spacing = w ∧ population = v) *Then* (*Decision* = e)	1.000			
	If (odor = n ∧ spore-print-color = y) *Then* (*Decision* = e)	1.000			
	If (odor = p) *Then* (*Decision* = p)	1.000			

predictions in rare exceptional conditions. Even if the proposed decision tree classifier augmented with exception does not improve the accuracy significantly, these are concise, semantically meaningful and more interesting.

5 Conclusion

In this paper, we have suggested an Enhanced Decision Tree Algorithm (EDTA) which accommodates exceptions hidden in small data disjuncts in the decision tree model. The advantage of the model is that it has rules that revise their conclusions in the presence of rare and exceptional conditions. Such a decision tree classifier is not only more accurate but more succinct and interesting. Knowing general rules for classification is essential but by knowing exceptions, a classifier becomes intelligent enough to avoid incorrect predictions in the manifestation of exceptional conditions. Such a classifier has a scope of applications in domains like robotics, fraud detection, and, fault and medical diagnosis where misclassifications are associated with high costs. In future, this work can be extended to accommodate intra- as well as inter-class exceptions.

References

1. Appavu alias Balamurugan, S., Rajaram, R.: Effective solution for unhandled exception in decision tree induction algorithms. Expert Syst. Appl. **36**(10), 12113–12119 (2009)
2. Bala, R., Ratnoo, S.: Discovering fuzzy censored classification rules (FCCRs): a genetic algorithm approach. Int. J. Artif. Intell. Appl. **3**(4), 175–188 (2012)
3. Bala, R., Ratnoo, S.: A genetic algorithm approach for discovering tuned fuzzy classification rules with intra- and inter-class exceptions. J. Intell. Syst. **25**, 263–282 (2016)
4. Bharadwaj, K.K., Al-Maqaleh, B.M.: Evolutionary approach for automated discovery of censored production rules **1**(10), 3230–3235 (2007)
5. Compton, P., et al.: Ripple down rules: turning knowledge acquisition into knowledge maintenance. Artif. Intell. Med. **4**(6), 463–475 (1992)
6. Carvalho, D.R., Freitas, A.A.: A hybrid decision tree/genetic algorithm method for data mining. Inf. Sci. **163**(1), 13–35 (2004)
7. Dietterich, T.G.: Ensemble methods in machine learning. In: Multiple Classifier Systems, pp. 1–15. Springer, Berlin Heidelberg (2000)
8. Gehrke, J., et al.: Rain forest—a framework for fast decision tree construction of large datasets. Data Min. Knowl. Disc. **4**(2–3), 127–162 (2000)
9. Geng, L., Hamilton, H.J.: Interestingness measures for data mining: a survey. ACM Comput. Surv. **38**(3), 9 (2006)
10. Gill, A.A., et al.: Improving decision tree performance through induction- and cluster-based stratified sampling. In: Intelligent Data Engineering and Automated Learning, pp. 339–344. Springer, Berlin, Heidelberg (2004) (LNCS, 7177)
11. Han, J., et al.: Data Mining: Concepts and Techniques, 3rd edn. Elsevier (2011)
12. Kotsiantis, S.B.: Decision trees: a recent overview. Artif. Intell. Rev. **39**(4), 261–283 (2013)
13. Lim, T.-S., et al.: A Comparison of prediction accuracy, complexity, and training time of thirty-three old and new classification algorithms. Mach. Learn. **40**(3), 203–228 (2000)
14. Mehta, M., et al.: SLIQ: a fast scalable classifier for data mining. In: Advances in Database Technology, pp. 18–32. Springer, Berlin, Heidelberg (1996) (LNCS, 1057)
15. Saroj, S., Bharadwaj, K.K.: A parallel genetic algorithm approach for automated discovery of censored production rules. In: Proceedings of the 25th International Multi-Conference: Artificial Intelligence and Applications, pp. 435–441. ACTA Press, Anaheim, CA, USA (2007)
16. Shafer, J.. et al.: SPRINT: a scalable parallel classifier for data mining. In: Proceedings of the 22nd International Conference on Very Large Databases, pp. 544–555. Morgan Kaufmann (1996)

17. Su, J., Zhang, H.: A fast decision tree learning algorithm. In: Proceedings of the 21st National Conference on Artificial Intelligence, vol. 1, pp. 500–505. AAAI Press, Boston, MA (2006)
18. Subramanian, A.A.B., et al.: Improving decision tree performance by exception handling. Int. J. Autom. Comput. **7**(3), 372–380 (2010)
19. Suzuki, E.: Discovering interesting exception rules with rule pair. In: Proceedings of the European Conference on Machine Learning and Principles and Practice of Knowledge Discovery in Databases Workshop on Advances in Inductive Rule Learning, pp. 163–178. (2004)
20. Suzuki, E., Zytkow, J.M.: Unified algorithm for undirected discovery of exception rules. In: Principles of Data Mining and Knowledge Discovery, pp. 169–180. Springer, Berlin, Heidelberg (2000) (LNCS, 1910)
21. Vashishtha, J., et al.: An evolutionary approach to discover intra– and inter-class exceptions in databases. Int. J. Intell. Syst. Technol. Appl. **12**(3–4), 283–300 (2013)
22. Yıldız, O.T., Dikmen, O.: Parallel univariate decision trees. Pattern Recogn. Lett. **28**(7), 825–832 (2007)

A Fuzzy Approach for Cost and Time Optimization in Agile Software Development

Anupama Kaushik, Devendra Kr. Tayal and Kalpana Yadav

Abstract Cost and time prediction is very crucial for software development. For many years conventional techniques to determine them were followed, but nowadays the software companies are moving toward adopting agile paradigm. It is due to its various characteristics such as iterative development, rapid delivery, and reduced risk. This work proposes the role of fuzzy logic in agile software cost and time optimization. Here, both fuzzy type-1 and type-2 are evaluated on agile dataset.

Keywords Agile cost estimation · Fuzzy type-1 · Fuzzy type-2 · Effort estimation · Time estimation

1 Introduction

Agile software development is a big leap from traditional plan-based approaches to software development. Since its inception through agile manifesto in 2001, many software development firms are sinking into it. The agile methodologies address the problem of requirement volatility as they allow replanning and changes in the requirement. It allows good communication between the developers and customers and also rapid delivery of the software.

Software cost estimation is a critical task in project development. There were many software cost estimation models present in the literature like COCOMO, SEER-SEM, SLIM, etc. These were the traditional models and later on they were refined

A. Kaushik (✉)
Department of IT, Maharaja Surajmal Institute of Technology, Janakpuri, New Delhi, India
e-mail: thisisanupama@gmail.com

Indira Gandhi Delhi Technical University for Women, Delhi, India

D. Kr. Tayal
Department of Computer Science, Indira Gandhi Delhi Technical University for Women, Delhi, India

K. Yadav
Department of IT, Indira Gandhi Delhi Technical University for Women, Delhi, India

B. Pati et al. (eds.), *Advanced Computing and Intelligent Engineering*,
Advances in Intelligent Systems and Computing 1082,
https://doi.org/10.1007/978-981-15-1081-6_53

by many researchers [1–3] by adding machine learning, evolutionary computation, metaheuristic techniques, etc.

Since predictability of required resources is the primary step for any project development and this implies even for the projects using agile techniques. But not much of work is done in software cost and time estimation for such projects. The present study is a step toward providing a fuzzy model of cost and time optimization for agile software development. In software development, cost and effort estimation is used interchangeably.

The paper is organized as follows: Sect. 2 presents the related research; Sect. 3 discusses in brief fuzzy type-1 and type-2; Sect. 4 describes the proposed methodology; Sect. 5 reflects upon the dataset, evaluation criteria and the results; Sect. 6 concludes the paper.

2 Related Research

Kang et al. [4] proposed a software cost estimation model for agile software development projects in which they used function points in addition to the story point. They used Kalman filter algorithm for observing the project progress in which they used function point as an input for providing cost estimation and velocity. The performance of the model is validated through a case study.

Coelho and Basu [5] illustrated the effort estimation in agile software development projects using story points and discussed the area for future work.

Choudhari and Suman [6] proposed a Software Maintenance Effort Estimation Model (SMEEM). But the model was effective for the maintenance environment of only the agile and extreme programming projects.

Ziauddin et al. [7] proposed a software effort estimation model for agile software projects and tested their approach using data taken from 21 software projects.

Hussain et al. [8] proposed the use of COSMIC which is an international standard to measure the functional size from the requirements of a software project, in agile processes which in turn supported the early effort estimation in it.

Popli and Chauhan [9] proposed a method of estimating cost, effort, and duration of small and medium-sized projects using agile. It also discusses various problems that the agile team faces.

Satapathy et al. [10] used Support Vector Regression (SVR) kernel methods to optimize the effort prediction in agile projects. They had also used story point approach.

Panda et al. [11] used different kinds of neural networks to enhance the prediction accuracy of their model based on story point approach and used Zia dataset [7] to validate the model.

Garg and Gupta [12] proposed a software cost estimation model using principal component analysis (PCA) and constraint programming approach. They used PCA to identify the attributes that directly affect the development costs and constraint programming is used to check whether the agile manifestos are taken care of.

Raslan et al. [13] used fuzzy logic in story point approach of effort estimation of projects using agile. They used trapezoidal membership functions to represent the input parameters.

Tanveer [14] provided a hybrid model to improve the effort estimation in agile projects using expert knowledge, change impact analysis and cost drivers as proposed by the experts.

Dragicevic et al. [15] proposed a Bayesian network model for effort estimation in projects using agile. Their model can be used for any type of agile method and they used the data of a single software company to validate the model.

Satapathy and Rath [16] used different machine learning algorithms such as decision tree, stochastic gradient boosting and random forest to analyse the effort estimation in agile software projects using story point approach.

Tanveer et al. [17] proposed a hybrid method based on boosted trees which used change impact analysis information for improving the effort estimation accuracy. They did a case study using their proposed approach with Insiders Technologies, a German software company and found their approach very effective.

The present study explores type-1 and type-2 fuzzy logic which have not been used earlier in agile software development.

3 Type-1 and Type-2 Fuzzy Systems

Prof. Zadeh introduced fuzzy logic in 1965 through his paper on fuzzy sets [18]. It is a technique to solve problems which are too difficult to be understood quantitatively. It models the human reasoning. These sets have degrees of membership lying in the interval [0, 1]. Formally, a fuzzy set F in V is expressed as:

$$F = \{(y, \mu_F(y))|y \text{ in } V\}$$

where, V is the universe of discourse. $\mu_F(y)$ is the membership function (MF) that defines the membership of y in F; There are different MFs like triangular MFs, trapezoidal MFs, Gaussian MFs, generalized bell MFs, etc.

In 1975, Zadeh first proposed Type-2 fuzzy sets (T2FSs), which were later extended by Mendel and Liang in 1999 [19]. These sets use the concept of footprint of uncertainty (FOU), and upper and lower MFs. If the type-1 MF is blurred both to the left and right, a type-2 MF is obtained.

If the boundaries of type-1 MF are blurred, FOU is obtained. The upper and lower boundaries of this FOU, represent the upper and lower MFs, respectively. A type-1 FS has crisp membership whereas type-2 FS has fuzzy membership.

A *T2 FS*, denoted $\tilde{F}$, [20] is characterized by a type-2 MF $\mu_{\tilde{F}}(y, u)$ where y $\in Y$ and $\in J_y \subseteq [0, 1]$, i.e.,

$$\tilde{\mathrm{F}} = \left\{\left((y, u), \mu_{\tilde{F}}(y, u)\right) \forall y \in Y, \forall u \in J_y \subseteq [0, 1]\right\}$$

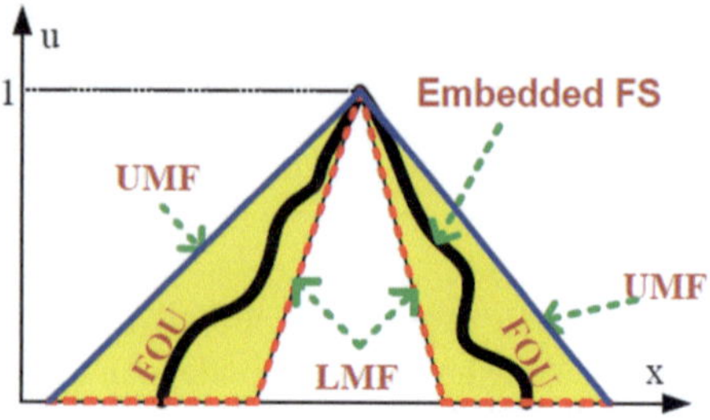

Fig. 1 Representation of triangular FMF in IT2FS

in which $0 \le \mu_{\tilde{F}}(y, u) \le 1$. When all $\mu_{\tilde{F}}(y, u) = 1$ then $\tilde{F}$ is an *interval T2 FS* (IT2 FS).

IT2 FS use only footprint of uncertainty (FOU). As there is a lot of computational complexity involved with general T2 FS, interval T2 FSs are more in use [20]. There are 20 different interval type-2 Fuzzy Membership Functions (FMF) produced in MATLAB environment. Figure 1 represents the triangular FMF in IT2FS. Here, the colored area is FOU, dots represent LMF, solid boundary line denotes UMF, and the wavy line represents embedded Fuzzy System for IT2FS. In this work type-1 and interval type-2 triangular FMF are used.

4 Proposed Methodology

Effort estimation in projects following agile methodology is very difficult in comparison to the traditional approach as there are lot of differences between agile approach and traditional approach. The story point approach is one of the most commonly used approaches for effort estimation in agile. Effort estimation using this approach depends upon two factors story size and complexity of the project. Certain values are assigned to each of these factors. The values assigned are unimportant, what matters are the relative values. For example, value 5 is assigned to an extremely large story, value 4 is assigned to a very large story, value 3 is assigned to a moderately large story, value 2 is assigned to the stories which can be completed in a day or two, and value 1 is assigned to the stories which can be completed within few hours. Similarly, for complexity value 5 is assigned for an extremely complex, with many unknowns and dependencies, value 4 is assigned to very complex, value 3 to moderately complex, value 2 to less complex with fewer unknowns and value 1 to very straightforward and clear stories.

So, the effort of a user story is given by (1):

$$\text{Effort}_{\text{userstery}} = \text{size} \times \text{complexity} \tag{1}$$

As the project consists of many individual user stories, the effort of a project is given as:

$$\text{Effort}_{\text{project}} = \sum_{i=1}^{n} \text{Effort}_{\text{userstory}} \quad (2)$$

In agile, velocity is calculated as:

$$V_i = \text{Units of effort completed/sprint time} \quad (3)$$

Here, units of effort relate to the effort of the individual user stories for the whole project, sprint is the time allotted for a specific work to get completed and reviewed.

According to Zia [7] there are two forces that reduce the project velocity. They are friction forces (FF) and dynamic forces (DF). Friction forces consistently reduce the productivity and dynamic forces slow down the project team members.

Deceleration (D) in a project using agile is given as:

$$D = FF \times DF \quad (4)$$

The final velocity is given as:

$$V_{\text{final}} = V^D \quad (5)$$

Total time to complete a project is given as:

$$T = \text{Effort}_{\text{project}} / V_{\text{final}} \quad (6)$$

The development cost of a project is:

$$\text{cost} = 1.681 \times TS \times T \quad (7)$$

where, *TS* is the monthly team salary and *T* is the total time to complete a project. All the above factors are described in Zia et al. [7].

The present study proposed to optimize the development cost and completion time of the project given in Eqs. (6) and (7) using fuzzy approach. Here, both type-1 and type-2 fuzzy approach is used and compared.

5 Experimental Evaluation

The proposed approach is evaluated on Zia dataset [7]. It is a collection of twenty-one software projects developed using agile software development methodology. The features of the dataset used in the study are "P.No, Effort, *V* (Velocity), Team Salary, Actual Time, Actual Cost." Here, salary and cost are in Pakistan Rupees and time is in months. Two type-1 fuzzy FIS is built. In the first FIS, effort and velocity are the input and they are fuzzified using triangular membership function and the output is the project completion time which is also fuzzified using triangular membership. Here, effort is divided into 14 fuzzy intervals from $e1$ to $e14$, velocity into 9 fuzzy intervals from $v1$ to $v9$ and time into 10 fuzzy intervals from $t1$ to $t10$. Figure 2 shows the type-1 triangular membership function for effort. In the same manner velocity and time are also represented using triangular membership function. 21 fuzzy rules are generated using fuzzy rule editor in this FIS to calculate project completion time. The fuzzy rules generated for this FIS are shown in Fig. 3. The second, type-1 FIS consists of team salary and time as inputs and cost as output. Here, team salary is divided into 11 fuzzy intervals from $s1$ to $s11$, time is divided into 10 fuzzy intervals from $t1$ to $t10$ and cost into 12 fuzzy intervals from $c1$ to $c12$. All these parameters are also mapped to triangular membership function and 21 fuzzy rules are generated.

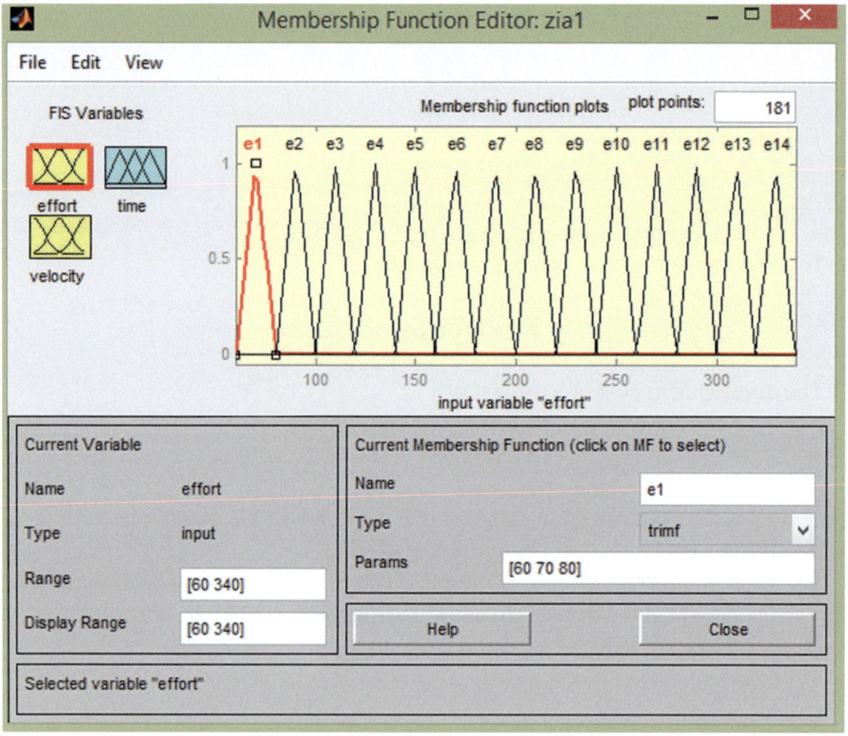

Fig. 2 Triangular membership function for effort

Fig. 3 Fuzzy rules for effort, velocity, and time

1. *If (effort is e5) and (velocity is v2) then (time is t5)*
2. *If (effort is e8) and (velocity is v1) then (time is t8)*
3. *If (effort is e6) and (velocity is v5) then (time is t4)*

----

20. *If (effort is e4) and (velocity is v2) then (time is t4)*
21. *If (effort is e2) and (velocity is v2) then (time is t2)*

Both these FISs are evaluated using "evalfis" function in MATLAB file to obtain the type-1 estimated time and cost.

The present study also evaluates the performance of type-2 fuzzy logic on cost and time optimization for agile software development projects. Here, also two interval type-2 FIS are developed, one for time optimization and another for cost optimization. All these parameters are mapped to interval type-2 triangular membership function. Figure 4 shows the type-2 membership editor for team salary generated. Similarly, the other two parameters, time and cost are also mapped to type-2 triangular membership function and fuzzy rules are generated. The second interval type-2 FIS consists of effort and velocity as inputs and time as output. These are also mapped to interval type-2 triangular membership function and fuzzy rules are generated. Here, also the same fuzzy rules are used. These FISs are evaluated using "evalifistype2" function in MATLAB file to obtain the type-2 estimated time and cost.

Table 1 shows the estimated time and cost obtained using type-1 and type-2 FIS. It also contains the estimated time and cost obtained using Zia model given by Zia et al. [7].

From Table 1, it can be deduced that estimated time and cost using type-2 FIS is more accurate and near to actual values of time and cost of the projects as given in Zia dataset [7]. So, the use of type-2 fuzzy logic in optimizing various parameters

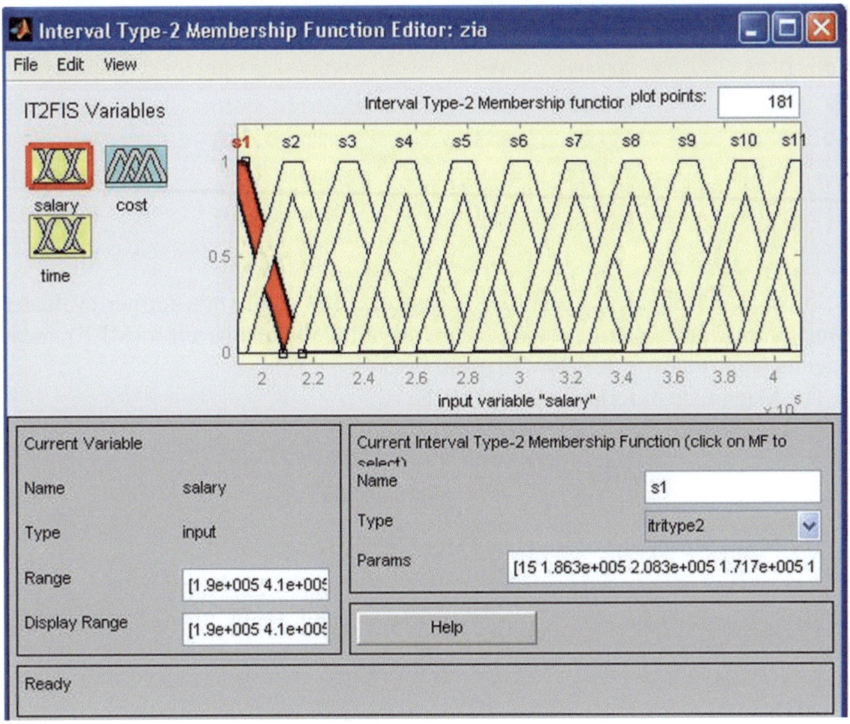

Fig. 4 Interval type-2 triangular membership function for team salary

Table 1 Estimated time and cost using type-1 and type-2 fuzzy logic

P.No	Type-1		Type-2		Zia's model	
	Estimated time	Estimated cost	Estimated time	Estimated cost	Estimated time	Estimated cost
1	70	1450000	60.9	1083800	58	1023207.14
2	70	1500460	90	1672600	81	1680663.89
3	72	1850000	53.3	1024800	52	992269.51
4	85	1799575	89.6	2336500	87	2002767.22
5	25	700110	31.2	755280	29	676081.32
6	85	3101142	90.3	3099000	95	2895132.85
7	39	750000	31.2	635140	29	540113.84
8	85	1850000	94.5	1756100	84	1614078.94
9	70	850000	34.1	516320	35	507264.58
10	55	1850000	60	1155400	66	1267179.55
11	40	700000	41.8	776100	41	786732.22
12	70	600490	34.7	621740	39	597142.61
13	70	500990	31.2	617030	35	538494.68
14	70	850000	29.8	357030	26	394545.65
15	65	501500	20.2	348220	22	330561.22
16	125	1850000	107.8	2181900	103	1971485.44
17	25	700000	35.3	785140	40	770857.32
18	60	1850000	55.7	955300	50	961866.44
19	85	1850000	82.5	1572600	76	1453032.29
20	70	700425	54.2	829760	51	854347.55
21	70	500345	34.1	595610	34	567484.33

of software development cannot be waved off. The results are further evaluated using three evaluation criteria's which are magnitude of relative error (MRE), mean magnitude of relative error (MMRE) and Pred (*l*).

The Magnitude of Relative Error (MRE) is:

$$\text{MRE} = \frac{|\text{Actual Data} - \text{Estimated Data}|}{\text{Actual Data}}$$

The Mean Magnitude of Relative Error (MMRE) is given as

$$\text{MMRE} = \frac{1}{N}\sum_{X=1}^{N} \text{MRE}$$

Smaller MMRE and MRE indicate less estimation error.

The Pred (*l*) is defined as

$$\text{Pred}(l) = \frac{k}{n}$$

where, n is the total number of projects and k is the number of projects whose MRE is less than or equal to l.

Table 2 shows the MRE values for type-1, type-2 and Zia model. It is found that MRE values for type-2 are very less in comparison to type-1 and Zia model. The smaller the MRE and MMRE values the better is the technique, the higher the PRED (l) values the better is the technique. The MMRE and PRED (l) for the estimated time and cost using type-1, type-2, and Zia model are depicted in Table 3. Here, the value of l in PRED (l) is taken as 7.19 for time and 5.76 for cost. These values are chosen in order to compare with the Zia model as they have used the same l values in their work Zia et al. [7]. It is found from Table 3 that, here also type-2 model defeats the type-1 and Zia model.

Table 2 MRE values for estimated time and cost

P.No	Type-1		Type-2		Zia model	
	MRE time	MRE cost	MRE time	MRE cost	MRE time	MRE cost
1	11.11	20.83	3.33	9.68	7.94	14.73
2	23.91	6.22	2.17	4.54	11.96	5.04
3	28.57	85.00	4.82	2.48	7.14	0.77
4	1.16	14.31	4.19	11.26	1.16	4.63
5	21.88	6.65	2.50	0.70	9.38	9.86
6	6.59	3.09	0.77	3.16	4.40	9.53
7	11.43	25.00	10.86	5.86	17.14	9.98
8	8.60	2.78	1.61	2.44	9.68	10.33
9	94.44	70.00	5.28	3.26	2.78	1.45
10	11.29	54.17	3.23	3.72	6.45	5.60
11	11.11	12.50	7.11	2.99	8.89	1.66
12	89.19	7.62	6.22	4.35	5.41	8.13
13	118.75	16.50	2.50	2.84	9.38	10.25
14	133.33	112.50	0.67	10.74	13.33	1.36
15	209.52	43.29	3.81	0.51	4.76	5.55
16	11.61	7.50	3.75	9.10	8.04	1.43
17	35.90	12.50	9.49	1.86	2.56	3.64
18	15.38	85.00	7.12	4.47	3.85	3.81
19	6.25	23.33	3.13	4.84	5.00	3.13
20	25.00	12.45	3.21	3.72	8.93	6.79
21	100.00	9.03	2.57	8.29	2.86	3.18

Table 3 MMRE and PRED values for estimated time and cost

Estimation models	MMRE time	MMRE cost	PRED (7.19) time	PRED (5.76) cost
Type-1	46.43	30.01	14.28	9.52
Type-2	4.20	4.79	90.47	71.42
Zia	7.19	5.75	57.14	61.90

6 Conclusion

Software cost and time optimization in projects using agile are very critical and in literature very few studies are present on it. This work evaluates the performance of type-1 and type-2 fuzzy logic on the projects using agile methodology. It uses the Zia dataset, on which most of the existing cost estimation studies are based upon. The study also compares type-1, type-2, and Zia model. It is found that type-2 outperforms type-1and Zia model at all the three evaluation criteria used, i.e., MRE, MMRE and PRED (*l*) for both time and cost optimization in projects using agile.

References

1. Kaushik, A., Verma, S., Singh, H.J., Chhabra, G.: Software cost optimization integrating fuzzy system and COA-Cuckoo optimization algorithm. Int. J. Syst. Assur. Eng. Manag. **8**, 1461–1471 (2017)
2. Kaushik, A., Tayal, D.K., Yadav, K., Kaur, A.: Integrating firefly algorithm in artificial neural network models for accurate software cost predictions. J. Softw. Evol. Process **28**, 665–688 (2016)
3. Dave, V.S., Dutta, K.: Neural network based models for software effort estimation: a Review. J. Artif. Intell. Rev. **42**, 295–307 (2014)
4. Kang, S., Choi, O., Baik, J.: Model based estimation and tracking method for agile software project. Int. J. Hum. Capital Inf. Technol. Professionals **3**(2), 1–15 (2012)
5. Coelho, E., Basu, A.: Effort estimation in agile software development using story points. Int. J. Appl. Inform. Syst. **3**(7), 7–10 (2012)
6. Choudhari, J., Suman, U.: Story points based effort estimation model for software maintenance. Procedia Technol. **4**, 761–765 (2012)
7. Ziauddin, Tipu, S.K., Zia, S.: An effort estimation model for agile software development. Adv. Comput. Sci. Appl. (ACSA) **2**(1), 314–324 (2012)
8. Hussain, I., Kosseim, L., Ormandjieva, O.: Approximation of COSMIC functional size to support early effort estimation in Agile. Data Knowl. Eng. **85**, 2–14 (2013)
9. Popli, R., Chauhan, N.: Cost and effort estimation in agile software development. In: International Conference on Reliability, Optimization and Information Technology-ICROIT, pp. 57–61, 6–8 Feb 2014
10. Satapathy, S.M., Panda, A., Rath, S.K.: Story point approach based agile software effort estimation using various SVR kernel methods. In: The Twenty-Sixth International Conference on Software Engineering and Knowledge Engineering, SEKE, pp. 304–307 (2014)
11. Panda, A., Satapathy, S.M., Rath, S.K.: Empirical validation of neural network models for agile software effort estimation based on story points. In: 3rd International Conference on Recent Trends in Computing 2015 (ICRTC-2015) Procedia Computer Science, vol. 57, pp. 772–781 (2015)

12. Garg, S., Gupta, D.: PCA based cost estimation model for agile software development projects. In: 2015 International Conference on Industrial Engineering and Operations Management (IEOM), pp. 1–7. IEEE (2015)
13. Raslan, A.T., Darwish, N.R., Hefny, H.A.: Towards a fuzzy based framework for effort estimation in agile software development. Int. J. Comput. Sci. Inform. Secur. **13**(1), 37–45 (2015)
14. Tanveer, B.: Hybrid effort estimation of changes in agile software development. In: Sharp, H., Hall, T. (eds.) Agile Processes, in Software Engineering, and Extreme Programming. XP 2016. Lecture Notes in Business Information Processing, vol. 251, pp. 316–320 (2016)
15. Dragicevic, S., Celar, S., Turic, M.: Bayesian network model for task effort estimation in agile software development. J. Syst. Softw. **127**, 109–119 (2017)
16. Satapathy, S.M., Rath, S.K.: Empirical assessment of machine learning models for agile software development effort estimation using story points. Innovations Syst. Softw. Eng. **13**:191–200 (2017)
17. Tanveer, B., Vollmer, A.M., Braun, S.: A hybrid methodology for effort estimation in Agile development: an industrial evaluation. In: ICSSP 18 Proceedings of the 2018 International Conference on Software and System Process, pp. 21–30 (2018)
18. Zadeh, L.A.: Fuzzy sets. Inf. Control **8**, 338–353 (1965)
19. Mendel, J.M., Liang, Q.: Pictorial comparison of type-1 and type-2 fuzzy logic systems. In: Proceedings of IASTED International Conference on Intelligent Systems and Control, Santa Barbara, CA, October (1999)
20. Liang, Q., Mendel, J.M.: Interval type-2 fuzzy logic systems: theory and design. IEEE Trans. Fuzzy Syst. **8**(5), 535–550 (2000)

Computing Articulation Points Using Maximal Clique-Based Vertex Classification

Sadhu Sai Kaushal, Mullapudi Aseesh and Jyothisha J Nair

Abstract The bridge vertices of an unweighted undirected graph are those vertices whose removal increases the number of connected components of the graph, i.e., the vertices whose removal disconnects the graph. However, not all the bridge vertices are equal. The removal of some of them might end in a single vertex disconnected from the graph, while in other cases, the graph can be split into several small pieces. This paper deals with the latter situation which ultimately finds vertices or articulation points among communities of a graph. A better method to find bridge vertices among a graph with maximal information termination is through maximal cliques and overlapping communities. The proposed method first divides the graph into cliques. These cliques are then merged into communities based on the similarity criteria. Vertices are classified into isolated, overlapping, and bridge vertices. Vertices which can be merged into overlapping communities are merged so that the remaining isolated vertices will form the articulation points or the pure bridge vertices, where removal of such vertices can affect the information passage between communities. After merging, we ignore a situation where removal of a bridge vertex in a graph which ends in a single vertex or a set of vertices which cannot be formed as a community as we focus on community-based vertex classification.

Keywords Clique · Maximal clique · Community · Overlapping communities · Classification of vertices

S. S. Kaushal · M. Aseesh · J. J. Nair (✉)
Department of Computer Science and Engineering,
Amrita Vishwa Vidyapeetham, Amritapuri, India
e-mail: jyothishaj@am.amrita.edu

S. S. Kaushal
e-mail: sadhu.saikaushal@gmail.com

M. Aseesh
e-mail: aseesh.mullapudi@gmail.com

B. Pati et al. (eds.), *Advanced Computing and Intelligent Engineering*,
Advances in Intelligent Systems and Computing 1082,
https://doi.org/10.1007/978-981-15-1081-6_54

1 Introduction

Community is a subpart of a real-world network that accounts for the system's functionality. Different communities, each of which performs a specific function with minimal differences are merged together to form a network. Therefore, these communities play a pivotal role in the structure and organization of networks. The amount of data to be processed is growing day by day and so is the complexity of the networks. Exchange of information between these communities in a network takes place either through a node that is present in both the communities or through an intermediate node connected to both the communities. However, the former case has been identified as the overlapping vertex and many methods have been found out to identify them. But, in the latter case removing the intermediate node from the network which is not a part of any community would greatly affect the organization of the network. There would not be any passage of information between the communities connected through the intermediate node. This concept is closely related to articulation points. However, considering this in real-world networks would give either positive or negative results depending on the network. For example in a terrorist cell organization blocking or removing the intermediate hierarchical level(node) would leave a void in the communication between the upper and lower hierarchy.

Similarly, in wireless sensor networks, this disruption will block data transmission from one network component to others. In the yeast protein– protein interaction network, lethal mutations enriched will be disrupted in the group of highly connected proteins. This has inspired us to find out the bridge vertices, their importance and how their inclusions and exclusions in a graph would affect that network. For these reasons, there is a growing importance for efficiently finding out the bridge vertices whose inclusion or exclusion would affect the properties of a network. In community detection, identifying central hubs that connect different groups can help isolate and identify communities. In epidemic diseases and tumors spreading, quarantining bridge vertices can stop the spread of infection and tumors into other communities. In viral marketing, the most influential bridge vertices can speed-up the new product marketing to different groups [1].

So, articulation points are viewed as vulnerabilities in networks. Different methods exist for detecting articulation points. But, we work on a method which finds those points or holes by using community detection techniques using maximal cliques. In this method, information contained in those communities is immense as the concept of maximal cliques is used. So, articulation points or bridge vertices have even more weight in this method as there is so much of information sharing among the communities which is being shared through those vertices.

In this paper, the following contributions are made:

1. Maximal cliques in the input graph are found and the vertices are segregated into overlapping, isolated, and bridge vertices that connect two maximal cliques.
2. Those maximal cliques are merged which gives the overlapping communities.
3. An algorithm is proposed to find out the articulation points in the graph whose exclusion divides the graph into separate components.

The rest of the paper is organized as follows: Sect. 2 mentions some of the related works. Section 3 presents preliminary concepts required. Section 4 presents our proposed methods. Section 5 gives the results, and Sect. 6 talks about conclusion for this research. The proposed algorithm works only for unweighted networks.

2 Related Work

Removal of articulation points in a graph either ends up with single vertex disconnection or a set of small pieces. Over the past several years many solutions have been coming into picture, where these articulation points are considered as nodes, which when removed make the graph into set of disjoint sets. These are found to be more vulnerable when present in a social networking community or in a reliable network.

Many algorithms are proposed for identifying the articulation points. One such algorithm is in [2], where the impact of articulation points is found in linear time, i.e., the number of disconnected vertices after the removal of articulation points. Similarly, in [3], strong articulation points are found by linear time algorithm. In this paper, we propose to find the articulation points in a network having different communities using overlapping community detection.

After observing the techniques presented in one of our reference papers [4], we found that finding maximal cliques is efficient using the Bron–Kerbosch algorithm which is not an output-sensitive algorithm and it is quite in contrast to other algorithms to clique problems. With the help of [5], we understood the use of Bron–Kerbosch algorithm in finding maximal cliques.

The above algorithms are redundant for finding articulation points. In cases, where, we need to find out the actual bridge between communities in a graph where information breakage between communities is crucial, the proposed algorithm shows a better performance.

From [6], the different advantages of community detection in graphs are understood and from [7], we understood the importance of overlapping communities in graphs. In [8–10], methods for finding communities are well demonstrated and in [11, 12] overlapping community detection methods are proposed.

The existing methods for finding articulation points stated in [13, 14], actually finds the articulation points in a network without considering communities present in the network. The importance of finding articulation points through communities using clique-based approach is that maximum information breakdown could be easily found and handled at these points. A clique is a community, handling maximum data sharing within itself. So, our proposed algorithm detects the communities using maximal cliques and then the articulation points are identified.

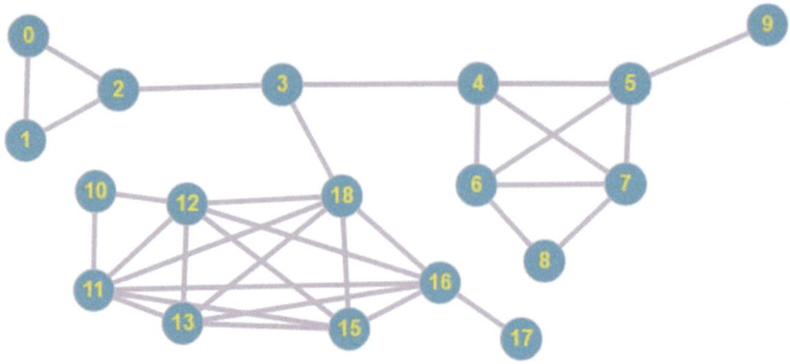

Fig. 1 Synthetic input 1

3 Preliminary Concepts

Clique: A subgraph in which all vertices are connected with all other vertices. In Fig. 1, (0, 1, 2) is a clique in which all the vertices are wholly connected. So are (4, 5, 6, 7), (11, 12, 18, 13, 15, 16), (10, 11, 12), (6, 7, 8).

Maximal Clique: A clique that cannot be extended by including one more adjacent vertex, i.e., it should not be a subset of a larger clique. That itself has to be the largest clique. One of the best-known algorithms to find maximal cliques is Bron–Kerbosch algorithm.

Community: It is a part of a network (graph) in which all the nodes are densely connected and the community as a whole is sparsely connected with other communities of the network.

Overlapping communities: A community that contains appropriate number of overlapping vertices.

Classification of Vertices:

1. Isolated Vertices: Vertices with low vertex degree, either 0 or 1 are considered to be isolated from the subgraph and does not belong to any maximal clique. In Fig. 1, vertices 9 and 17 are considered to be isolated.
2. Overlapping Vertices: Some vertices belong to more than one community in a network. In the literature, there are many methods for detecting nonoverlapping communities. But the real networks consist of overlapping communities. In Fig. 1, vertices 12, 11, 6, and 7 belong to more than one maximal clique and are considered to be overlapping vertices, i.e., 12 and 11 belong to (10, 12, 11) and (12, 13, 11, 15, 18, 16) maximal cliques. Similarly, 6 and 7 belong to (4, 5, 6, 7) and (6, 7, 8) maximal cliques.
3. Bridge Vertices: Some real networks can be divided into two or more communities because of a bridge vertex between them which connect two or more communities and acts as a messenger between them. So, these vertices play an important role in passage of information and they do not belong to any maximal clique. In

Fig. 1, node 3 connects three communities and hence termed as a bridge vertex. In recent years, there is a growing importance to find the pure bridge vertices or the articulation points in a community which inspired us to take up the task of finding them.

4 Proposed Work

In this paper, the algorithm proposed consists of two main procedures. One is finding out the overlapping communities through maximal cliques and classification of vertices. Second is that the possible vertices from the vertex sets are merged into the communities. The remaining vertices present in the bridge vertex set are the articulation points.

A. Extract Maximal Cliques:
Any network is represented as a Graph G(V, E) where V and E are the number of vertices and edges in graph G. The algorithm which we implemented finds that the maximal cliques are based on Bron–Kerbosch algorithm. Given a graph, this algorithm searches for all the maximal cliques using a recursive backtracking approach. Let R, P, X be the sets given as input to the graph where R and X are initially empty, P is the vertex set of the graph. Within each recursive call, the algorithm considers the vertices in P, in turn, if there are no such vertices, it either reports R as a maximal clique (if X is empty), or backtracks. For each vertex v chosen from P, it makes a recursive call in which v is added to R and in which P and X are restricted to the neighbor set N(v) of v, which finds and reports all clique extensions of R that contain v. Then, it moves v from P to X to exclude it from consideration in future cliques and continues with the next vertex in P.

B. Expand Maximal Cliques into communities:

1. N_m is the minimum of total number of vertices of the two maximal cliques, and N_o is the total number of the common vertices between these two maximal cliques. These two maximal cliques can be merged into a larger subgraph only if $N_m/2 <= N_o$ [4].
2. If an isolated vertex connects with only one nonoverlapping vertex, this isolated vertex and vertex connected with it can be combined into a same community. Otherwise, isolated nodes will be not merged into any community.
   The above points are mentioned briefly in the form of a pseudo code in Algorithm 1.

C. Find out pure Bridge vertices:

1. Finding Bridge Vertices: To find out the bridge vertices we will consider the isolated vertics, overlapping vertices, and the vertices present in all the maximal cliques.
    a. The union of isolated vertices, overlapping vertices, and vertices from all the maximal cliques are found and arranged in an ordered set.
    b. In another set, consider all the vertices of the input graph.
    c. Now, subtracting the two sets gives the bridge vertex set.

    These are the pure Bridge vertices or the Articulation points present in the given graph or a network.

    Articulation points are those which cannot be merged into any other community and their exclusion from the network would divide the network into different components. The information loss because of the exclusion of those articulation points is maximal.
2. Finding Pure Bridge Vertices by refining Bridge vertices:

    Consider Fig. 2. In this figure, vertex 7 is initially in the bridge vertex set. But after the graph is processed with below procedure, vertex 7 is merged into the left community because the degree of connectedness of this vertex to the left community is more. Below proposed algorithm finds out the pure bridge vertices which cannot be merged into any other community. The algorithm is as follows:
    a. The bridge vertex set is iterated and the corresponding connectedness of each vertex in the bridge vertex set to all the communities in the graph is checked.
    b. Now, the communities are sorted from largest to smallest based on their degree of connectedness with the particular bridge vertex.
    c. Now after step 2, the first and second community's degree's are checked. If they are different, then the bridge vertex is merged into the first community and if they are not, then it is left alone.
    d. The vertices remaining in the bridge vertex set are the pure bridge vertices or the articulation points.

    So, these pure bridge vertices are almost similar to articulation points whose exclusion will divide the network into several components. Similarly, considering the bridge vertices that are merged into any community, breaking the link between the bridge vertex and the community to which it is not merged would also affect the amount of information passage between the two communities.

```
Input: Set of : [maximal_cliques], [isolated_vertices];
Output:Set of : [overlapping_communities], [overlapping_vertices],
    [isolated_vertices] ;
N_m is the minimum of total number of vertices of the two maximal cliques,
    and N_o is the total number of the common vertices between these two maximal
    cliques.
for clique1, clique2 in [maximal_cliques] do
    if Nm / 2 ≤ No then
        community ← merge two maximal cliques
        Add community to [overlapping_communities]
    else
        do nothing;
    end
end
for vertex in [overlapping_communities] do
    if vertex found in more than two communities then
        Add vertex to [overlapping_vertices]
    else
        do nothing;
    end
end
for vertex in [isolated_vertices] do
    if edge(Vertex, non overlapping vertex) ∈ [overlapping_communities] then
        Add vertex to [overlapping_communities]
        Remove vertex from [isolated_vertices]
    else
        do nothing;
    end
end
```

Algorithm 1: Expand Maximal cliques

5 Results

The proposed work of detecting overlapping communities and pure bridge vertices is implemented in Python programming language running on a personal computer with 3.0 GHz processor, 8.0 GB memory, and Win8 operating system. We tested our algorithm with various test case networks and synthetic inputs. We have imported the Networkx library in python for studying and manipulating complex graphs and networks.

Comparison of our algorithm and the general articulation points algorithm is made with synthetic inputs 1–4 [Figs. 1, 2, 3, 4]. Table 1, shows the differences in finding articulation points. In Fig. 1, Tarjan's algorithm gives six articulation points

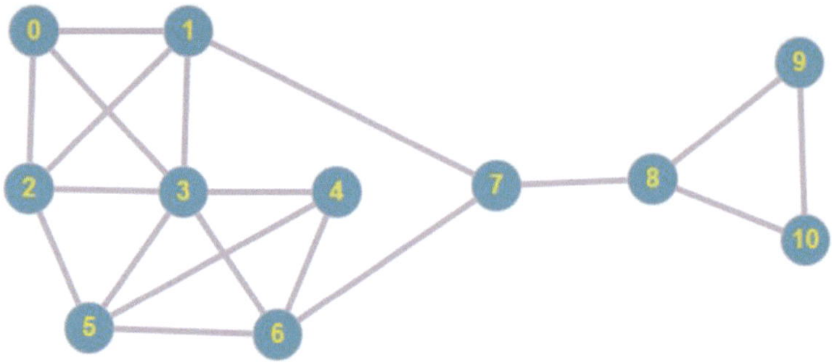

Fig. 2 Synthetic input 2

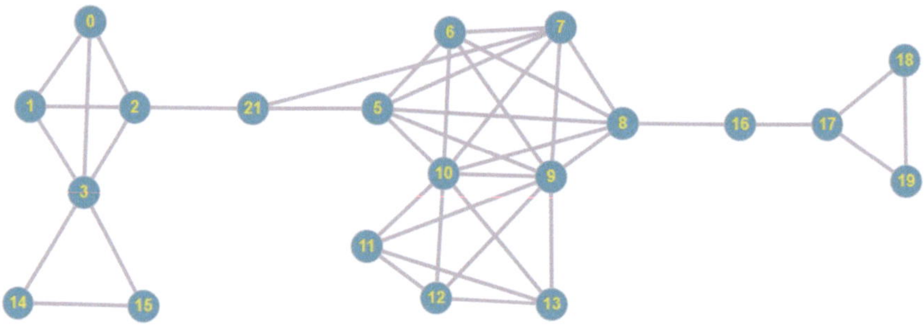

Fig. 3 Synthetic input 3

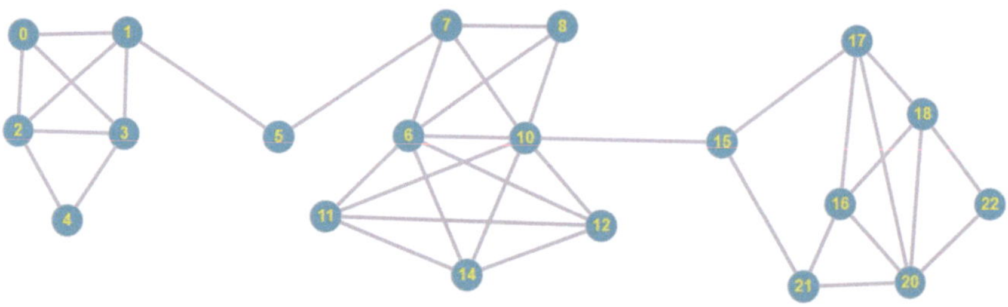

Fig. 4 Synthetic input 4

which are (2, 3, 4, 5, 18, 16) nodes, whereas with our algorithm there is only one articulation point which is node 3. In Fig. 2, Tarjan's algorithm gives two articulation points which are (7, 8) nodes, where as with our algorithm there are no articulation points. In Fig. 3, there are six articulation points with Tarjan's algorithm which are (2, 3, 21, 8, 16, 17) nodes, whereas our algorithm gives only one articulation point which is node 16. In Fig. 4, there are six articulation points with Tarjan's method which are (1, 5, 7, 10, 15, 17) nodes, whereas our algorithm gives only node 5. Our main aim is to merge as many nodes into the communities which leave with lesser

Table 1 Comparison of Tarjan's algorithm and proposed algorithm to find articulation points

Tested network	No. of articulation points using Tarjan's algorithm	No. of articulation points using proposed algorithm
Figure 1	6	1
Figure 2	2	0
Figure 3	6	1
Figure 4	6	1

Table 2 Test datasets

Name	No. of nodes	Overlapping vertices	Isolated vertices	Bridge vertices	Articulation points
Zachary karate	34	14	1	1	1
Marknewman-adjnoun	112	67	8	23	22
Marknewman-celegens	297	268	15	4	4

Table 3 Synthetic inputs

Name	No. of nodes	Overlapping vertices	Isolated vertices	Bridge vertices	Articulation points
Figure 1	18	4	2	1	1
Figure 2	11	3	0	1	0
Figure 3	20	3	0	2	1
Figure 4	20	7	0	2	1

number of articulation points. The left out articulation points hold the central area for maximum passage of information.

Some of the test case networks include zachary karate dataset with 34 vertices, marknewman-adjnoun dataset which has 112 vertices and marknewman-colegenes dataset which has 297 vertices. Table 2 shows the results of different test datasets and Table 3 shows the results of Synthetic inputs using our proposed algorithm. Since our algorithm works only for unweighted networks, all the test case networks were unweighted. Testing our algorithm on different data sets gave expected results with an average execution time of 5 seconds. The results which are mentioned above can be found in Table 2 [Test Datasets] table.

As shown in the Synthetic inputs table [Table 3], if we consider Fig. 3, there are three overlapping vertices (3, 9, 10) five maximal cliques, and two bridge vertices (16, 21). After implementing it using the proposed algorithm, the bridge vertex set now contains just (16). It is because the degree of connectedness of the bridge vertex (21) with the community (5, 6, 7, 8, 9, 10) is 2 and hence it is merged into the community. So, the number of articulation points is 1. Similarly, in Fig. 4, there are seven overlapping vertices (2, 3, 6, 10, 16, 20, 18), zero isolated nodes and two bridge vertices (5, 15). However, after implementing the algorithm on the bridge vertex set, bridge vertex (15) gets merged into the community because of the degree it possesses with the respective community, i.e., 2. So, the number of articulation points, in this case, is also 1. So, the concept behind this is to include maximum number of bridge vertices into communities where ever possible. Ultimately removing those articulation points from the network would divide it into different communities and hence lack of information passage between communities.

6 Conclusion

The bridge vertices of an unweighted undirected graph are those vertices whose removal increase the number of connected components of the graph, i.e., the vertices whose removal disconnects the graph.

An algorithm to detect pure bridge vertices is proposed and implemented effectively. First step is to find all the maximal cliques in the graph using the Bron–Kerbosch algorithm. Then, expand it to communities and finally, the actual bridge vertices were found whose inclusion or exclusion would greatly affect the information passage between the communities. The possible future work for this research could be finding the bridge edges using edge classification among the communities of a given graph.

References

1. Wenzheng, X., Rezvani, M., Liang, W., Yu, J.X., Liu, C.: Efficient algorithms for the identification of top-k structural hole spanners in large social networks. IEEE Trans. Knowl. Data Eng. **29**(5), 1017–1030 (2017)
2. Farina, G.: A linear time algorithm to compute the impact of all the articulation points (2015)
3. Firmani, D., Italiano, G.F., Laura, L., Orlandi, A., Santaroni, F.: Computing strong articulation points and strong bridges in large scale graphs. In: Experimental Algorithms, SEA (2012)
4. Li, J., Wang, X., Cui, Y.: Uncovering the overlapping community structure of complex networks by maximal cliques. Physica A Stat. Mech. Appl. **415**, 398–406 (2014)
5. Cazalsa, F., Karande, C.: A note on the problem of reporting maximal cliques. Theor. Comput. Sci. **407**, 564–568 (2008)
6. Fortunato, S.: Community detection in graphs. Phys. Rep. **486**, 75–174 (2010)
7. Xie, J., Kelley, S., Szymanski, B.K.: Overlapping community detection in networks: the state-of-the-art and comparative study. ACM Comput. Surv. (CSUR) **45**(4) (2013)
8. Girvan, M., Newman, M.E.J.: Community structure in social and biological networks. PNAS **99**(12) (2002)
9. Clauset, A., Newman, M.E.J., Moore, C.: Finding community structure in very large networks. Phys. Rev. E **70** (2004)
10. Wang, X.T., Chen, G.R., Lu, H.T.: A very fast algorithm for detecting community structures in complex networks. Physica A Stat. Mech. Appl. **384**(2), 667–674 (2007)

11. Evans, T.S.: Clique graphs and overlapping communities. J. Stat. Mech. Theory Exp. **2010** (2010)
12. Lancichinetti, A., Fortunato, S., Kertész, J.: Detecting the overlapping and hierarchical community structure in complex networks. New J. Phys. **11** (2009)
13. Farach-Colton, M., Hsu, T.-S., Li, M., Tsai, M.-T.: Finding articulation points of large graphs in linear time. In: Lecture Notes in Computer Science, vol. 9214. Springer, Cham (2015)
14. Italiano, G.F., Laura, L., Santaroni, F.: Finding strong bridges and strong articulation points in linear time. Theoret. Comput. Sci. **447**

MIX
Papier aus verantwortungsvollen Quellen
Paper from responsible sources
FSC® C105338

If you have any concerns about our products,
you can contact us on
ProductSafety@springernature.com

In case Publisher is established outside the EU,
the EU authorized representative is:
Springer Nature Customer Service Center GmbH
Europaplatz 3, 69115 Heidelberg, Germany

Printed by Libri Plureos GmbH
in Hamburg, Germany